WORKBOOK FOR ORGANIC CHEMISTRY
SUPPLEMENTAL PROBLEMS AND SOLUTIONS

Jerry A. Jenkins
Otterbein College

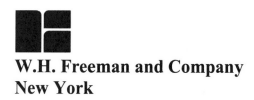

W.H. Freeman and Company
New York

© 2010 by W.H. Freeman and Company

All rights reserved.

Printed in the United States of America

ISBN-13: 978-1-4292-4758-0
ISBN-10: 1-4292-4758-4

First printing

W.H. Freeman and Company
41 Madison Avenue
New York, NY 10010
Houndmills, Basingstoke
RG21 6XS England

www.whfreeman.com/chemistry

TABLE OF CONTENTS

PREFACE v
*About the author **vi** | Acknowledgments **vi** | Selected concepts/reactions locator **vii***
TIPS* viii** | *Common abbreviations **ix

CHAPTER 1   THE BASICS 1
1.1 Hybridization, formulas, physical properties **1** | *1.2 Acids and bases* **4** | *1.3 Resonance* **7**

CHAPTER 2   ALKANES 11
2.1 General **11** | *2.2 Nomenclature* **12** | *2.3 Conformational analysis, acyclic* **13**

CHAPTER 3   CYCLOALKANES 15
3.1 General **15** | *3.2 Nomenclature* **16** | *3.3 Conformational analysis, cyclic* **18**

CHAPTER 4   REACTION BASICS 21

CHAPTER 5   ALKENES AND CARBOCATIONS 27
5.1 General **27** | *5.2 Reactions* **30** | *5.3 Syntheses* **36** | *5.4 Mechanisms* **39**

CHAPTER 6   ALKYNES 49
6.1 Reactions **49** | *6.2 Syntheses* **50** | *6.3 Mechanisms* **53**

CHAPTER 7   STEREOCHEMISTRY 55
7.1 General **55** | *7.2 Reactions and stereochemistry* **61**

CHAPTER 8   ALKYL HALIDES AND RADICALS 65
8.1 Reactions **65** | *8.2 Syntheses* **66** | *8.3 Mechanisms* **67**

CHAPTER 9 S_N1, S_N2, E1, AND E2 REACTIONS 69
9.1 General **69** | *9.2 Reactions* **71** | *9.3 Syntheses* **76** | *9.4 Mechanisms* **78**

CHAPTER 10   NMR 87

CHAPTER 11   CONJUGATED SYSTEMS 93
11.1 Reactions **93** | *11.2 Syntheses* **96** | *11.3 Mechanisms* **98**

CHAPTER 12   AROMATICS 103
12.1 General **103** | *12. Reactions* **105** | *12.3 Syntheses* **109** | *12.4 Mechanisms* **111**

CHAPTER 13   ALCOHOLS 117
13.1 Reactions **117** | *13.2 Syntheses* **120** | *13.3 Mechanisms* **124**

CHAPTER 14   ETHERS 129
14.1 Reactions **129** | *14.2 Syntheses* **133** | *14.3 Mechanisms* **134**

CHAPTER 15   ALDEHYDES AND KETONES 139
15.1 Reactions **139** | *15.2 Syntheses* **149** | *15.3 Mechanisms* **154**

CHAPTER 16   CARBOXYLIC ACIDS 167
16.1 Reactions **167** | *16.2 Syntheses* **169** | *16.3 Mechanisms* **172**

CHAPTER 17   CARBOXYLIC ACID DERIVATIVES 177
17.1 Reactions **177** | *17.2 Syntheses* **186** | *17.3 Mechanisms* **193**

CHAPTER 18   CARBONYL α-SUBSTITUTION REACTIONS AND ENOLATES   201
18.1 Reactions 201 | *18.2 Syntheses* 204 | *18.3 Mechanisms* 207

CHAPTER 19   CARBONYL CONDENSATION REACTIONS   209
19.1 Reactions 209 | *19.2 Syntheses* 217 | *19.3 Mechanisms* 219

CHAPTER 20   AMINES   229
20.1 Reactions 229 | *20.2 Syntheses* 233 | *20.3 Mechanisms* 236

SOLUTIONS TO PROBLEMS   241

CHAPTER 1	THE BASICS	243
CHAPTER 2	ALKANES	251
CHAPTER 3	CYCLOALKANES	255
CHAPTER 4	REACTION BASICS	261
CHAPTER 5	ALKENES AND CARBOCATIONS	263
CHAPTER 6	ALKYNES	281
CHAPTER 7	STEREOCHEMISTRY	287
CHAPTER 8	ALKYL HALIDES AND RADICALS	295
CHAPTER 9	S_N1, S_N2, E1, AND E2 REACTIONS	299
CHAPTER 10	NMR	315
CHAPTER 11	CONJUGATED SYSTEMS	319
CHAPTER 12	AROMATICS	327
CHAPTER 13	ALCOHOLS	341
CHAPTER 14	ETHERS	351
CHAPTER 15	ALDEHYDES AND KETONES	357
CHAPTER 16	CARBOXYLIC ACIDS	379
CHAPTER 17	CARBOXYLIC ACID DERIVATIVES	387
CHAPTER 18	CARBONYL α-SUBSTITUTION REACTIONS AND ENOLATES	405
CHAPTER 19	CARBONYL CONDENSATION REACTIONS	413
CHAPTER 20	AMINES	427

PREFACE
WORKBOOK FOR ORGANIC CHEMISTRY
SUPPLEMENTAL PROBLEMS AND SOLUTIONS

Organic Chemistry is mastered by *reading* (textbook), by *listening* (lecture), by *writing* (outlining, notetaking), and by *experimenting* (laboratory). But perhaps most importantly, it is learned by *doing, i.e.*, solving problems. It is not uncommon for students who have performed below expectations on exams to explain that they honestly *thought* they understood the text and lectures. The difficulty, however, lies in *applying, generalizing,* and *extending* the specific reactions and mechanisms they have "memorized" to the solution of a very broad array of related problems. In so doing, students will begin to "internalize" Organic, to develop an intuitive feel for, and appreciation of, the underlying logic of the subject. Acquiring that level of skill requires but goes far beyond rote memorization. It is the ultimate process by which one learns to manipulate the myriad of reactions and, in time, gains a predictive power that will facilitate solving new problems.

Mastering Organic is challenging. It demands memorization (*an organolithium reagent will undergo addition to a ketone*), but then requires application of those facts to solve real problems (*methyllithium and androstenedione dimethyl ketal will yield the anabolic steroid methyltestosterone*). It features a highly logical structural hierarchy (like mathematics) and builds upon a cumulative learning process (like a foreign language). The requisite investment in time and effort, however, can lead to the development of a sense of self-confidence in Organic, an intellectually satisfying experience indeed.

Many excellent first-year textbooks are available to explain the theory of Organic; all provide extensive exercises. Better performing students, however, consistently ask for additional exercises. It is the purpose of this manual, then, to provide *Supplemental Problems and Solutions* that reinforce and extend those textbook exercises.

Workbook organization and coverage. Arrangement is according to classical functional group organization, with each group typically divided into *Reactions*, *Syntheses*, and *Mechanisms*. To emphasize the vertical integration of Organic, problems in later chapters heavily draw upon and integrate reactions learned in earlier chapters.

It is desirable, but impossible, to write a workbook that is *completely* text-independent. Most textbooks will follow a similar developmental sequence, progressing from alkane/alkene/alkyne to aromatic to aldehyde/ketone to carboxylic acid to enol/enolate to amine chemistry. But within the earlier domains placement of stereochemistry, spectroscopy, S_N/E, and other functional groups (*e.g.*, alkyl halides, alcohols, ethers) varies considerably. The sequence is important because it establishes the concepts and reactions that can be utilized in subsequent problems. It is the intent of this workbook to follow a *consensus* sequence that complements a broad array of Organic textbooks. Consequently, instructors utilizing a specific textbook may on occasion need to offer their students guidance on workbook chapter and problem selection.

Most Organic textbooks contain later chapters on biochemical topics (proteins, lipids, carbohydrates, nucleic acids, *etc.*). This workbook does not include separate chapters on such subjects. However, consistent with the current trend to incorporate biochemical relevance into Organic textbooks, numerous problems with a bioorganic, metabolic, or medicinal flavor are presented throughout all chapters.

To produce an error-free manual is certainly a noble, but unrealistic, goal. For those errors that remain, I am solely responsible. I encourage the reader to please inform me of any inaccuracies so that they may be corrected in future versions.

Jerry A. Jenkins
Otterbein College
Westerville, OH 43081
jjenkins@otterbein.edu

Grindstones sharpen knives; problem-solving sharpens minds!

ABOUT THE AUTHOR

Jerry A. Jenkins received his BA degree *summa cum laude* from Anderson University and PhD in Organic from the University of Pittsburgh (*T Cohen*). After an NSF Postdoctoral Fellowship at Yale University (*JA Berson*), he joined the faculty of Otterbein College where he has taught Organic, Advanced Organic, and Biochemistry, and chaired the Department of Chemistry & Biochemistry. Prof. Jenkins has spent sabbaticals at Oxford University (*JM Brown*), The Ohio State University (*LA Paquette*), and Battelle Memorial Institute, represented liberal arts colleges on the Advisory Board of Chemical Abstracts Service, and served as Councilor to the American Chemical Society. He has published in the areas of oxidative decarboxylations, orbital symmetry controlled reactions, immobilized micelles, chiral resolving reagents, nonlinear optical effects, and chemical education. Prof. Jenkins has devoted a career to challenging students to appreciate the logic, structure, and aesthetics of Organic chemistry through a problem-solving approach.

ACKNOWLEDGMENTS

I wish to express gratitude to my students, whose continued requests for additional problems inspired the need for this book; to Mark Santee, Director of Marketing, WebAssign, for encouraging and facilitating its publication; to Dave Quinn, Media and Supplements Editor, W. H. Freeman, for invaluable assistance in bringing this project to completion; to the production team at W.H. Freeman, specifically Jodi Isman, Project Editor, for all their assistance with the printing process; to Diana Blume, Art Director, and Eleanor Jaekel for their assistance in the cover design; and to my wife Carol, for her endless patience and support.

SELECTED CONCEPTS/REACTIONS LOCATOR

The location of problems relating to the majority of concepts and reactions in most Organic textbooks will be generally predictable: pinacol rearrangements will be found under *ALCOHOLS*, benzynes under *AROMATICS*, acetals under *ALDEHYDES AND KETONES*, etc. Placement of others, however, may vary from one text to another: diazonium ions may be under *AROMATICS* or *AMINES*, thiols may be under *ALCOHOLS* or *ETHERS*, the Claisen rearrangement may be under *ETHERS* or *AROMATICS*, etc. The following indicates where problems on several of these often variably placed concepts or reactions are *initially* encountered in *Workbook for Organic Chemistry*.

Selected concept/reaction	Chapter
Active methylene chemistry (*e.g.*, malonic/acetoacetic ester syntheses)	18
Brønsted-Lowry/Lewis equations	1
Carbocation rearrangements	5
cis-, *trans*- (geometric) isomers	3
Claisen, Cope, oxy-Cope rearrangements	14
Conformational analysis	2, 3
Curved arrow notation	*vi*, 1
Degrees of unsaturation (units of hydrogen deficiency)	5
Diazonium ions	20
Diels-Alder reaction	11
Enamines, synthesis of	15
Enamines, reactions of	19
Epoxides, synthesis of	5
Epoxides, reactions of	14
Free radical additions	5
Free radical substitutions	8
Hydrogens, distinguishing different	2
Isocyanates, ketenes	17
Kinetic isotope effects	9
Kinetics, thermodynamics	4
Neighboring group participation	9
Nitriles	16
Organometallics (Grignard, organolithium, Gilman), synthesis of	8
Phenols	12
Polymers	5
Reaction coordinate diagrams	4
Reaction types/mechanisms	4
Resonance	1
Thiols, (di)sulfides	14
UV/VIS spectroscopy	11

# TIPS (TO IMPROVE PROBLEM SOLVING)

Mechanism arrows. All reactions (except nuclear) involve the flow of electrons. Arrows are used to account for that movement. They originate at a site of *higher* electron density (*e.g.*, lone pairs, π bond) and point to an area of *lower* electron density (*e.g.*, positively or partially positively charged atoms).

Equilibrium vs. resonance arrows. Equilibrium arrows interrelate real species (as above). Resonance arrows interrelate imaginary valence bond structures. Do not interchange them.

Hydrogen nomenclature. The word "hydrogen" is commonly misused. Be more specific.

A proton ($H^\oplus$) is removed by hydride ($H:^\ominus$) to form hydrogen (H_2).

A hydrogen atom (H·) is removed by a free radical species.

State of association/dissociation. Correct identification of the appropriate charge state on a species in a particular environment is important. Generally speaking, alkoxides (hydroxide), carboxylates, carbanions, enolates, amines, *etc.*, exist under *alkaline* conditions. Protons, carboxylic acids, carbocations, enols, *etc.*, exist under *acidic* conditions. For example, hydroxide does not exist in an acidic solvent

and a proton is not *directly* available in base.

COMMON ABBREVIATIONS

The following abbreviations and symbols are used throughout this workbook:

Ac	acetyl (CH_3CO-)
AcOH	acetic acid
*	chiral center or isotopic label
B:	base
Bn	benzyl ($PhCH_2-$)
Bu	butyl (C_4H_9-)
CA	conjugate acid
CB	conjugate base
Δ	heat energy
D-A or (4+2)	Diels-Alder
DB	double bond(s)
DCC	dicyclohexylcarbodiimide
DIBAH	diisobutylaluminum hydride
DMF	dimethylformamide
DMSO	dimethyl sulfoxide
EAS	electrophilic aromatic substitution
ee	enantiomeric excess
equiv	equivalent(s)
Et	ethyl (CH_3CH_2-)
F-C	Friedel-Crafts
[H]	reduction
$\sim H^+$	proton shift
HMPA	hexamethylphosphoramide
HSCoA	coenzyme A
$h\nu$	light energy
H-V-Z	Hell-Volhard-Zelinsky reaction
inv	inversion of configuration
L	leaving group
LDA	lithium diisopropylamide
mCPBA	m-chloroperbenzoic acid
Me	methyl (CH_3-)
NAS	nucleophilic acyl (or aryl) substitution
NBS	N-bromosuccinimide
NGP	neighboring group participation
NR	no reaction
Nu:	nucleophile
[O]	oxidation
PCC	pyridinium chlorochromate
Ph	phenyl (C_6H_5-)
Pr	propyl (C_3H_7-)
py	pyridine
Ra-Ni	Raney nickel
ret	retention of configuration
rds	rate determining step
taut	tautomerization
THF	tetrahydrofuran
TMS	tetramethylsilane or trimethylsilyl
Ts	tosyl (p-toluenesulfonyl)
TsOH	tosyl acid (p-toluenesulfonic acid)
TS	transition state
W-K	Wolff-Kishner reduction
X	halogen
(XS)	excess

PROBLEMS

CHAPTER 1
THE BASICS

1.1 Hybridization, formulas, physical properties

1. Seldane™ is a major drug for seasonal allergies; Relenza™ is a common antiviral.

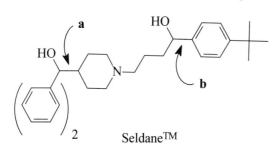

Seldane™

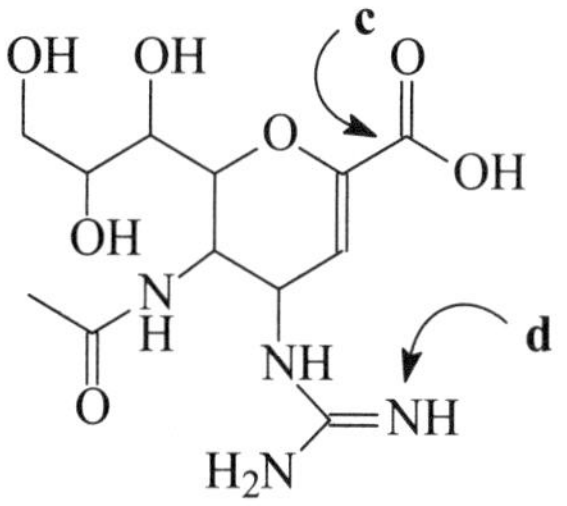

Relenza™

a. Complete the molecular formula for each. Seldane™: C___H___NO$_2$ Relenza™: C___H___N$_4$O$_7$

b. Draw all the lone electron pairs in both structures.

c. Which orbitals overlap to form the covalent bonds indicated by arrows **a, b,** and **c**?

    a _____   b _____   c _____

d. What is the hybridization state of both oxygens in Seldane™ and of nitrogen **d** in Relenza™?

2. Place formal charge over any atom that possesses it in the following structures:

a. :C≡C: b. H–C≡O: c. :O=N=O: d. the *conjugate base* of :NH$_2$CH$_3$

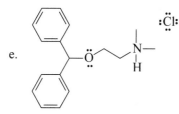

Benadryl™ (antihistamine)   zingerone (a constituent of the spice ginger)

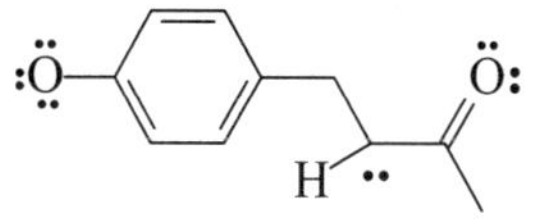

3. a. One type of carbene, [:CH$_2$], a very reactive species, has the two unshared electrons in the *same* orbital and is called "singlet" carbene. Identify the orbital and predict the HCH bond angle.

b. Another type of carbene is called "triplet" carbene and has a linear HCH bond angle. Identify the orbitals housing the two lone electrons.

4. a. Which has the *higher* bp? ⟋N–H or –N–   b. *lower* mp? 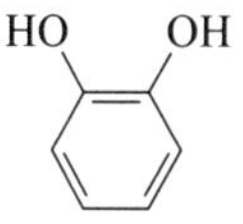   catechol   or   hydroquinone

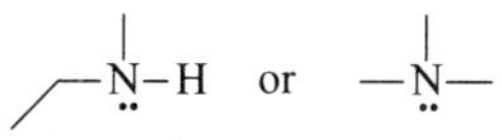

1.1 Hybridization, formulas, physical properties

2 • Chapter 1 The Basics

5. Must the indicated carbon atoms in each of the following structures lie in the same plane?

a. 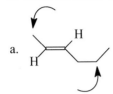   b. 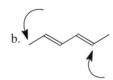   c. 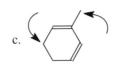   d.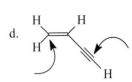

e. $(CH_3)_3C^{\oplus}$ all four carbons

f.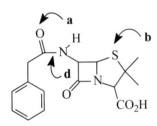

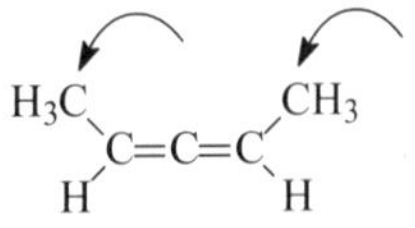

g.

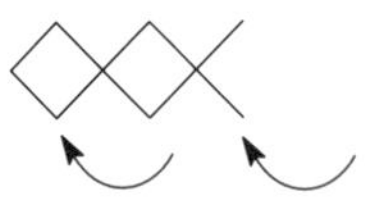

h.

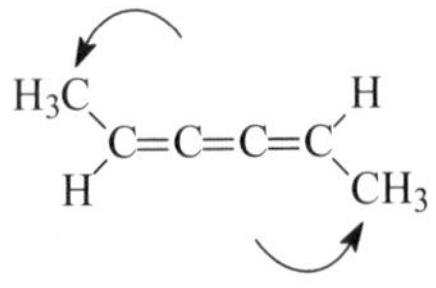

6. Which species in each pair has the *higher* molecular dipole moment (μ)?

a. $CHCl_3$ or $CFCl_3$

b. CH_3NH_2 or CH_3NO_2

c. CO_2 or SO_2

7. *Penicillin V* and the antiulcerative *cimetidine* (Tagamet^{TM} – the first billion dollar ethical drug) have the structures below:

penicillin V 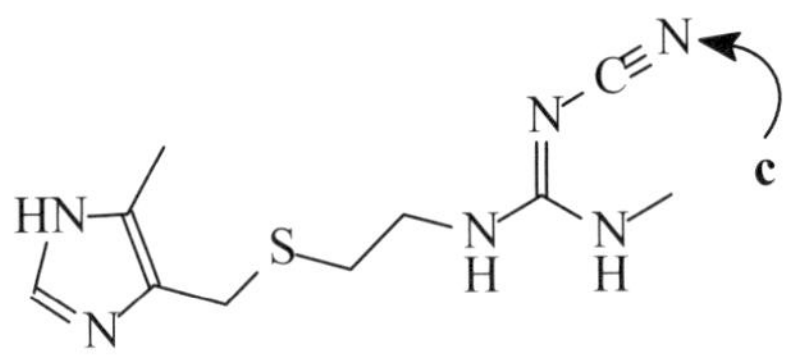   *cimetidine*

a. Complete the molecular formulas for each.

penicillin V: C____H____N____O____S____   cimetidine: C____H____N____S____

b. Identify the type of orbital (s, p, sp, sp^2, sp^3) that houses the lone electron pairs on the atoms indicated by arrows **a**, **b**, and **c** in the above structures.

a _____   b _____   c _____

c. The bond between the carbonyl carbon and nitrogen (indicated by arrow **d**) is somewhat *stronger* than a single but *weaker* than a double bond. Given that fact, what type of orbital houses the lone pair of electrons on that nitrogen? (Suggestion: do this problem after studying *resonance*.)

d. How many lone pairs of electrons are in each structure?

penicillin V: _____   cimetidine: _____

1.1 Hybridization, formulas, physical properties

8. *Sumatriptan* is often prescribed for the treatment of migraines. *Prostacyclin* is a platelet aggregation inhibitor.

sumatriptan

prostacyclin

a. Complete the molecular formulas for each.

   sumatriptan: C____H____N____O____S   prostacyclin: C____H____O____

b. *Sumatriptan* contains _____ sp² and _____ sp³ carbons; *prostacyclin* contains _____ sp² and _____ sp³ carbons.

c. *Sumatriptan* and *prostacyclin* possess _____ and _____ lone pairs of electrons, respectively.

9. Rozerem™ is prescribed for the treatment of insomnia, Chantix™ for smoking cessation, and Ritalin™ for ADHD.

Roserem™   Chantix™   Ritalin™

a. What is the molecular formula for each?

Rozerem™ _____   Chantix™ _____   Ritalin™ _____

b. How many *lone pairs* of electrons are there in each?

Rozerem™ _____   Chantix™ _____   Ritalin™ _____

10. *Theobromine* (Greek *theobroma* – "food of the gods") is a constituent of cocoa. How many lone pairs of electrons are in its structure? How many lone pairs of electrons are in the plasticizer *melamine*?

theobromine   melamine

1.1 Hybridization, formulas, physical properties

11. Which functional groups are present in each of the following medicines?

a. Tamiflu™ (antiviral)

b. Cipro™ (antibiotic)

c. Yasmin™ component (OCP)

1.2 Acids and bases

1. What is the *strongest* base that can exist in ammonia?

Sodium hydride (NaH) is, in fact, a stronger base than the above answer. Write a reaction to describe what happens when NaH is added to NH_3. Use arrows to show the flow of electrons.

2. Which is the *stronger* base: $(CH_3)_2NH$ or $CH_3\text{-}O\text{-}CH_3$?

3. Using curved arrow notation, write Lewis acid/base equations for each of the following. Remember to place formal charge on the appropriate atoms.

a. cyclohexanone =Ö: + $AlCl_3$ ⟶

b. $Ph_3P:$ + BF_3 ⟶

c. —N: :Ö: + BH_3 ⟶

4. Place formal charge on all appropriate atoms. Label the reactants on the left of the arrow as Lewis acids (LA) or Lewis bases (LB) and draw curved arrows to show the movement of electron pairs in each reaction.

a. $H_3C\text{-}\ddot{\text{O}}:$ + $CH_3CH_2\text{-}\ddot{\text{Cl}}:$ ⟶ $CH_3\text{-}\ddot{\text{O}}\text{-}CH_2CH_3$ + $:\ddot{\text{Cl}}:$

b. $H_2C=CH_2$ + BF_3 ⟶ $CH_2\text{-}CH_2\text{-}BF_3$

1.2 Acids and bases

c. $H_3C-\ddot{O}-H$ + $:CH_2-CH_3$ ⟶ $H_3C-\ddot{\ddot{O}}:$ + H_3C-CH_3

d. $:\ddot{\ddot{Cl}}-\ddot{\ddot{Cl}}:$ + $AlCl_3$ ⟶ $:\ddot{\ddot{Cl}}$ + $AlCl_4$

e. $CH_3-\ddot{N}=C=\ddot{\ddot{S}}:$ + $:NH_3$ ⟶ $CH_3-\ddot{N}-C\overset{:S:}{\underset{NH_3}{\diagup}}$

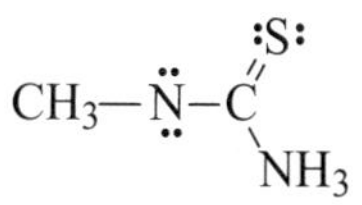

5. *Lynestrenol*, a component of certain oral contraceptives, has the structure

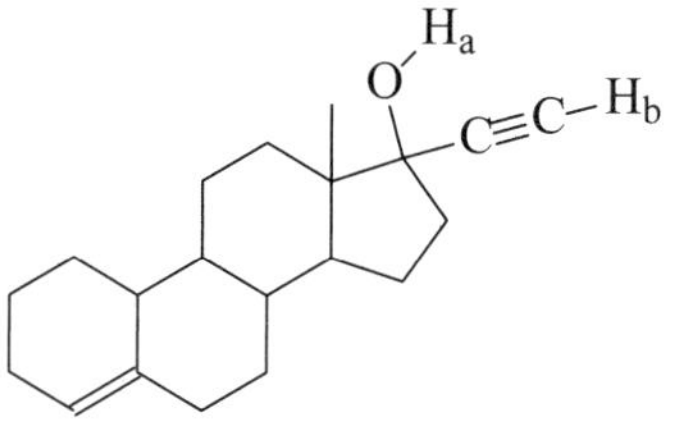

a. Calculate the molecular formula: C___H___O.

b. The pK$_a$s of hydrogens *a* and *b* are about 16 and 25, respectively, and the pK$_a$ of ammonia is about 35. Write a Brønsted-Lowry equation for the reaction of the *conjugate base* of lynestrenol with ammonia.

c. Is the K$_{eq}$ for the above reaction about equal to, greater than, or less than 1?

6. The structure of *ibuprofen* (**A**) and *acetaminophen* (**B**) are drawn below.

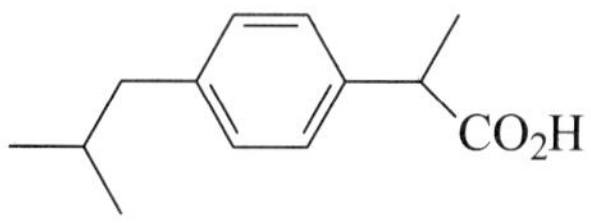

A 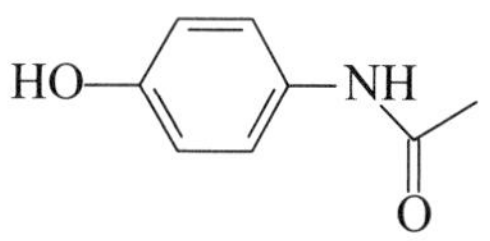   **B**

a. Write a reaction for the *conjugate base* of **A** with **B**.

1.2 Acids and bases

b. Identify the weak and strong acids and bases.

c. Is K_{eq} about equal to, less than, or greater than 1?

7. Which compound has the *lowest* pK_a?

   a. EtOH   b. HOAc   c. H_2O d. PhOH   e. H_2 f. NH_3

8. Which species has the ability to quantitatively (completely) remove the proton H_a (pK_a 22) from
   R−C≡C−H_a ?

   a. hydroxide   b. *CB* of NH_3 c. *CA* of hydride   d. *CB* of EtOH

9. Stress levels in horses may be monitored by measuring urine *estradiol*. Comment on the K_{eq} for the reaction of the *conjugate base* of nitromethane (pK_a 10.3) with estradiol.

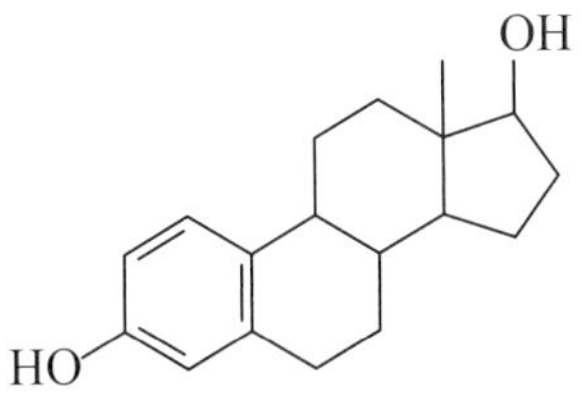

estradiol

CH_3NO_2

nitromethane

10. *Pyridinium chloride* is drawn below.
a. Place the appropriate formal charge on the atoms that bear it.

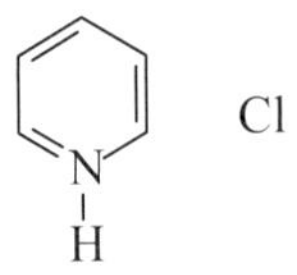

b. The pK_as for pyridinium chloride and sodium bicarbonate ($NaHCO_3$) are 5.2 and 10.2, respectively. Write a Brønsted-Lowry equation for the reaction of pyridinium chloride with the *conjugate base* of bicarbonate. Use curved arrow notation to show the flow of electrons.

c. Is K_{eq} greater than, less than, or about one?

1.2 Acids and bases

1.3 Resonance

1. Identify the type of orbital housing the electrons specified by the arrows.

2. Which species has the lower pK$_a$, H−C≡N: or H−O−C≡N: ?

3. How many nuclei can *reasonably* bear the charge in each of these ions?

a. HO−CH⁺−NH$_2$ ⟷

b. 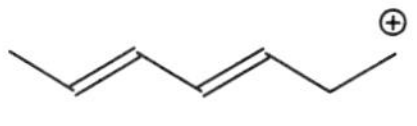 [pentadienyl cation] ⟷

c. 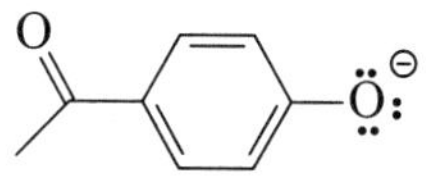 [4-acetylphenoxide] ⟷

d. 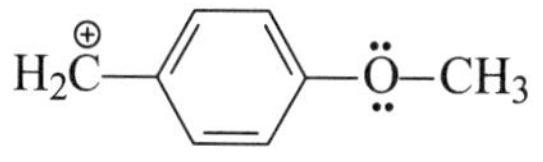 [H$_2$C⁺−C$_6$H$_4$−Ö−CH$_3$] ⟶

4. The compound below can be protonated at any of the three nitrogen atoms to give a *guanidinium* ion derivative (creatine phosphate and the amino acid arginine possess this moiety). One of these nitrogens is much more basic than the others, however. Draw the conjugate acids resulting from such protonation, then identify the conjugate acid which is most stable. Why?

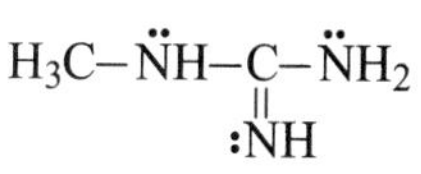

H$_3$C−N̈H−C(=N̈H)−N̈H$_2$

8 • Chapter 1 The Basics

5. Draw a resonance structure that is *more stable* than the one given. Use curved arrows to derive.

   a. :Ö–Ö–Ö: ⟷
      ozone

   b. [bicyclic structure with N-H and ⊕ charge] ⟷

   c.   ⟷

   d. [cyclohexenyl cation with :ÖH] ⟷

6. How many nuclei can *reasonably* bear the charge in each of the following ions?

   a. [pyrroline with ⊕N-H]

   b. [oxepine cation with ⊕]

   c. [cyclohexenyl-CH$_2$⊕]

   d. [cyclohexenone anion]

7. Recalling that resonance is a stabilizing force, explain why the pK$_a$ of H$_a$ in **A** is (only!) about 10.

   [diketone structure with H and H$_a$]
   A

8. Either oxygen in acetic acid (HOAc) could, in theory, be protonated to produce two different conjugate acid forms. Draw each and explain which is more favored.

1.3 Resonance

9. How many nuclei can *reasonably* bear the charge or odd electron in each of the following?

a. 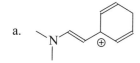   b.   c.

d.   e.   f.

g.   h.   i.

10. **B**'s molecular dipole moment (μ) is larger than **A**'s. Explain.

A   **B**

11. Bioluminescence in fireflies is a result of the conversion of chemical energy (in ATP) to light energy. Specifically, ATP, O_2, and the enzyme luciferase cause *luciferin* (~ 9 mg can be collected from about 15,000 fireflies!) to be oxidatively decarboxylated to an electronically excited *oxyluciferin*. Relaxation of the latter to its ground state is accompanied by the emission of light (fluorescence). Subsequent regeneration reactions then recycle oxyluciferin back to luciferin. Draw the two resonance structures of the CB of oxyluciferin in which either oxygen bears the negative charge.

luciferin → (ATP, O_2, luciferase, $-CO_2$) → *oxyluciferin* + hν

1.3 Resonance

CHAPTER 2
ALKANES

2.1 General

1. Which compound has the *highest* mp?

    1. *n*-octane
    2. 2,5-dimethylhexane
    3. 2,3,4-trimethylpentane
    4. bicyclo[2.2.2]octane
    5. all have the same number of carbons and would melt at the same T

2. Which compound has the *highest* bp?

    1. *n*-pentane
    2. neopentane (dimethylpropane)
    3. isopentane

3. Dodecahedrane, one of the three Platonic solids (tetrahedron, hexahedron, and dodecahedron), is a regular polyhedron consisting of twelve cyclopentane rings (think soccer ball). Eicosane is a straight-chain compound. Although both are C_{20} hydrocarbon alkanes, one melts at 420^0 and the other at 37^0. Explain.

4. How many *constitutional* (structural) isomers exist for

    a. C_6H_{14}?
    b. C_7H_{16}?

5. How many *different kinds* (constitutional) of hydrogens are in

    a. 2,3-dimethylpentane?
    b. 2,4-dimethylpentane?
    c. 3-ethylpentane?
    d. 2,2,4-trimethylpentane?
    e. 2,5,5-trimethylheptane?
    f. 4-ethyl-3,3,5-trimethylheptane?

2.2 Nomenclature

Give the IUPAC name for each of the following. Be certain to specify stereochemistry when relevant.

1. 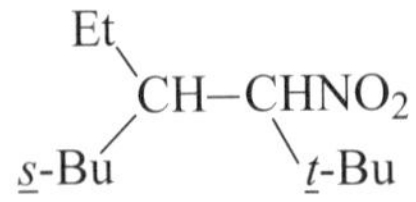
```
    Et
     \
      CH—CHNO₂
     /      \
   s-Bu     t-Bu
```

2. 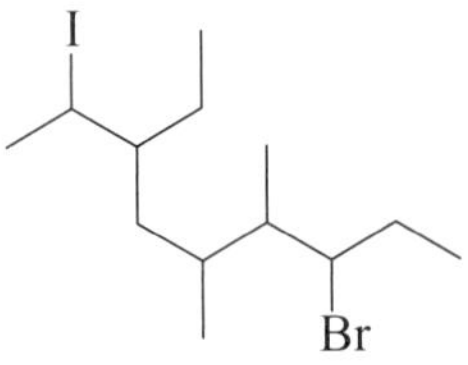 (structure with I and Br substituents)

3. 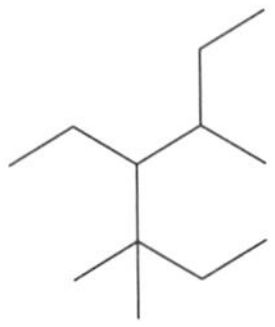 (branched alkane structure)

4. 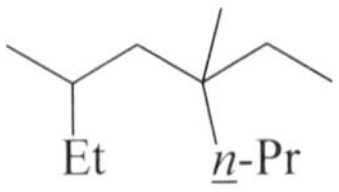 (structure with Et and n-Pr substituents)

5. 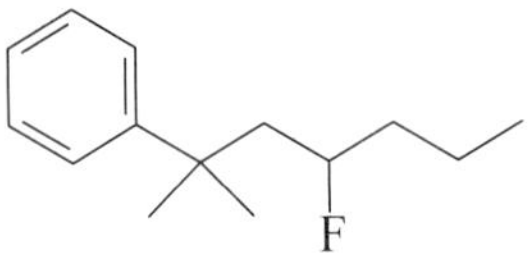 (structure with phenyl and F)

6. 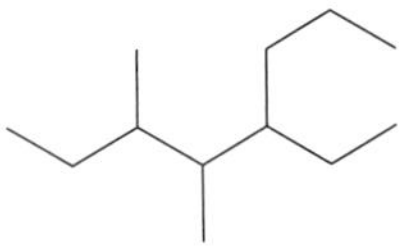 (branched alkane structure)

7. 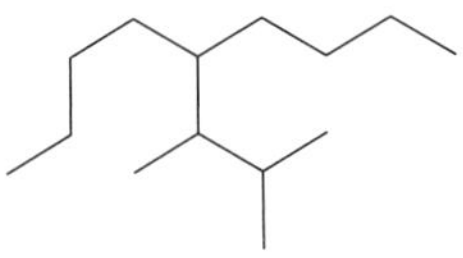 (branched alkane structure)

8. 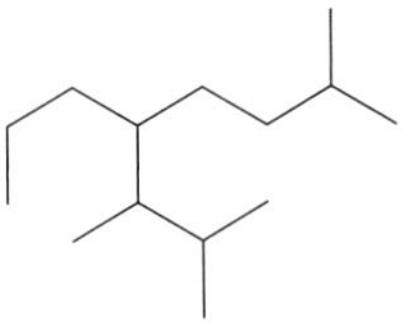 (branched alkane structure)

9. 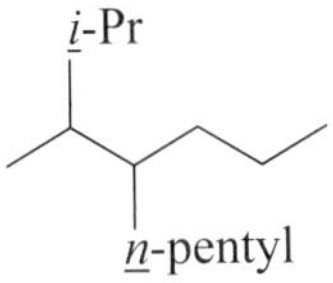
```
    i-Pr
     |
   ———————
     |
   n-pentyl
```

10. isohexyl iodide

11. 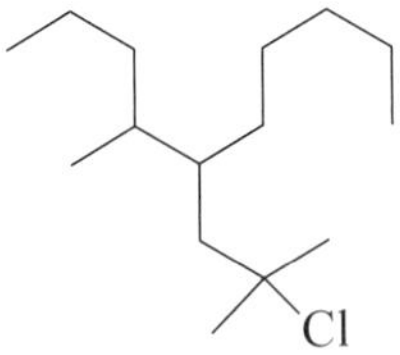 (structure with Cl)

12. 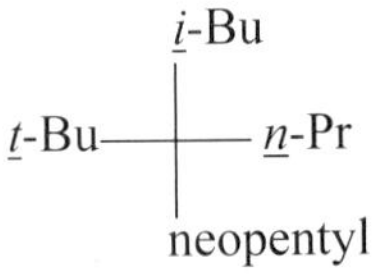
```
            i-Bu
             |
    t-Bu————————n-Pr
             |
          neopentyl
```

Give the *correct* IUPAC names for problems 13 – 16.

13. 2-isopropyl-4-methylheptane

14. 3-(1-methylbutyl)octane

15. 3-*s*-butyl-7-*t*-butylnonane

16. tetraethylmethane

17. Draw structural formulas, using bond line notation, for the following:

   a. neopentyl alcohol (R-OH)   b. isobutyl *n*-pentyl ether (R-O-R')   c. allyl bromide (R-X)

2.3 Conformational analysis, acyclic

1. The rotational energy barrier about the C-C bond in EtBr is 3.7 kcal/mole. What is the energy cost of eclipsing a C-H and C-Br bond?

2. Draw Newman projections of the
   a. *most stable* conformer, looking down the C_2-C_3 bond, of 2-cyclopentyl-6-methylheptane

   b. *gauche* conformer of 1-phenylbutane, looking down the C_1-C_2 bond (use two-letter abbreviations for R groups).

3. Give the *common* name for (a) and the *IUPAC* name for (b).

   a. [Newman projection with OH, Me, H on front carbon and H, H, Me on back carbon]
   (R-OH = alkyl alcohol)

   b. [Newman projection with *s*-Bu, H, Et on front carbon and H, H, *t*-Bu on back carbon, with phenyl group]

4. Draw the *conformer* of isopentane that corresponds to the *highest minimum* in a plot of the potential energy *vs.* rotation about the C_2-C_3 bond (use a Newman projection).

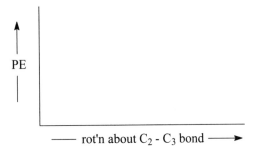

5. The molecular dipole moment (μ) for FCH_2CH_2OH is much larger than that for FCH_2CH_2F. Use conformational analysis to explain.

2.3 Conformational analysis, acyclic

CHAPTER 3
CYCLOALKANES

3.1 General

1. Which compound has the *highest* molecular dipole moment (u)?

   a.   b. *anti* conformer of 2,3-dichlorobutane   c. C$_2$Cl$_2$ d. *cis*-1,3-dichlorocyclobutane

2. How many *constitutional* (structural) isomers exist for

   a. dichlorocyclopentane?

   b. C$_6$H$_{12}$ that have a cyclopropyl ring in their structure?

3. How many *cis/trans* stereoisomers exixt for

   a. dichlorocyclopentane?

   b. diphenylcyclohexane?

   c. 2-chloro-4-ethyl-1-methylcyclohexane?

4. How many *different kinds* [constitutional and geometric (*cis/trans*)] of hydrogens are there in

   a. 1-ethyl-1-methylcyclopropane?

   b. allylcyclobutane?

   c. methylcyclobutane?

d. chlorocyclopentane?

e. vinylcyclopentane?

5. Which bicyclic compound is *least* strained?

a. 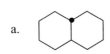 b. 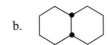 c. 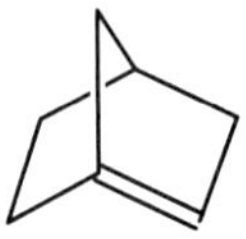   d.

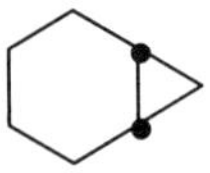

6. Three structural isomers are possible for methylbicyclo[2.2.1]heptane. One of them has two stereoisomeric forms. Draw structures for all four isomers.

7. In view of the previous problem, how many structural and geometric isomers exist for methylbicyclo[2.2.2]octane?

3.2 Nomenclature

Give the IUPAC name for each of the following. Be certain to specify stereochemistry when relevant.

1. 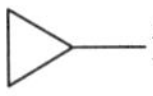 ▷—isoamyl

2.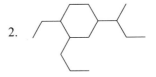

3.  (three names!)

4.

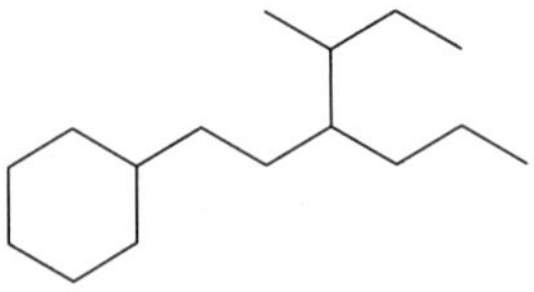

Problems • 17

5.

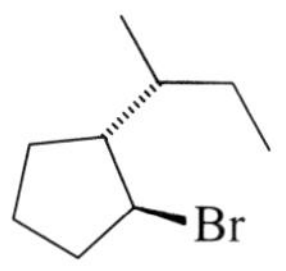

6.

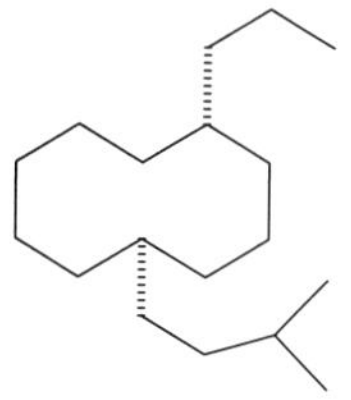

7.

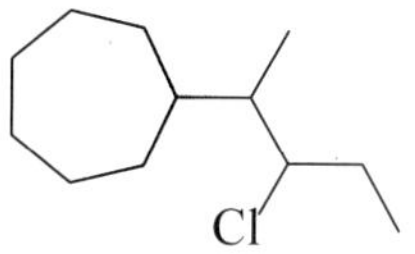

8.

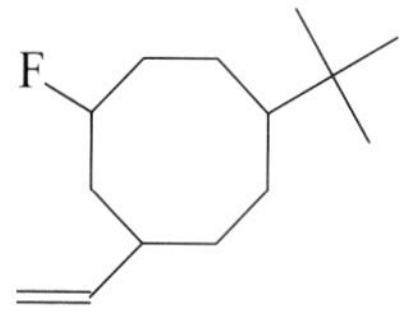

9.

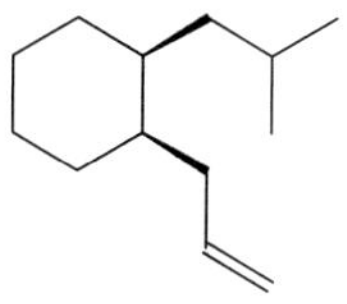

10. 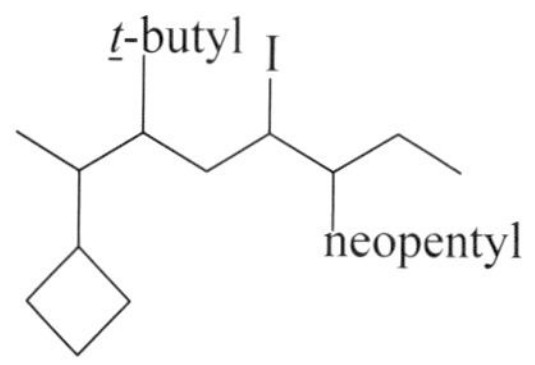 *t*-butyl, I, neopentyl

11.  F, Ph

12.

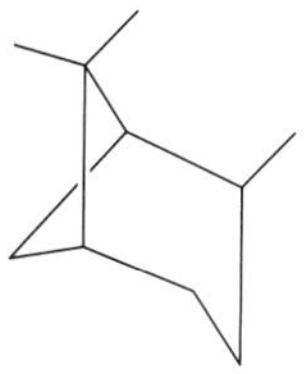

13.

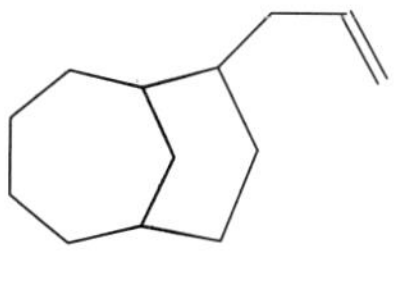

14.

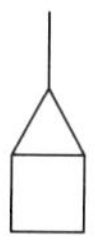

roof-methylhausane (!)

15.

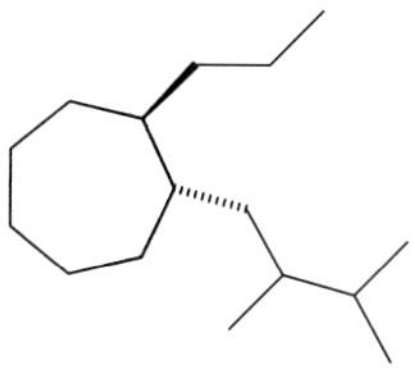

16.

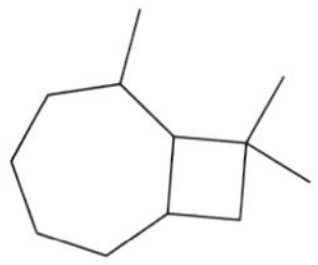

3.2 Nomenclature

3.3 Conformational analysis, cyclic

1. Draw the *most stable* conformer of

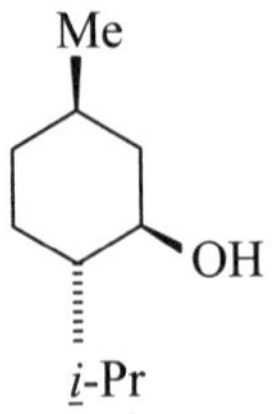

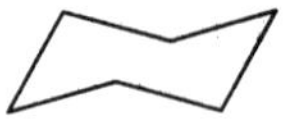

menthol

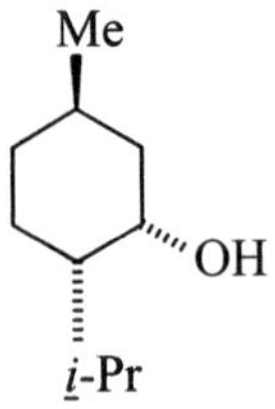

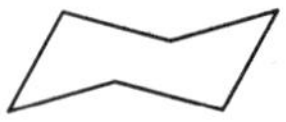

neomenthol

2. In each of the following predict whether K_{eq} is about *equal to*, *greater than*, or *less than* one:

a. *trans*-1,3-diphenylcyclohexane ⇌ "flipped" conformer

b. 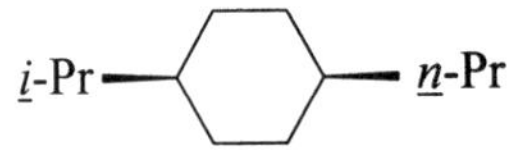 *i*-Pr—⟨⟩—*n*-Pr ⇌ "flipped" conformer

(if *i*-Pr is equatorial)

c. 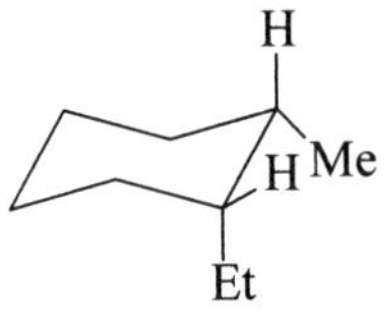 ⇌ "flipped" conformer

3. Which has the *most* negative heat of combustion (ΔH_{comb}) in each of (a), (b), or (c)?

a.

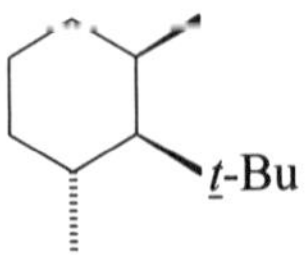

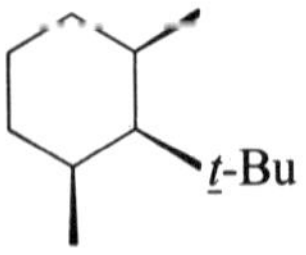

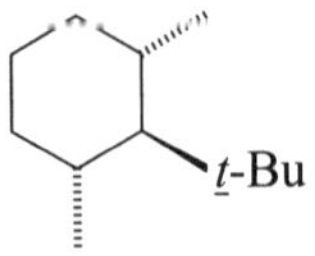

b.

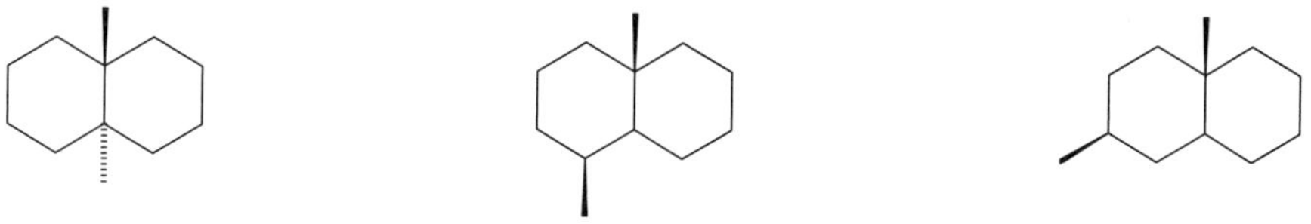

c.

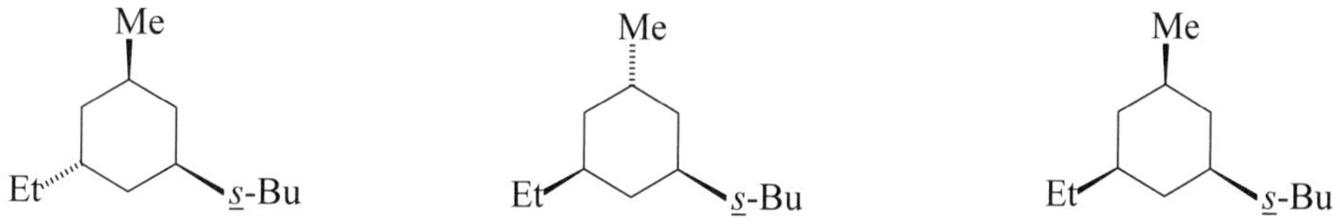

4. a. Which has the *least* negative heat of combustion (ΔH_{comb})?

b. Which two structures in (a) are the same compound?

5. Many alkyl halides undergo loss of HX in the presence of base. For example, chlorocyclohexane gives cyclohexene when treated with sodium hydroxide. The reaction mechanism generally requires both the leaving proton and halide to occupy axial positions, a process known as a *trans*-diaxial elimination. Therefore, which do you think would react faster, *cis*-1-chloro-2-*t*-butylcyclohexane or *trans*-1-chloro-2-*t*-butylcyclohexane?

6. *Trans*-4-fluorocyclohexanol exists largely in a chair conformation, whereas the *cis*-isomer favors a twist-boat conformation. Explain.

3.3 Conformational analysis, cyclic

7. Glucose, like cyclohexane, exists in a chair conformation. Two configurations of glucose are possible; they are drawn below:

1   and   **2**

a. Complete the chair conformations below to show the *most stable* conformer of **1** and **2**.

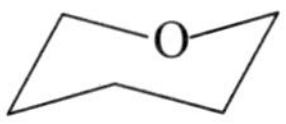

1 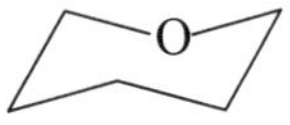   **2**

b. Which configuration would you predict would be *less* stable, *i.e.*, burn with a *more negative* heat of combustion?

8. One of the chair conformations of *cis*-1,3-dimethylcyclohexane is 5.4 kcal/mol more stable than the other. If the steric strain of 1,3-diaxial interactions between hydrogen and methyl is 0.9 kcal/mol, what is the strain cost of a 1,3-diaxial interaction between the two methyl groups?

9. a. How many *cis/trans* stereoisomers exist for 1,2,3,4,5,6-hexamethylcyclohexane?

b. For three of those stereoisomers, $K_{eq} = 1$ for conformational chair-chair flipping. Draw them.

c. Of those three, which is the *least* stable?

d. Which stereoisomer would be *least* likely to undergo conformational flipping?

3.3 Conformational analysis, cyclic

CHAPTER 4
REACTION BASICS

1. Which type of reaction – *addition, elimination, rearrangement, substitution, reduction* [H], or *oxidation* [O] – best describes each of the following?

a.  MeLi + (acetone) ⟶ (2-methyl-2-lithoxypropane)

b. 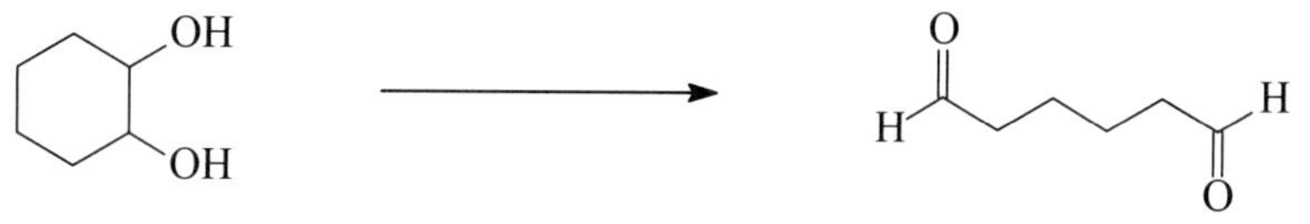 (1,2-cyclohexanediol) ⟶ (hexanedial)

c. RCO$_2$R + NH$_2$R ⟶ RCONHR + HOR

d. 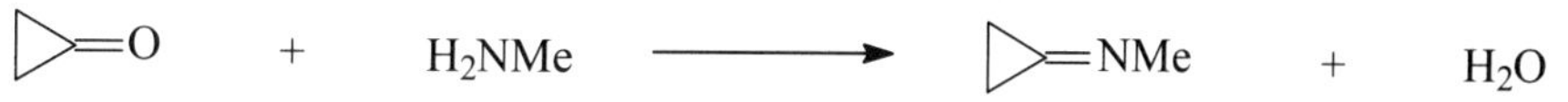 (cyclopropanone) + H$_2$NMe ⟶ (cyclopropanimine N-Me) + H$_2$O

e. 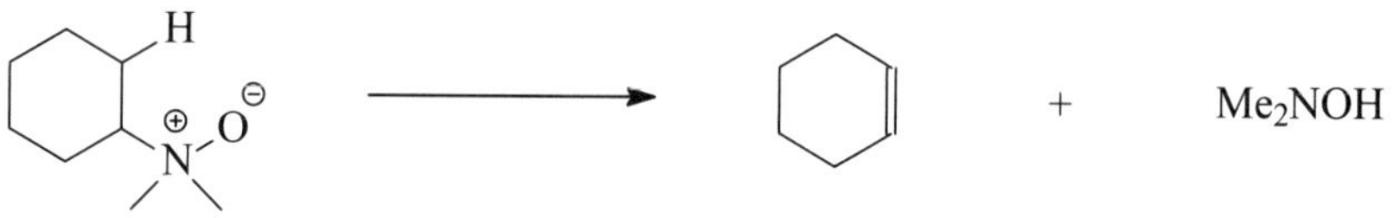 (cyclohexyl-NMe$_2$ N-oxide) ⟶ (cyclohexene) + Me$_2$NOH

f. 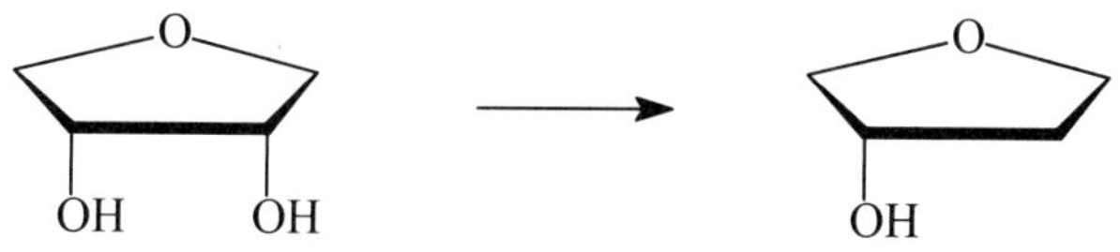 (furanose diol) ⟶ (furanose mono-ol)

g. RHC=CHR ⟶ RC≡CR

h. 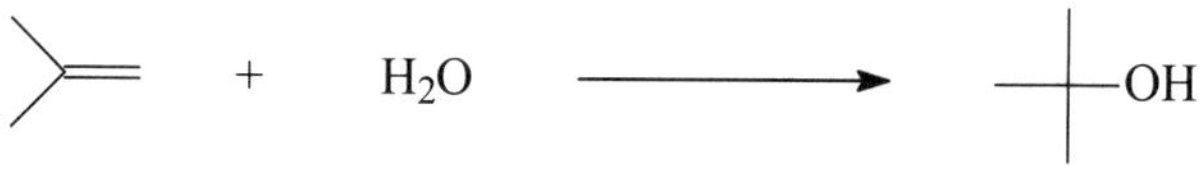 (isobutylene) + H$_2$O ⟶ (t-butanol)

i. PhCO$_2$H ⟶ PhCHO

j.  (cyclobutene) ⟶ (butadiene)

k. BnOH ⟶ PhCHO

4. Reaction Basics

l. Ac₂O + H₂O ⟶ 2 AcOH

m. CHCl₃ + KO-t-Bu ⟶ :CCl₂ + t-BuOH + KCl

n. [cyclopentadiene] + HC≡CH ⟶ [norbornadiene]

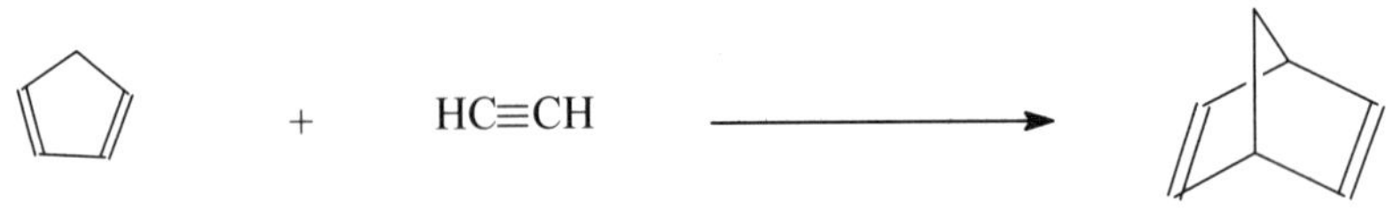

o. Br₂ ⟶ 2 Br⁻

p. Ph—≡—Ph ⟶ Ph\\=/Ph (cis-stilbene)

q. [cyclohexenol] —OH ⟶ [cyclohexanone] =O

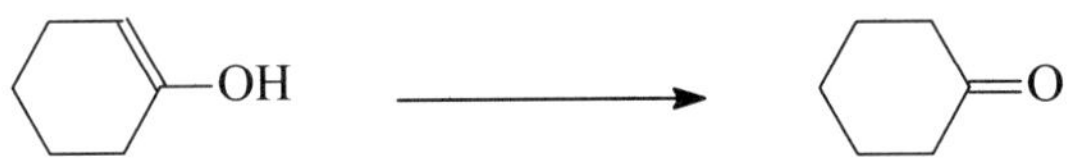

r. vinyl chloride ⟶ C₂H₂

s. [benzene-1,2-dicarboxylic acid] ⟶ [phthalic anhydride] + H₂O

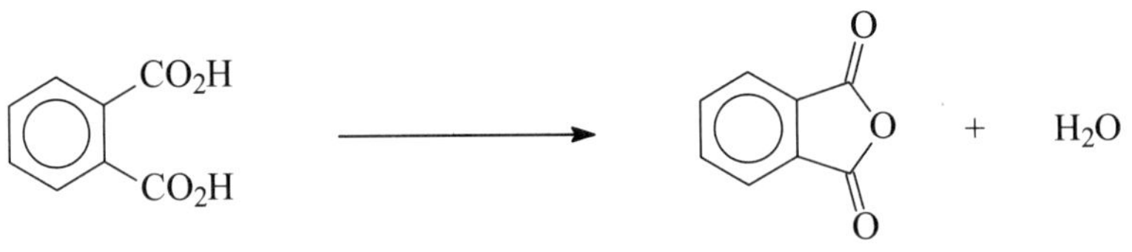

t. isohexyl alcohol ⟶ isohexane

2. Imagine a 2-step (**A** to **B** and **B** to **C**) endothermic reaction for which ΔG° values for each step are, respectively, +3 and +7 kcal/mole. The ΔG‡ value for the rate determining step is 11 kcal/mole. (a) Draw a potential energy diagram for this reaction. (b) What is the ΔG‡ value for the conversion of **C** to **B**?

3. A simplified mechanism for the exothermic substitution reaction below involves two steps:

a. Draw an overall energy diagram and label the transition state(s), intermediate, $\Delta G^{\ddagger}$ for the rate determining step, and $\Delta G°$.

b. The overall K_{eq} for the conversion of RCOCl to RCO$_2$R' could be calculated from $\Delta G°$ according to the equation:

$K_{eq} =$ _____

c. If $\Delta G^{\ddagger}$ is known, the rate of the reaction could be calculated according to the equation:

rate = _____

4. Bromoform (**A**) in the presence of base (:B$^-$) can form a very reactive intermediate, dibromocarbene (**B**), which can rapidly add to olefins to produce *gem*-dibromocyclopropane derivatives. The following summarizes the two-step mechanism:

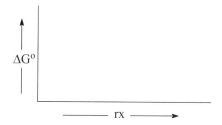

a. Assuming that $\Delta G°$ for the overall reaction is +2.5 kcal/mol and that step (2) is rate-determining, draw a reaction energy diagram that depicts all three steps.

b. Calculate K_{eq} for this reaction (R = 2 *cal*/mol·K, T = 300).

5. Consider the following reaction mechanism for **A** in equilibrium with **B**:

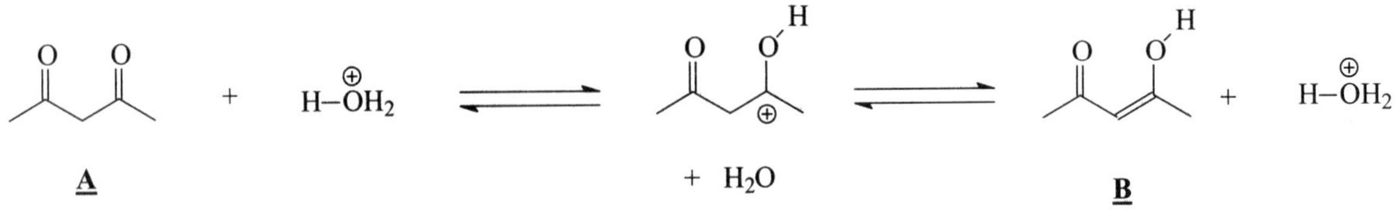

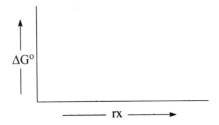

A   + H$_2$O   **B**

                                                                                               75%

a. The overall reaction is an example of a(n) _____ (type) reaction that occurs by a(n) _____ mechanism.

b. Draw curved arrows to show the electron flow that has occurred in each step.

c. Calculate K_{eq}, assuming only **A** and **B** are present (note: **B** is formed in 75% yield).

                  K_{eq} = _____

If K_{eq} is known, then $\Delta G°$ = _____ .

d. Which species is (are) nucleophilic in this reaction?

e. Draw a qualitative energy diagram for the reaction (assume the first step is slower than the second). Label the transition state(s) and intermediate.

6. Consider the following reaction:

          cyclohexyl-I   + MeOH   ⟶   cyclohexyl-OMe +   HI

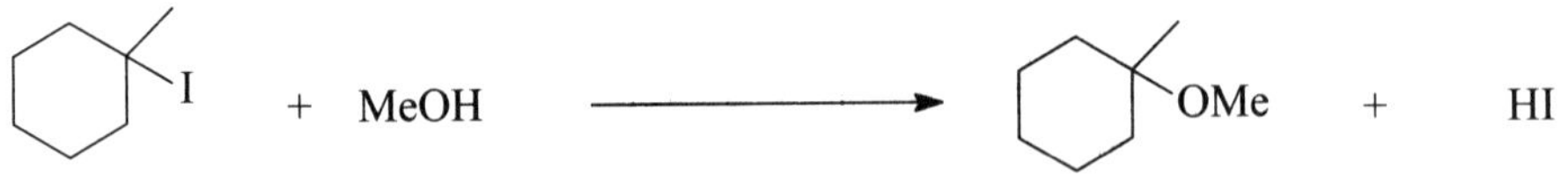

A

The rate law for the reaction may be expressed as: rate = $k[\mathbf{A}]$. Given that methyl alcohol is *not* in the rate law, propose a reaction for the rate determining step.

7. Below are reactions we shall examine in more detail later. Classify the mechanisms as *polar/ionic, free radical,* or *pericyclic* (concerted).

a.

b. hausene

c.

d.

e.

f.

g.

h.

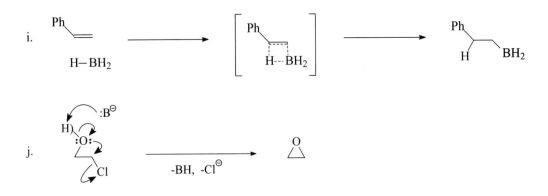

CHAPTER 5
ALKENES AND CARBOCATIONS

5.1 General

1. Nomenclature. Give the complete IUPAC name for the following:

   a.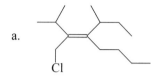

   b. [structure: phenyl CH with propenyl group]

   c. 4-vinyldecane (an *incorrect* name!)

   d. [bicyclic structure with methyl group]

2. Identify each of the olefins below as (E)- or (Z)-:

   a. [cyclopentylidene with CO₂H and CH₂OH]

   b. [alkene: NC, vinyl, H₂NH₂C, t-Bu]

   c. [Ph, cyclohexyl, =NH, NH₂ substituted alkene]

   d. [tetrasubstituted alkene with acyl, ester, dioxolane groups]

   e. [alkene with NH₂, CH₂F, isobutyl, sec-butyl groups]

   f. [Ph, cyclopentadienyl, SH, CHO substituted alkene]

3. a. How many alkenes, C_7H_{12}, could you treat with H_2 / Pt to prepare methylcyclohexane?

   b. Which would have the *least negative* heat of hydrogenation?

4. How many *geometric* isomers exist for 2,4-heptadiene?

5. Which carbocation is the most stable?

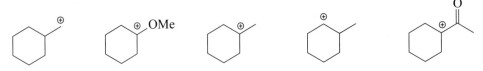

6. Degrees of unsaturation (units of hydrogen deficiency).
a. The antidepressant *fluoxetine* (Prozac™), $C_{17}H_{18}F_3NO$, when treated with H_2 / Ni gives a structure with molecular formula $C_{17}H_{30}F_3NO$. It contains no triple bonds. How many rings are in fluoxetine?

b. Cipro™ is an antibacterial that is used to treat anthrax. Its molecular formula is $C_{17}H_{18}FN_3O_3$. The drug has four rings and no triple bonds. How many double bonds does it contain?

c. *RU 486* is an abortion medication. Its molecular formula is $C_{28}H_{35}NO_2$. Its structure contains five double bonds and one triple bond. How many rings are in *RU 486*?

d. The COX-2 inhibitor *rofecoxib* (Vioxx™), an anti-inflammatory agent, has been taken off the market because of potential increased cardiovascular risk. Its molecular formula is $C_{17}H_{14}O_4S$. There are three rings and no triple bonds in rofecoxib. How many double bonds are there? (Note: for each sulfur atom, subtract four hydrogen atoms to arrive at the equivalent hydrocarbon formula.)

e. The antibiotic *floxacillin*, $C_{19}H_{17}ClFN_3O_5S$, contains eight double bonds. How many rings are present? (In this case, treat sulfur as you would oxygen.)

f. The antidepressant Paxil™ has the molecular formula $C_{19}H_{20}FNO_3$. Upon exhaustive hydrogenation (H_2/Pt) a compound $C_{19}H_{32}FNO_3$ is formed. How many double bonds and how many rings are in Paxil™?

5.1 General

7. How many stereoisomers exist for 2,4-hexadiene?   for 2-chloro-2,4-hexadiene?

8. Draw structural formulas for each of the following:

   a. (Z)-3-methyl-2-phenyl-2-hexene   b. propylene dichloride

   c. styrene bromohydrin   d. *trans*-cyclohexene glycol   e. isobutylene epoxide

9. Draw an energy *vs.* progress of reaction diagram for the *exothermic* reaction of vinylcyclobutane with HCl to yield 1-chloro-1-methylcyclopentane. Be certain the number of intermediates is clearly indicated.

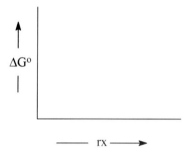

10. Draw the *most*

    a. *important* contributing resonance structure of the conjugate acid of 6-methyl-1,3,5-heptatriene

    b. *stable intermediate* in the following reaction:

11. The following 1,2-hydride shift does not occur. Why?

    adamantyl carbocation

5.1 General

30 • Chapter 5 Alkenes and Carbocations

12. Which reaction demonstrates NEITHER *regiospecificity* nor *stereospecificity*?

a. *trans*-2-pentene $\xrightarrow{\text{HF}}$

b. 1-pentene $\xrightarrow[\text{(XS) NaBr}]{\text{Cl}_2}$

c. cyclobutene $\xrightarrow[\text{H}_2\text{O}]{\text{Cl}_2}$

d. 1-ethylcyclopropene $\xrightarrow{\text{D}_2/\text{Pt}}$

13. Why, and how, does β-pinene readily isomerize to α-pinene in the presence of an acid catalyst?

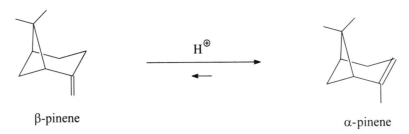

β-pinene   α-pinene

5.2 Reactions
Draw the structural formula of the major organic product(s). Show stereochemistry where appropriate.

1. Ph-CH(CH₃)-CH=CH₂ $\xrightarrow{\text{HCl}}$

2. (cyclohexenyl-NO₂) $\xrightarrow{\text{HI}}$

3. (1-methyl-1-isopropenylcyclopentane) $\xrightarrow{\text{H}_3\text{O}^\oplus}$

4. cyclopentane $\xrightarrow{\begin{array}{l}\text{1. Cl}_2/\Delta \\ \text{2. KOMe, MeOH} \\ \text{3. Br}_2, \text{CHCl}_3\end{array}}$

5.2 Reactions

5. 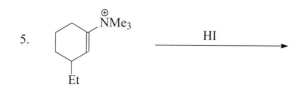 HI

6.  HF

7. vinylcyclohexane — DCl →

8. 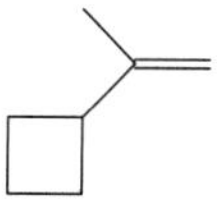  HBr

9. 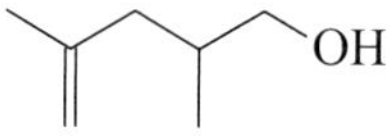  $H^{\oplus}$  → (complete)

10. 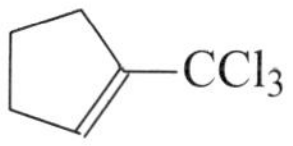 HBr

11. 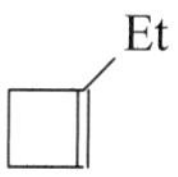 DBr

12. cyclopentene
   1. H_2 / Pd
   2. Br_2 / $h\nu$

5.2 Reactions

32 • Chapter 5 Alkenes and Carbocations

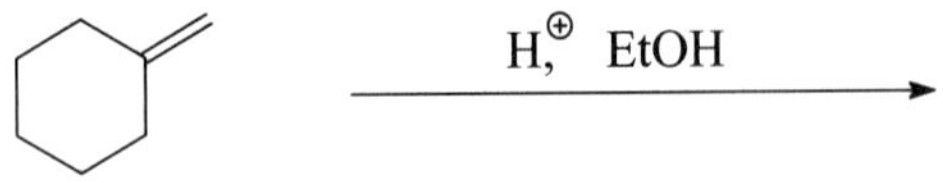

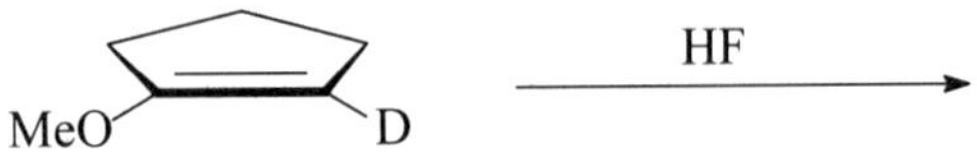

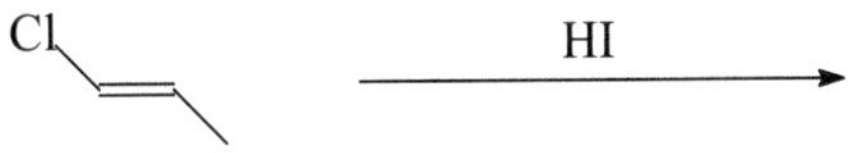

13. [methylenecyclohexane] → H⁺, EtOH

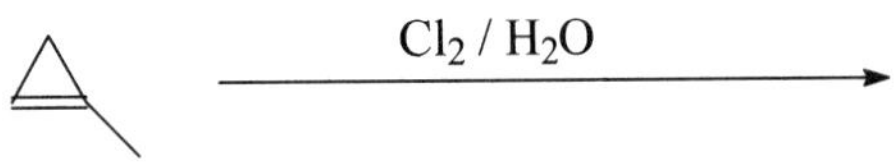

14. [1-methoxy-2-deutero-cyclopentene, MeO and D on sp² carbons] → HF

15. [1-chloropropene, Cl on CH=CH-CH₃] → HI

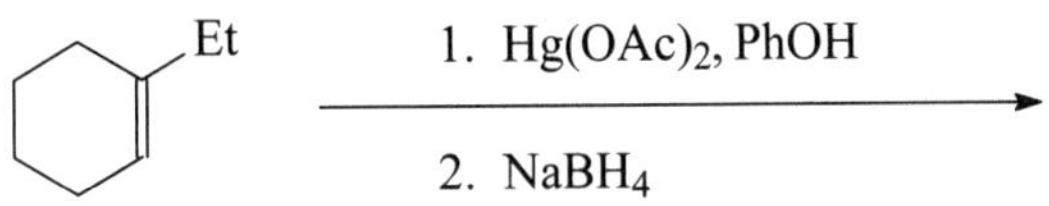

16. [methylenecyclopropane] → Cl₂ / H₂O

17. [methylenecyclopentane] → 1. B₂D₆ 2. H₂O₂, ⁻OH

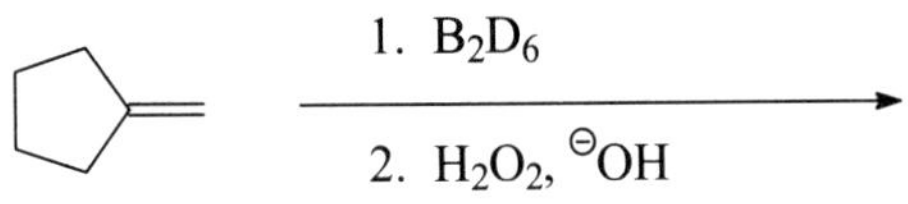

18. propylene → Cl₂ / (XS) NaI

19. [1-ethylcyclohexene] → 1. Hg(OAc)₂, PhOH 2. NaBH₄

20. H₂C=C=CH₂ (allene) → (XS) CH₂I₂ / Zn(Cu)

5.2 Reactions

21. 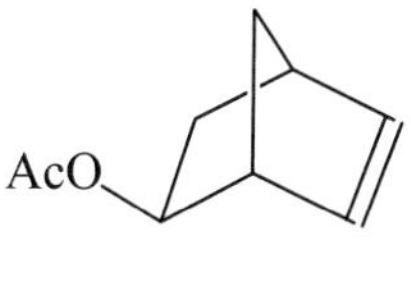 AcO-[norbornene with OAc] → 1. KMnO₄, ⁻OH 2. HIO₄

22. 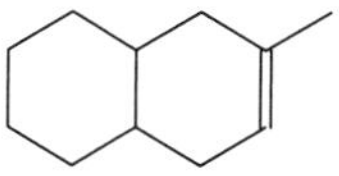 [decalin with methylidene substituent] → 1. O₃ 2. H₃O⁺, Zn

23. 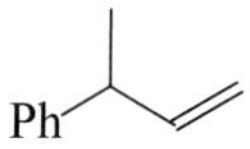 Ph-CH(CH₃)-CH=CH₂ → HBr, di-*t*-butyl peroxide

24. (E)-3-hexene → diazomethane, hν

25. cyclopentyl bromide → 1. base 2. OsO₄ 3. NaHSO₃

26. 3-methyl-1-butene → IN₃

27. 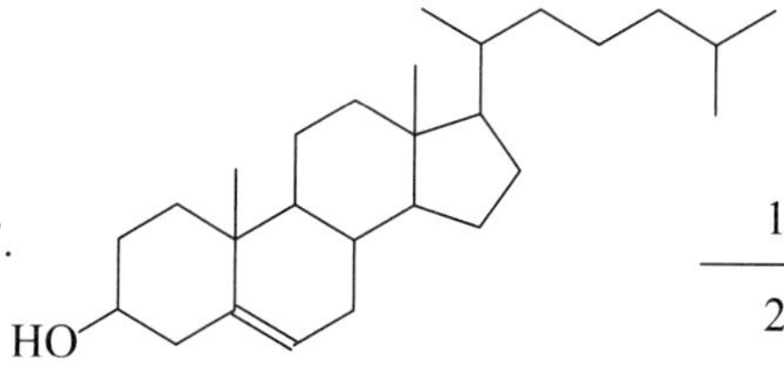 cholesterol → 1. BD₃, THF 2. H₂O₂, ⁻OH

28. 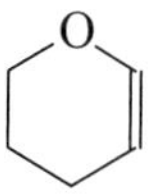 [3,4-dihydro-2H-pyran] → H⁺, MeOH

29. 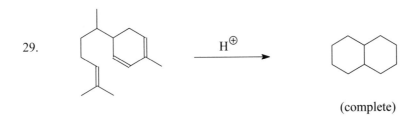 (complete)

30. styrene glycol —HIO₄→

31. [structure with OH on decalin] 1. H₂SO₄ 2. KMnO₄, ⁻OH

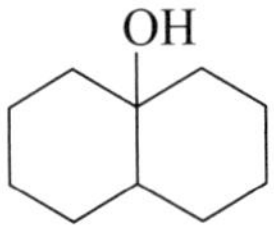

32. H₂C=C=O —HCl→

33. [structure with OR] 1. BH₃, THF 2. H₂O₂, ⁻OH → [ ] ⟹ *artemisinin* (antimalarial)

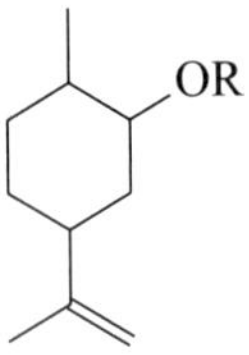

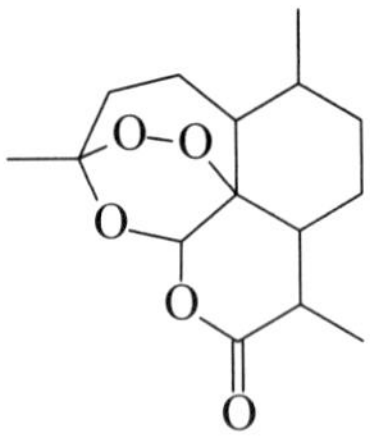

34. [estradiene with HO] 1. Br₂, H₂O 2. base → [ ] ⟹ *estrone*

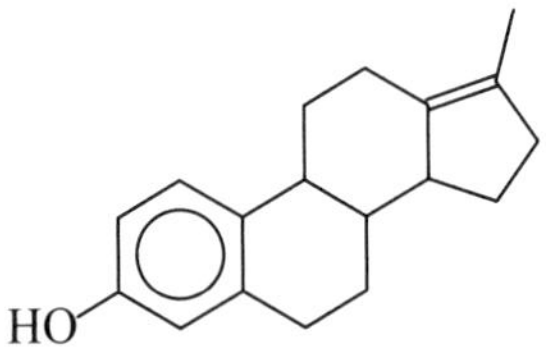

35. [epoxide terpene] —H⁺→ [decalin] (complete)

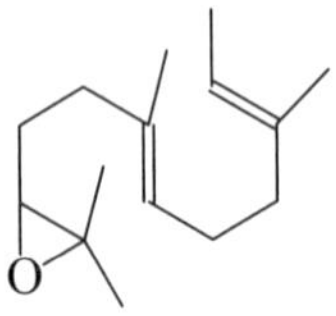

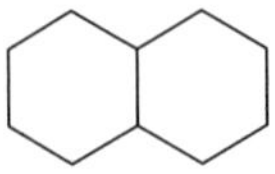

5.2 Reactions

36. chlorocyclopentane $\xrightarrow[\text{3. EtOH, H}^\oplus]{\text{1. }^\ominus\text{OR} \\ \text{2. mCPBA}}$

37. 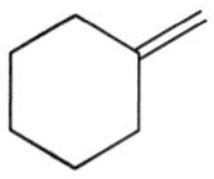 [methylenecyclohexane] $\xrightarrow{\text{Br}_2, \text{ }s\text{-BuOH}}$

38. 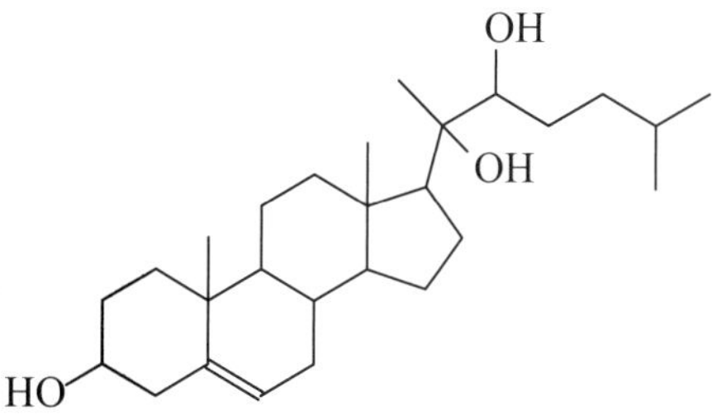 [steroid with diol side chain, HO at C3, Δ5] $\xrightarrow{\text{HIO}_4}$ pregnenolone

39. Draw the structure of the *largest* carbon-containing product in the following reaction:

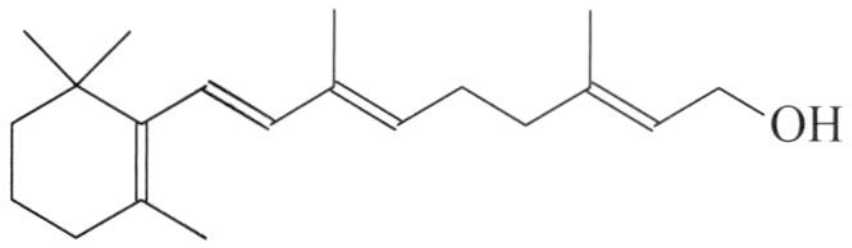

[vitamin A] $\xrightarrow{\text{KMnO}_4, \text{ H}^\oplus}$

vitamin A

40. 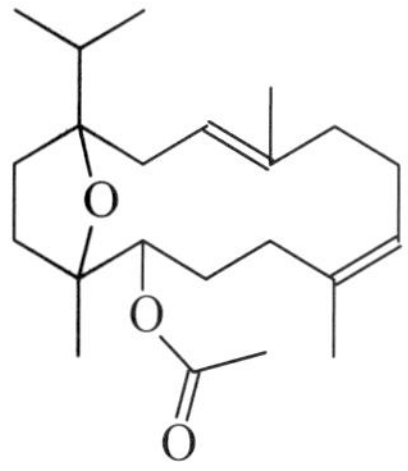 [incensole acetate structure] $\xrightarrow[\text{3. HIO}_4]{\text{1. OsO}_4 \\ \text{2. NaHSO}_3}$

incensole acetate
(found in frankincense)

41. 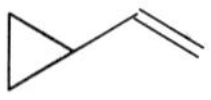 [vinylcyclopropane] $\xrightarrow[\text{two 1,2-shifts}]{\text{HCl}}$ [cyclobutane]

(complete)

5.2 Reactions

42. PhCH₂Cl + n-BuLi $\xrightarrow[-HCl]{1,1\text{-elimination}}$ [ ] $\xrightarrow{\text{cyclohexene}}$
   (*a very strong base*)

43. 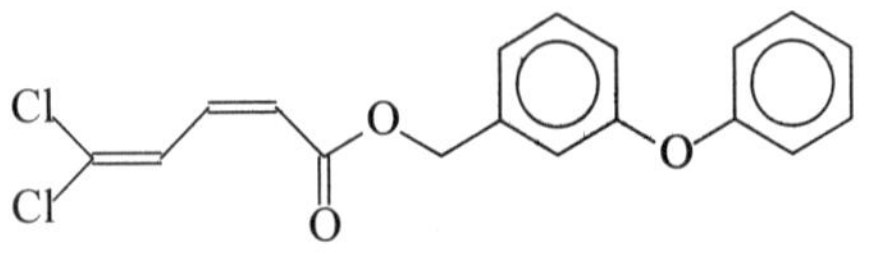 (dichlorovinyl acrylate ester of 3-phenoxybenzyl alcohol) $\xrightarrow[\text{Zn(Cu)}]{(CH_3)_2CI_2 \text{ (1 equiv)}}$

   permethrin (insect repellent)

5.3 Syntheses
Supply a reagent or sequence of reagents that will effect the following conversions.

1. 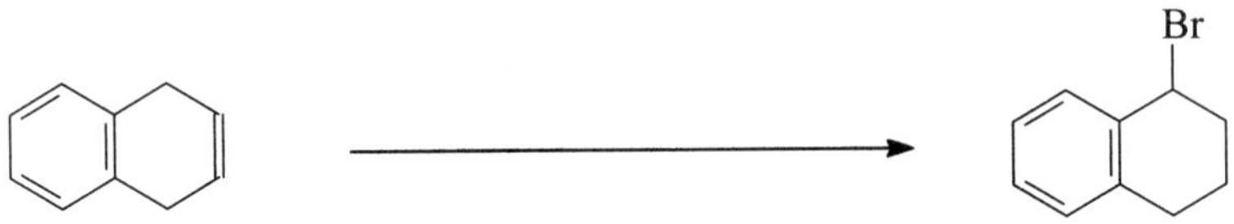 dihydronaphthalene ⟶ 1-bromo-tetrahydronaphthalene

2. cyclohexyl alcohol ⟶ cyclohexyl chloride

3. *t*-BuCl ⟶ *t*-BuF

4.  cyclobutyl chloride ⟶ cyclobutane

5.

6. cycloheptane ⟶ HOOC-(CH₂)₅-COOH

7. (3-methyl-1-pentene type alkene) ⟶ 2-bromo-2-methyl-1-chloro product

8. allylbenzene ⟶ phenylacetaldehyde (PhCH₂CHO)

9. ethylene ⟶ bromocyclopropane

10. cyclopentyl alcohol ⟶ bicyclo[2.1.0]pentane

11. 1,2-dihydronaphthalene ⟶ 2-tetralol (OH)
    ⟶ 1-tetralol (OH)

5.3 Syntheses

38 • Chapter 5 Alkenes and Carbocations

12.  [bicyclic compound with Br at ring junction] ⟶ [cyclooctane-1,5-dione]

13. cyclohexane ⟶ [trans-1-bromo-2-deuteriocyclohexane]

14.  [3-bromo-2,2-dimethylbutane type structure] ⟶ [1-bromo-3,3-dimethylbutane structure]

15. isobutane ⟶ isobutyl alcohol

16.  [cyclohexane with two adjacent CH(Cl)CH₃ groups] ⟶ [cyclohexane with two adjacent C(CH₃)CO₂H groups]

17. *t*-butyl chloride ⟶ isobutylene chlorohydrin

18.  [1-methyl-1-vinylcyclohexane] ⟶ [open-chain diketone]

5.3 Syntheses

19. 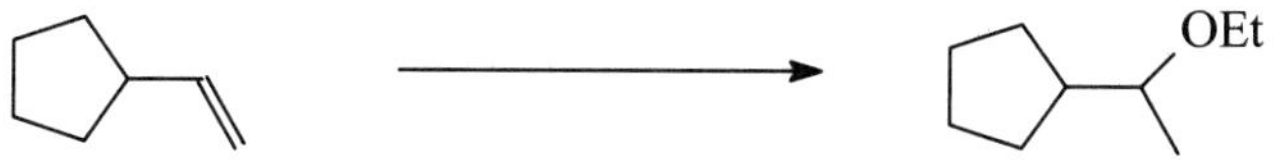

20. 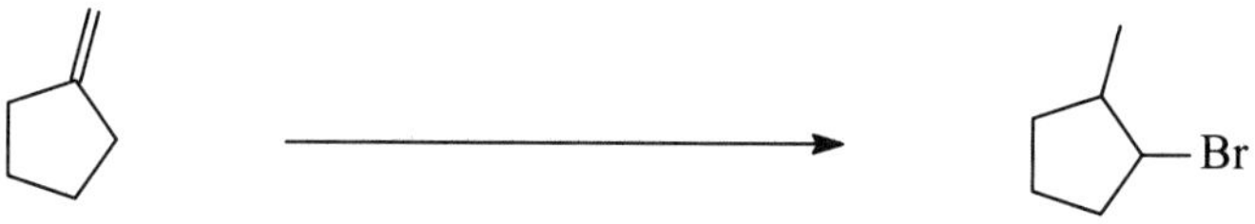

21. ethylene 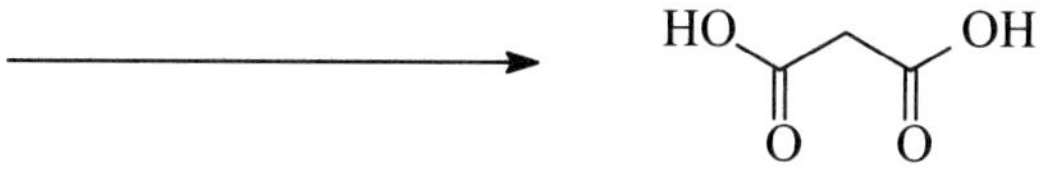

22. *t*-butyl bromide (*only source of carbon*) → di-*t*-butyl ether

23. cyclobutyl alcohol → hausane (bicyclo[2.1.0]pentane)

5.4 Mechanisms

Outline a detailed mechanism for each of the following. No other reagents than those given are necessary. Use arrows to explain the flow of electrons and show all intermediates. NO WORDS!

1. 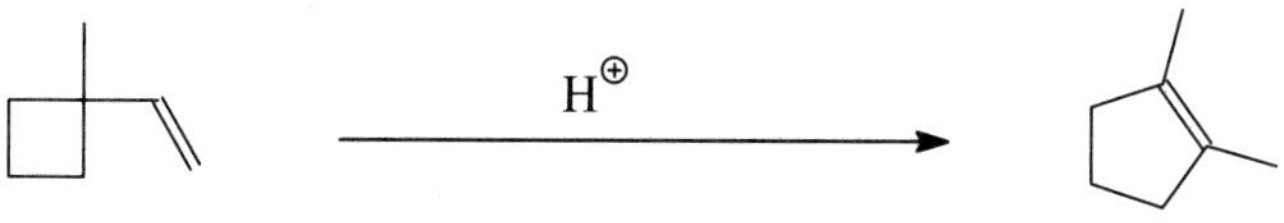

2. 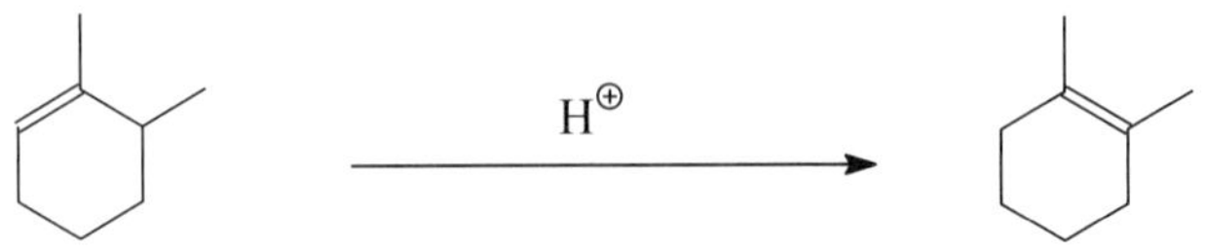 [alkene] →(H⁺) [isomerized alkene]

3.  isoprene →(H⁺, MeOH) [product with OMe]

4. 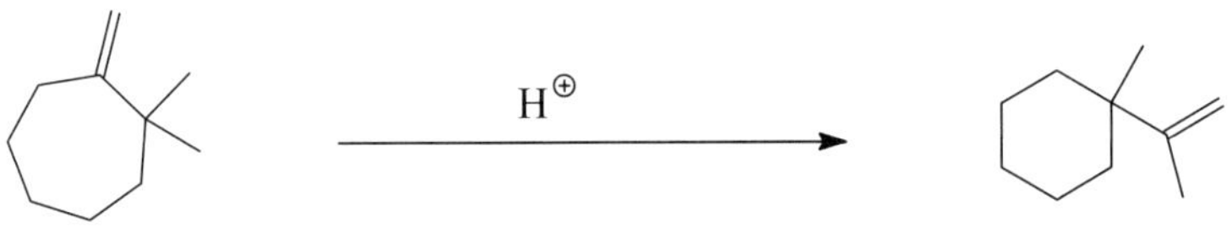 [methylenecycloheptane derivative] →(H⁺) [product]

5. 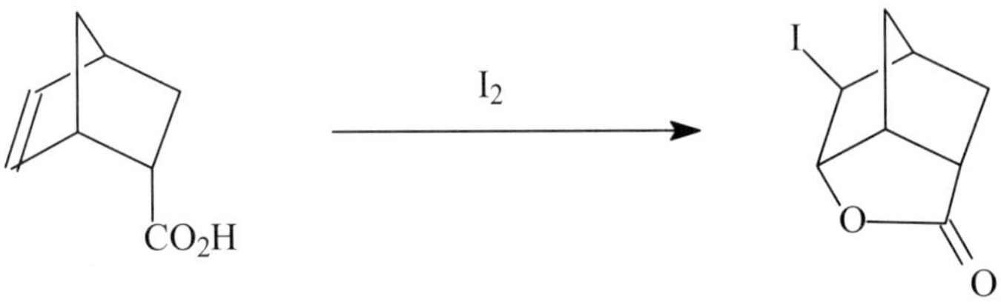 [norbornene carboxylic acid] →(I₂) [iodolactone product]

6. Isobutylene in the presence of excess propylene and a trace of acid yields C_7H_{14}. Deduce this product.

5.4 Mechanisms

7.

8. [structure with OH] → 1. Hg(OAc)₂ 2. NaBH₄ → [2,2-dimethyltetrahydropyran]

9. [methylcyclohexene] + [2-butene] → (a C_{11} olefin)
   $H^{\oplus}$

10. [amine with alkene] → I₂, −HI → [bicyclic amine with I]

11. [2-ethyl-1-butene] + $H_2C=\overset{\oplus}{N}=\overset{\ominus}{\underset{..}{N}}:$ (diazomethane) → [pyrazoline product]

5.4 Mechanisms

12. The reaction of 3-bromocyclohexene with HBr yields only *trans*-cyclohexene dibromide, i.e., no *cis*-product is formed. In contrast, 3-methylcyclohexene reacts with HBr to yield a mixture of *cis*- and *trans*-stereoisomers, as well as a tertiary alkyl halide. Explain with appropriate structures and arrows.

13. The natural products *caryophyllene* and *isocaryophyllene* (odor somewhere between cloves and turpentine) are stereoisomers that differ in the configuration of a double bond. They have the molecular formula $C_{15}H_{24}$. Catalytic hydrogenation of either yields the same compound, $C_{15}H_{28}$. Ozonolysis, followed by zinc and aqueous acid, yields **A** and an other aldehyde. Suggest structures for the caryophyllenes.

A

14. Treatment of an unknown alkene with $Hg(OAc)_2$ in H_2O/THF, followed by a $NaBH_4$ workup, produces an alcohol isomeric to one obtained by hydroboration-oxidation of the same alkene. Reduction of the alkene affords the compound C_5H_{12}, while ozonolysis yields an aldehyde, CH_3CHO, as one of the products. Deduce the structure of the alkene.

15. *Partial* catalytic hydrogenation of C_5H_8 (**A**) yields a mixture of **B**, **C**, and **D**. Ozonolysis, followed by a reductive work-up (Zn, H_3O^+), of **B** gives no new products. When treated in the same way, **C** gives formaldehyde and 2-butanone and **D** gives formaldehyde and isobutyraldehyde. Provide structures for compounds **A** through **D**. What is the *common* name of **A**?

formaldehyde   2-butanone   isobutyraldehyde

5.4 Mechanisms

16. β-Myrcene, C₁₀H₁₆, found in bayleaves and hops, is an intermediate in the manufacture of perfumes. When treated with H₂/Pt, 2,6-dimethyloctane is formed (β-myrcene has no triple bonds). Treatment of β-myrcene with ozone, followed by an acidic zinc work-up, yields **A** (C₅H₆O₃), acetone (Me₂CO), and two equivalents of formaldehyde. What are the structures of β-myrcene and **A**?

17. Reaction of **A**, C₁₀H₁₆, with H₂/Pd yields **B**. When treated with KMnO₄, a brown precipitate forms. When **A** is treated with ozone followed by zinc in acid, compound **C** and another product are produced. What are the structures of **A** and the other ozonolysis product?

B

C

18. Draw
a. the structure of the *monomer* that would give the following polymer by an addition mechanism:

b. a segment (three or four repeating units) of poly(styrene).

19. *t*-Butyl vinyl ether is polymerized commercially by a cationic process for use in adhesives. Show the mechanism for linking three monomeric units.

20. 2 CH₂N₂ —Δ→ ethylene + 2 N₂

5.4 Mechanisms

44 • Chapter 5 Alkenes and Carbocations

21.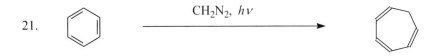

22.

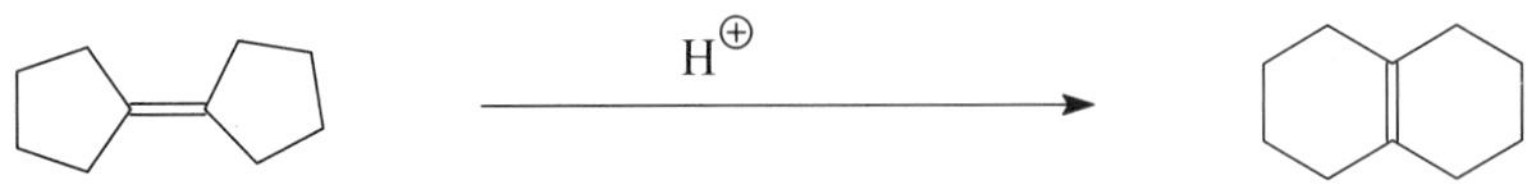

23.

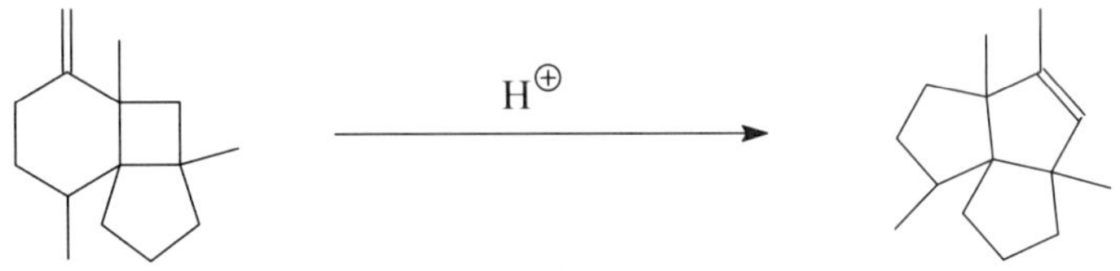

isocomene (from goldenrod)

24. Hydride shifts and alkyl migrations occur in many enzyme-catalyzed reactions in all living species – *including you as you are working these problems!* Below is one such biochemical reaction (see 14.3, 6 for perhaps the very best example). Account for the formation of all intermediates leading to the product. (Hint: positive sulfur, like positive oxygen, is a good leaving group, *i.e.*, it easily leaves a carbon to which it is attached, taking with it both bonding electrons.)

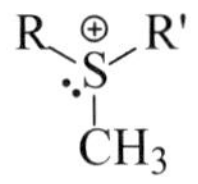

SAM (S-adenosylmethionine, a common methylating agent in all of us)

oleic acid (a fatty acid)

5.4 Mechanisms

25.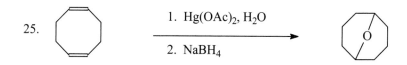

26.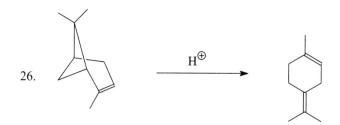

27. Elaidic acid ($C_{18}H_{34}O_2$), a fatty acid, is present in processed foods such as margarine and *may* contribute to elevated levels of cholesterol. Reaction of elaidic acid with Simmons-Smith reagent produces compound **I**, whereas reaction with acidic permanganate yields **II** and **III**. What is the structure of elaidic acid? Indicate stereochemistry.

28. Compound **A** ($C_{10}H_{18}O$) reacts with H_2SO_4 to give **B** ($C_{10}H_{16}$) and an isomer **C**. Ozonolysis of **B** yields a diketone; ozonolysis of **C** yields **D**. (a) Draw structures for **A**, **B**, and **C**. (b) Describe a simple chemical color test that would differentiate **A** from **B** or **C**.

5.4 Mechanisms

46 • Chapter 5 Alkenes and Carbocations

29. 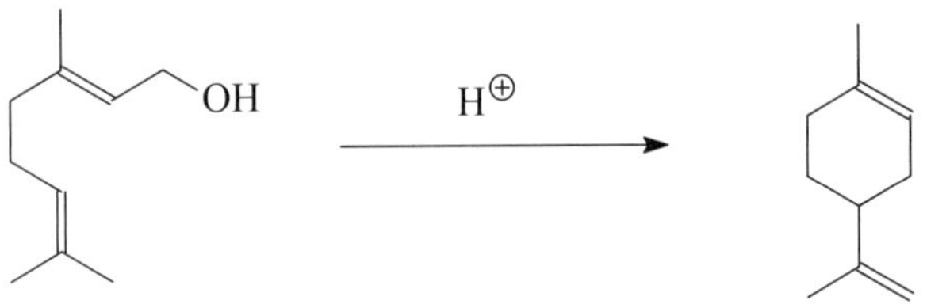 [structure of geraniol-like alcohol] —H⊕→ limonene (volatile in lemons and oranges)

Hint: some alcohols can be protonated to form oxonium ions which may then "leave" as water to give a carbocation.

30. Aziridines (**B**) are nitrogen analogs of epoxides and can be made from azides (**A**) by the following reaction:

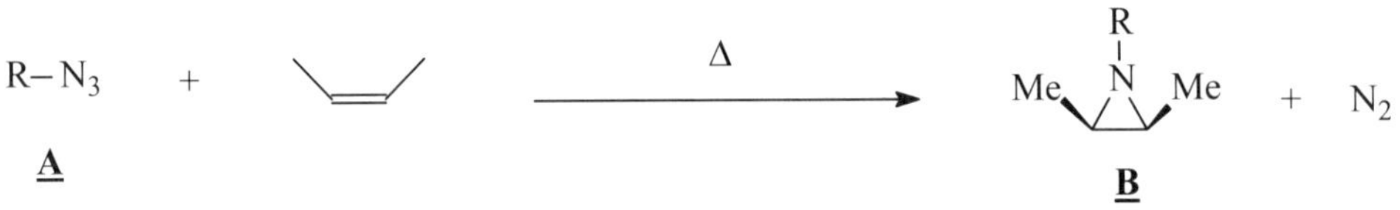

R–N₃ + [cis-2-butene] —Δ→ [aziridine with R on N, Me and Me on ring] + N₂
A   **B**

Recalling the mechanism of generating carbene from diazomethane, and the fact that nitrogen is an excellent "leaving group," (a) draw the resonance structure of **A** that *best* illustrates how it can decompose to extrude N₂ and (b) supply electron flow arrows to show the *structure of the reactive intermediate* derived from **A** that reacts with *cis*-2-butene to give **B**.

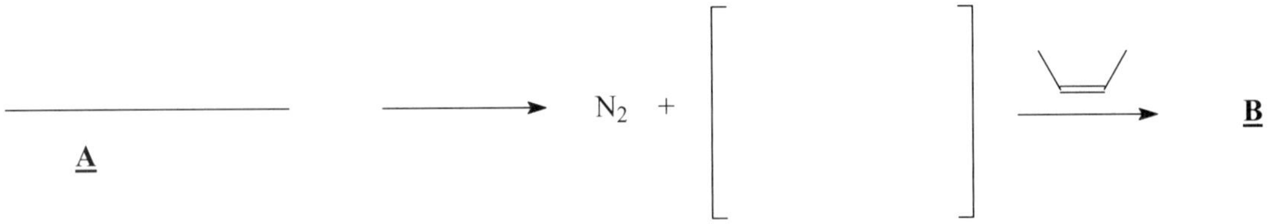

_____ ———→ N₂ + [_____] ——→ **B**
A

(c) Given the observed stereochemistry, what *type* of mechanism does this addition reaction illustrate?

31. 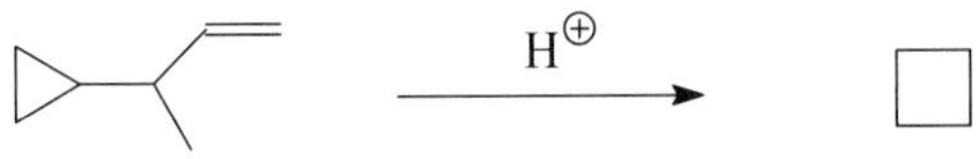 [cyclopropyl-substituted alkene] —H⊕→ ☐

(an olefin - complete)

32. styrene (vinylbenzene)   1. Cl₂, H₂O   2. base   3. *dry* HCl →

5.4 Mechanisms

33. 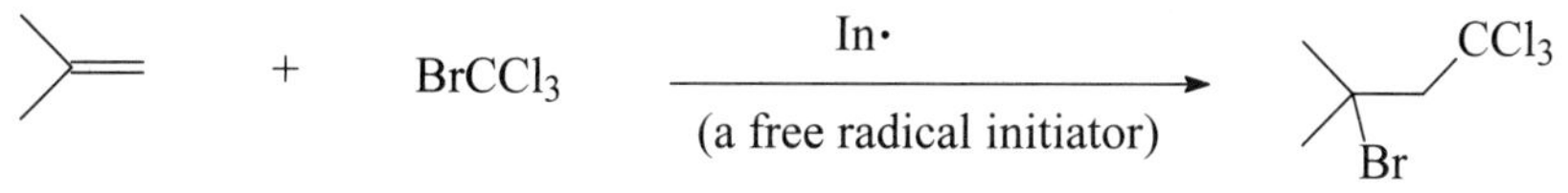 (CH₃)₂C=CH₂ + BrCCl₃ →(In•, a free radical initiator)→ (CH₃)₂C(Br)CH₂CCl₃

34. Compound **A**, $C_{16}H_{30}O$, is a sex attractant (pheremone) for the male silkworm moth. Given the data from the following three experiments, deduce the structure of **A**, clearly showing its stereochemistry.

a. Catalytic hydrogenation of **A** yields $C_{16}H_{34}O$.

b. Ozonolysis of **A**, followed by treatment with zinc and acid, yields compounds **B**, **C**, and **D**.

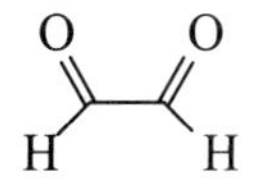

B: OHC–CHO (glyoxal)

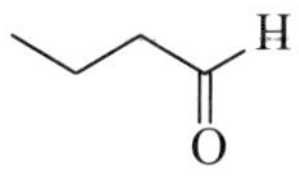

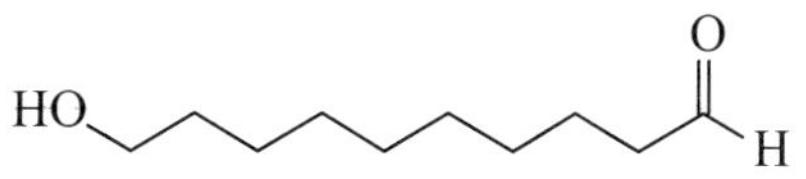

C: CH₃CH₂CH₂CHO (butanal)

D: HO–(CH₂)₈–CHO

c. *Incomplete* reaction of **A** with diazomethane (CH_2N_2) gives a mixture of **E** and **F** (the C_{11} and C_9 substituents contain one oxygen atom). Note: this experiment establishes the stereochemistry of **A**.

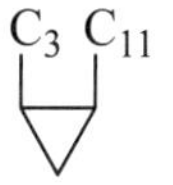

E: cyclopropane with C_3 and C_{11} substituents

F: cyclopropane at C_5 with C_9 substituent

A = _____

35. 1,4-Cyclohexadiene undergoes isomerization to 1,3-cyclohexadiene in the presence of acid. Two mechanisms are possible: protonation followed by deprotonation (path a) vs. protonation followed by a 1,2-hydride shift and subsequent deprotonation (path b):

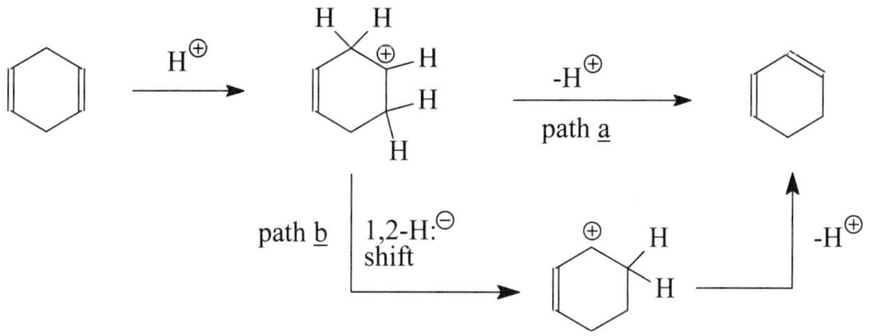

If 3,3,6,6-tetradeuterio-1,4-cyclohexadiene is treated with acid, 1,2,5,5-tetradeuterio-1,3-cyclohexadiene is formed. Which path is favored? (Note: C-H bonds are *slightly* weaker than C-D bonds.)

36. Carbonyl groups greatly affect the acidity of nearby (α-) protons. For example, the pK$_a$ of cyclohexane is about 60, but the pK$_a$ of H$_a$ in cyclohexanone is about 20. This dramatic increase in acidity is largely a consequence of resonance stabilization of the conjugate base of the latter (for an example of the additive effect on pK$_a$s of 1,3-dicarbonyls, see problem 1.3, 7), and allows an easy exchange of α-hydrogen for deuterium atoms by the following mechanism:

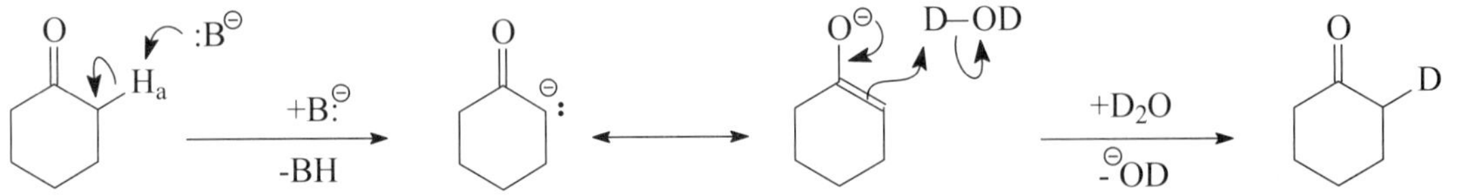

cyclohexanone

Under the same conditions, however, species **A** does not undergo hydrogen-deuterium exchange. Explain. Hint: consider the geometric constraints of olefinic moieties.

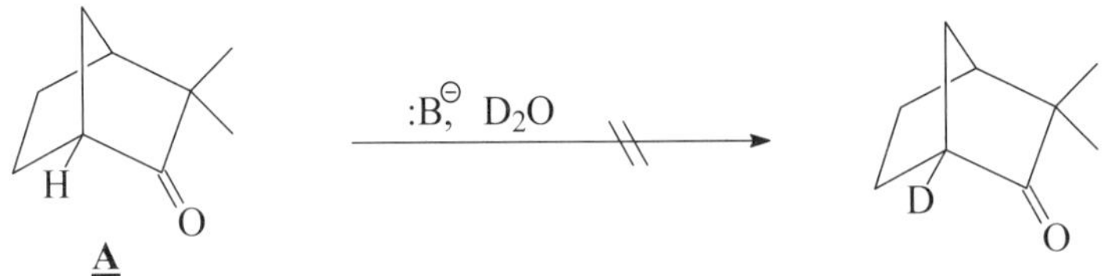

5.4 Mechanisms

CHAPTER 6
ALKYNES

6.1 Reactions
Draw the structural formula of the major organic product(s). Show stereochemistry where appropriate.

1. 3-penten-1-yne $\xrightarrow{\text{1. NaH} \quad \text{2. D}_2\text{O}}$

2. 1-octyne $\xrightarrow{\text{H}^\oplus, \text{HgSO}_4, \text{PhOH}}$

3. phenylacetylene $\xrightarrow{\text{1. B}_2\text{H}_6 \quad \text{2. H}_2\text{O}_2, {}^\ominus\text{OH}}$

4. *n*-BuCl
   1. $^\ominus$OMe, HOMe
   2. Cl$_2$
   3. (XS) NaNH$_2$
   4. BH$_3$·THF
   5. H$_2$O$_2$, $^\ominus$OH

5. (1-chloro-1-methylcyclopentane) $\xrightarrow{\text{RC}\equiv\text{C:}^\ominus}$

6. (cyclohexylacetylene)
   1. Li / NH$_3$
   2. HBr, di-*t*-butyl peroxide

7. isopropylacetylene
   1. H$_2$ / Pd(Pb)
   2. BH$_3$
   3. H$_2$O$_2$, $^\ominus$OH

8. 1-decyne
   1. NaH
   2. CH$_3$(CH$_2$)$_{12}$Cl
   3. Lindlar catalyst

   muscalure (pheromone for house fly)

9. 1,1-dichlorobutane
   1. (XS) NaNH$_2$
   2. H$_3$O$^\oplus$, HgSO$_4$

6.1 Reactions

50 • Chapter 6 Alkynes

10. 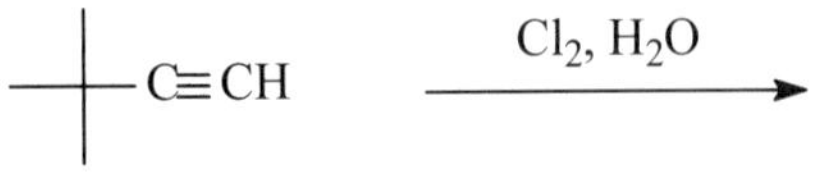 (CH₃)₃C—C≡CH $\xrightarrow{\text{Cl}_2, \text{H}_2\text{O}}$

11. PhC≡CH $\xrightarrow[\text{2. Zn(Cu), cyclopentene}]{\text{1. (XS) HI}}$

12.

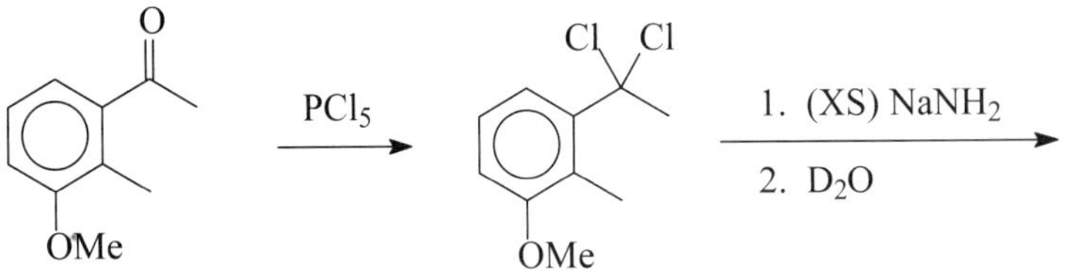

[2-methoxy-6-methylphenyl methyl ketone] $\xrightarrow{\text{PCl}_5}$ [2-methoxy-6-methylphenyl-CCl₂-CH₃] $\xrightarrow[\text{2. D}_2\text{O}]{\text{1. (XS) NaNH}_2}$

13. H—C≡C—CH₂OH $\xrightarrow[\substack{\text{2. }\underline{n}\text{-C}_5\text{H}_{11}\text{Br (1 equiv)} \\ \text{3. H}^\oplus}]{\text{1. LiNH}_2 \text{ (2 equiv)}}$

14. acetylene $\xrightarrow[\substack{\text{3. NaNH}_2 \\ \text{4. }\underline{t}\text{-Bu-Cl}}]{\substack{\text{1. NaNH}_2 \text{ (1 equiv)} \\ \text{2. }\underline{n}\text{-Pr-I}}}$

6.2 Syntheses
Supply a reagent or sequence of reagents that will effect the following conversions.

1.  [3,4-dibromohexane] ⟶ [propanoic acid, CH₃CH₂COOH]

2. acetylene ⟶ $\underline{n}$-pentyl bromide

6.2 Syntheses

3. vinyl chloride $\longrightarrow$ methyl vinyl ether

4. acetylene $\longrightarrow$ (E)-3-octene

5. *t*-butylacetylene $\longrightarrow$ 2-chloro-2,3-dimethylbutane

6. 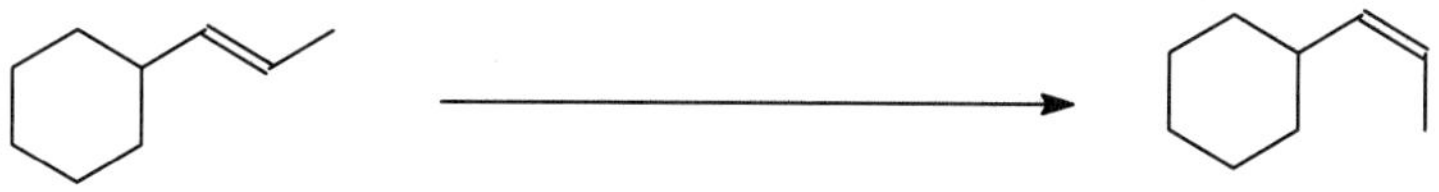 [cyclohexyl-CH=CH-CH₃ (E)] $\longrightarrow$ [cyclohexyl-CH=CH-CH₃ (Z)]

7. propyne $\longrightarrow$ [methylcyclopropane]

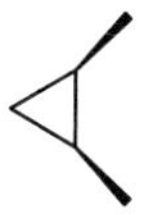

8. propyne $\longrightarrow$ *n*-propyl bromide

9. 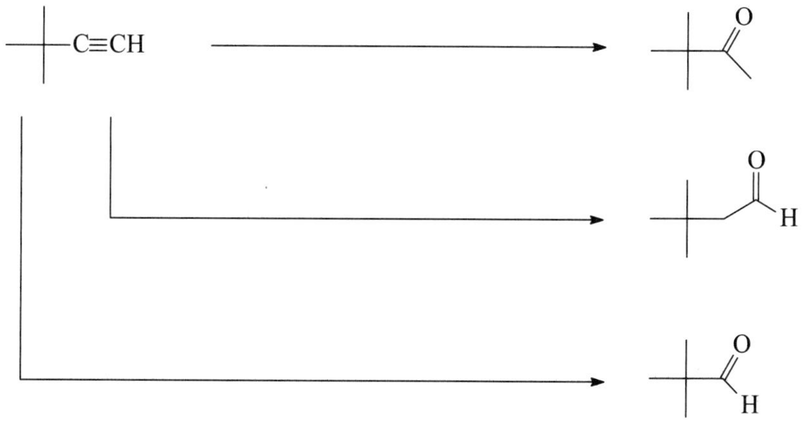 (CH₃)₃C—C≡CH $\longrightarrow$ (CH₃)₃C—C(=O)—CH₃

    $\longrightarrow$ (CH₃)₃C—CH₂—CHO

    $\longrightarrow$ (CH₃)₃C—CHO

6.2 Syntheses

10. styrene ⟶ (E)-1-phenyl-1-butene

11. diphenylacetylene ⟶ cis-1,2-diphenylcyclopropane

    diphenylacetylene ⟶ PhCHO

12. 3-hexyne ⟶

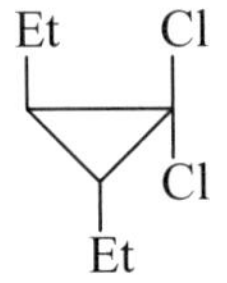

    (cyclopropane with Et, Cl on one carbon and Et, Cl on another)

13. CH₃(CH₂)₄CH₂Cl ⟶

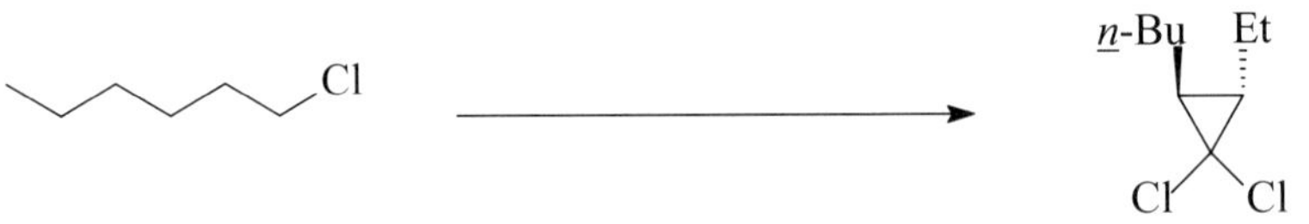

    (cyclopropane with n-Bu, Et substituents and CCl₂)

14. acetylene ⟶ 2-heptanone (odor of cheddar cheese)

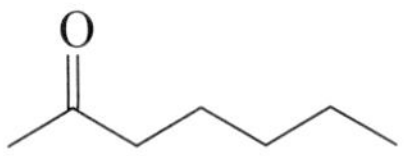

15. acetylene ⟶ *disparlure* (pheremone for female gypsy moth)

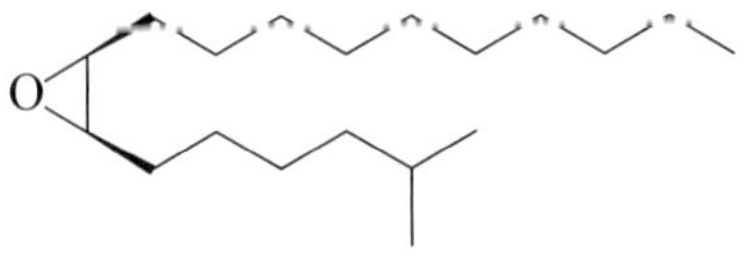

6.2 Syntheses

16. 1-pentyne ⟶ 2 CH₃CH₂CH₂COOH

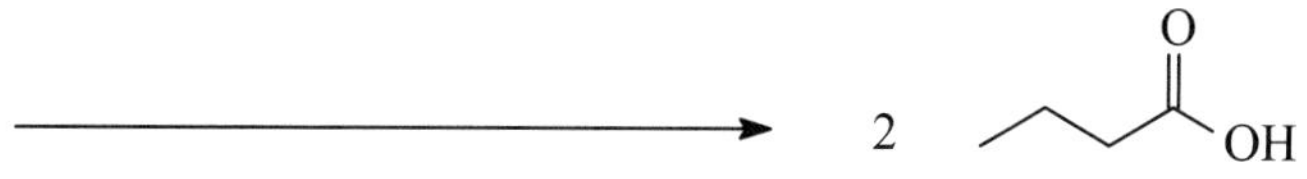

6.3 Mechanisms

1. 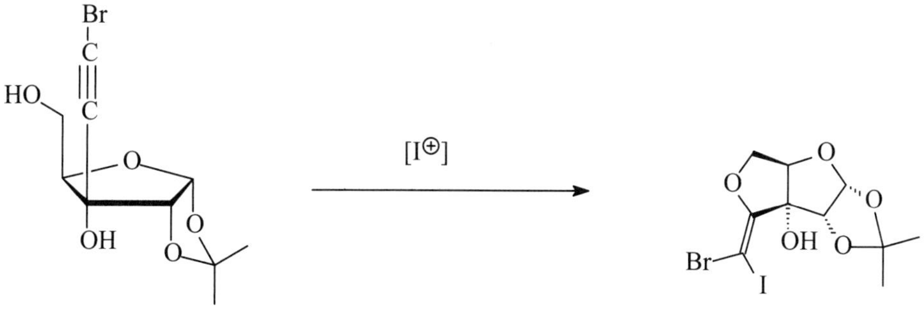 [sugar substrate with HO, Br-C≡C-, OH, and isopropylidene acetal groups] —[I⁺]→ [bicyclic product with =CBrI, OH, and isopropylidene acetal]

2. 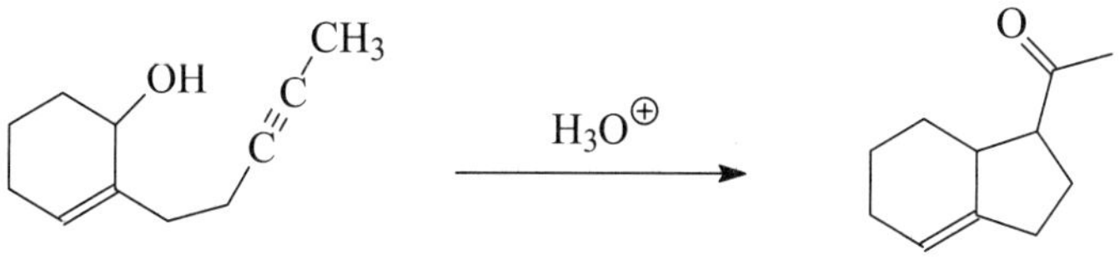 [cyclohexenol with pendant 2-butynyl chain] —H₃O⁺→ [bicyclic hydrindanone with acetyl group]

3. 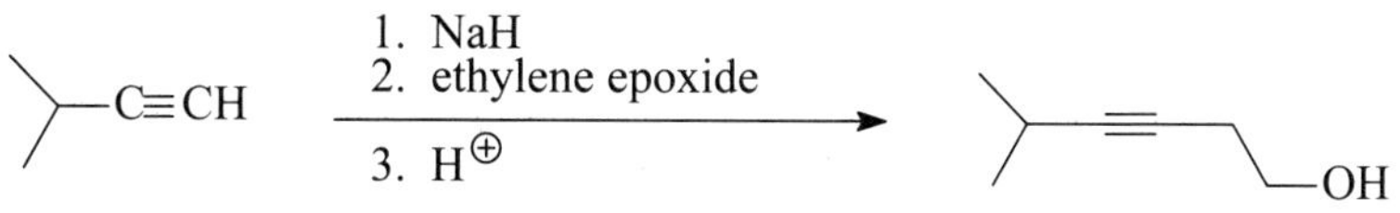 (CH₃)₂CH−C≡CH
   1. NaH
   2. ethylene epoxide
   3. H⁺
   ⟶ (CH₃)₂CH−C≡C−CH₂CH₂−OH

4. [Structure of steroid with Me₂N-aryl group and ketone] → 1. H₃CC≡C:⁻   2. H⁺ → mifepristone (RU-486)

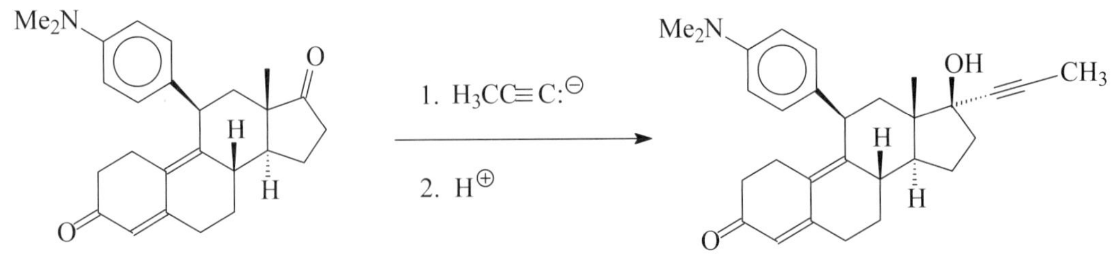

5. In the presence of *very* strong base an internal triple bond in *any* position of a straight chain alkyne will shift to the terminus of the chain, a process known as the *acetylene zipper reaction*:

$$R-\!\!\equiv\!\!-CH_3 \xrightarrow{\text{strong base}} RCH_2-C\equiv C:^{\ominus}$$

6.3 Mechanisms

CHAPTER 7
STEREOCHEMISTRY

7.1 General

1. Which of the following molecules are chiral?

a., 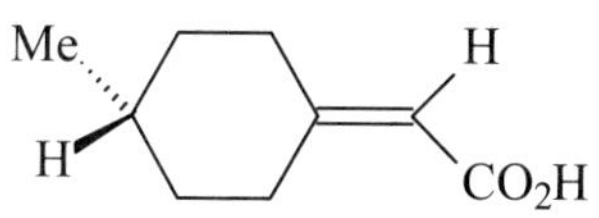 b., 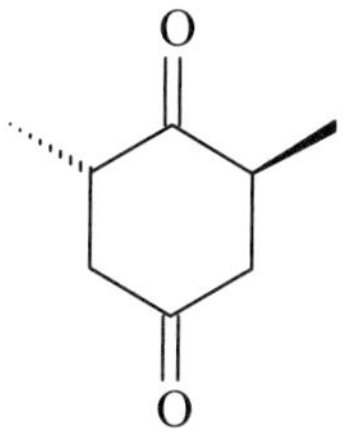 c.

d., 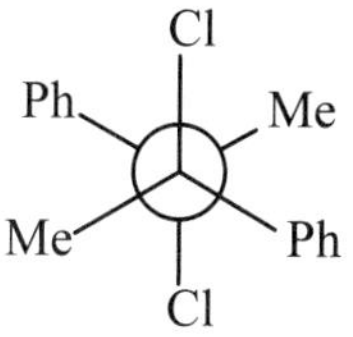 e., 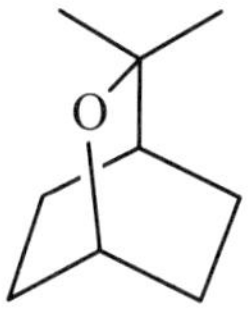 f.

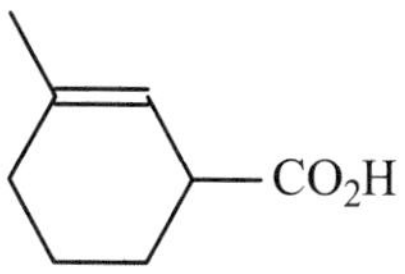

g., 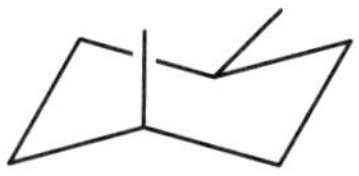 h. HC≡C—C≡C—CH=C=CH—CH=CHCH$_2$CO$_2$H
(an antibiotic)

i., 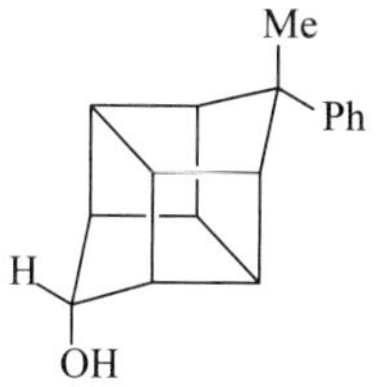 j. 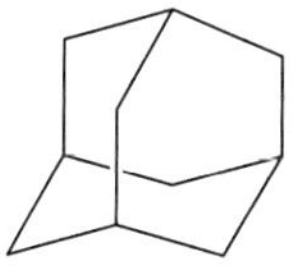 adamantane (an antiviral agent), k.

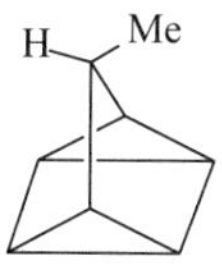

l., 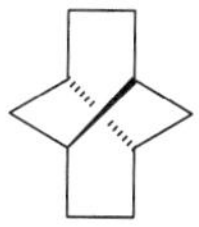 m. the C$_2$-epimer of 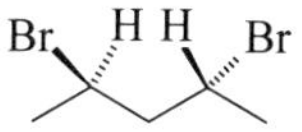 [Br-CH-CH$_2$-C(CH$_3$)$_2$-Br with H H], n. 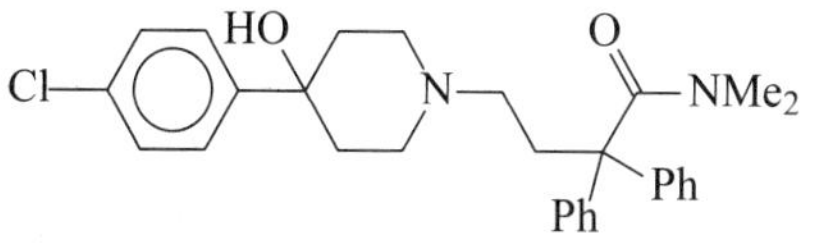 loperamide (Imodium^{TM} - antidiarrheal)

2. How many chiral carbons are there in each of the following molecules?

a. 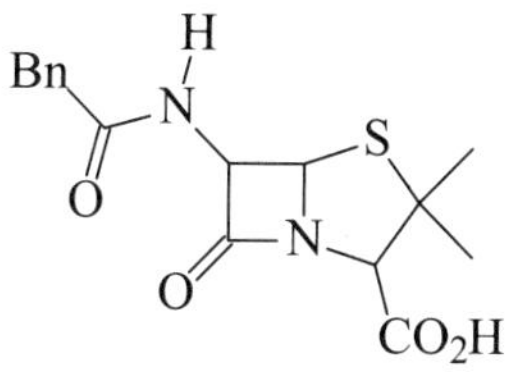 penicillin G

b. 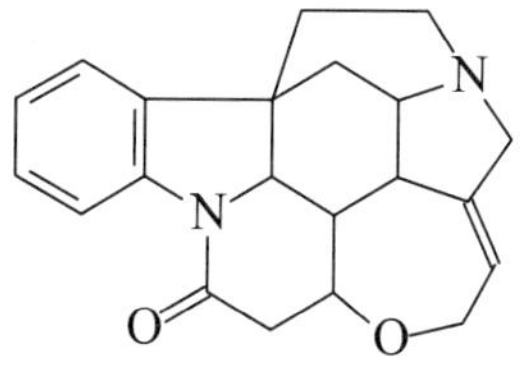 strychnine

56 • Chapter 7 Stereochemistry

c. 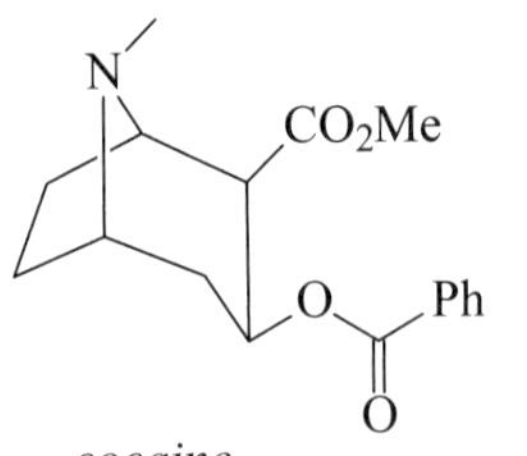 cocaine

d. 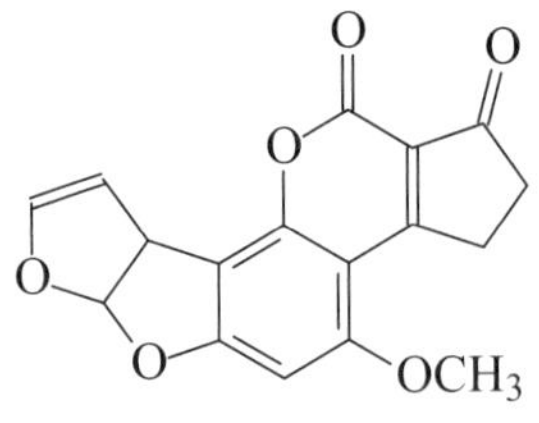 aflatoxin B$_1$

3. Identify each chiral center as (R)- or (S)-.

a. 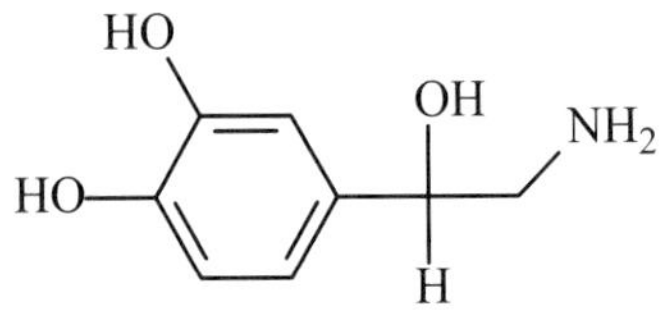 (−)-norepinephrine

b.

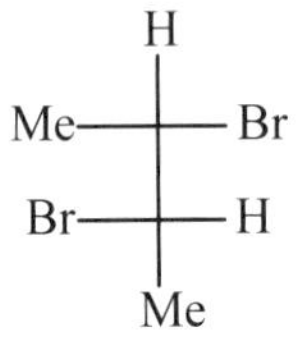

c.

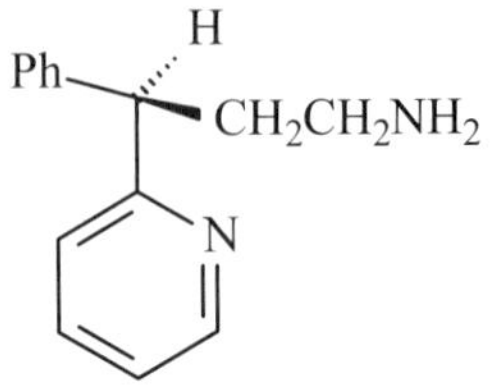

d.

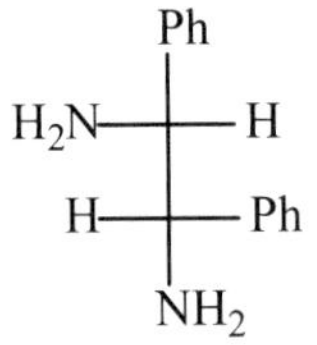

e.

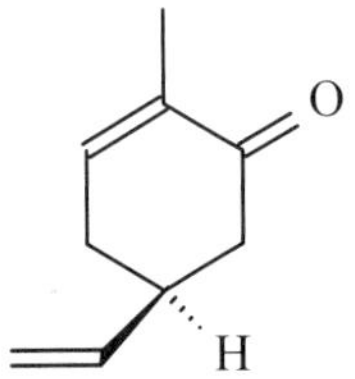

f. 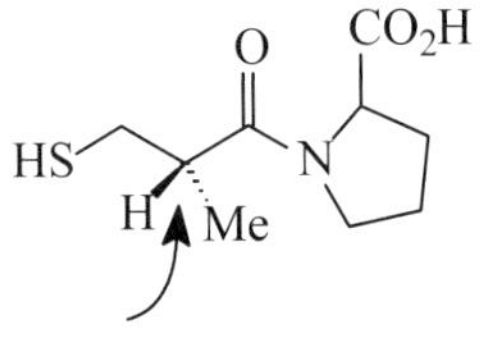 captopril (antihypertensive)

g. 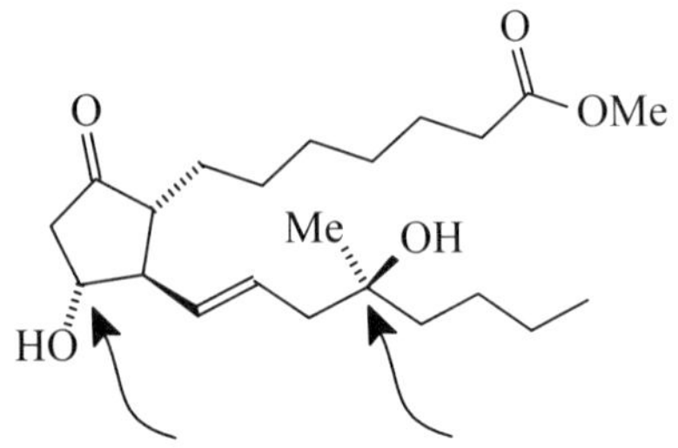 misoprostol (Cytotec™ - promotes cervical ripening)

4. Identify each of the following pairs of structures as identical, enantiomers, or diastereomers.

a.

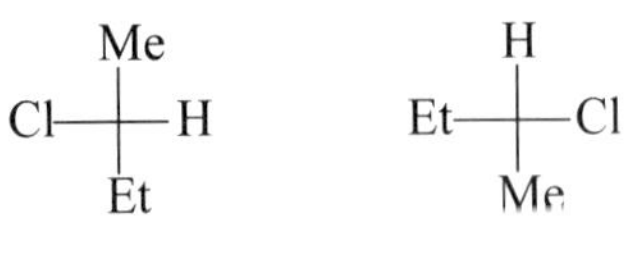

b.

c.

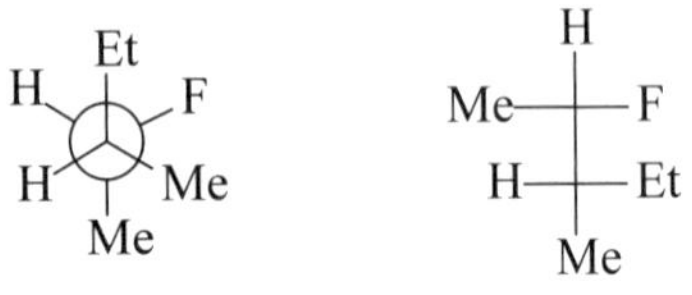

d.

7.1 General

e. 
```
     CHO              H
H─┬─OH         HO─┬─CHO
    CH₃             CH₃
```

f. 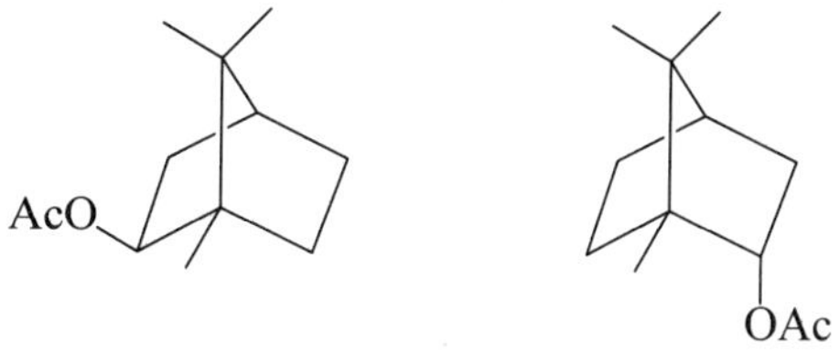 (structures with AcO and OAc on bicyclic framework)

g. 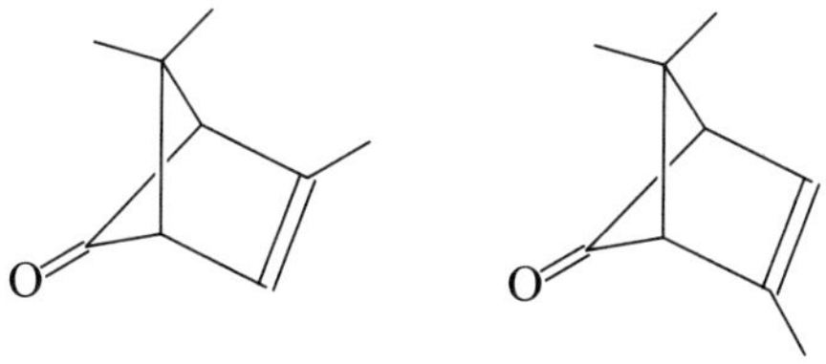 α-pinene (from pine resin)

h. 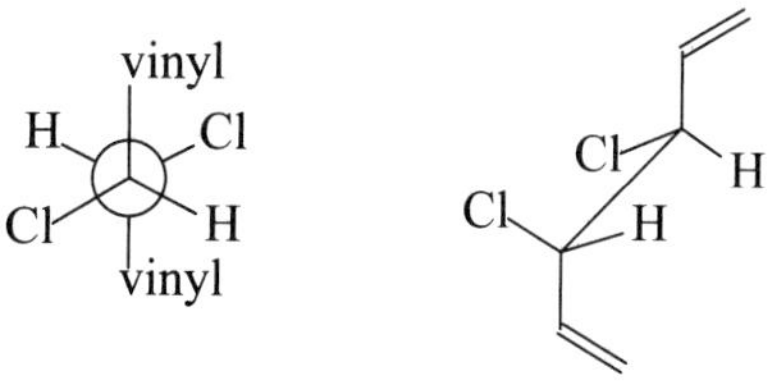 (Newman projection with vinyl, Cl, H / Cl, H, vinyl) and (allyl chain with Cl, H substituents)

i. (two spiro bicyclic acetal structures)

j. 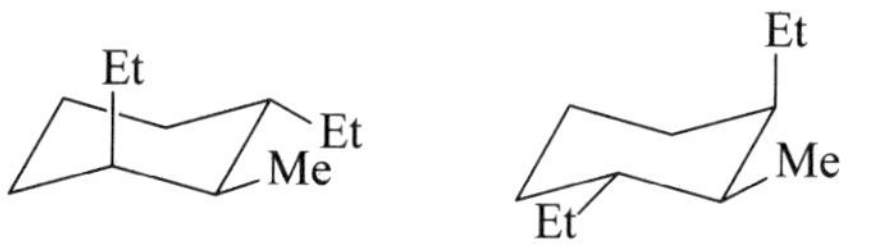 (two cyclohexane structures with Et, Me, Et substituents)

5. How many "kinds" of hydrogens (enantiomeric and diastereomeric hydrogens are different!) are there in

a. isohexane?   b. (R)-2-chloropentane?   c. (S)-4-chloro-1-pentene?

6. Nomenclature. Give the complete IUPAC name for the following:

a. 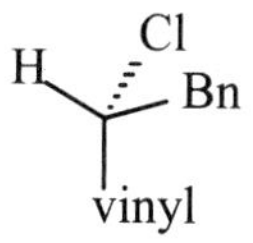 (stereocenter with H, Cl, Bn, vinyl)

b. 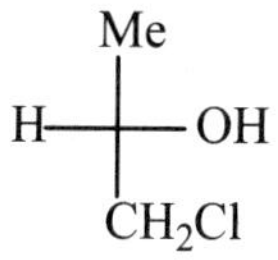
```
      Me
H─┬─OH
   CH₂Cl
```

c. 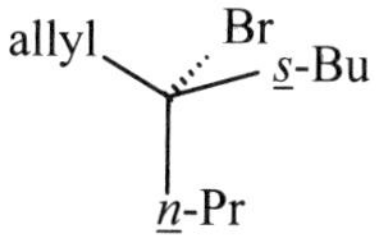 (stereocenter with allyl, Br, s-Bu, n-Pr)

d. 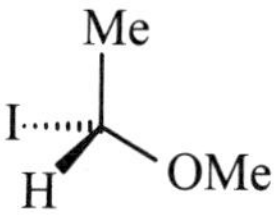 (stereocenter with Me, I, H, OMe)

e. 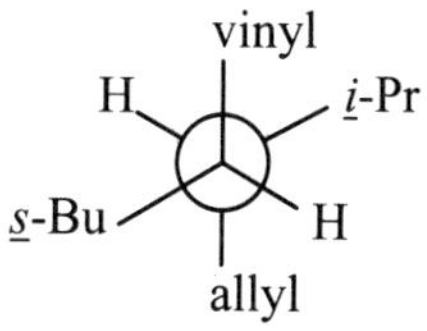 (Newman projection with vinyl, i-Pr, H / s-Bu, allyl, H)

f. 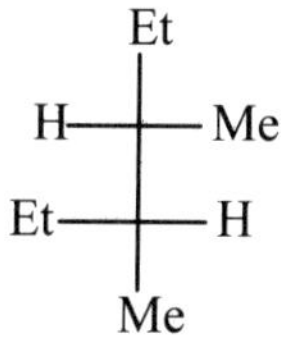
```
     Et
H─┬─Me
Et─┴─H
     Me
```

7.1 General

7. How many

a. *pairs of enantiomers* exist for bromochlorocyclopentane?

b. *geometric diastereomers* exist for 1,3-dichloro-2,4-dimethylcyclobutane?

c. *pairs of enantiomers* are possible for chlorofluorocyclobutane?

d. *meso* stereoisomers and how many *enantiomeric pairs* exist for

$$\text{CH}_3\text{CH(OH)CH(Cl)CH(OH)CH}_2\text{CH}_3$$?

e. *meso* stereoisomers exist for 2,3,4,5-tetrachlorohexane?

8. a. *D-Xylose* is a common sugar found in maple trees. Because it is much less likely to cause tooth decay than sucrose, D-xylose is often used in the manufacture of candy and gum. D-Xylose is the C_4-epimer of the enantiomer of **A**. Draw its structure.

Structure **A** (Fischer projection):
- CHO (top)
- HO—H
- H—OH
- H—OH
- CH$_2$OH (bottom)

A

(-)-ephedrine (Fischer projection):
- H (top)
- MeHN—Me
- H—OH
- Ph (bottom)

b. *Ephedrine*, a very potent dilator of the air passages in the lungs, has been used to treat asthma. The naturally occurring stereoisomer, isolable from the plant *Ephedra sinica*, is levorotatory ($[\alpha] = -40°$) and has the configuration above.

(i) Assign (R)- or (S)- configuration to each chiral center.

(ii) If a solution of (+) and (-) ephedrine has a specific rotation of $+10°$, what percentage of the mixture is *dextrorotatory* enantiomer?

7.1 General

9. Optically pure *quinine* has a specific rotation of -170°. What percent of levorotatory form is present in an optically impure sample whose [α] is +68°? How many chiral carbons are there in quinine?

quinine

10. (*S*)-*Naproxen* is an active non-steroidal anti-inflammatory drug (NSAID), but the (R)-enantiomer is a harmful liver toxin. Assign the configuration for the (S)-enantiomer.

naproxin

11. For each of the molecules below, indicate whether it is capable of *enantiomerism only* (**E**), *diastereomerism only* (**D**), or *both enantiomerism and diastereomerism* (**ED**).

a.   b.   c.   d.   e.

f.   g.   h.

12. *Thalidomide* was used as a sedative and anti-nausea drug for pregnant women in Europe (1959-62). Unfortunately, it was sold as a racemate and each enantiomer has a different biochemical activity. One enantiomer, the (S)-form, is a teratogen that was responsible for thousands of serious birth defects. Which of the following is (R)-thalidomide?

vs.

7.1 General

13. Another example of different enantiomers having remarkably different biochemical activities is *penicillamine*. The (S)-form has anti-arthritic properties, whereas the (R)-form is toxic. Which form is the following configuration?

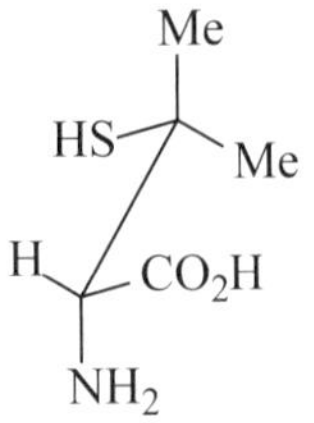

penicillamine

14. *Taxol* is an anticancer agent active against ovarian and breast tumors. (a) How many chiral carbons are in taxol? (b) If the specific rotation of optically pure taxol is -120°, and a synthetic preparation of taxol containing only its two enantiomers shows a specific rotation of +24°, what is the percentage of *dextrorotatory* enantiomer in the mixture?

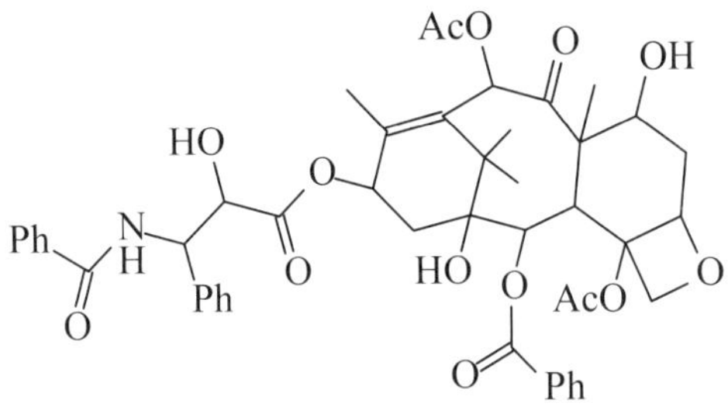

taxol

15. Compound **A** below has _____ chiral carbons, _____ *meso* stereoisomers, and _____ pair(s) of enantiomers.

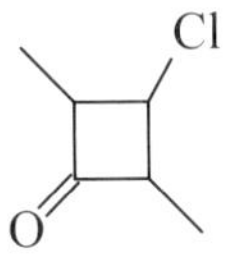

A

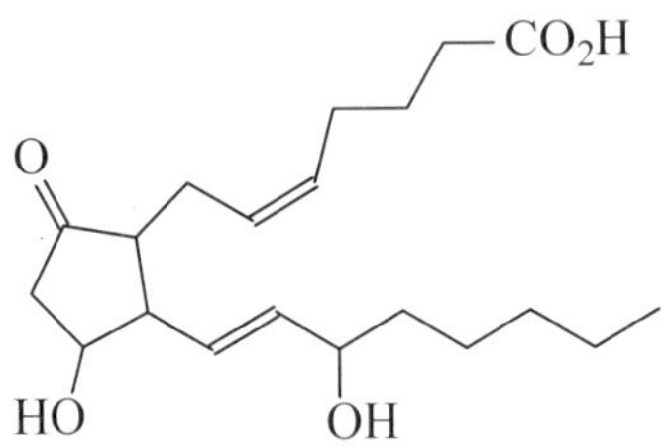

B PGE$_2$ (a prostaglandin)

The number of stereoisomers possible for **B** is _____ (do not change *cis/trans* configurations of the olefins).

7.1 General

16. The antibiotic *cephalosporin C* has a specific rotation of +103° in water.

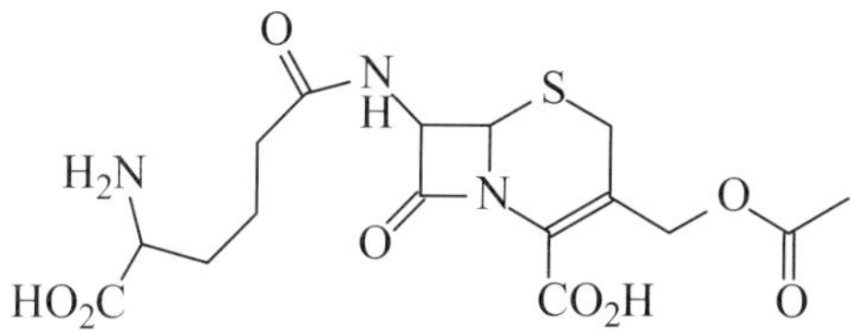

cephalosporin C

a. What is the maximum number of stereoisomers for the above structure?

b. If a synthetic sample of cephalosporin C has an optical rotation of +82°, what percent of the enantiomers is levorotatory?

7.2 Reactions and stereochemistry

1. Draw the stereochemical formula for the major organic product(s) in the following reactions by completing the Fisher projections.

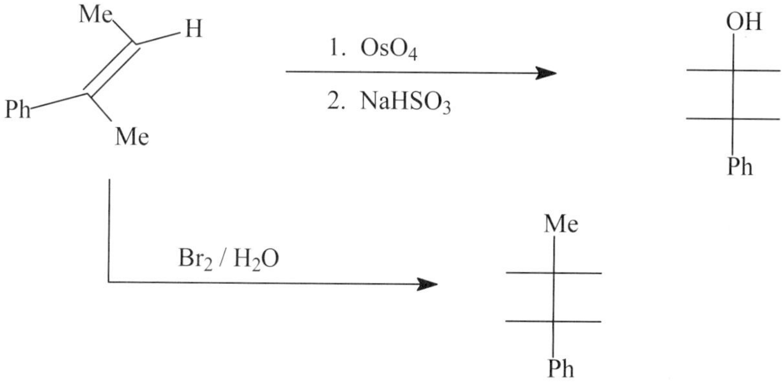

2. Have the following reactions proceeded with *syn-* or *anti-* stereochemistry?

a. *cis*-2-butene

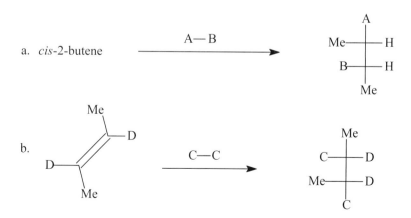

b.

c. *Fumarase* catalyzes the following reaction in mitochondria:

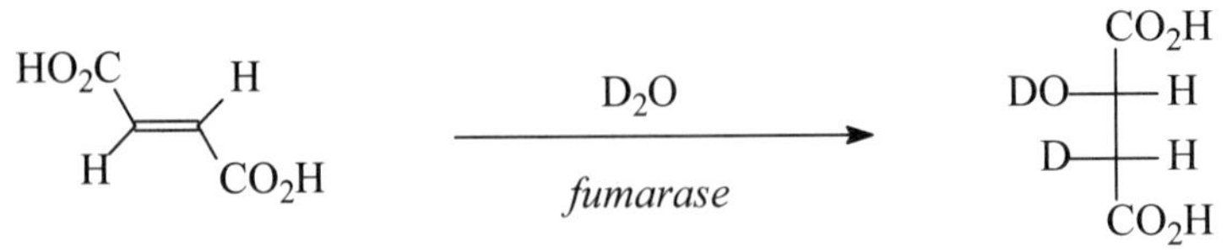

$$\text{HO}_2\text{C-CH=CH-CO}_2\text{H} \xrightarrow[\text{fumarase}]{\text{D}_2\text{O}} \text{malic acid}$$

(Fischer projection: CO$_2$H / DO—H / D—H / CO$_2$H)

3. For each of the following reactions, (a) <u>how many fractions</u> could be collected by fractional distillation or recrystallization, and (b) <u>for each fraction</u> describe whether it is one enantiomer (**E**), a racemate (**R**), or a meso compound (**M**).

a. 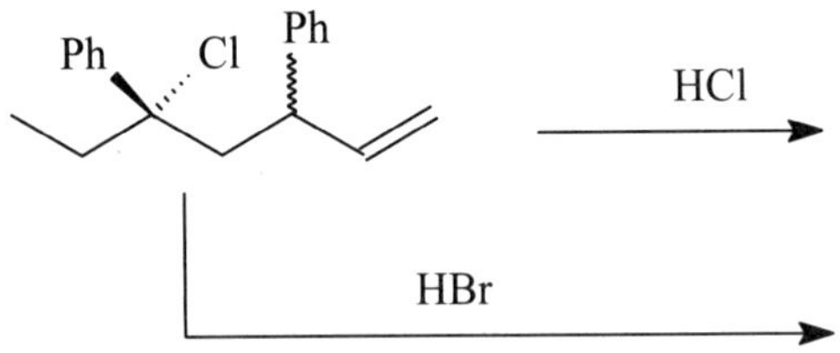 [structure with Ph, Cl, Ph substituents] → HCl
   → HBr

b. (Z)-3-hexene $\xrightarrow{\text{Br}_2}$

c. 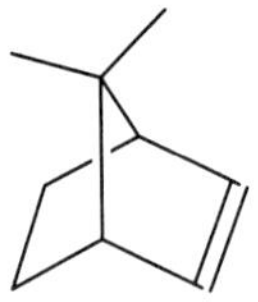 [bicyclic alkene structure] $\xrightarrow{\text{KMnO}_4,\ \text{H}^\oplus}$

d. (S)-3-phenyl-1-butene $\xrightarrow{\text{HI}}$

e. 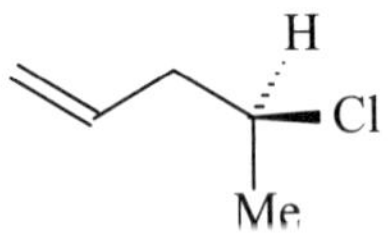 [structure with H, Cl, Me] $\xrightarrow{\text{HF}}$

f.  [cyclohexene with two methyl groups] $\xrightarrow{\text{D}_2\ /\ \text{Ni}}$

7.2 Reactions and stereochemistry

g. 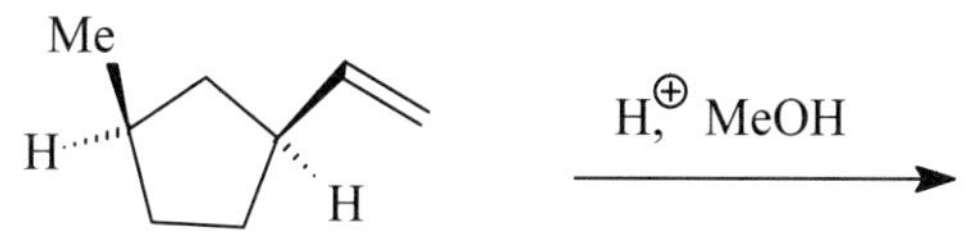 [structure: (1S,3R)-1-methyl-3-vinylcyclopentane, trans] → H⁺, MeOH

h. [1,2-dimethylcyclohexene] → 1. BH₃ 2. H₂O₂, ⁻OH

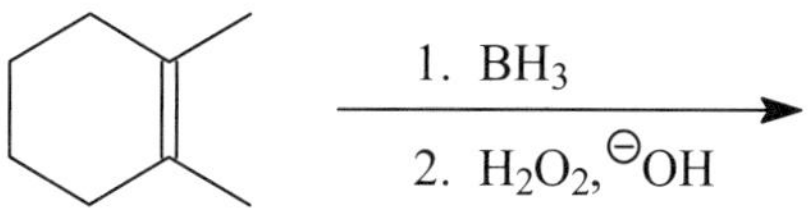

i. 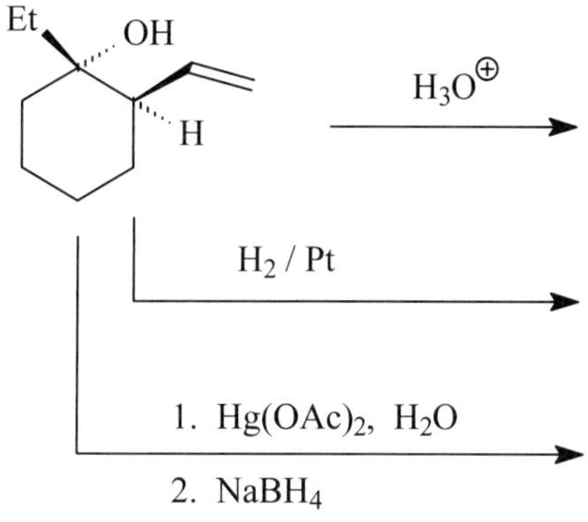 [1-ethyl-2-vinylcyclohexan-1-ol, Et and OH on same carbon, H trans] → H₃O⁺

   → H₂ / Pt

   → 1. Hg(OAc)₂, H₂O   2. NaBH₄

j. 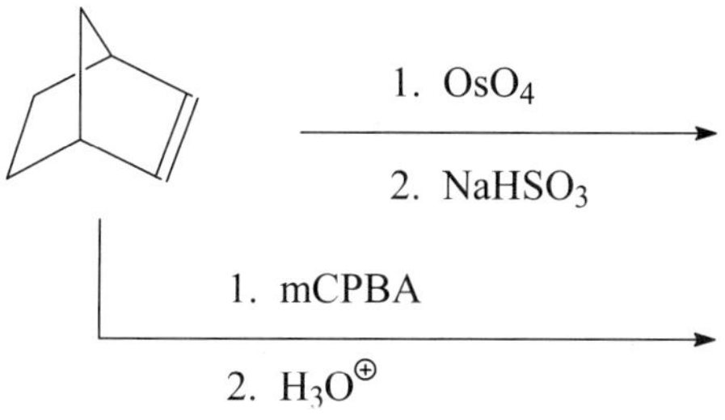 [norbornene] → 1. OsO₄   2. NaHSO₃

   → 1. mCPBA   2. H₃O⁺

k. 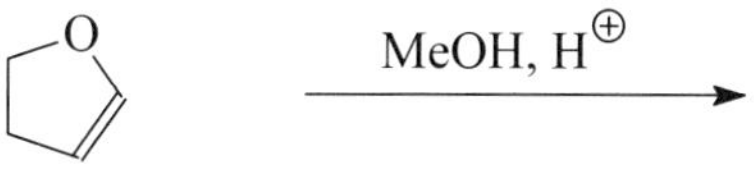 [2,3-dihydrofuran] → MeOH, H⁺

4. Outline syntheses for the following conversions that ensure the indicated stereochemical outcomes.

a. 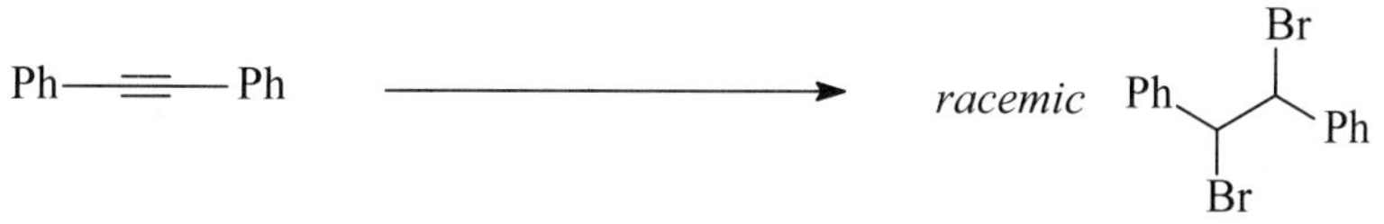 Ph—≡—Ph → racemic Ph-CHBr-CHBr-Ph

7.2 Reactions and stereochemistry

b. 2-butyne $\longrightarrow$ *meso*-2,3-dibromobutane

c. [alkyne structure] $\longrightarrow$ *racemic* [diol structure with OH groups]

d. *trans*-2-butene $\longrightarrow$ *meso*-glycol only

5. Consider the structure [Ph-CH(OH)-CH(Me)-NHMe] , and answer the following:

    a. How many stereoisomers are possible?

    b. Two of the structures are the decongestants *ephedrine* and *pseudoephedrine*:

    ephedrine   *pseudoephedrine*

    Which stereochemical term best describes their structural relationship?

    c. The HCl salt of ephedrine has a specific rotation of -34°. What would you predict for the specific rotation of the HCl salt of pseudoephedrine?

    d. Both ephedrine and pseudoephedrine can be dehydrated to an olefin, which upon hydrogenation produces methamphetamine ("speed," "meth").

       i. How many stereoisomers exist for the olefin?

       ii. How many stereoisomers are possible for "meth"?

7.2 Reactions and stereochemistry

CHAPTER 8
ALKYL HALIDES AND RADICALS

8.1 Reactions

Draw the structural formula of the major organic product(s). Show stereochemistry where appropriate.

1. How many different dichlorides could be isolated by ordinary physical methods (*e.g.*, fractional distillation) from the following reaction? Would each, as collected, be optically active or inactive?

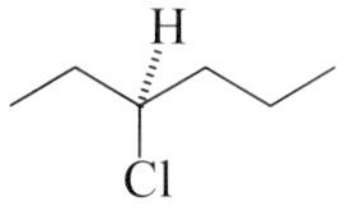

   (S)-2-chloropentane (with H wedge and Cl) → Cl_2, $h\nu$

2. Calculate the maximum % of (R)-2-bromopentane that could be formed from the reaction of bromine with *n*-pentane.

3. 2,3-dimethylbutane
   1. Br_2, Δ
   2. Mg
   3. D_2O

4. 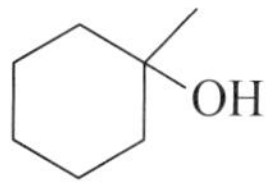 1-methylcyclohexanol
   1. conc HCl
   2. Li
   3. CuI
   4. allyl iodide

5. cyclobutane
   1. Cl_2, $h\nu$ 2. Li
   3. CuI
   4. vinyl iodide   5. HI

6. propane
   1. Br_2, Δ
   2. Mg
   3. phenylacetylene

7. bromobenzene
   1. Li   2. CuI
   3. *n*-PrBr   4. NBS, peroxides
   5. KOH   6. Br_2 / H_2O

8.2 Syntheses
Supply a reagent or sequence of reagents that will effect the following conversions.

1.  cyclopentane → 3-bromocyclopentene

2.  bicyclo[3.3.0] → cyclooctane-1,5-dione

3. isopentane → 2-bromo-3-methylbutane

4. iodobenzene → biphenyl

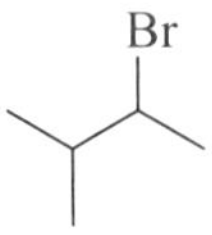

5. cyclopentane → 1-methyl-1-benzylcyclopentane

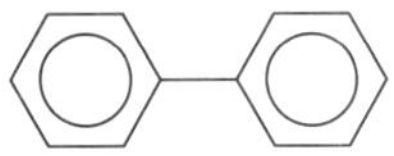

6. chlorobenzene → benzaldehyde

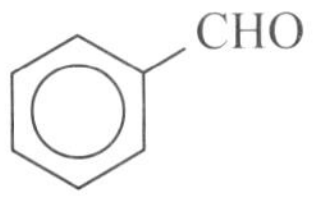

7. cyclohexene $\xrightarrow{\text{3 ways!}}$ deuteriocyclohexane

8. 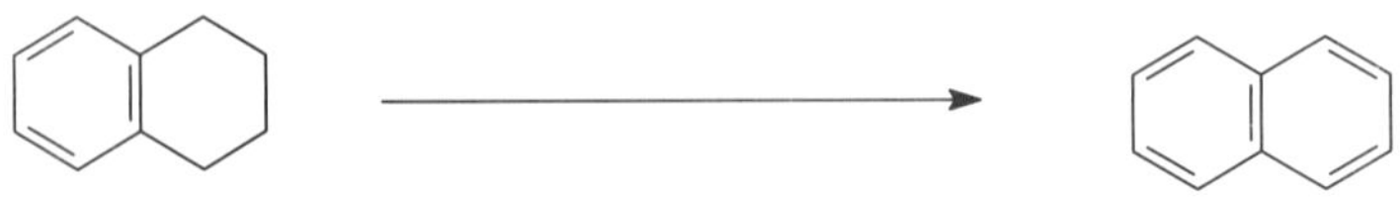 tetralin → naphthalene

8.3 Mechanisms

Outline a detailed mechanism for each of the following. No other reagents than those given are necessary. Use arrows to explain the flow of electrons and show all intermediates.

1. methylcyclohexene $\xrightarrow{O_2,\ ROOR}$ 3-methylcyclohex-2-enyl hydroperoxide (O–OH)

2. $H_2C=C=CH_2$ (allene) $\xrightarrow{1.\ CH_2N_2,\ h\nu}$ spiropentane $\xrightarrow{2.\ Cl_2,\ h\nu}$ (1,1-dichlorospiropentane) + (1,1-bis(chloromethyl)cyclopropane) + Cl–CH₂CH₂–C(=CH₂)–CH₂Cl

   (propose a mechanism for step 2)

3. *Bergman reaction:*

   (Z)-hexa-3-ene-1,5-diyne-1,6-d₂ $\xrightarrow{\Delta}$ 1,4-dideutero-benzene-1,2-diyl diradical (drawn as 1,2-diethynyl D-substituted alkene form)

4. Alkyl nitrite esters (RO–NO) readily undergo photolytic homolysis. The *Barton reaction* utilizes this fact to functionalize the remote δ-position of steroids. Use conformational analysis to explain.

   AcO–steroid–O–N=O (with δ and α positions labeled) $\xrightarrow{h\nu}$ AcO–steroid with O=N–CH₂ at angular position and OH at 6β

5. a. The *vinylcyclopropane – cyclopentene rearrangement* proceeds by a free radical mechanism. Explain. Hint: the cyclopropyl C-C bond is easily homolyzed.

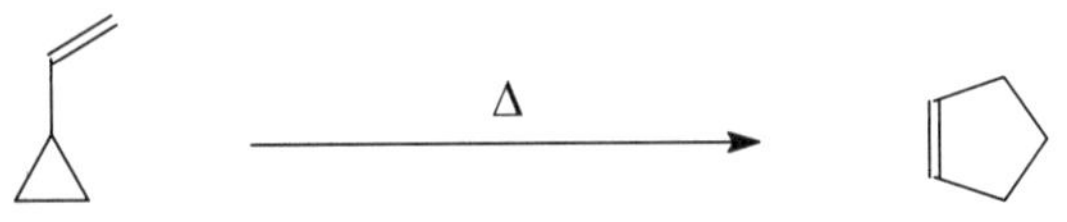

b. Predict the product:

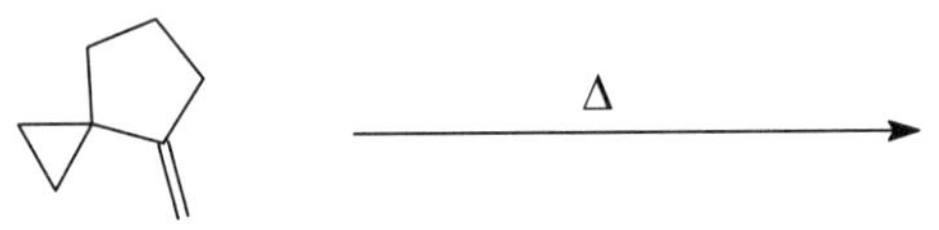

6. Aspirin, as well as other non-steroidal anti-inflammatory drugs (NSAIDS), blocks the synthesis of certain inflammation-mediating prostaglandins by inhibiting the enzyme cyclooxygenase (COX – see 5.1, 6d). COX converts arachidonic acid to the prostaglandin PGG_2, which subsequently undergoes reduction to give PGH_2. Other prostaglandins derive from the latter. Outline a mechanism for the synthesis of PGG_2. Hint: begin by a free radical removal of one of the doubly allylic hydrogen atoms.

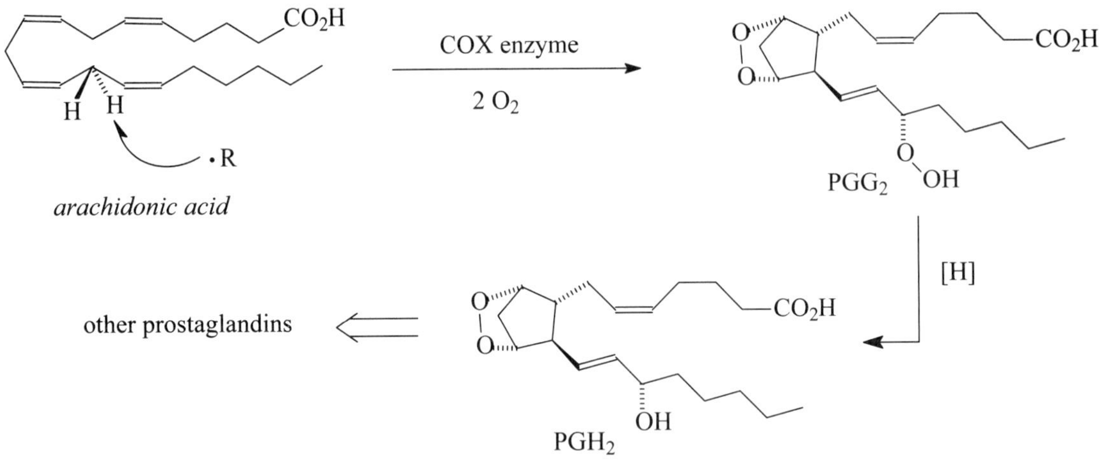

8.3 Mechanisms

CHAPTER 9
S$_N$1, S$_N$2, E1, AND E2 REACTIONS

9.1 General
For problems 1 – 9, circle the

1. reaction that will go *faster*:

    a. AcO$^\ominus$ + allyl chloride $\xrightarrow{\text{ethanol}}$

    b. AcO$^\ominus$ + allyl chloride $\xrightarrow{\text{HMPA}}$

2. structure with the *poorest* leaving group:

    a. R-SH b. R-NH$_2$ c. R-OAc d. R-OH

3. *stronger* nucleophile:

    a.   b. Et$_3$N

4. alkyl halide most reactive by an S$_N$2 pathway:

    a.  b.   c.

5. solvent that will *maximize* the rate of the reaction of Et$_3$N with *n*-BuBr:

    a. DMSO b. MeOH c. PhH d. chloroform

6. halide that will react *more rapidly* by an E2 pathway:

    a.   b.

7. approximate value of k_H / k_D when PhCHBrCH$_3$, *vs.* PhCDBrCD$_3$, is allowed to react with potassium *t*-butoxide:

    a. 1 b. <1 c. >1

9.1 General

8. reaction that will yield the more stereochemically pure product(s):

a. (or diastereomer) — methanolysis (S_N) →

b. (R)-2-bromopentane (or enantiomer) — $MeO^\ominus$, MeOH (S_N) →

9. *change* in rate of reaction if the concentration of Ph_2CHBr is *tripled* and the concentration of ethanol is *doubled*:

a. rate is unaffected b. rate triples c. rate doubles
d. rate increases 5-fold e. rate increases 6-fold

---

10. Which would be the *reaction of choice* (higher yielding) for each of these syntheses?

a. [propyl bromide] —$^\ominus$OMe→ [propene] ←$^\ominus$O-$\underline{t}$-Bu— [isopropyl bromide]

b. [cyclohexyl bromide] —$^\ominus$OH→ [methylenecyclohexane/methylcyclohexene] ←$^\ominus$OH— [1-bromo-1-methylcyclohexane]

c. [isopropyl bromide] + $^\ominus$OR vs. $^\ominus$SR (which?) to maximize S_N

11. Which reaction would be expected to show a *primary hydrogen kinetic isotope effect*?

a. [structure with Cl and H(D)] —KO-$\underline{t}$-Bu / $\underline{t}$-BuOH→

b. [structure with Cl, H(D), H(D)] —KOH / MeOH→

c. [structure with Cl, (D)H, H(D)] —KOMe / MeOH→

9.1 General

12. The following reaction might be envisioned as occurring by an intramolecular S_N2 process. However, kinetic evidence indicates a bimolecular mechanism. Explain.

9.2 Reactions
Identify (if not already stated) each reaction as largely S_N1, S_N2, $E1$, or $E2$ – then draw the structural formula of the major organic product(s). Show stereochemistry where appropriate.

1. *n*-octyl bromide + KOCH$_3$ $\xrightarrow{\text{MeOH}}$

2. 3-iodo-3-methylpentane + sodium ethoxide / EtOH $\longrightarrow$

3. potassium *t*-butoxide + *sec*-BuCl $\longrightarrow$

4. 2-bromo-3-methylbutane + lithium diisopropylamide $\longrightarrow$

5. *n*-hexyl iodide $\xrightarrow{\text{KCN / DMF}}$

6. 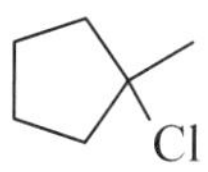 [cyclopentyl chloride with methyl] $\xrightarrow{\text{methanolysis (RT)}}$

7. 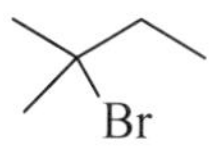 [neopentyl-type with Br] $\xrightarrow{\text{refluxing EtOH}}$

8. 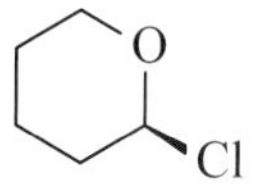 [2-chlorotetrahydropyran] $\xrightarrow{\text{acetolysis } (S_N)}$

9. *n*-propyl bromide + Me$_2$NH $\longrightarrow$

10. isopropyl bromide + sodium *t*-butoxide $\longrightarrow$

11. 3-iodopentane $\xrightarrow{\text{sodium acetate / DMF}}$

12. 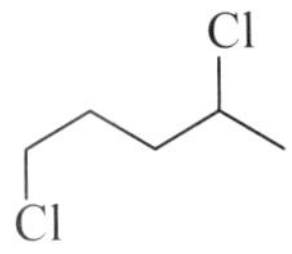 ClCH$_2$CH$_2$CH$_2$CH(Cl)CH$_3$ $\xrightarrow{\text{NaSH (1 equiv)}}$

13. 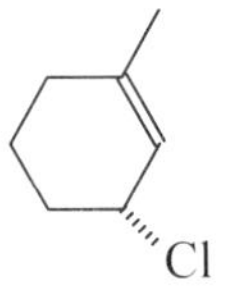 (3-chloro-1-methylcyclohexene, Cl on wedge) $\xrightarrow{\text{ethanolysis (S}_N\text{)}}$

14. 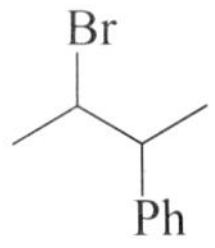 (2-bromo-3-phenylbutane) $\xrightarrow{^{\ominus}\text{OAc / Ag}^{\oplus}}$

15. 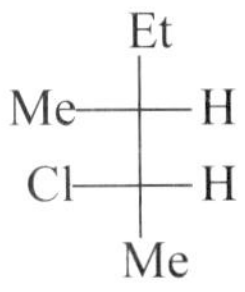 Fischer projection: Me—C(Et)(H)—C(Cl)(H)(Me) ... Et top, Me-H, Cl-H, Me bottom $\xrightarrow{^{\ominus}\text{OEt (E)}}$

16. 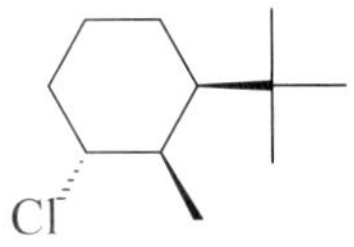 (1-chloro-2-methyl-3-*t*-butylcyclohexane) $\xrightarrow{^{\ominus}\text{OMe / MeOH}}$

17. 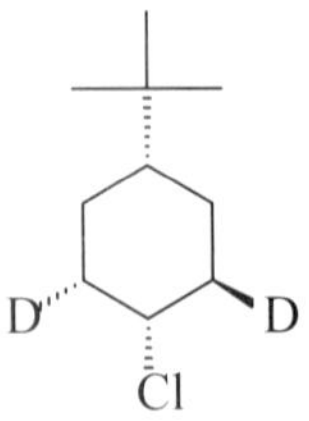 (4-*t*-butyl-1-chloro-cyclohexane with D's at 2,6 positions) $\xrightarrow{\text{E2}}$

9.2 Reactions

18.

       CH₃ / H—D / Br—H / Ph → E2

19. [cyclohexyl with Br, CH₃, and CH₂CH₂OH substituents] → acetone

20. 4-iodo-1-pentane → triphenylphosphine
            ↓ t-butyl alcohol (S$_N$)

21. CH₃CH₂—S—CH(CH₃)—Cl → methanolysis

22. I—CH₂CH₂CH₂—CH(NMe₂)—CH₃ → Δ

23. conjugate base of H₂Se + [cyclobutyl chloride] →

24. Ph—CH(OH)—CH₂—NHMe → PhCH₂Cl (1 equiv)
        ephedrine

25. HO—CH₂CH₂—⁺NMe₃ → Me₃O⁺ BF₄⁻
        choline

26.

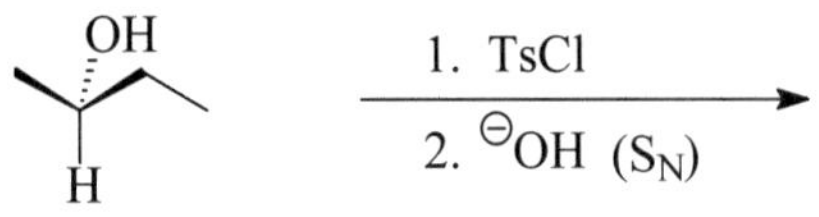

![structure: (S)-butan-2-ol with OH up and H down]
1. TsCl
2. ⁻OH (S_N)

27.

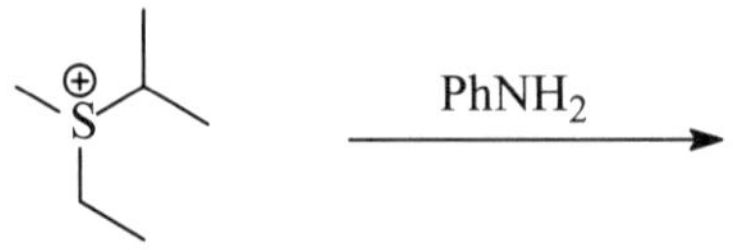

[ethyl methyl isopropyl sulfonium cation] —PhNH₂→

28. 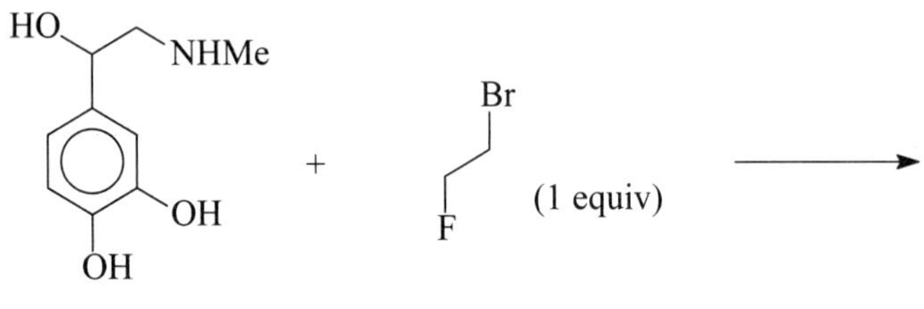 epinephrine (HO-CH(aryl)-CH₂-NHMe with catechol OH, OH) + BrCH₂CH₂F (1 equiv) →

29. 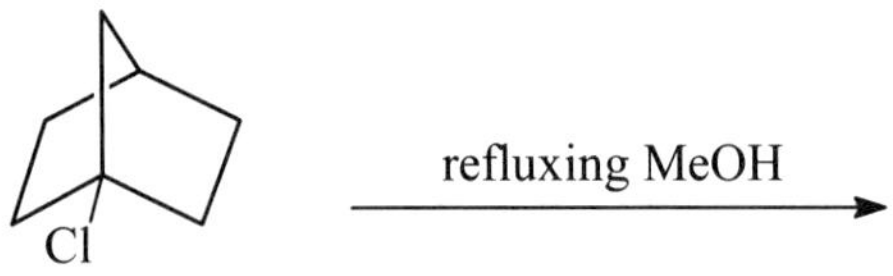 [1-chloronorbornane] —refluxing MeOH→

30. 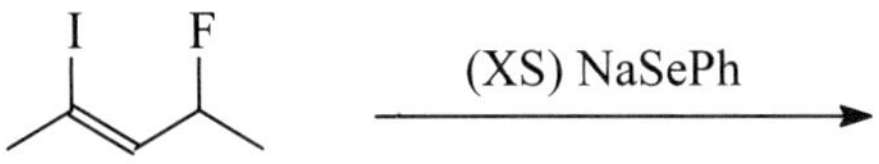 [(CH₃)(I)C=C(F)(CH₃)] —(XS) NaSePh→

31. 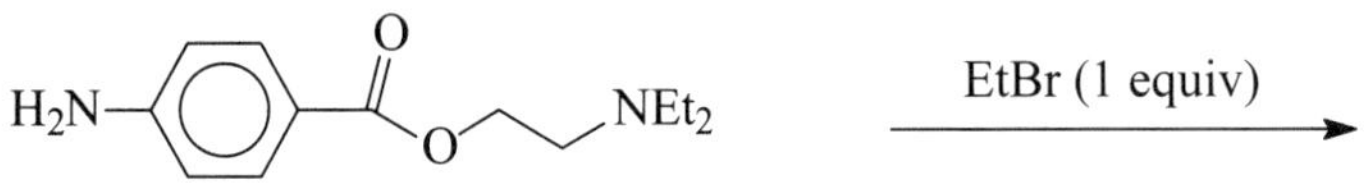 H₂N–C₆H₄–C(=O)–O–CH₂CH₂–NEt₂ (Novocaine™) —EtBr (1 equiv)→

32. 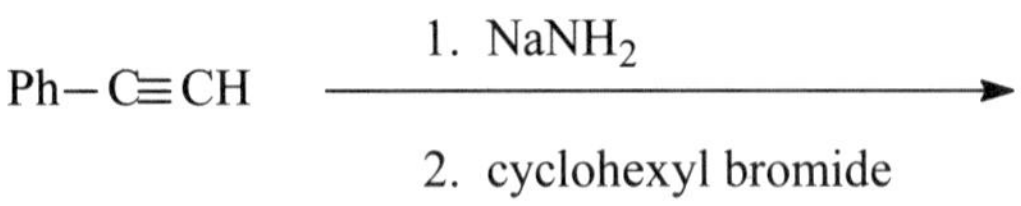 Ph–C≡CH
1. NaNH₂
2. cyclohexyl bromide

33. 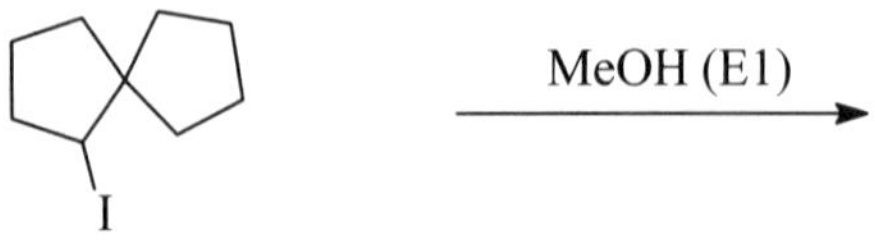 [1-iodospiro[4.4]nonane] —MeOH (E1)→

9.2 Reactions

34. Ph—OTs $\xrightarrow{\text{KSCN}}$

35. 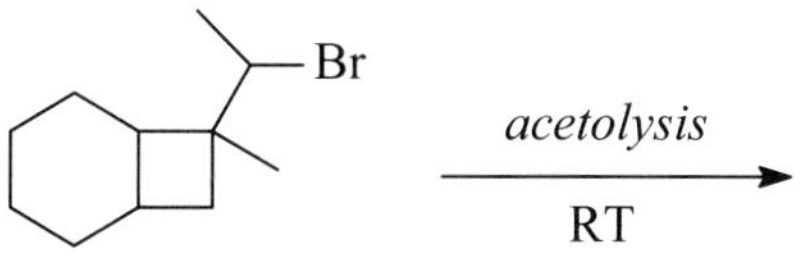 [bicyclic structure with CHBr-Me and Me substituents] $\xrightarrow[\text{RT}]{acetolysis}$

36. 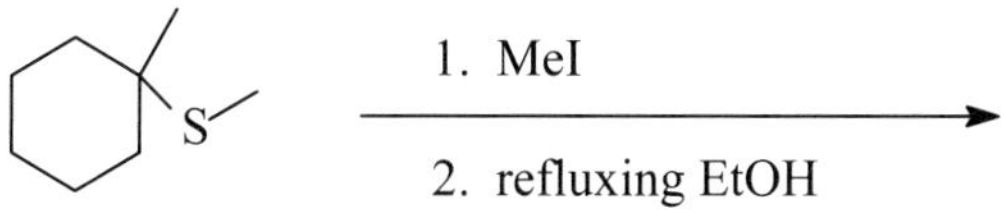 [1,1-dimethylcyclohexyl-SMe] $\xrightarrow{\text{1. MeI} \quad \text{2. refluxing EtOH}}$

37. 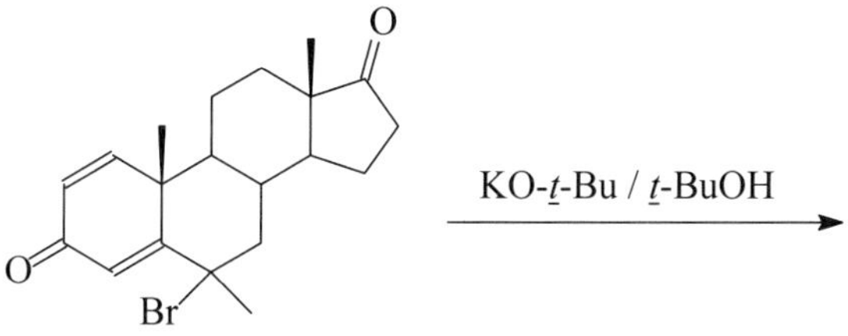 [steroid structure with Br and Me substituents] $\xrightarrow{\text{KO-}t\text{-Bu / }t\text{-BuOH}}$

Aromasin™ (an aromatase inhibitor used in breast cancer therapy)

38. 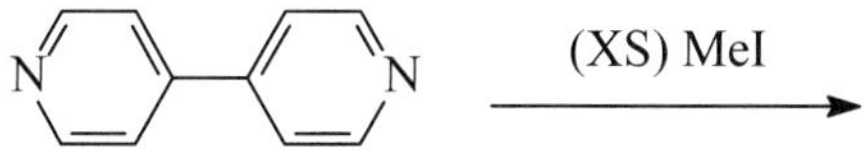 4,4'-bipyridine $\xrightarrow{\text{(XS) MeI}}$

paraquat (an herbicide)

39.

a. 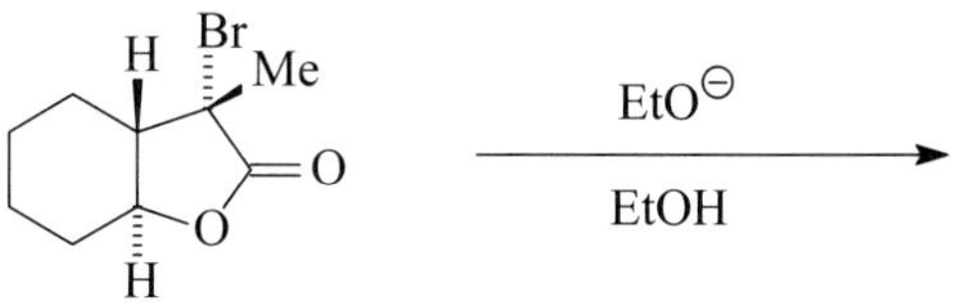 [bicyclic lactone with H, Br, Me stereochemistry] $\xrightarrow[\text{EtOH}]{\text{EtO}^\ominus}$

b. 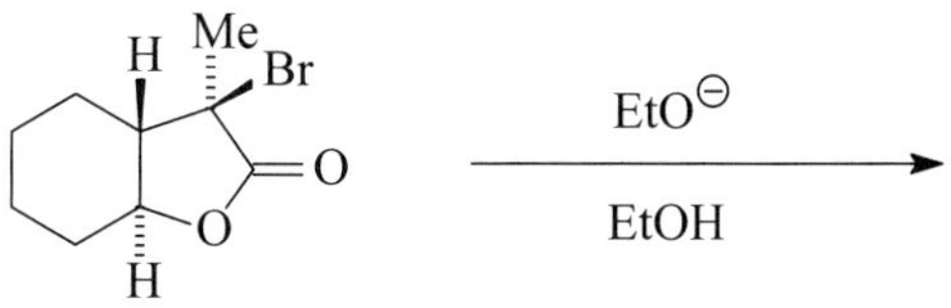 [bicyclic lactone with H, Me, Br stereochemistry] $\xrightarrow[\text{EtOH}]{\text{EtO}^\ominus}$

9.2 Reactions

9.3 Syntheses
Supply a reagent or sequence of reagents that will effect the following conversions.

1. 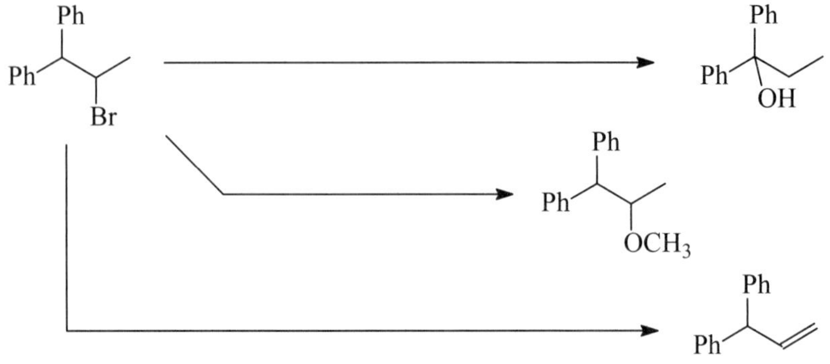 Ph-CH(Ph)-CH(Br)-CH3 → Ph-C(Ph)(OH)-CH2CH3

   → Ph-CH(Ph)-CH(OCH3)-CH3

   → Ph-CH(Ph)-CH=CH2

2.  t-Bu, Me, Br-Me cyclohexane → t-Bu-substituted cyclohexene with dimethyl

3.  (cyclohexane with Me, Br, H, D stereochem) → 3-methylcyclohexene

4. 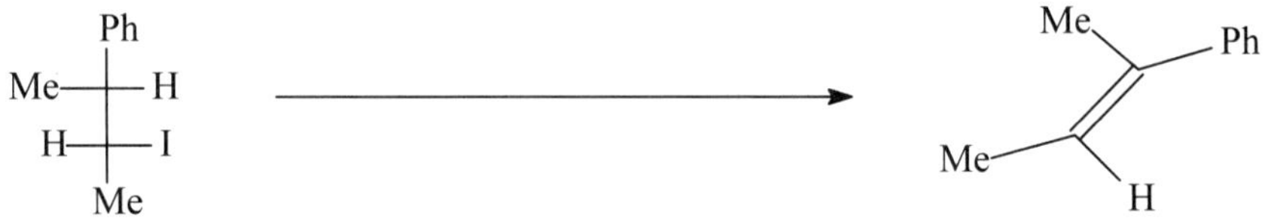 Fischer projection: Me—H, Ph top; H—I, Me bottom → (E/Z)-alkene: Me/Ph and Me/H

5.  1-methyl-1-(iodomethyl)cyclohexane → 1-ethylcyclohexene

6. 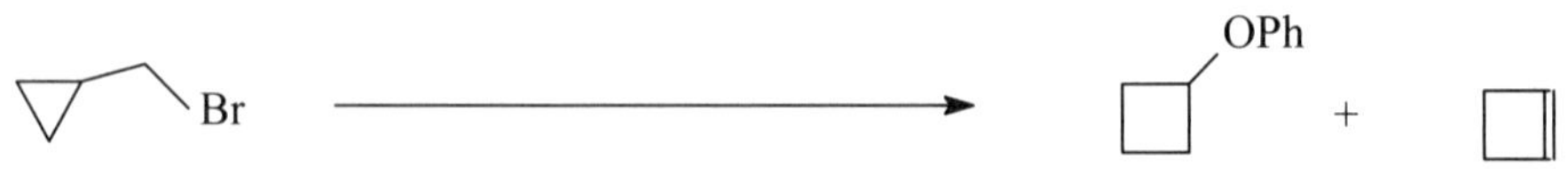 cyclopropyl-CH2-Br → cyclobutyl-OPh + cyclobutyl cation/product

7.  [1-methylcyclohexyl chloride] ⟶ [1-methylcyclohexyl benzoate]

8. 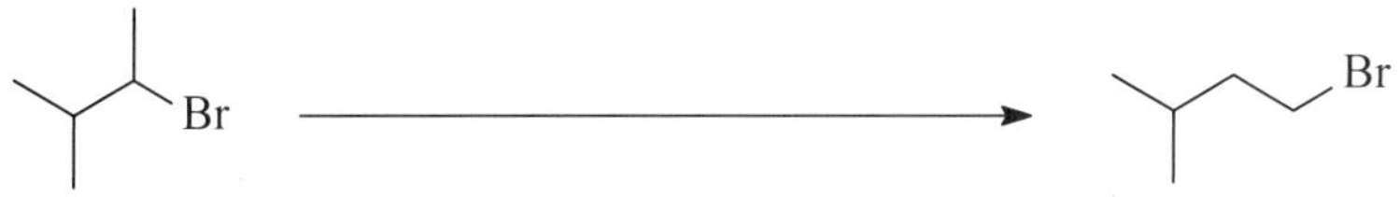 [2-bromo-3-methylbutane] ⟶ [1-bromo-3-methylbutane]

9.  [2-methylbutane] ⟶ [2-ethoxy-2-methylbutane (EtO on tertiary carbon of 2-methylbutane)]

10. 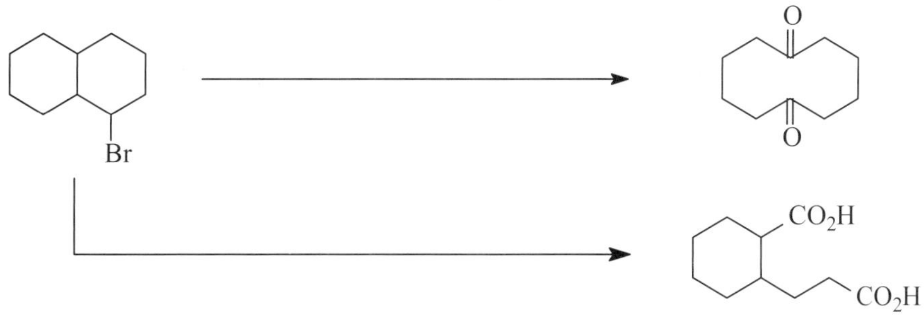 [decahydronaphthalene with Br on ring] ⟶ [cyclodecane-1,6-dione]

    [decahydronaphthalene with Br on ring] ⟶ [cyclohexane with CO$_2$H and CH$_2$CH$_2$CO$_2$H substituents]

11. 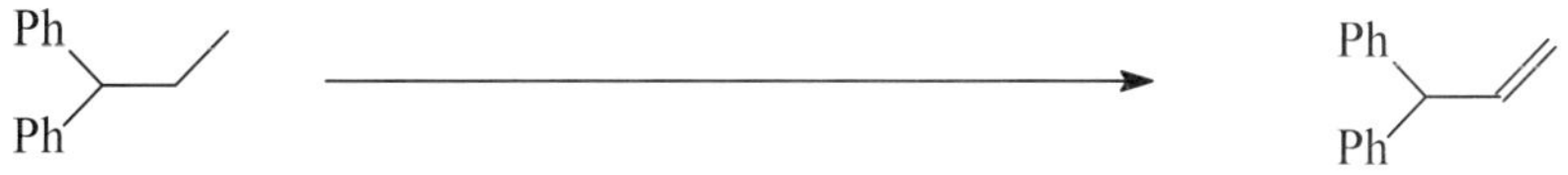 [1,1-diphenylpropane, Ph$_2$CH–CH$_2$CH$_3$] ⟶ [3,3-diphenylpropene, Ph$_2$CH–CH=CH$_2$]

12.  [PhCH$_2$–CH(OTs)CH$_3$] ⟶ [Ph–CH(OTs)–CH$_2$CH$_3$]

13.  [methylcyclopentane] ⟶ [trans-2-methylcyclopentanol]

9.3 Syntheses

78 • Chapter 9 S_N1, S_N2, E1, and E2 Reactions

14.

15. ethylene ⟶ [1,4-dithiane structure]

16. [cyclohexene] —via an alkyne→ [cyclohexyl-CH2-CHO]

17. [1-iodobicyclo[2.2.2]octane] ⟶ [1-bromobicyclo[2.2.2]octane]

18. [trans-1,2-D, 3-D, Br cyclohexane] ⟶ [cyclohexene with D at C1 and C3] and [cyclohexene with H at C1, D at C3]

9.4 Mechanisms

Outline a detailed mechanism for each of the following. No other reagents than those given are necessary. Use arrows to explain the flow of electrons and show all intermediates.

1. Br-(CH2)5-Br —NaSH / HCO3⁻→ [thiane]

9.4 Mechanisms

9.4 Mechanisms

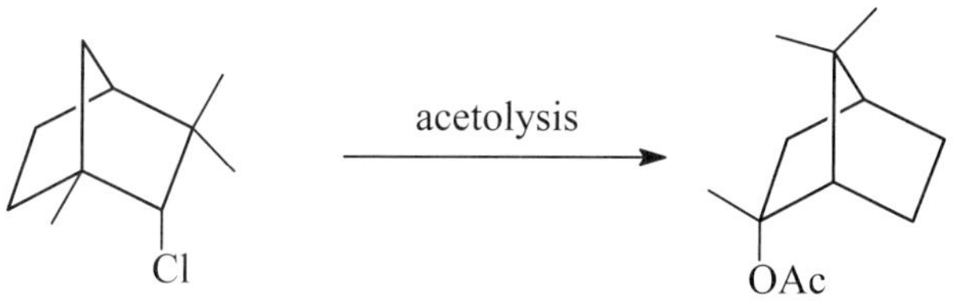

2. [bicyclic chloride with gem-dimethyl] —acetolysis→ [rearranged bicyclic OAc product]

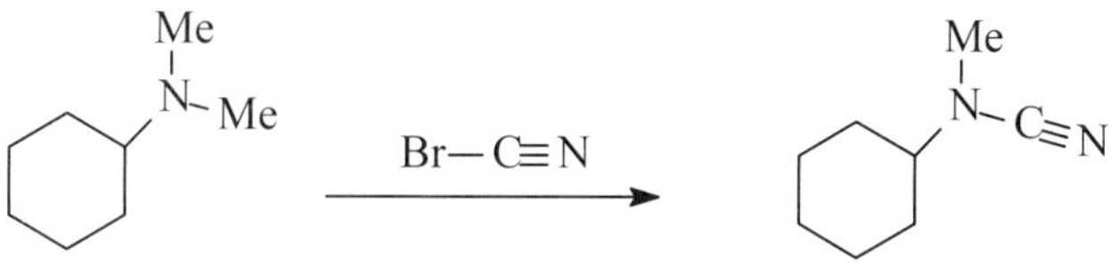

3. Cyclohexyl-N(Me)₂ + Br—C≡N → Cyclohexyl-N(Me)-C≡N

4.

$$\underset{\text{NEt}_2}{\text{CH}_2}-\underset{\text{Cl}}{\text{CH}}-\text{CH}_2-\text{CH}_2-\text{Ph} \xrightarrow{^{\ominus}\text{OH}} \underset{\text{OH}}{\text{CH}_2}-\underset{\text{NEt}_2}{\text{CH}}-\text{CH}_2-\text{CH}_2-\text{Ph}$$

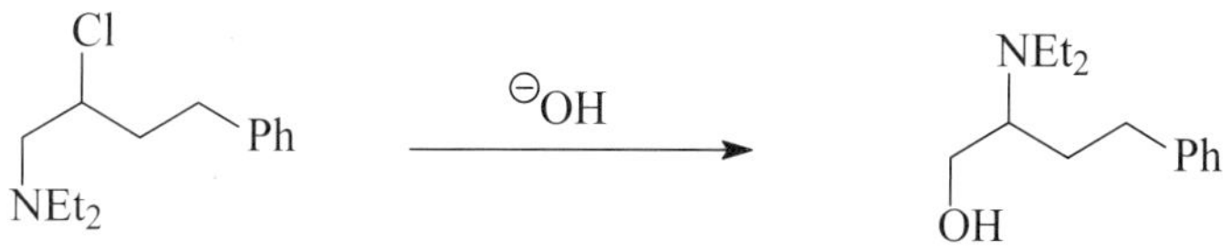

5.

CH₃-CHCl-COO⁻ —dil ⁻OH→ CH₃-CH(OH)-COO⁻

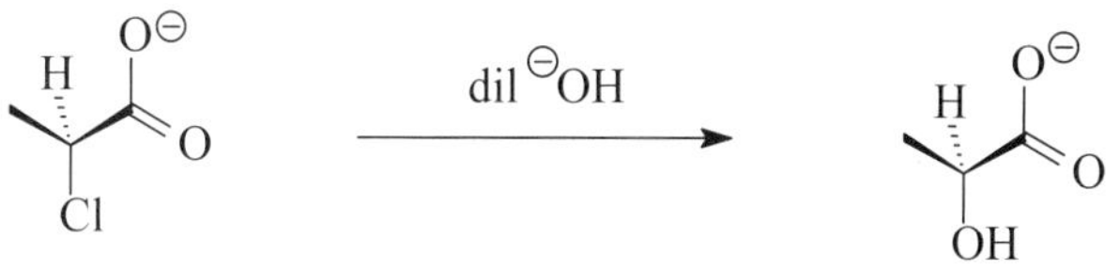

(Note: *retention* of configuration!)

6.  [HS-CH(CH₃)-CH₂-C(CH₃)=CH-CH=CH-CH₂-I] —DMF→ [six-membered sulfur ring with vinyl and methyl substituents]

7. 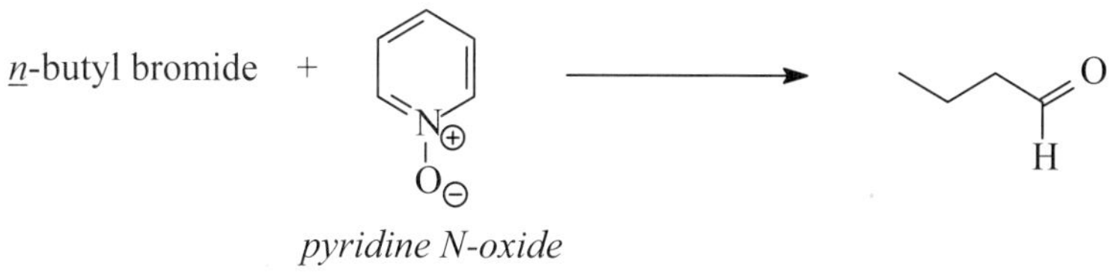 *n*-butyl bromide + pyridine N-oxide → CH₃CH₂CH₂CHO

9.4 Mechanisms

8. When treated with hydroxide, *trans*-**A** yields **B**. However, when *cis*-**A** is treated with hydroxide, no **B** is observed. Explain.

A → **B**

Problems 4 and 5 above illustrate the concept of "*neighboring group participation*" (NGP), wherein an *internal* nucleophilic atom (*e.g.*, N and O, respectively, in those examples) facilitates the ejection of the leaving group by an intramolecular S_N2 attack to form an unstable intermediate. This type of mechanism is often evidenced by (1) *rearrangement* (problem 4), (2) *stereochemistry* (problem 5), or (3) *kinetic data* (problem 9 below). Problems 9 – 16 are additional examples. Account for the observations mechanistically.

9. Unlike most primary alkyl halides the molecules below, types of sulfur and nitrogen mustard gases, do NOT undergo second order hydrolysis, but rather first order: $-d[RX]/dt = k[RX]$. Yet their rates of hydrolysis are *enormously faster* than those of most primary alkyl halides.

10. Compound **II** undergoes acetolysis at 75° about 10^3 times more rapidly than **I** and yields a racemate. Explain. What stereochemical outcome would you predict for the product from **I**?

I   **II**   HOAc → a racemate

9.4 Mechanisms

11.

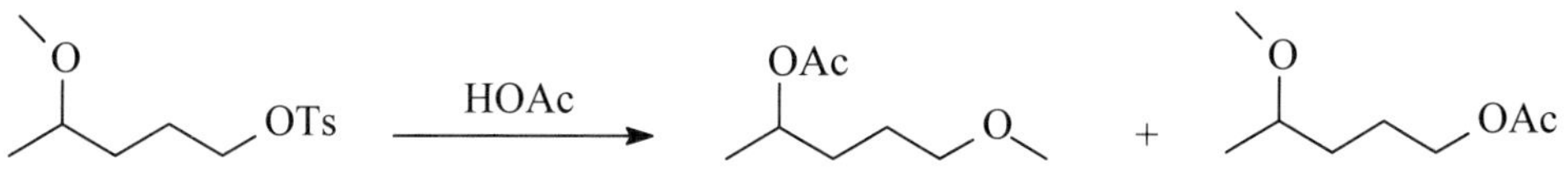

12. Paquette (*OSU*) observed that **II** undergoes solvolysis, *e.g.*, acetolysis about 10^4 times more rapidly than **I**.

I   **II**

13. 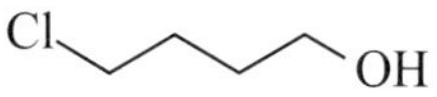 Cl~~~~~OH undergoes ethanolysis 5,700 times more rapidly than Cl~~~OH.

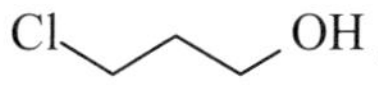

14. Sometimes a carbon-carbon double bond can act as a neighboring group nucleophile. For example, **II** undergoes acetolysis ~ 10^{11} times faster than **I** and does so with *retention of configuration*. Explain.

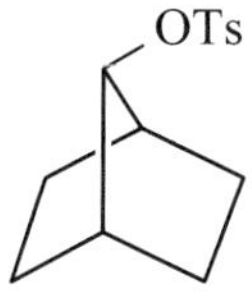

I 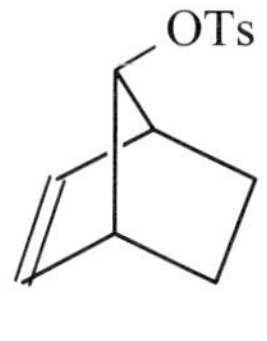   **II**

15. In view of the previous problem, account for the following:

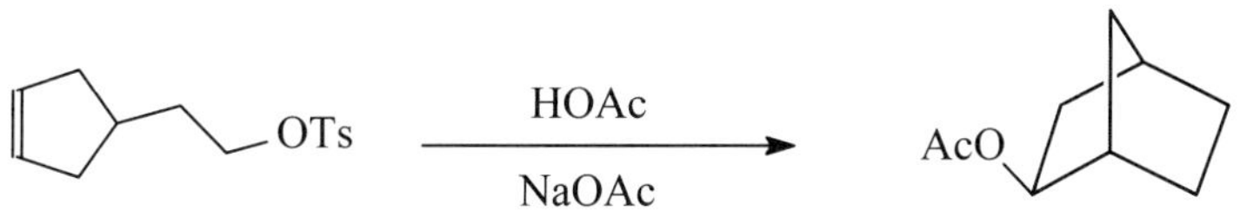

[cyclopentenyl-CH₂CH₂-OTs] →(HOAc, NaOAc)→ AcO-norbornyl

9.4 Mechanisms

82 • Chapter 9 S_N1, S_N2, E1, and E2 Reactions

16. DNA is stable in dilute aqueous hydroxide solution, but RNA rapidly hydrolyzes. A mechanistic clue is provided in the observation that hydrolysis of the latter yields not only 3'-phosphates but also 2'-phosphates. Explain.

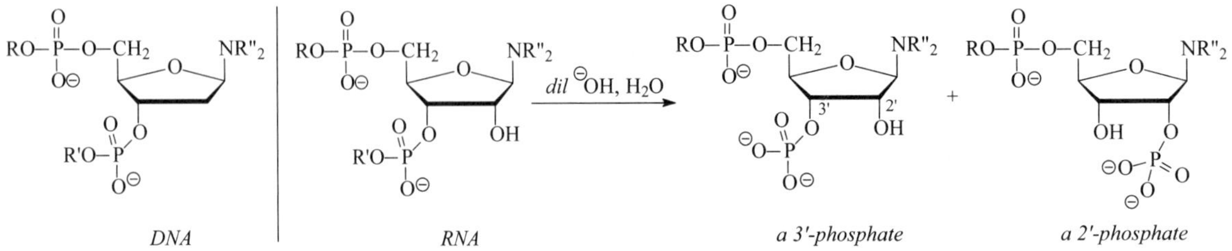

17. 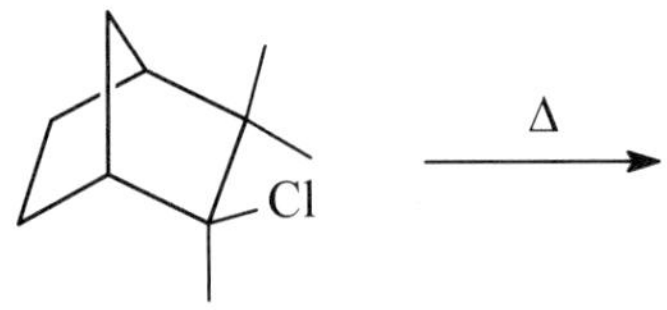 camphene hydrochloride $\xrightarrow{\Delta}$ racemic camphene + HCl

18. 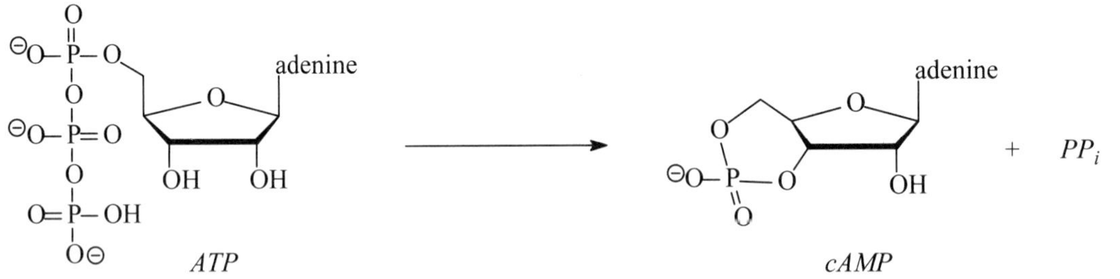 ATP $\longrightarrow$ cAMP + PP_i

9.4 Mechanisms

Some terpene chemistry...

19. The biosynthesis of *terpenes* (natural products constructed from the essence of *n* units of isoprene) begins with a "head-to-tail" coupling of two derivatives of isoprene, dimethylallyl pyrophosphate (*DMA-PP*) and isopentenyl pyrophosphate (*I-PP*) to form geranyl pyrophosphate (*G-PP*):

a. DMA-PP + I-PP $\xrightarrow{\text{base}}$ G-PP $\xrightarrow{H_2O}$ geraniol (a monoterpene)

$$PP = -\overset{O}{\underset{O^\ominus}{P}}-O-\overset{O}{\underset{O^\ominus}{P}}-OH \quad \text{(-OPP is a good leaving group)}$$

b. geraniol $\xrightarrow{H^\oplus}$ terpineol

c. A similar coupling of *G-PP* with *I-PP* yields the C_{15}-sesquiterpene farnesyl pyrophosphate (*F-PP*) to produce a C_{20}-diterpene:

F-PP $\xrightarrow{I\text{-}PP}$ (a C_{20}-diterpene)

F-PP $\xrightarrow{\times 2}$ triterpenes (C_{30}) (*e.g.*, squalene => cholesterol)

C_{20}-diterpene $\xrightarrow{[O], H_3O^\oplus}$ **A** $\xrightarrow{H^\oplus}$ vitamin A (retinol)

C_{20}-diterpene $\xrightarrow{\times 2}$ tetraterpenes (C_{40}) (*e.g.*, lycopene, β-carotene)

Outline a mechanism for the coupling and for the conversion of the diterpene **A** to vitamin A.

9.4 Mechanisms

d. *F-PP* can isomerizes to nerolidol pyrophosphate (*N-PP*). *F-PP* and *N-PP* undergo a "head-to-head" reductive coupling by an E1 reaction to form the C_{30}-triterpene *squalene*. Outline the mechanisms for each of these events. Hint: reductive coupling is initiated by hydride attack on *N-PP* as shown below.

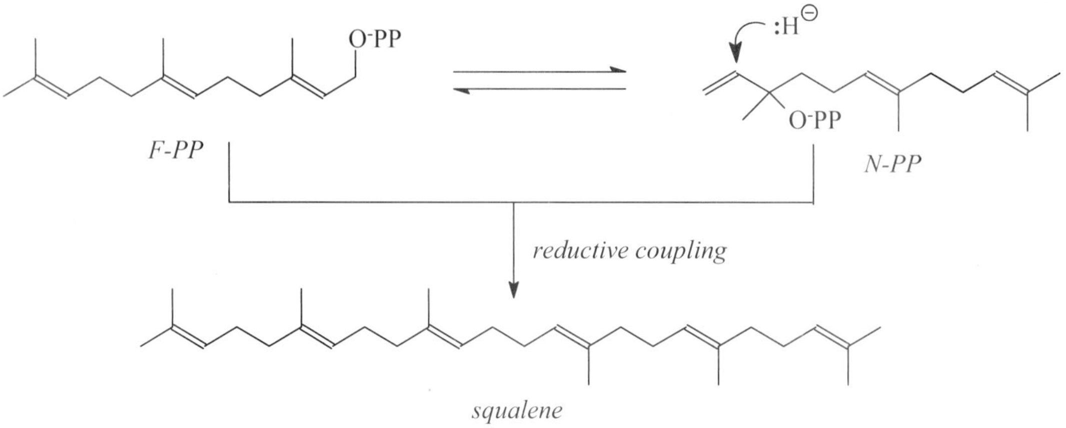

20. The most common methylating agent in biochemistry is *SAM* (S-adenosylmethionine), formed by an S_N reaction between the amino acid methionine and ATP. An example of a metabolic methylation is the conversion of *norepinephrine* (the prefix "*nor*" means one-less-carbon-than) to *epinephrine*. Formulate a mechanism for producing SAM and draw the structure of epinephrine.

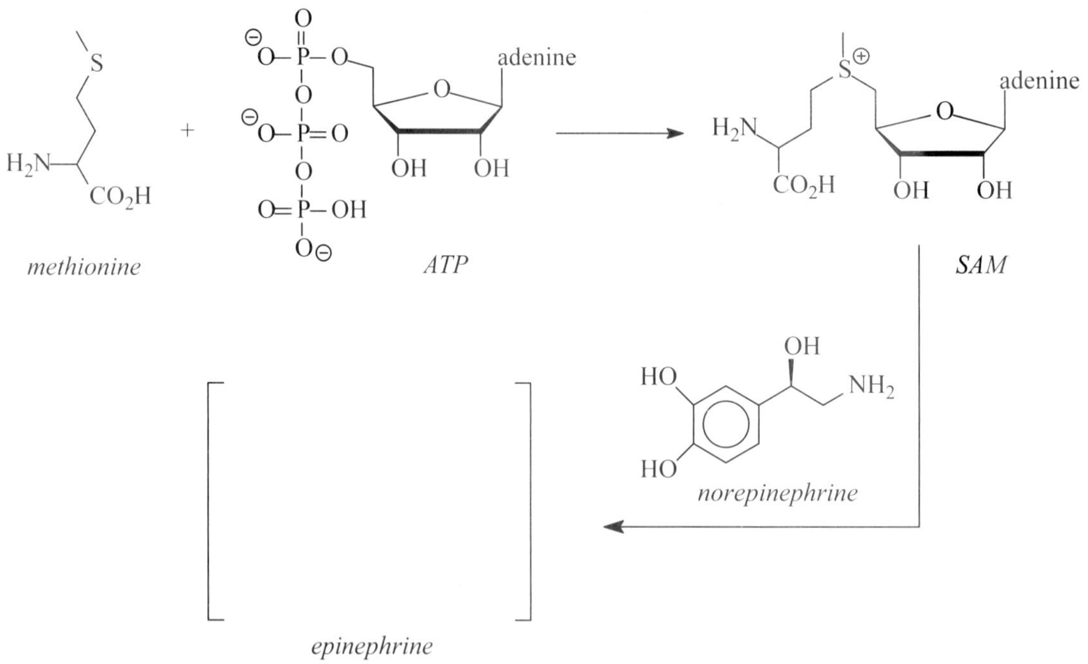

9.4 Mechanisms

21. $Ph_2C=\overset{\oplus}{N}=\overset{..}{\underset{..}{N}}{}^{\ominus}$ $\xrightarrow{\text{TsOH}}$ [ ] $\xrightarrow[-N_2]{\text{EtOH}}$ Ph_2CHOEt

22. $PhCH_2Cl$ + :P(OMe)$_3$ $\longrightarrow$ $PhCH_2-\overset{\overset{O}{\|}}{P}(OMe)_2$ + MeCl

23. $HCCl_3$ + KI $\xrightarrow{\text{KOH, H}_2\text{O}}$ $HCCl_2I$ (Note: reaction does NOT occur in the absence of KOH!)

9.4 Mechanisms

CHAPTER 10
NMR

Deduce the structures in problems 1 - 17 from the ^{1}H NMR and IR information.

1. C_6H_{12}: δ 0.9 (t, 3H), 1.6 (s, 3H), 1.7 (s, 3H), 2.0 (p, 2H), 5.1 (t, 1H); no long-range coupling evident.

2. $C_6H_{12}Cl_2O_2$: δ 1.3 (t, 6H), 3.6 (q, 4H), 4.4 (d, 1H), 5.4 (d, 1H).

3. $C_8H_{18}O_2$: IR (3405 cm^{-1}). ^{1}H NMR δ 1.3 (s, 12H), 1.5 (s, 4H), 1.9 (s, 2H).

4. $C_{10}H_{14}O$: IR (3200 cm^{-1}). ^{1}H NMR δ 1.2 (s, 6H), 1.6 (s, 1H), 2.7 (s, 2H), 7.2 (s, 5H).

5. $C_5H_{10}O_4$: δ 3.2 ((s, 6H), 3.8 (s, 3H), 4.8 (s, 1H).

6. C_8H_9BrO: δ 1.4 (t, 3H), 3.9 (q, 2H), 6.7 (d, 2H), 7.4 (d, 2H).

7. $C_3H_5ClF_2$: δ 1.75 (t, 3H), 3.63 (t, 2H).

8. C_9H_{10}: δ 2.04 (m, 2H), 2.91 (t, 4H), 7.17 (s, 4H).

9. **C₈H₉Br**: δ 2.0 (d, 3H), 5.3 (q, 1H), 7.6 (m, 5H).

10. **C₄H₆Cl₂**: δ 2.18 (s, 3H), 4.16 (d, 2H), 5.71 (t, 1H).

11. **C₉H₁₁Br**: δ 2.15 (m, 2H), 2.75 (t, 2H), 3.38 (t, 2H), 7.22 (s, 5H).

12. **C₉H₁₀O₃**: δ 2.3 (t, 2H), 4.1 (t, 2H), 7.3 (m, 5H), 11.0 (br s, 1H).

13. **C₆H₁₁Br**: δ 1.0 (s, 9H), 5.5 (d, 1H, J = 17 Hz), 6.6 (d, 1H, J = 17 Hz).

14. **C₈H₁₄**: δ 1.7 (s, 6H), 1.8 (s, 6H), 6.0 (s, 2H).

15. **C₆H₁₁FO₂**: IR (3412 cm^{-1}). ¹H NMR δ 1.2 (s, 6H), 2.2 (s, 3H), 3.8 (d, 1H), 4.1 (s, 1H).

16. **C₇H₁₄O₂**: IR (1610 cm^{-1}). ¹H NMR δ 1.0 (s, 9H), 2.1 (m, 2H), 3.8 (br s, 1H), 4.0 (t, 1H), 8.6 (t, 1H).

10. NMR

17. $C_{11}H_{12}O_2$: IR (1705 cm^{-1}). ^{1}H NMR δ 2.2 (s, 3H), 2.5 (s, 3H), 5.8 (m, 1H), 7.1 (d, 2H), 7.9 (d, 2H), 9.8 (s, 1H).

---

18. What is the *maximum* multiplicity for either of the methylene protons in the proton NMR for

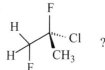 ?

19. The structure below represents two diastereomeric compounds, <u>A</u> and <u>B</u>. Compound <u>A</u> gives a *singlet* proton NMR for the methylene group, but <u>B</u> gives a *multiplet* for the same group. What are the structures of <u>A</u> and <u>B</u>?

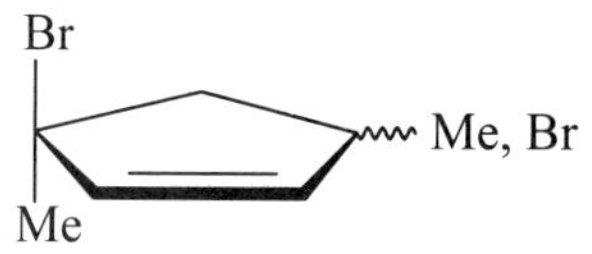

20. *Trans*-3-bromo-1-phenyl-1-propene shows a spectrum in which the vinylic proton at C_2 is coupled with the C_1 proton (J = 16 Hz) and the C_3 protons (J = 8 Hz). What is the expected multiplicity for that proton? Use a spin tree diagram to explain.

21. a. What is the multiplicity of the chemical shift at *highest* field in the proton NMR of (R)-1,2-dichloro-2-fluoropropane?

b. Use a spin tree diagram to explain why the *lowest* field chemical shift appears as a *triplet*.

22. What is the *maximum* multiplicity for H_a in the amino acid phenylalanine?

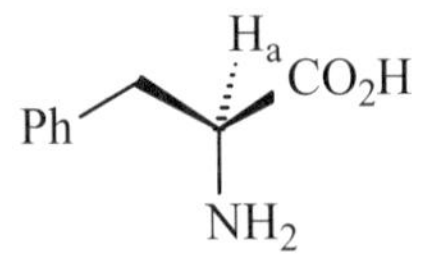

Ph–CH$_2$–C(H$_a$)(NH$_2$)–CO$_2$H

phenylalanine

23. A compound has only two singlets in its ^{1}H NMR spectrum: δ 1.4 and 2.0 with relative intensities of 3:1. Its ^{13}C NMR spectrum has chemical shifts at δ 22, 28, 80, and 170. A strong absorption in its IR occurs at 1740 cm^{-1}. Draw a possible structure for the compound.

24. The following questions relate to deuterated cholesterol, drawn below:

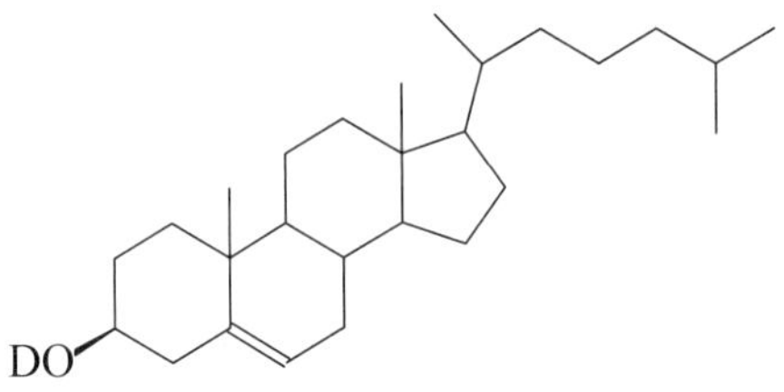

a. Predict the theoretical multiplicity of the *lowest field* proton.

b. What is the maximum number of ^{13}C chemical shifts that would be expected for the C_8H_{17} alkyl side chain?

25. Treatment of 2,3-dibromo-2,3-dimethylbutane with SbF$_5$ (a very strong Lewis acid) in SO$_2$ at -60° yields SbF$_6^-$ and a substance whose ^{1}H NMR shows only a singlet at δ 2.9. Draw the structure of that substance.

26. What is the *multiplicity* of the methylene group in the following compound?

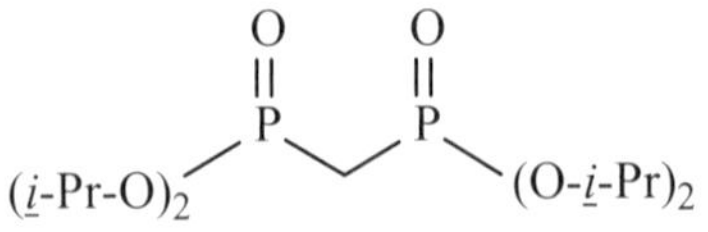

(i-Pr-O)$_2$P(=O)–CH$_2$–P(=O)(O-i-Pr)$_2$

27. Below is the structure and partial ^{1}H NMR for an organoplatinum compound. Platinum has three isotopes: ^{195}Pt (I = ½, 34% natural abundance), ^{194}Pt (I = 0), and ^{196}Pt (I = 0) – the latter two account for the remaining 66% natural abundance. (Note: aromatic proton resonances are not shown.)

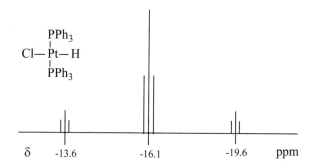

a. Explain the relative *amplitude* and *multiplicity* of the signal at δ -16.6. Clearly explain $J_{H,?}$ by using a spin tree diagram.

b. Explain the *amplitude* and *multiplicity* of the two signals at δ -13.6 and -19.6. Again, clearly explain $J_{H,?}$ by using a spin tree diagram.

c. What do the very negative chemical shift values of the signals suggest about the magnetic environment of the resonating proton?

10. NMR

28. Pettit (*UT*) observed that the protonation of cyclooctatetraene (*COT*) yields a carbocation (*homotropylium ion*) that possesses *homoaromatic* stabilization. (*Homoaromatics* refers to π systems that are interrupted by a saturated center but in which the geometry still permits significant overlap of the *p* orbitals across a gap.)

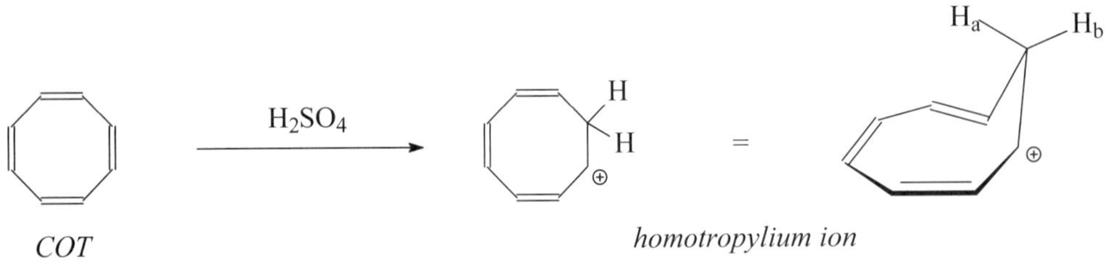

COT → H₂SO₄ → homotropylium ion

The ¹H NMR of the homotropylium ion shows a remarkable chemical shift difference of 5.5 ppm for geminal protons H_a (δ 0.5) and H_b (δ 5.0). Each appears as a pseudoquartet. Explain both the location of the chemical shifts and multiplicities of these protons.

29. The ¹H NMR spectrum of NaBH₄ is shown below. Boron has two isotopes: ¹⁰B (I = 3) and ¹¹B (I = 3/2) whose natural abundances are 20% and 80%, respectively. Interpret the spectrum.

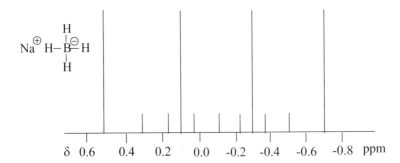

CHAPTER 11
CONJUGATED SYSTEMS

11.1 Reactions
Draw the structural formula of the major organic product(s). Show stereochemistry where appropriate.

1. 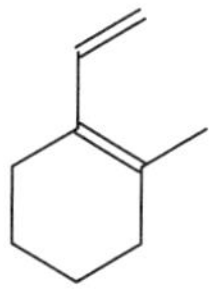 [1-vinyl-2-methylcyclohexene] $\xrightarrow[\text{(1,4-addition)}]{\text{HBr (1 equiv)}}$

2. α-farnesene $\xrightarrow[\text{(1,4-addition)}]{\text{DCl (1 equiv)}}$

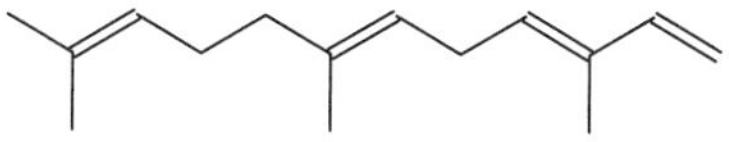

   α-farnesene (in waxy coatings of apple skins)

3. isoprene + MeO$_2$C—≡—CO$_2$Me ⟶

4. 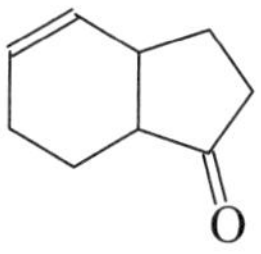 [bicyclic enone] $\xrightarrow[\Delta]{\text{retro D-A}}$

5. 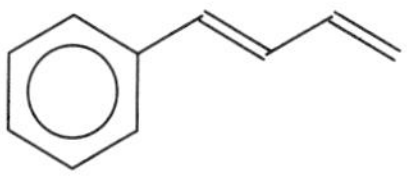 [PhCH=CH–CH=CH$_2$] $\xrightarrow{\text{HBr (1 equiv)}}$

   (product of thermodynamic control)

6. 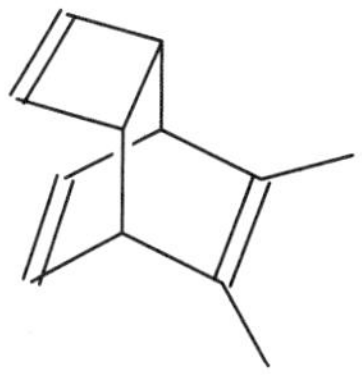 [bicyclic diene with methyls] $\xrightarrow{\Delta}$ 2-butyne +

7. 3-methyl-1,3,5-hexatriene $\xrightarrow[\text{ROOR (1,4-addition)}]{\text{DBr (1 equiv)}}$

8. 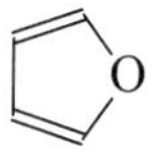 [furan] + [N-phenyl triazoline-3,5-dione] ⟶

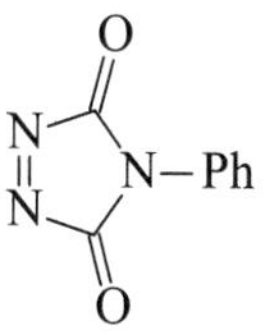

   Cookson's dienophile

Chapter 11 Conjugated Systems

9. 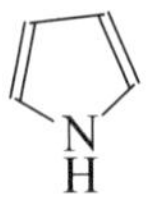 pyrrole + (Z)-1,2-diphenylethene ⟶

10. cyclohexene
   1. NBS, ROOR
   2. KOMe (E2)
   3. phenylacetylene
   ⟶

11. 4-vinylcyclohexene
   1. Δ (retro D-A)
   2. *trans*-2-butene
   ⟶

12.  (cyclopentadiene with -(CH$_2$)$_3$-C≡CH side chain) $\xrightarrow{\text{intra D-A}}$

13.  furan
   1. vinyl chloride
   2. KO-*t*-Bu
   ⟶

14. 1,3-cyclohexadiene
   1.  CH$_2$=CH-CO$_2$Me
   2. O$_3$ 3. Zn, H$^\oplus$
   ⟶

15. 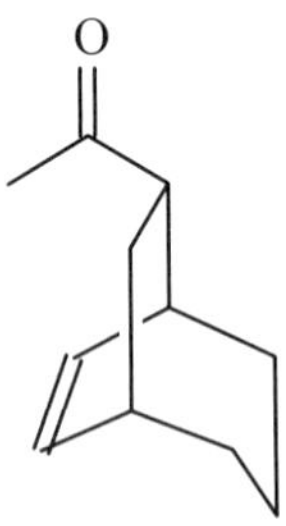 (bicyclic ketone structure) $\xrightarrow{\Delta \text{ (retro 4+2)}}$

11.1 Reactions

16. 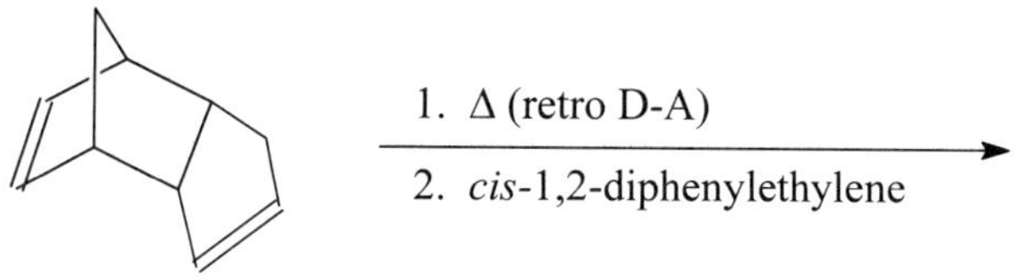 [dicyclopentadiene] $\xrightarrow[\text{2. }cis\text{-1,2-diphenylethylene}]{\text{1. }\Delta\text{ (retro D-A)}}$

17. 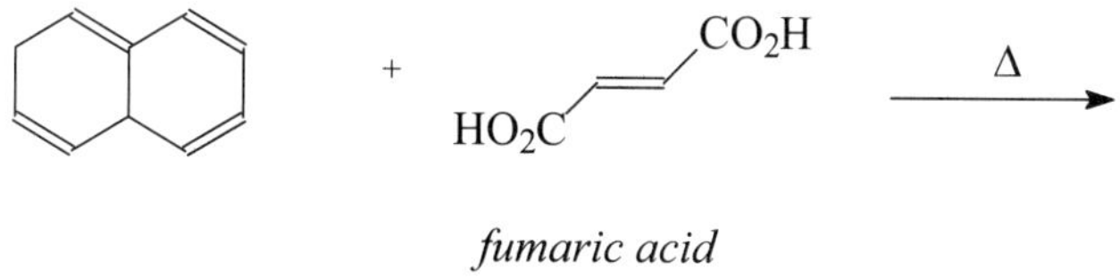 [dihydronaphthalene] + [fumaric acid: HO₂C-CH=CH-CO₂H] $\xrightarrow{\Delta}$

    fumaric acid

18. 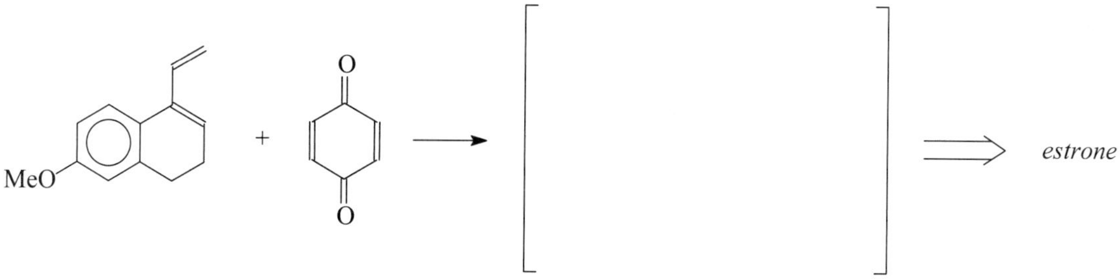 [1-vinyl-6-methoxy-3,4-dihydronaphthalene] + [1,4-benzoquinone] → [ ] ⟹ *estrone*

19. 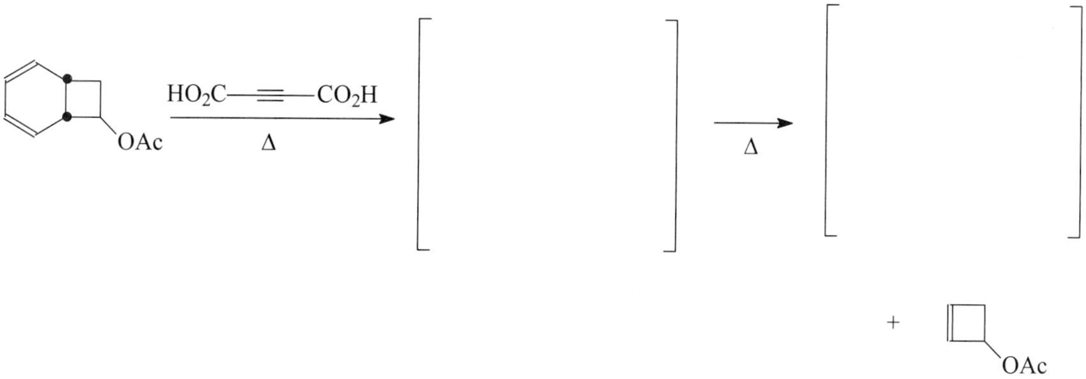 [benzocyclobutene-OAc] $\xrightarrow[\Delta]{\text{HO}_2\text{C}-\equiv-\text{CO}_2\text{H}}$ [ ] $\xrightarrow{\Delta}$ [ ] + [cyclobutenyl-OAc]

20. 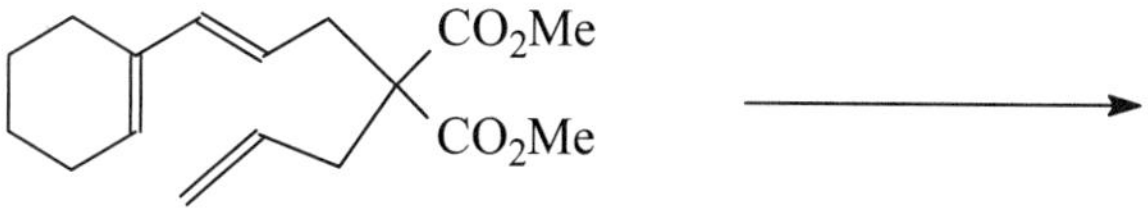 [cyclohexenyl-CH=CH-C(CO₂Me)₂-CH₂-CH=CH₂] →

21. 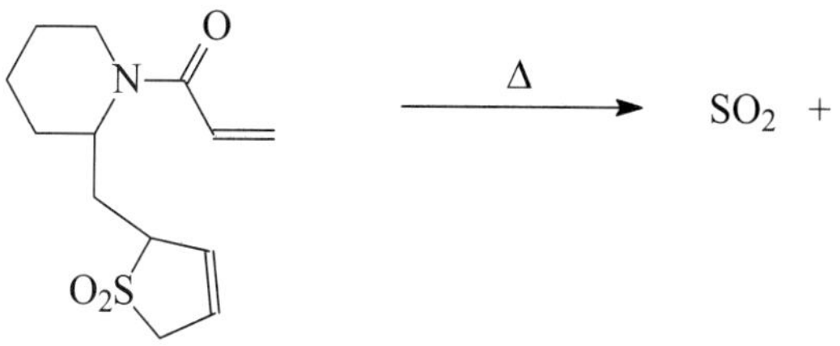 [N-acryloyl-2-(sulfolenylmethyl)piperidine] $\xrightarrow{\Delta}$ SO₂ +

11.1 Reactions

22.

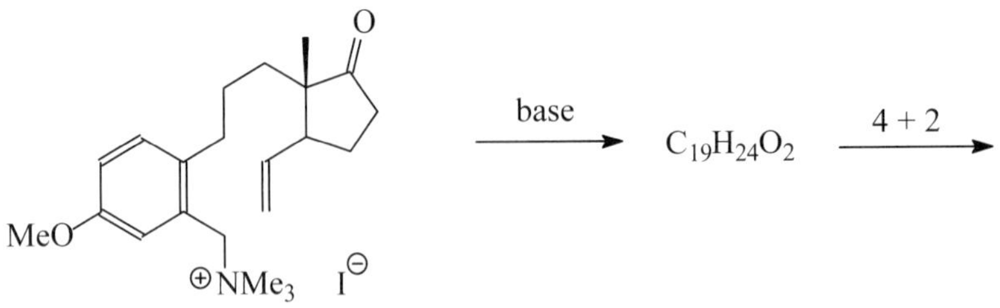
O
MeO
⊕NMe3 I⊖
base
C19H24O2
4 + 2

23.

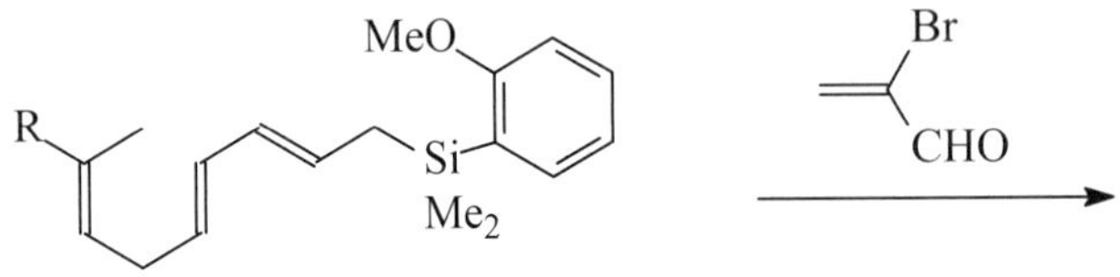
R
MeO
Si
Me2
Br
CHO

24.

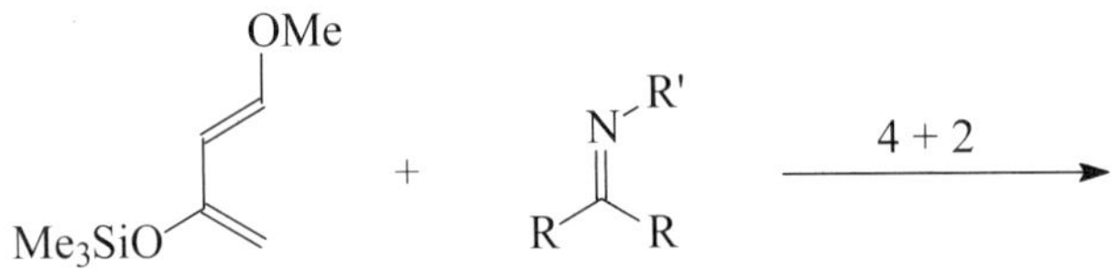
OMe
Me3SiO
+
N
R'
R
R
4 + 2

Danishefsky's diene

11.2 Syntheses

Supply a reagent or sequence of reagents that will effect the following conversions.

1. cyclohexane →

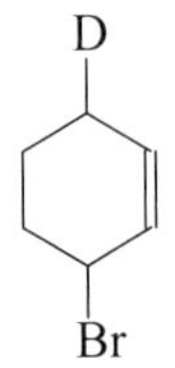
D
Br

via a conjugated diene

2. cyclohexane → bicyclo[2.2.2]octane

3. cyclohexene →

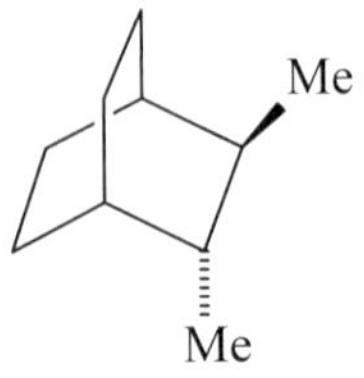
Me
Me

4. vinylcyclohexane ⟶ [bicyclic diene product]

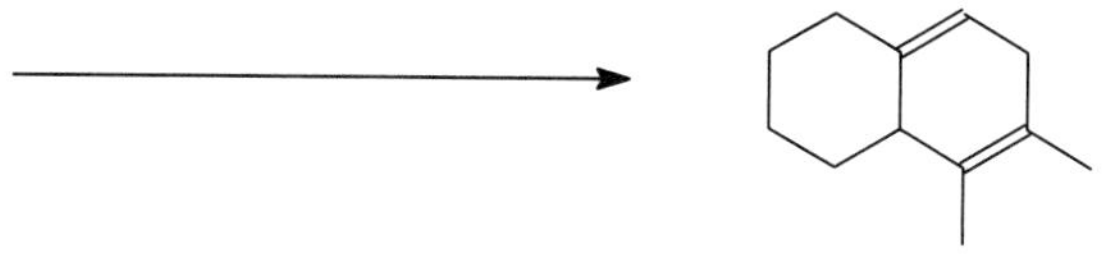

5. A Diels-Alder dimerization of **A** gives the indicated product. Draw the structure of **A**.

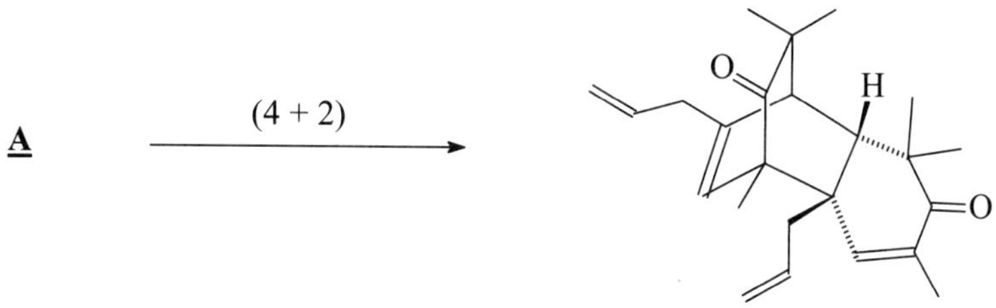

A $\xrightarrow{(4+2)}$ [product shown]

6. Draw the structures of the starting materials that may be used to synthesize the following product:

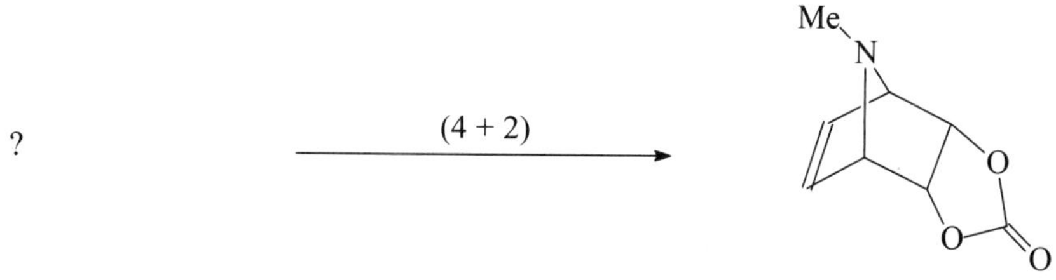

? $\xrightarrow{(4+2)}$ [product shown]

7. The *Alder-ene reaction*, like the Diels-Alder, is a concerted (pericyclic) reaction:

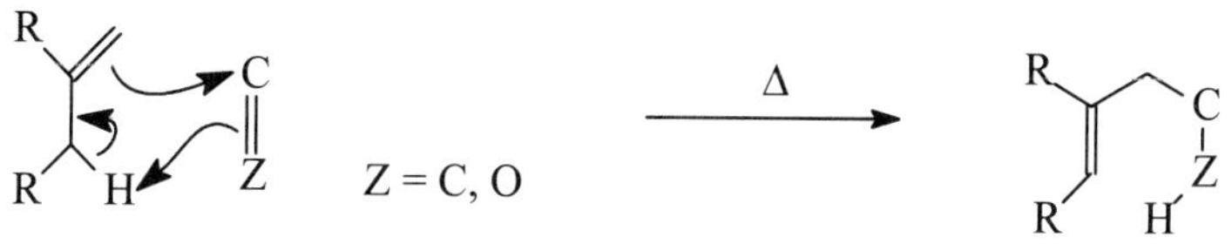

[scheme shown]   Z = C, O

How could the following compound be prepared by an ene reaction?

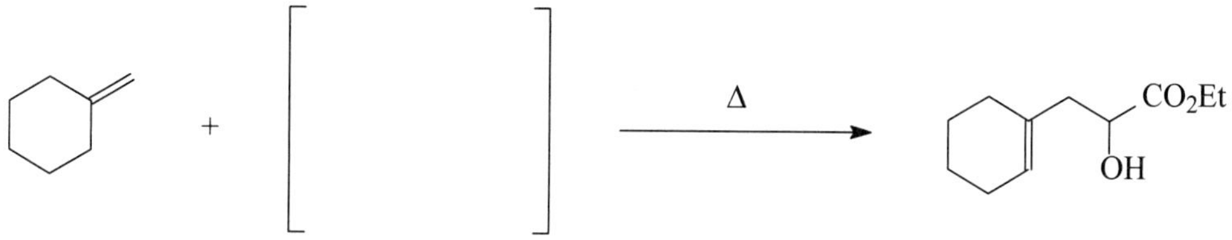

[methylenecyclohexane] + [ ] $\xrightarrow{\Delta}$ [product with CO$_2$Et and OH]

11.2 Syntheses

11.3 Mechanisms

Outline a detailed mechanism for each of the following. No other reagents than those given are necessary. Use arrows to explain the flow of electrons and show all intermediates.

1. *Phenolphthalein* in solutions below pH 8.5 is colorless, but in solutions above pH 8.5 is a deep red-purple color. Explain.

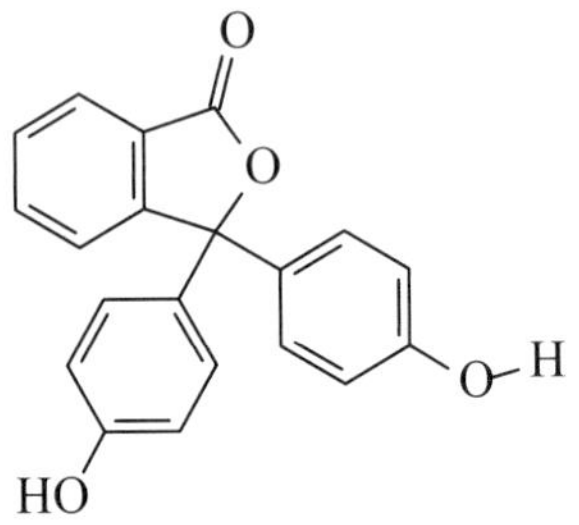

phenolphthalein

2.

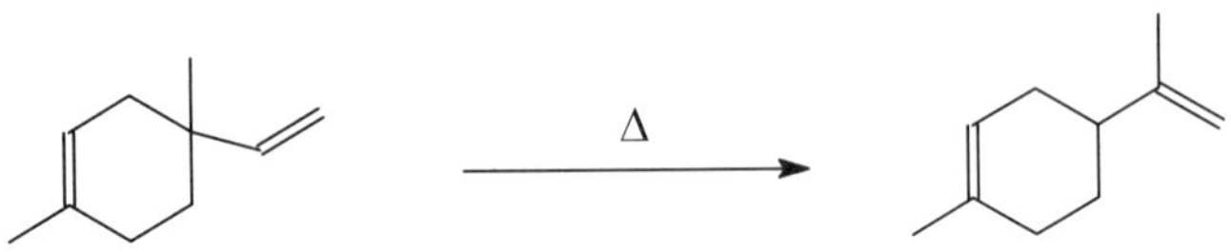

3. Similar to the Diels-Alder the following *electrocyclic reaction* is generally concerted (pericyclic) and readily reversible.

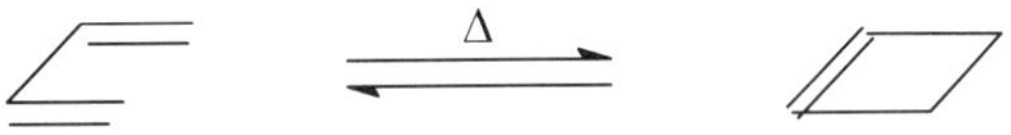

Explain the observed conversions:

a.

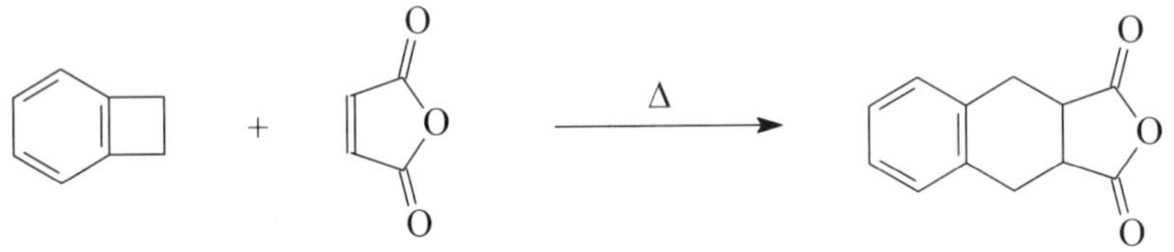

b.

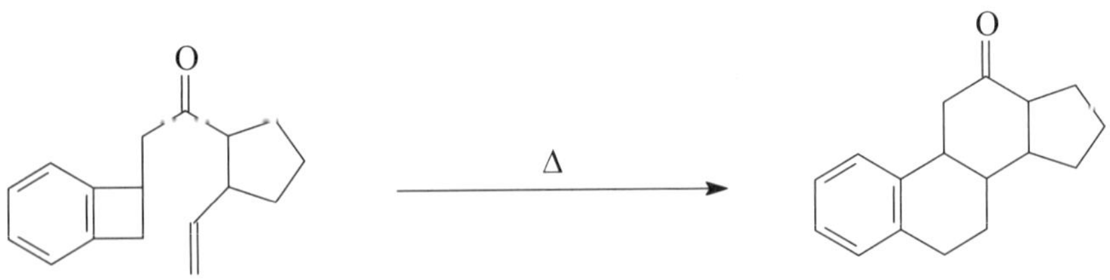

4. The structure of *pyridine* is shown below:

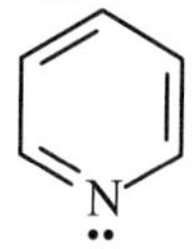

pyridine

a. Describe the *longest* wavelength λ_{max} electronic transition in terms of σ, σ*, π, π*, or n.

b. Comment on the *probability* of that transition. What term in the Beer-Lambert equation reflects this probability?

c. Draw the conjugate acid of pyridine. How would that transition in (a) be affected?

5. Compound **A**, upon standing in acid, yields a new isomeric compound **B** whose ^{1}H NMR is δ 1.7 (s, 3H), 1.8 (s, 3H), 2.3 (br s, 1H), 4.1 (d, J = 8 Hz, 2H), 5.5 (t, J = 8 Hz, 1H). Draw the structure of compound **B** and give its mechanism of formation.

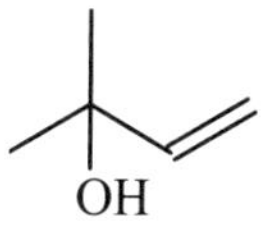

A

6. One approach to synthesizing the sesquiterpene *occidentalol*, found in New England white cedar trees, begins with a forward Diels-Alder reaction, followed by a retro-Diels-Alder, to form **A**. Explain.

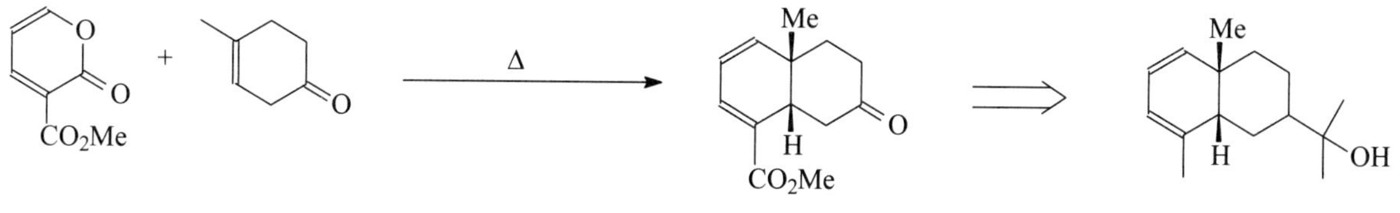

A

occidentalol

11.3 Mechanisms

7. An early stage reaction in Paquette's (*OSU*) total synthesis of dodecahedrane employed the following "domino" Diels-Alder:

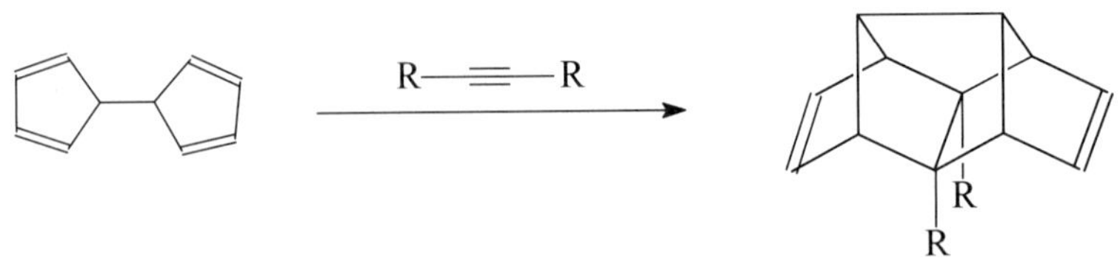

8.

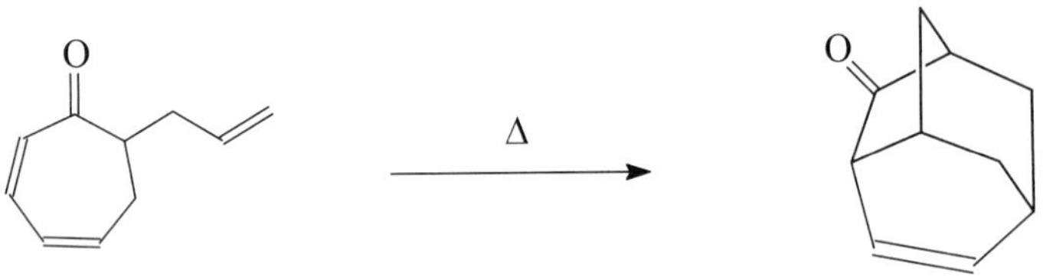

9.

10.

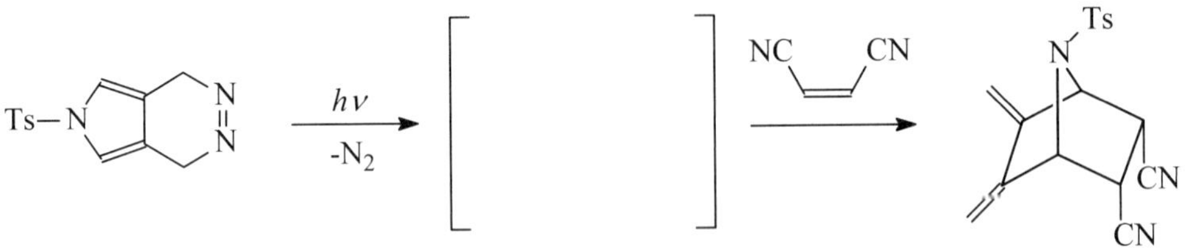

11.3 Mechanisms

11. The degradation of heme proceeds by way of the bile pigments *biliverdin* and *bilirubin*, green and red, respectively. Elevated levels of the latter produce jaundice. Bilirubin, a principal antioxidant in blood plasma, is formed by reducing biliverdin. Label the structures below as biliverdin or bilirubin and identify the site of reduction in the former. Explain the difference in color of the two pigments.

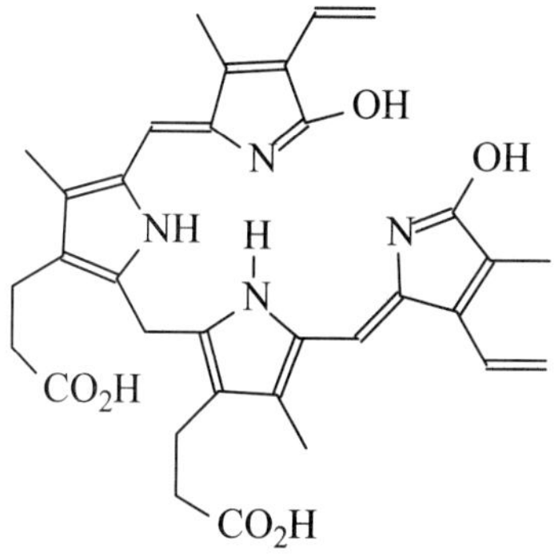

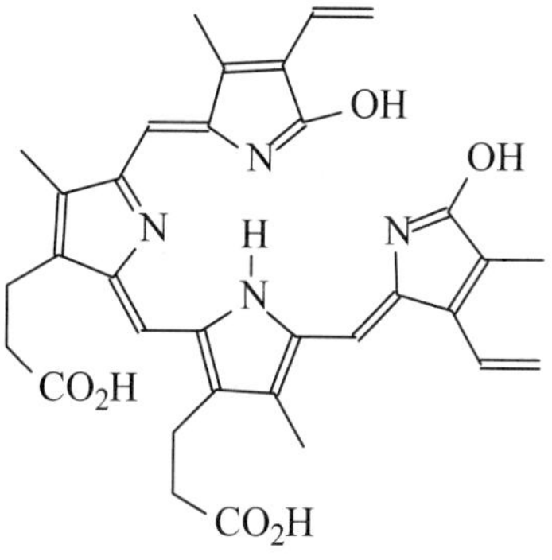

12. Depending upon the number of π electrons in a pericyclic process, reversible cycloaddition reactions may be classified as *thermally* "allowed" or "forbidden" (a theoretical prediction of the probability that such a reaction will occur). The Diels-Alder reaction is the most common example of a *thermally allowed* (4+2) cycloaddition. Examples of *thermally forbidden* reactions include (2+2) and (4+4) cycloadditions (they do occur, however, under photochemical conditions). Formation of the dibenzenes below could be envisioned by a cycloaddition mechanism. Identify each as (2+2), (4+2), or (4+4). Which would be expected to undergo a thermal *retro*-cycloaddition to benzene most rapidly?

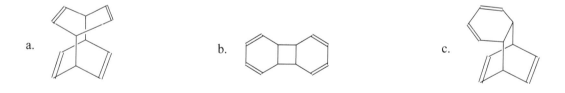

11.3 Mechanisms

CHAPTER 12
AROMATICS

12.1 General

1. Circle the compounds that would be expected to have aromatic character.

a.   b.   c.   d.

e.   f.   g.   h.

i.   j.   k.

l. carbocation in the reaction of [structure] SbF₅ →

m. product in the reaction of [cyclobutadiene] $\xrightarrow{2\ Li}$ 2 Li⁺ +

n. product in the reaction of [structure] $\xrightarrow{2\ MeLi}$ 2 MeH +

o. product in the reaction of [dibromide structure] $\xrightarrow[-ZnBr_2]{Zn}$

calicene

2. Which would have the *largest* molecular dipole moment (μ)?

a. 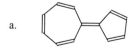   b. 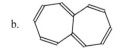   c.

3. Which nitrogen atom is *least basic* in purine and *most basic* in Zofran™?

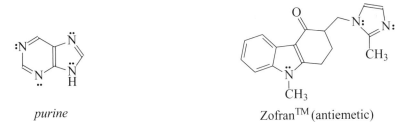

purine   Zofran™ (antiemetic)

4. One of the following ketones is *unstable* and undergoes a Diels-Alder reaction rapidly. Which?

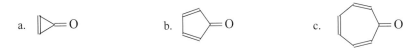

5. Which of the following compounds would most easily form its conjugate base?

6. Which would undergo an S_N1 reaction most readily?

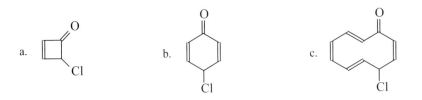

7. Circle the *more(most) basic* electron pair in each of the following:

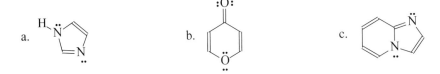

8. Use a Frost mnemonic to explain why 7-chloro-1,3,5-cycloheptatriene gives a *singlet* ^{1}H NMR spectrum when dissolved in a solvent containing a Lewis acid.

12.1 General

9. Which ketone has the *largest* molecular dipole moment (μ)?

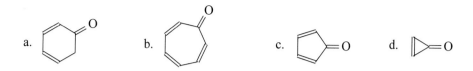

10. **A** in the presence of HBF₄ forms a salt. Explain.

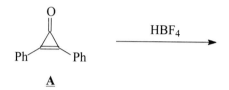

11. Explain the regioselectivity of the following addition:

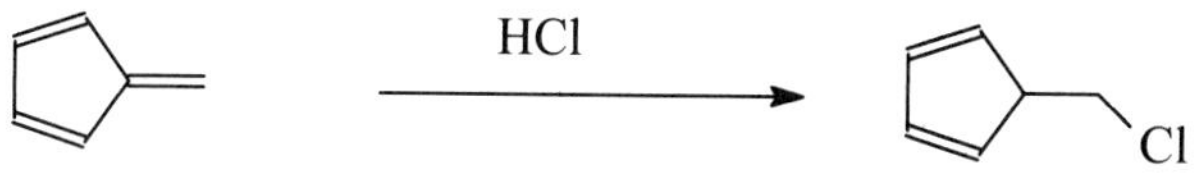

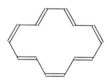

12. The ¹H NMR spectrum for the following [14]annulene compound shows two major chemical shifts. Simulate their approximate location and predict the integration of each.

a [14]annulene

12.2 Reactions

Draw the structural formula of the major organic product(s). Show stereochemistry where appropriate.

1. 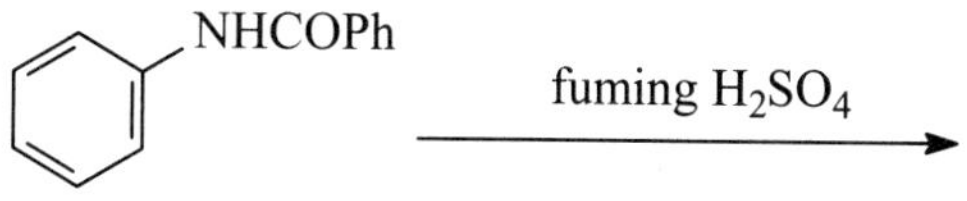 [phenyl-NHCOPh] → fuming H₂SO₄

2. *o*-methylphenol → HONO₂ / H₂SO₄

12.2 Reactions

106 • Chapter 12 Aromatics

3. 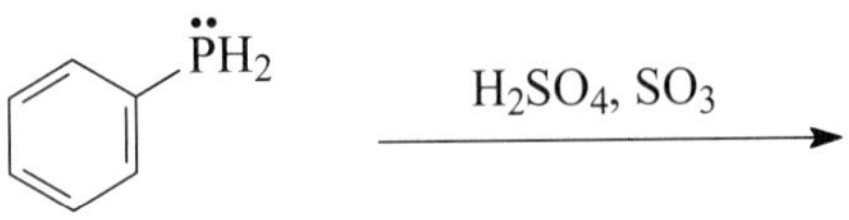 Ph-PH₂ → H₂SO₄, SO₃

4. 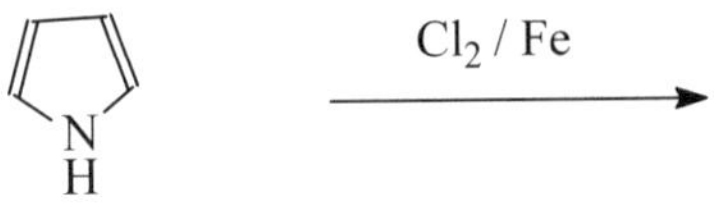 pyrrole → Cl₂ / Fe

5. benzene
   1. PhCH₂CH₂Br, AlCl₃
   2. NBS, R₂O₂
   3. KOMe, MeOH

6. 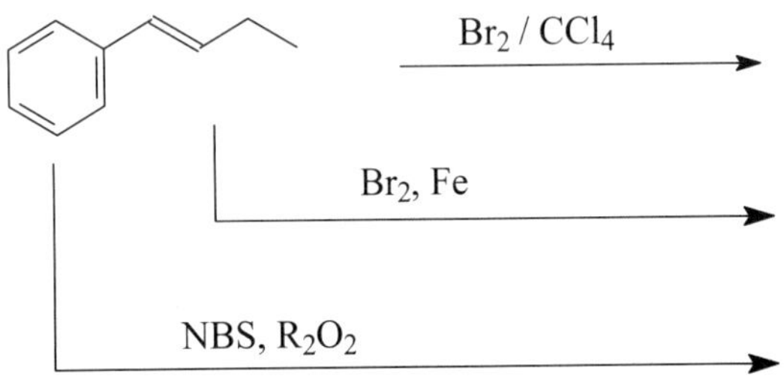 PhCH=CHCH₂CH₃
   Br₂ / CCl₄
   Br₂, Fe
   NBS, R₂O₂

7. 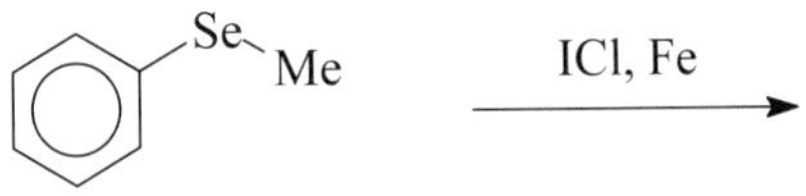 Ph-Se-Me → ICl, Fe

8. Ph—N=O → Br₂, FeBr₃

9. 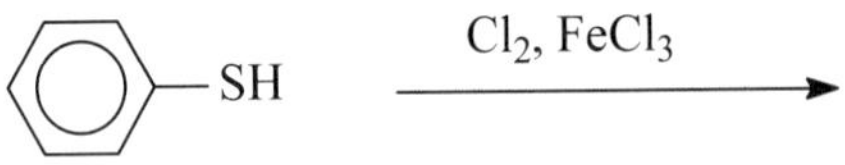 Ph-SH → Cl₂, FeCl₃

10. 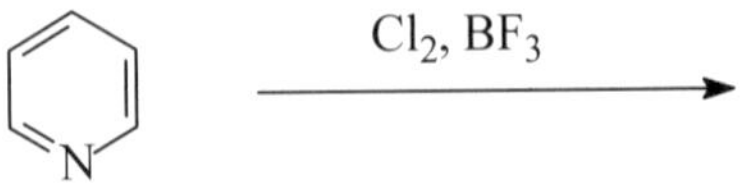 pyridine → Cl₂, BF₃

12.2 Reactions

11. 2 (2,4,5-trichlorophenol) + formaldehyde $\xrightarrow{H_2SO_4}$

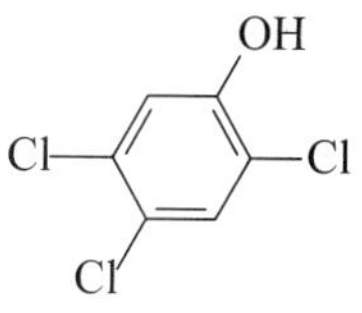

($C_{13}H_6Cl_6$ - *hexachlorophene*, a disinfectant)

12. 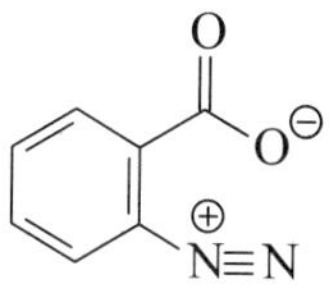 (2-carboxylatobenzenediazonium) $\xrightarrow{\text{1. }\Delta\text{ }(-CO_2, -N_2)}_{\text{2. 1,3-cyclohexadiene}}$

13. 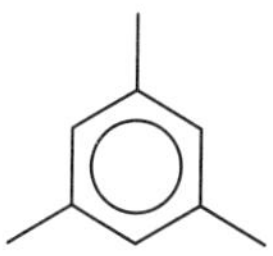 (1,3,5-trimethylbenzene) $\xrightarrow{\text{picric acid}}$

14. 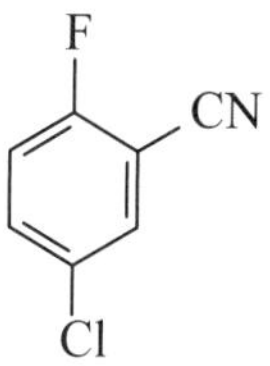 (2-fluoro-5-chlorobenzonitrile) $\xrightarrow{NH_3}$

15. 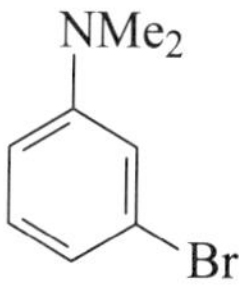 (3-bromo-N,N-dimethylaniline) $\xrightarrow{\text{1. MeLi}}_{\text{2. }H^\oplus}$

16. (1-chloro-4-trifluoromethylbenzene) $\xrightarrow{\text{1. }HNO_3, H_2SO_4}_{\text{2. NaOMe, MeOH}}$

17. toluene + (2,5-dimethyl-2,5-dihydrofuran) $\xrightarrow{H^\oplus}_{-H_2O}$

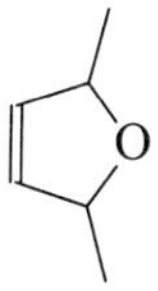

(*a bicyclic C_{13} compound*)

12.2 Reactions

18.

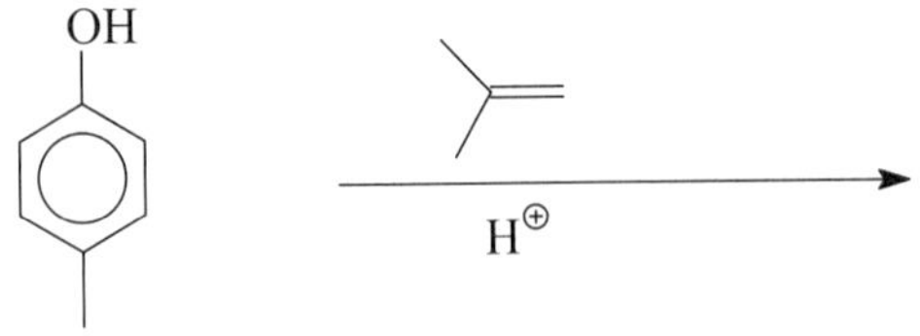

[p-cresol] + [isobutylene] →(H⁺) BHT ($C_{15}H_{22}O$ - a food preservative)

19. 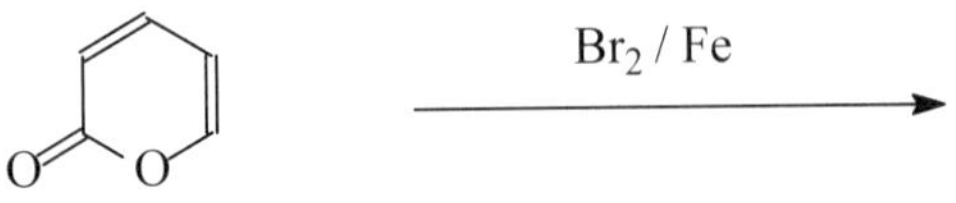 α-pyrone →(Br₂ / Fe)

20. 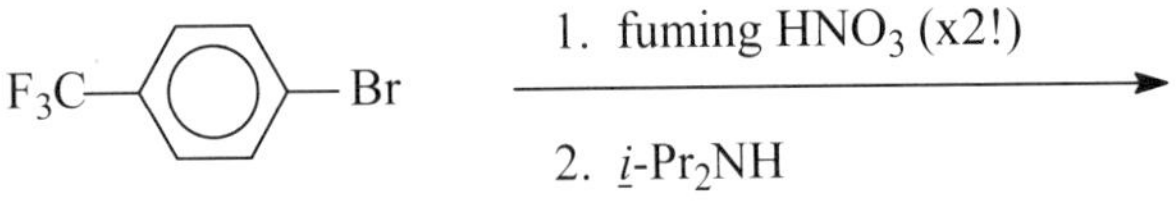 F_3C—[benzene]—Br →(1. fuming HNO_3 (x2!); 2. i-Pr_2NH) Trifluralin B™ (a pre-emergent herbicide)

21. anisole →(1. MeI, $AlCl_3$; 2. NBS, ROOR; 3. KOH)

(flavor in licorice)
partial ¹H NMR: δ 7.2 (d, 2H), 7.9 (d, 2H)

22. Although iodination of aromatic rings does not occur as readily as bromination, it can be observed when activating substituents are present, *e.g.*, in the biosynthesis of the hormone *thyroxine*:

tyrosine →(I_2, catalyst) [ ] → thyroxine

12.2 Reactions

23. 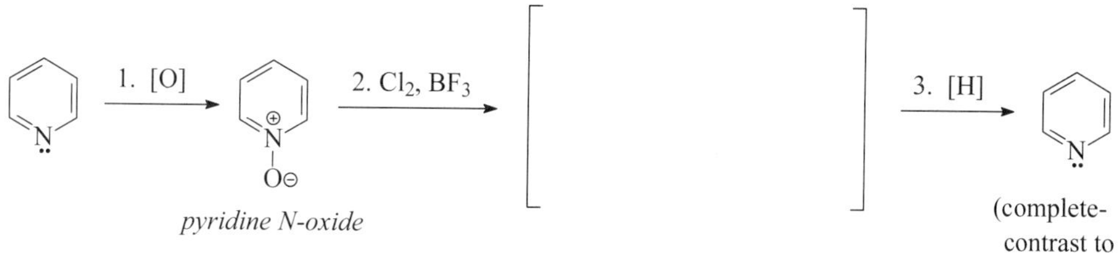 pyridine →(1. [O])→ pyridine N-oxide →(2. Cl₂, BF₃)→ [ ] →(3. [H])→ pyridine (complete-contrast to 12.2, 10)

12.3 Syntheses

Supply a reagent or sequence of reagents that will effect the following conversions.

1. benzene ⟶ 1-chloro-2-methylcyclohexane

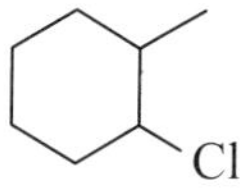

2. benzene ⟶ 1,3,5-trideuterobenzene

3. benzene ⟶ 1,4-dideuterobenzene

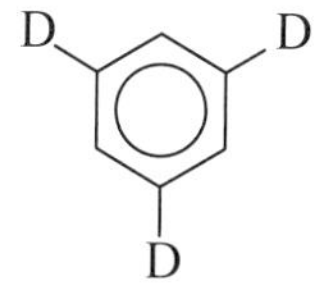

4. benzene ⟶ 1,2-diphenylethane

5. 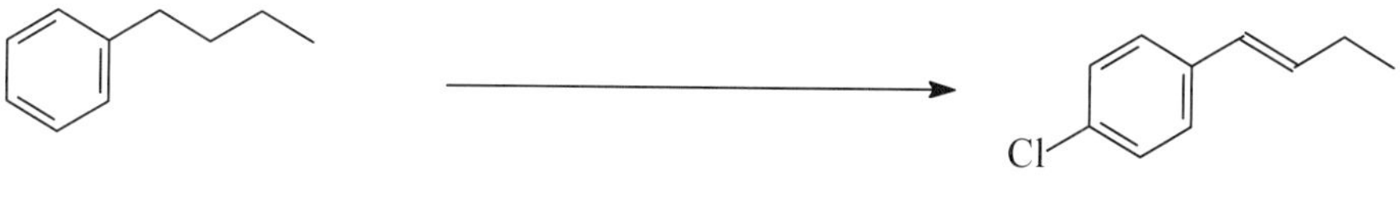 butylbenzene ⟶ 1-(4-chlorophenyl)-1-butene

6. PhH ⟶ [phenylcyclohexene]

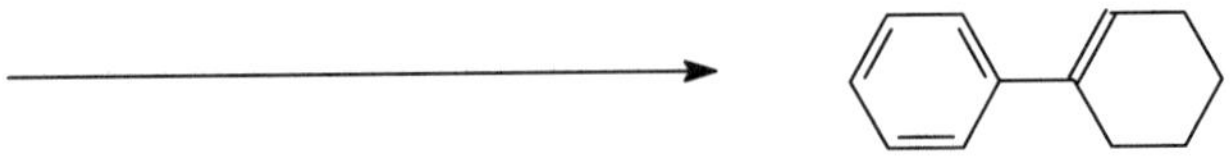

7. [4-phenyl-2-methylpent-1-ene / 5-methyl-1-phenyl-hex-4-ene structure] ⟶ [α-tetralone]

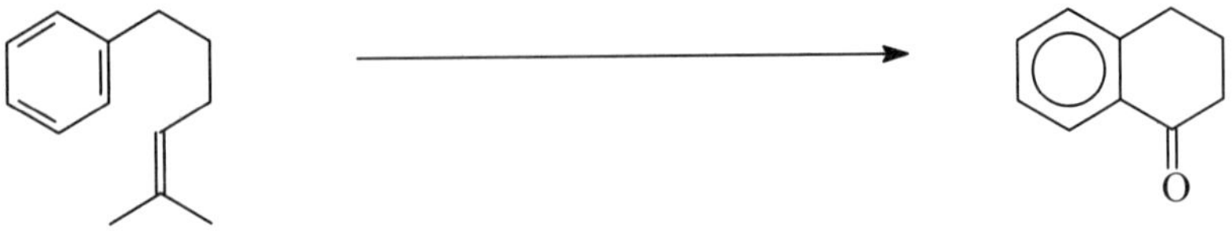

8. benzene ⟶ o-nitrobenzoic acid

9. benzene ⟶ [4-methylbiphenyl]

10. [PhCH₂CH₂NHMe] ⟶ [N-methylindoline]

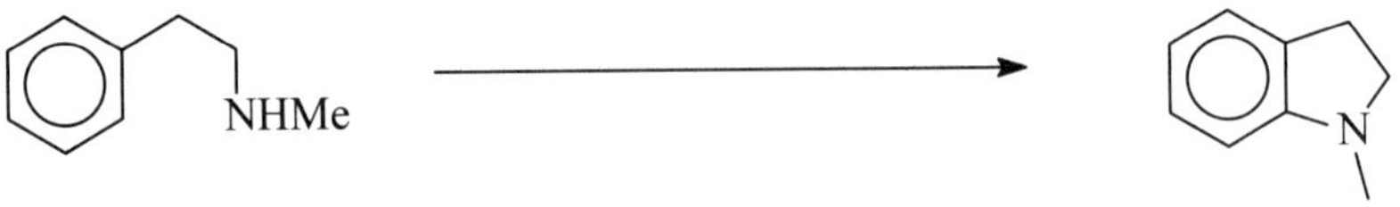

11. benzene ⟶ [4-ethyl-isobutylbenzene] ⟹ [ibuprofen structure with CO₂H]

    ibuprofen

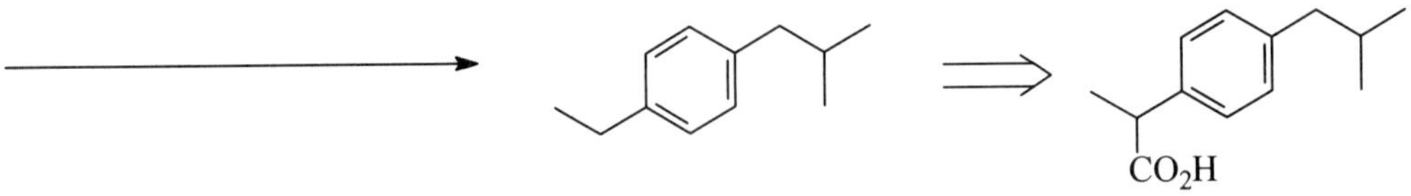

12. acetone, phenol ⟶ HO–C₆H₄–C(CH₃)₂–OH ⟶ [2,6-di-tert-butyl-4-(2-phenylpropan-2-yl)phenol]

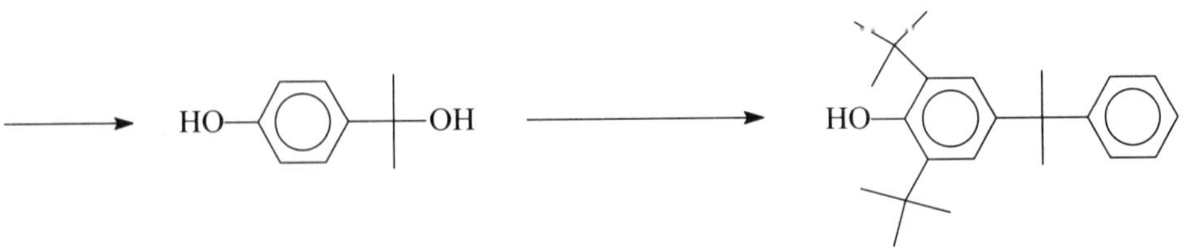

12.3 Syntheses

13. 2,4-D and 2,4,5-T are the active agents in the defoliant Agent Orange™. How could they be prepared from the indicated starting materials?

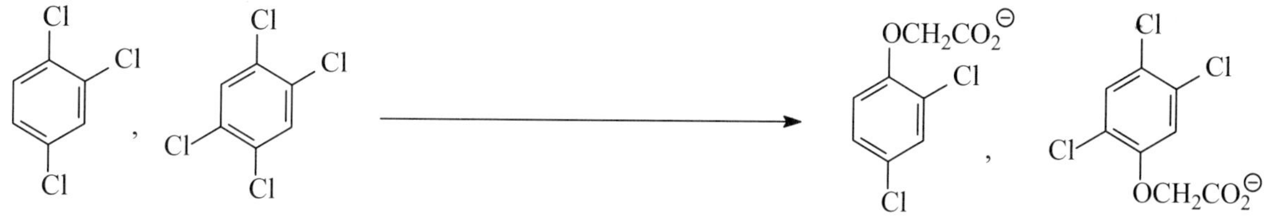

14. [4-bromo-nitrobenzene] ⟶ Tylenol™ (acetamido-phenol)

15. *p*-hydroxybenzoic acid ⟶ [3-amino-4-propoxy benzoic acid] ⟹ proparacaine (a local anesthetic)

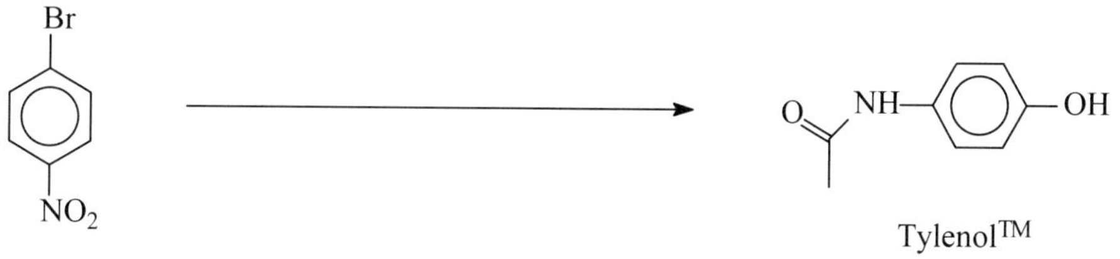

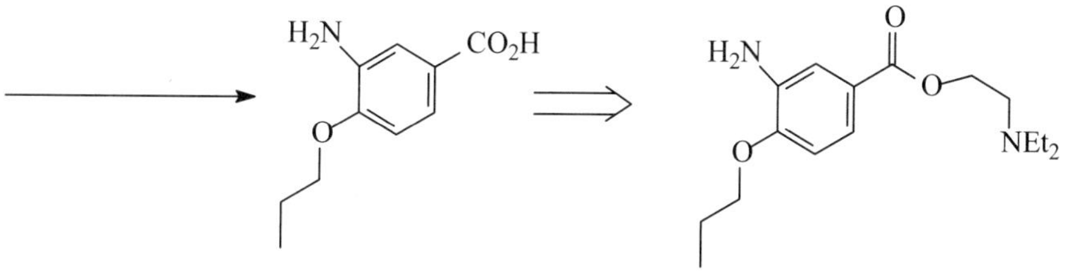

12.4 Mechanisms

Outline a detailed mechanism for each of the following. No other reagents than those given are necessary. Use arrows to explain the flow of electrons and show all intermediates.

1. styrene $\xrightarrow{H^{\oplus}}$ [1-methyl-3-phenylindane]

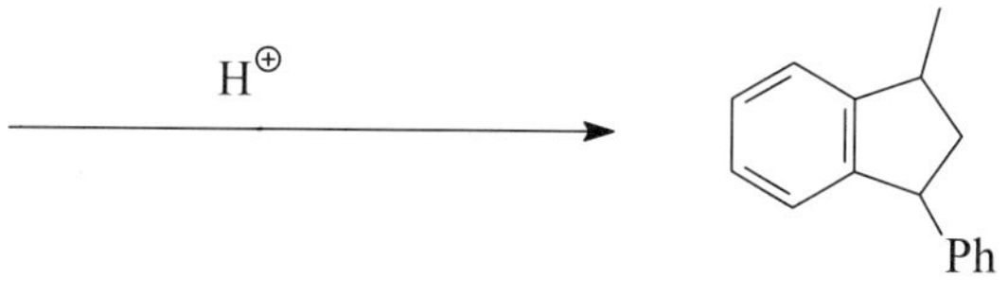

2. Epoxides, because of ring strain, are much more reactive than most ethers. Account for the following:

   anisole $\xrightarrow{\text{ethylene oxide, } H^{\oplus}}$ [4-methoxyphenyl-CH(CH$_3$)-CH$_2$OH... actually product: 4-MeO-C$_6$H$_4$-CH(CH$_3$)-OH with CH$_2$OH]

12.4 Mechanisms

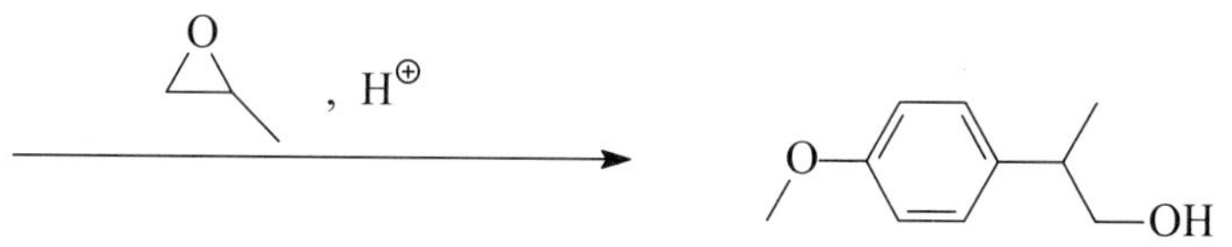

3. (XS) N,N-dimethylaniline + COCl₂ (*phosgene*) →[AlCl₃] *Michler's ketone* (4,4'-bis(dimethylamino)benzophenone)

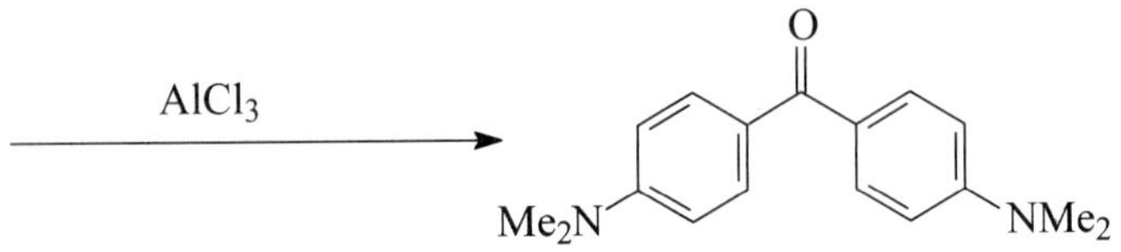

4. The *Kolbe reaction* is used industrially to convert phenol to salicylic acid, an immediate precursor to aspirin.

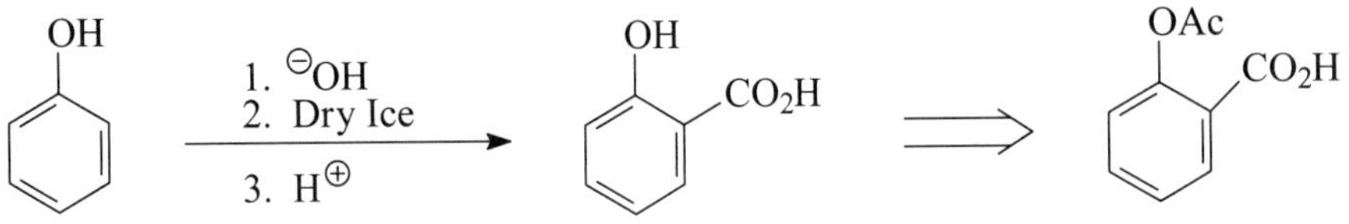

phenol →[1. ⁻OH; 2. Dry Ice; 3. H⁺] *salicylic acid* ⇒ *aspirin*

5. *tert*-butylbenzene →[Br₂ / AlCl₃] bromobenzene + isobutylene

6. diphenylmethane →[1. R–C(=O)–X, AlCl₃; 2. HBr] 9-R-anthracene

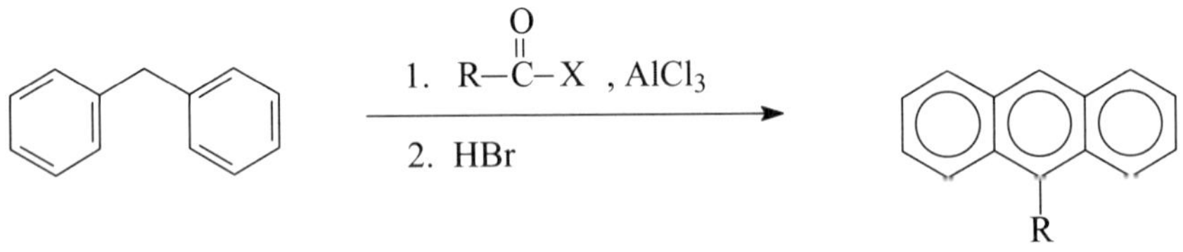

12.4 Mechanisms

7. 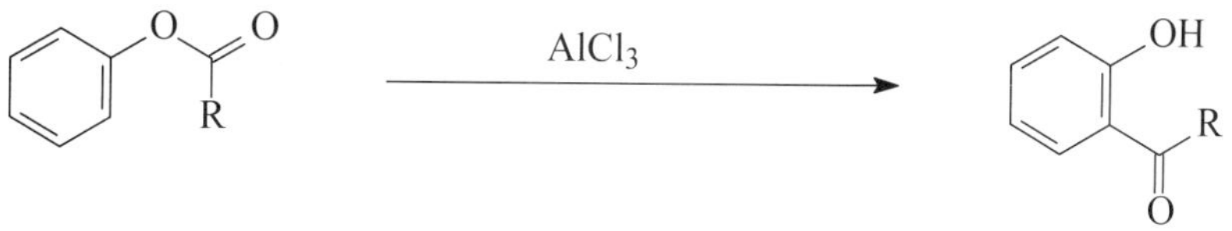 [phenyl ester with R group] —AlCl₃→ [2-hydroxyaryl ketone with R]

8. 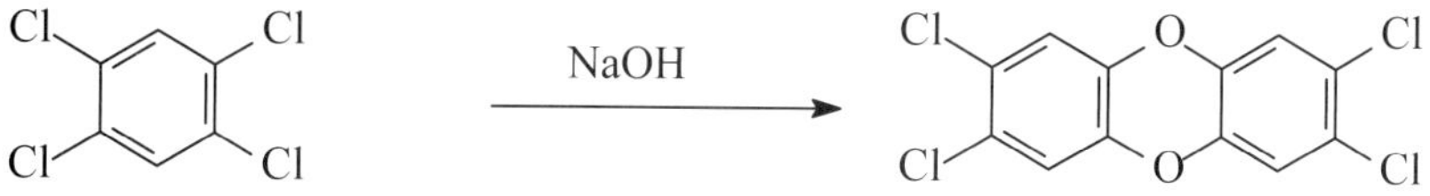 [tetrachlorobenzene] —NaOH→ [tetrachlorodibenzodioxin]

   dioxin

9. 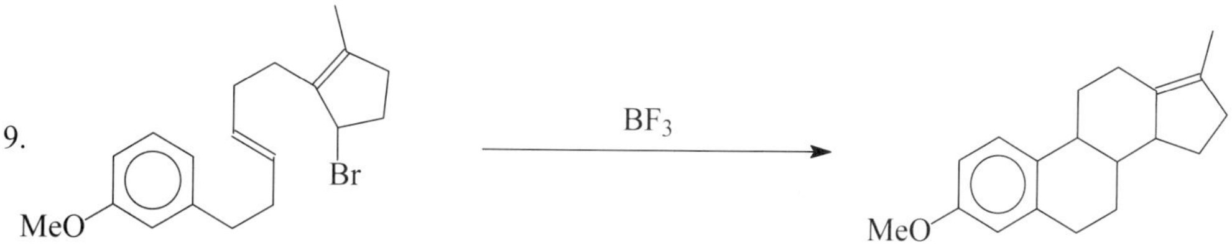 [MeO-aryl with pendant alkene and bromocyclopentene] —BF₃→ [methoxy steroid skeleton]

10. Formyl chloride, **A**, does NOT exist; therefore, one cannot do a Friedel-Crafts type acylation to produce benzaldehyde. However, the latter can by synthesized by the reaction of benzene with carbon monoxide and HCl (a process known as the *Gatterman-Koch reaction*). Outline a mechanism.

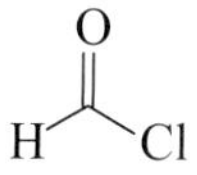

    H–C(=O)–Cl   +   benzene   —CO / HCl→   PhCHO
    A

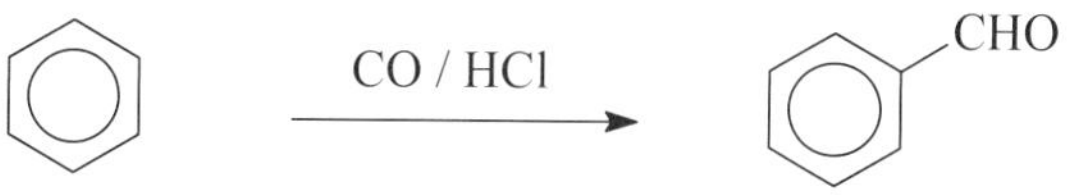

11. toluene —H₂O₂ / CF₃SO₃H→ 4-methylphenol–OH (*a convenient way to substitute an hydroxyl group onto an aromatic ring*)

12.4 Mechanisms

114 • Chapter 12 Aromatics

12. Dyes such as indigo blue (see 19.1, 34) do not bond well to cotton and tend to wash off after repeated laundering; they are known as *surface* dyes. On the other hand, *reactive* dyes bind covalently to cotton, resulting in greater color retention ('fastness'). The following process illustrates the latter. An amino-containing dye is initially bound to cyanuryl chloride to give a product that subsequently is allowed to react with the hydroxyl groups of cotton. Show a mechanism for this process that illustrates how cyanuryl chloride serves to crosslink the dye with cotton. What type of reaction describes each step?

cyanuryl chloride → 1. dye-NH$_2$ 2. cotton-OH

cyanuryl chloride

13. Malaria, which claims over one million lives *per year*, mostly children and largely in Africa, could be eradicated with the judicious use of *DDT*. Banned in the US in 1972, in large part because of Rachael Carson's 1962 book *The Silent Spring*, exhaustive scientific review has since shown DDT, *in moderation*, not only to be safe for humans and the environment, but also the single most effective anti-malarial agent ever formulated. Although the World Health Organization and the US have now reversed their anti-DDT stance, emotional opposition to the pesticide remains so fierce that its use continues to be resisted – at the cost of millions of unnecessary deaths.

DDT is easily prepared as follows:

Cl$_3$C-CHO + 2 chlorobenzene →[H$_2$SO$_4$] DDT

14. benzene + (3-chloro-2-methyl-2-butene) →[1. BF$_3$][2. H$^\oplus$] product

12.4 Mechanisms

15. [structure: 2-(2'-(3-hydroxyphenyl)phenyl)acetaldehyde] $\xrightarrow{H^\oplus}$ [structure: hydroxyphenanthrene]

16. When poly(styrene) is treated with chloromethyl methyl ether and SnCl₄ (a strong Lewis acid), Merrifield resin (named after Nobel laureate Bruce Merrifield who pioneered *in vitro* peptide syntheses) is formed.

poly(stryene) $\xrightarrow[\text{SnCl}_4]{\text{ClCH}_2\text{—O—CH}_3}$ Merrifield resin

17. [cycloheptatriene] $\xrightarrow{\text{1. Br}_2}$ $C_7H_8Br_2$ $\xrightarrow{\text{2. :B}}$ C_7H_7Br $\xrightarrow{\text{3. H}_2\text{O}}$ *ditropyl ether*

18.

$$\begin{array}{c} \text{Me} \\ \text{H}—\!\!\!-\!\!\!-\text{OTs} \\ \text{Ph}—\!\!\!-\!\!\!-\text{H} \\ \text{Me} \end{array} \xrightarrow[S_N \ (acetolysis)]{\text{HOAc}}$$ (a *racemate*)

Hint: recall the concept of neighboring group participation (9.4, 9-16) in some nucleophilic substitution reactions; even aromatic rings are sometimes capable of acting as a "neighboring group."

12.4 Mechanisms

116 • Chapter 12 Aromatics

19.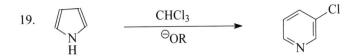

20.

21.

12.4 Mechanisms

CHAPTER 13
ALCOHOLS

13.1 Reactions
Draw the structural formula of the major organic product(s). Show stereochemistry where appropriate.

1. benzyl methyl ketone
   - 1. NaBH$_4$
   - 2. NH$_4$Cl (a weak acid)
   - 3. PCl$_3$
   - 4. KO-t-Bu

   ↓
   - 1. H$_2$ / Pd
   - 2. H$_2$SO$_4$

2. 3-ethyl-3-pentanol
   - 1. i-PrMgBr
   - 2. 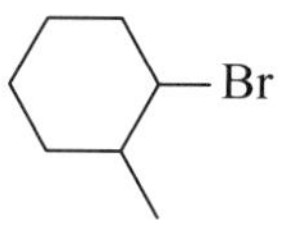 (2-methylcyclohexyl bromide)

3. 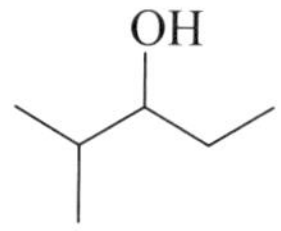 (3-methyl-2-pentanol structure, OH on C3)
   - 1. H$_2$SO$_4$
   - 2. H$_3$O$^\oplus$

   ↓
   - HCl

4. 5-hydroxy-2-heptanone
   - 1. TsCl
   - 2. NaOAc

5. 1-hexen-3-ol
   - 1. NaH
   - 2. 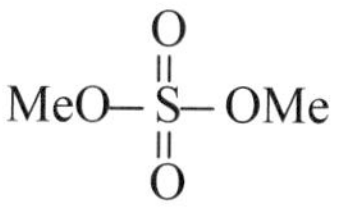 MeO–S(=O)$_2$–OMe

6. 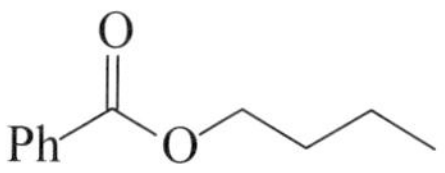 PhC(=O)O-butyl
   - 1. PhMgCl
   - 2. H$^\oplus$

   ↓
   - 1. LiAlH$_4$
   - 2. H$^\oplus$

7. iodomethane $\xrightarrow{\begin{array}{l}\text{1. Li}\\\text{2. diisopropyl ketone}\\\text{3. H}^\oplus\end{array}}$

8. 2-butanol $\xrightarrow{\begin{array}{l}\text{1. HBr}\\\text{2. LDA}\\\text{3. BH}_3\cdot\text{THF}\\\text{4. H}_2\text{O}_2,\text{HO}^\ominus\end{array}}$

9. 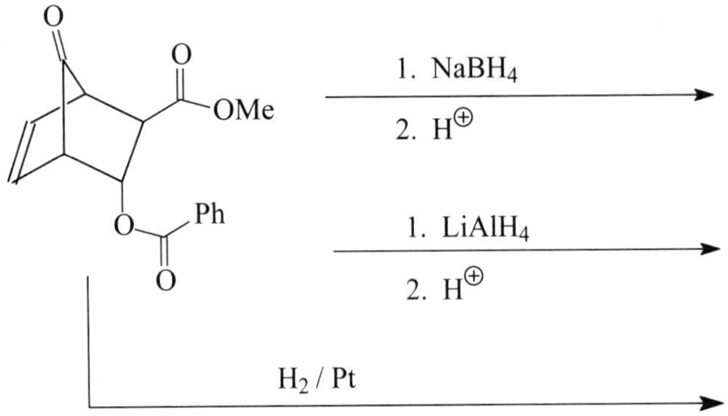 [bicyclic structure with ketone, CO₂Me, and OC(O)Ph groups]
   - $\xrightarrow{\text{1. NaBH}_4;\ \text{2. H}^\oplus}$
   - $\xrightarrow{\text{1. LiAlH}_4;\ \text{2. H}^\oplus}$
   - $\xrightarrow{\text{H}_2/\text{Pt}}$

10. 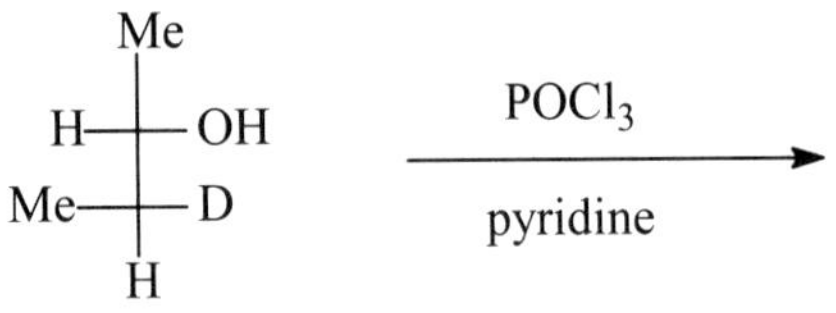 [stereochemical structure: H–C(Me)–OH / Me–C–D / H] $\xrightarrow{\text{POCl}_3,\ \text{pyridine}}$

11. morphine $\xrightarrow{\text{1. NaOH;\ 2. CH}_3\text{I (1 equiv)}}$ codeine

12. 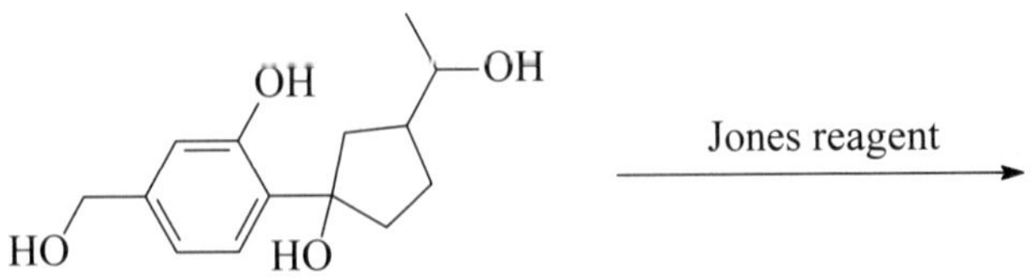 [cyclopentane-aromatic diol structure with CH(OH)Me, OH, HO, and HOCH₂ groups] $\xrightarrow{\text{Jones reagent}}$

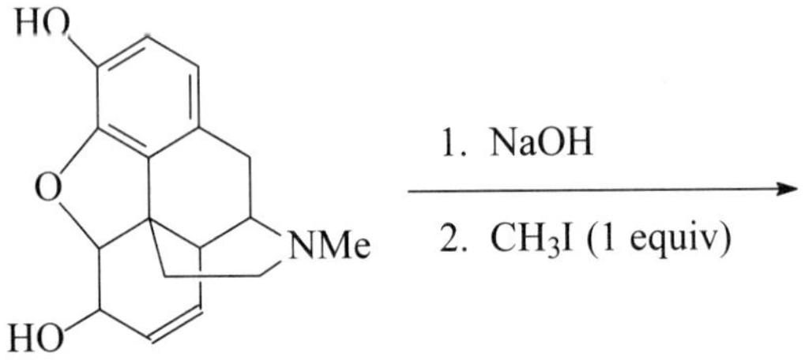

13.1 Reactions

13. 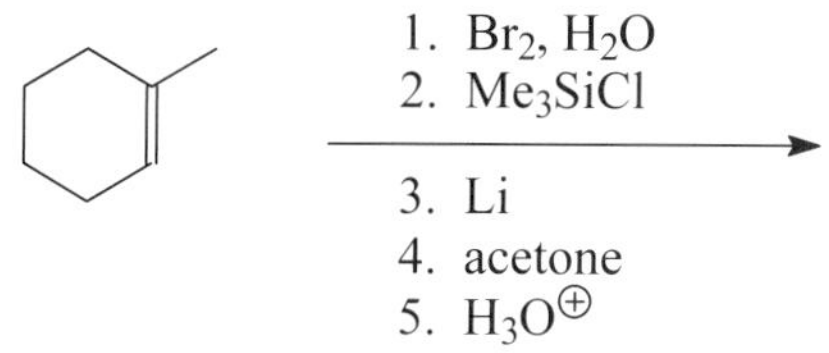 [1-methylcyclohexene]
   1. Br$_2$, H$_2$O
   2. Me$_3$SiCl
   3. Li
   4. acetone
   5. H$_3$O$^\oplus$

14. 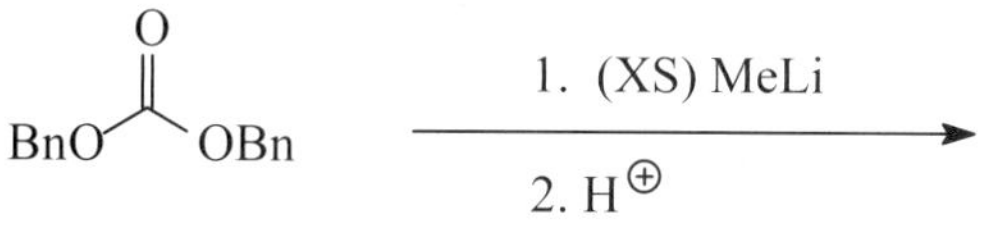 BnO–C(=O)–OBn
   1. (XS) MeLi
   2. H$^\oplus$

15. *p*-hydroxybenzoic acid
   1. LiAlH$_4$
   2. (XS) HBr

16. 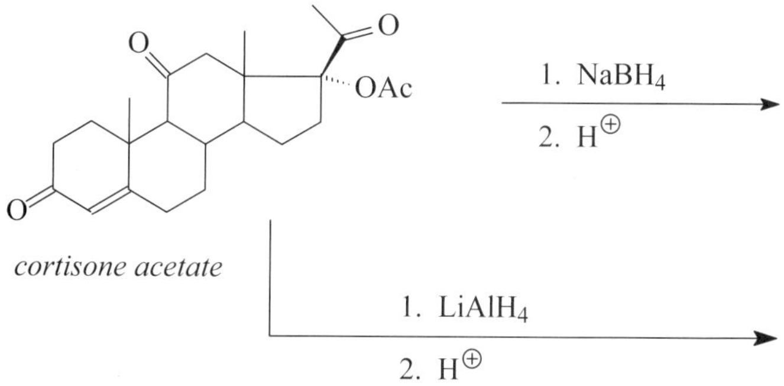 cortisone acetate
   1. NaBH$_4$
   2. H$^\oplus$

   1. LiAlH$_4$
   2. H$^\oplus$

17. 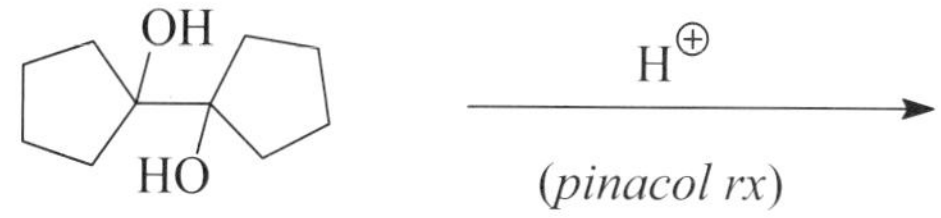 [1,1'-bicyclopentyl-1,1'-diol] H$^\oplus$ (*pinacol rx*)

18.  [1-(1-hydroxycyclobutyl)-1-phenyl-1-hydroxyethane] H$^\oplus$

19. 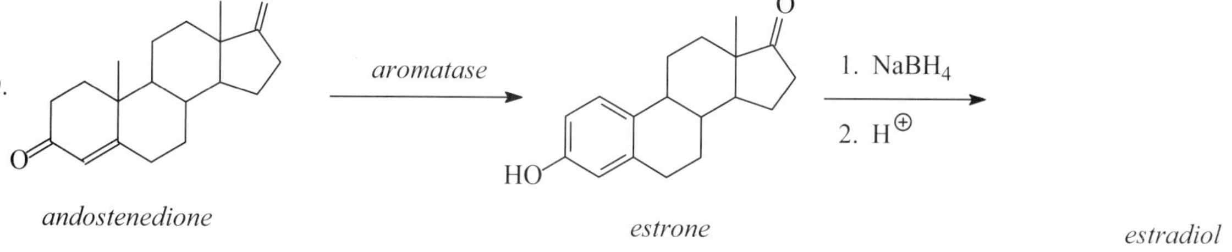 androstenedione →(aromatase)→ estrone →(1. NaBH$_4$, 2. H$^\oplus$)→ estradiol

13.1 Reactions

20.

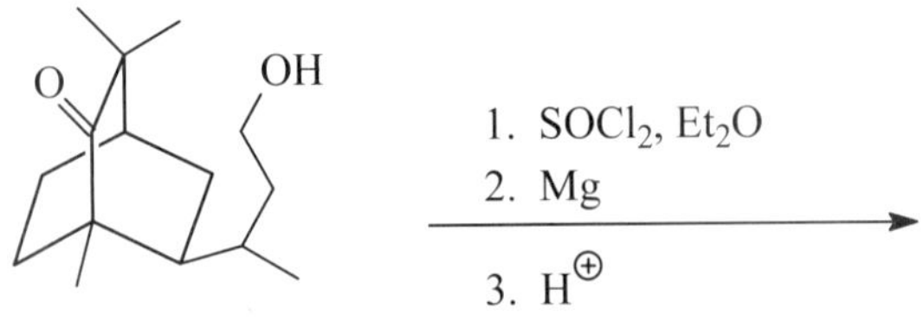

[Structure with ketone and OH group] → 1. SOCl₂, Et₂O   2. Mg   3. H⊕ → *patchouli alcohol* (used as a fragrance)

21.

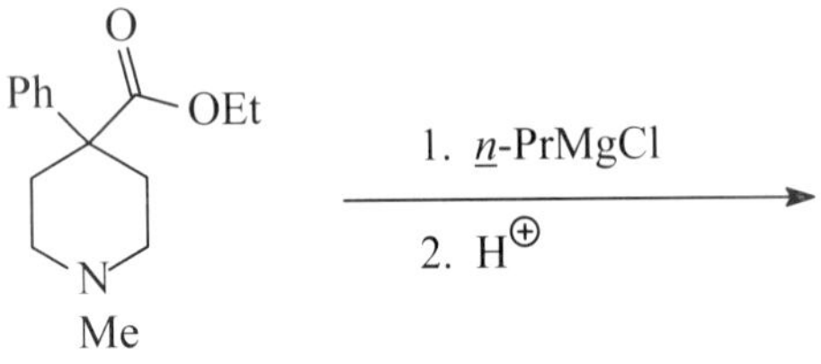

[Ph-substituted piperidine ester with N-Me] → 1. *n*-PrMgCl   2. H⊕ →

Demerol™ (narcotic analgesic)

22.

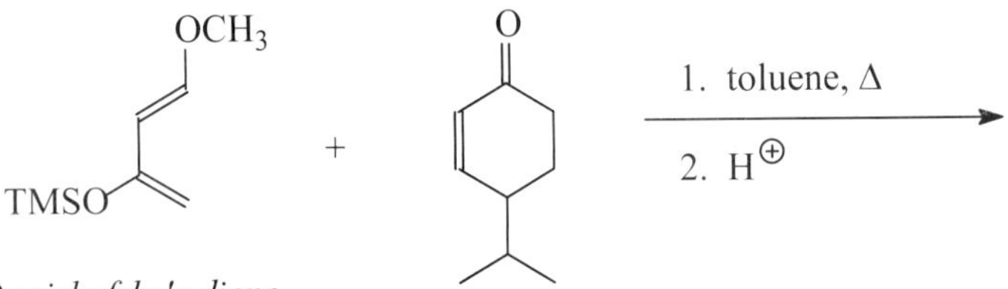

[Danishefsky's diene with OCH₃ and TMSO groups] + [cyclohexenone with isopropyl substituent] → 1. toluene, Δ   2. H⊕ →

Danishefsky's diene

13.2 Syntheses
Supply a reagent or sequence of reagents that will effect the following conversions.

1. 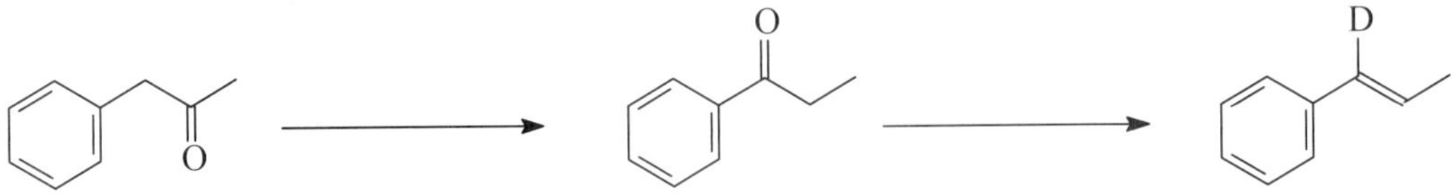 PhCH₂C(O)CH₃ → PhC(O)CH₂CH₃ → PhC(D)=CHCH₃

2. cyclohexanol → cyclohexyl-CHO → cyclohexyl-C(O)CH(CH₃)₂

13.2 Syntheses

3. 1-butene

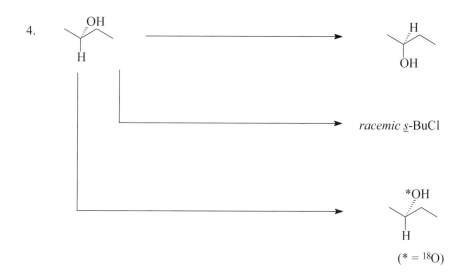

5. n-butane ⟶ (2-methyl-1-propanol) ⟶ (2-methyl-2-butanol, HO)

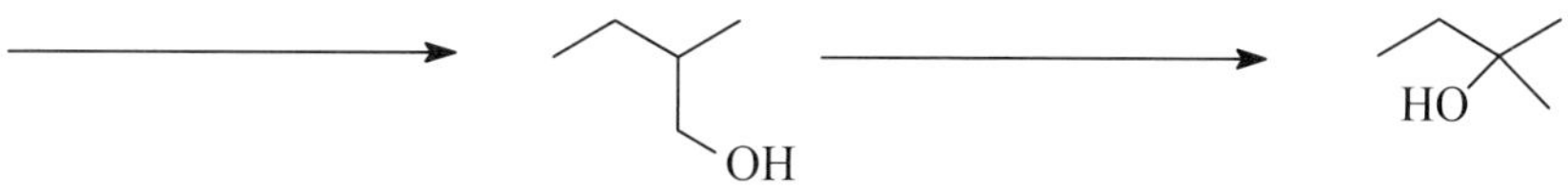

6. cyclohexane ⟶ (cyclohexenone) ⟶ (methylcyclohexadiene)

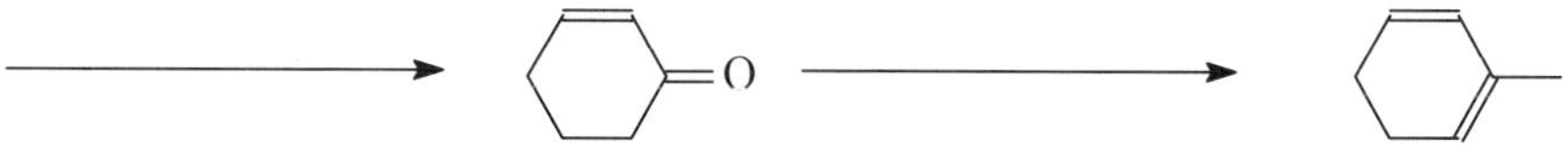

7. vinyl chloride ⟶ 1,3,5-hexatriene

8. n-hexyl alcohol ⟶ (hexyl cyanide, C≡N)

13.2 Syntheses

122 • Chapter 13 Alcohols

9.  [cyclobutane-CO₂H] ⟶ [cyclobutane-CH(D)(CH₃)?]

10. *p*-chlorophenol ⟶ *p*-hydroxybenzaldehyde (*via a Grignard*)

11.  estradiol ⟶ [progesterone-like steroid with ketone at C17, OH on aromatic ring]

12. 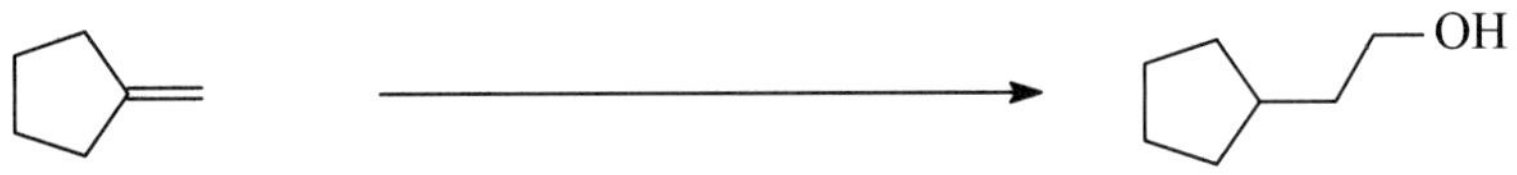 methylenecyclopentane ⟶ 2-cyclopentylethanol

13. *n*-butane ⟶ butanoic acid

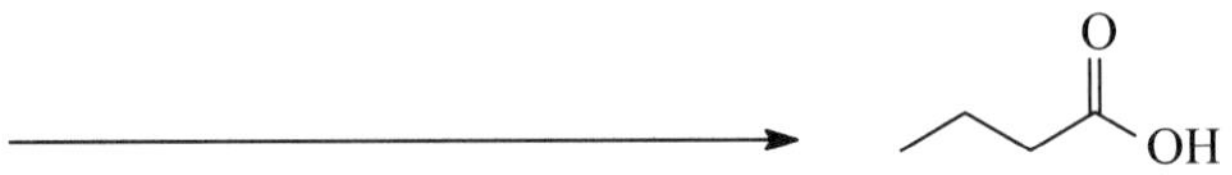

14. 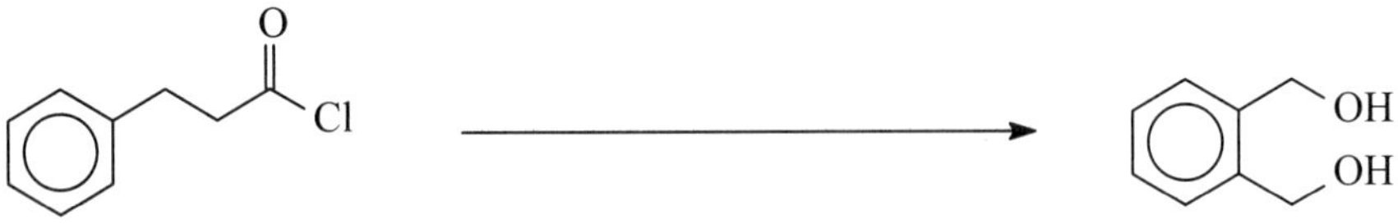 3-phenylpropanoyl chloride ⟶ 1,2-bis(hydroxymethyl)benzene

15. 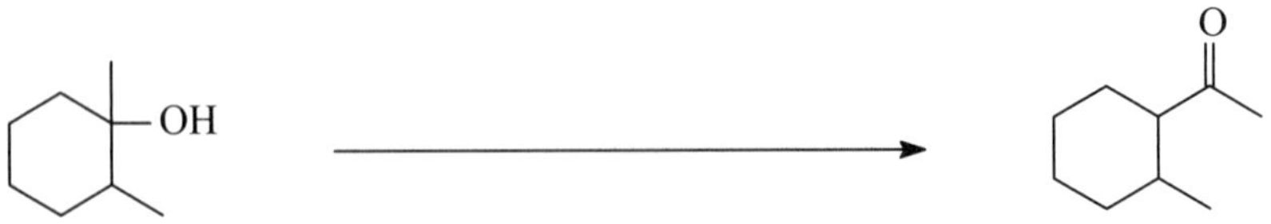 1-methyl-2-methylcyclohexan-1-ol ⟶ 1-(2-methylcyclohexyl)ethan-1-one

13.2 Syntheses

16. 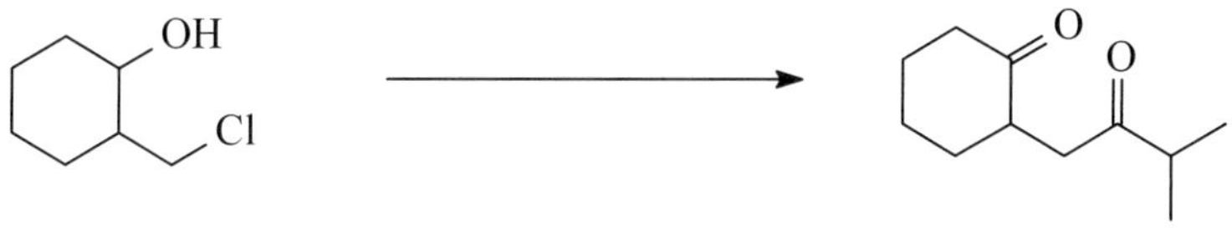 [structure: 2-(chloromethyl)cyclohexan-1-ol] → [structure: 2-(3-methyl-2-oxobutyl)cyclohexan-1-one]

17. 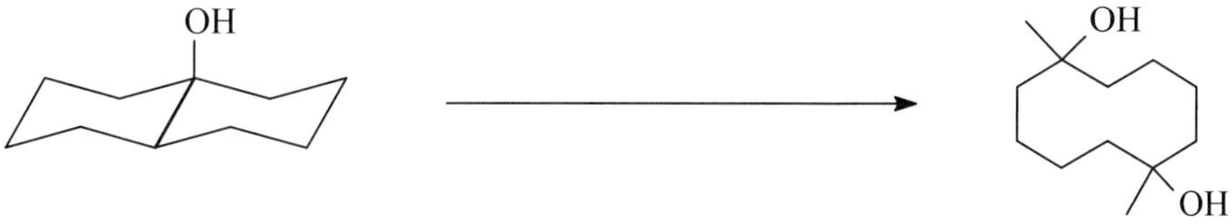 [structure: bicyclic tertiary alcohol] → [structure: 1,6-dimethylcyclodecane-1,6-diol]

18.  [structure: bicyclic alkene diol] → [structure: ethynyl bicyclic diol with methyl]

19. 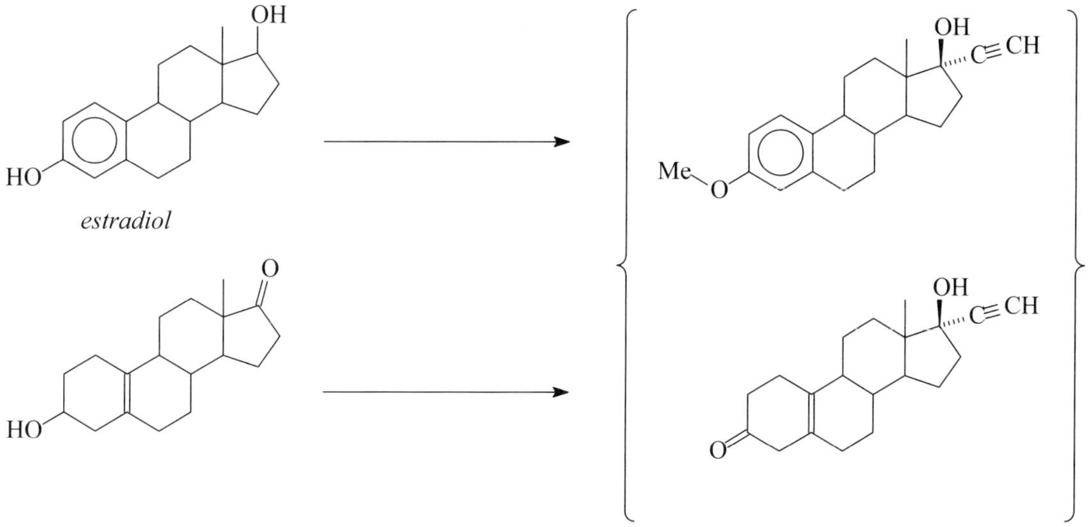 estradiol → { Enovid™ constituents (OCPs) }
    [structure: estradiol] → [structure: mestranol (methyl ether ethynyl)]
    [structure: estrone-like enol] → [structure: norethynodrel-like]

20.  [structure: 9-bromofluorene] → [structure: phenanthrene]

13.2 Syntheses

21. Berson (*Yale*) discovered that the bicyclic carbocation **A** undergoes a clever rearrangement in which the cyclopropyl ring circumambulates around the cyclopentenium ring. Beginning with **B**, synthesize a deuterium-labeled species that would support this observation.

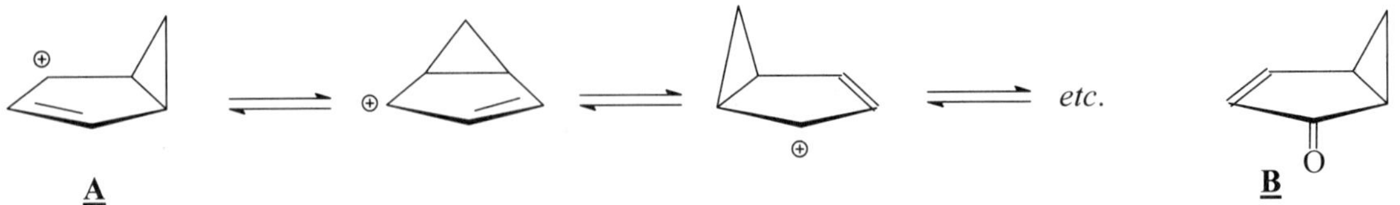

13.3 Mechanisms
Outline a detailed mechanism for each of the following. No other reagents than those given are necessary. Use arrows to explain the flow of electrons and show all intermediates.

1. (diol) $\xrightarrow{\text{dil } H_2SO_4}$ (*a cyclic ether*)

2. retinal $\xrightarrow[\text{2. } H_2SO_4]{\text{1. NaBH}_4}$ (product shown)

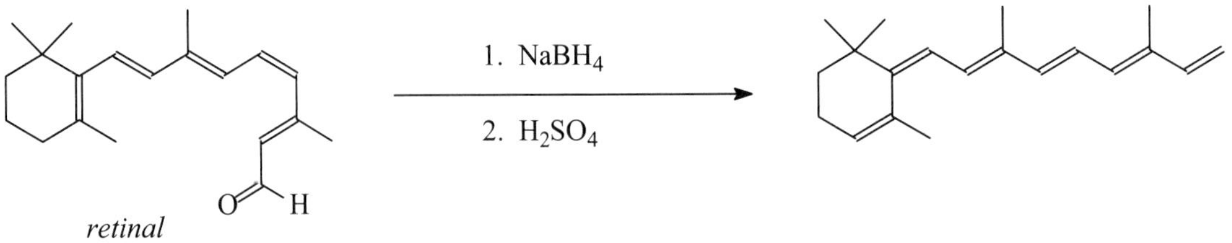

3. cycloheptene glycol $\xrightarrow{H^\oplus}$

$C_7H_{12}O$
IR: 1729 cm^{-1}
^{1}H NMR: δ 9.5 (d, 1H),
plus other chemicals shifts

13.3 Mechanisms

4.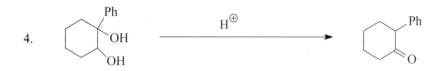

5. [reaction scheme: 1,1-dimethyl-2,2-dimethylcycloheptanol with H⁺ → tert-butylcyclohexene + isopropenyl-methylcyclohexane]

6. glycerol —H⁺→ CH₂=CH–CHO (acrolein)

7. [2-ethyl-1,3-hexanediol] —H⁺→ (*an aldehyde*)

8. [diphenylethane-carboxylic acid]
   1. H₃PO₄ → [dibenzosuberone]
   2. cyclopropyl–MgCl
   3. H₃O⁺
   4. HBr
   5. Me₂NH
   → amitriptyline (an antidepressant)

13.3 Mechanisms

126 • Chapter 13 Alcohols

9. Aflatoxin B_1 is one of the most potent carcinogens known. In the presence of water and acid, compound **A** is formed.

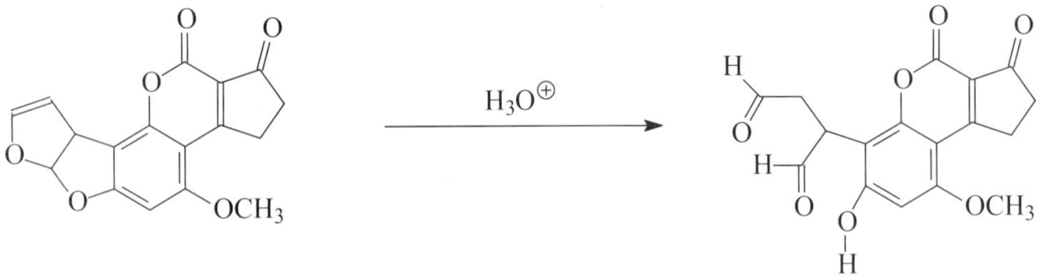

aflatoxin B_1 — $H_3O^{\oplus}$ → **A**

10. (Fischer projection: CH₃ / H–Br / HO–H / CH₃) — HBr → (H–Br / Br–H / CH₃ on both ends) + (Br–H / H–Br / CH₃ on both ends) } racemate

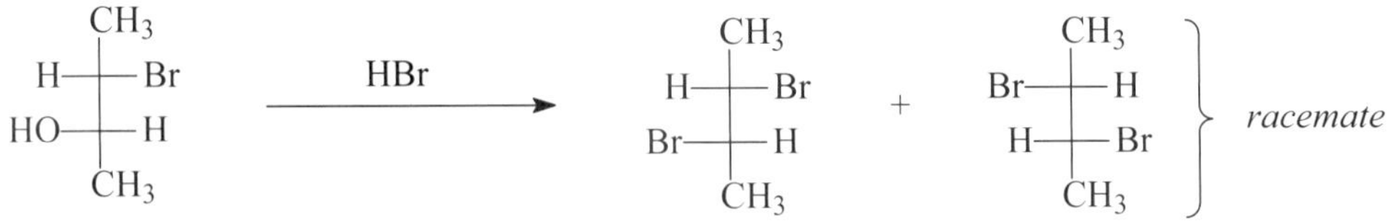

Note: *retention at $C_{2,3}$*   *inversion at BOTH $C_{2,3}$!*

This observation by Winstein (*UCLA*) provided stereochemical support for the concept of neighboring group participation (see 9.4, 9-16).

Similarly,

11. (trans-2-bromocyclohexanol) *or* (cis-2-bromocyclohexanol) — HBr → only *trans*-product is formed!

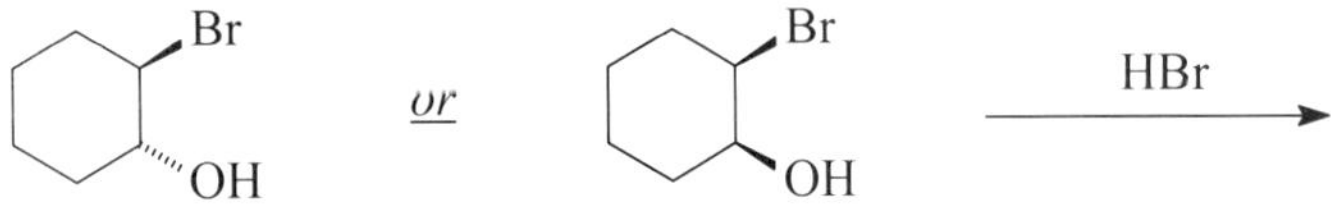

12. A step in the biosynthesis of the amino acid *valine*:

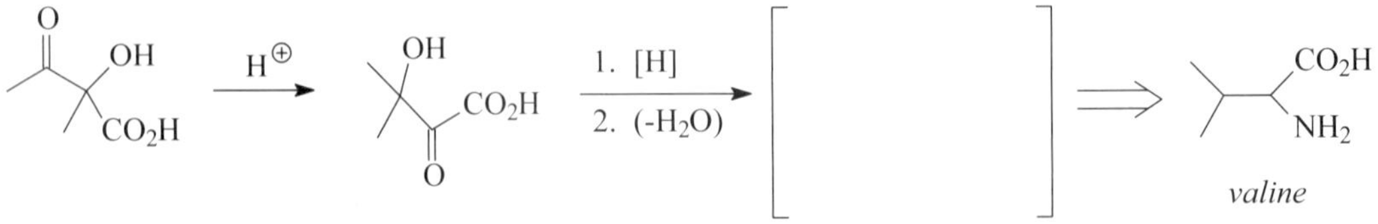

(α-hydroxy-β-keto acid with OH and CO_2H) — $H^{\oplus}$ → (diol with OH, CO_2H, and C=O) — 1. [H], 2. (-H_2O) → [ ] ⇒ (valine: CO_2H, NH_2)

valine

13.3 Mechanisms

13. 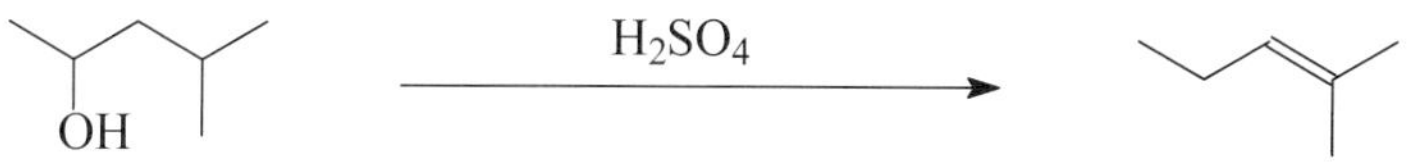 [reaction: 4-methyl-2-pentanol (with OH) → H₂SO₄ → 2-methyl-2-butene type alkene]

14. The conversion of ethylene glycol to acetaldehyde under acidic conditions could occur by one of two pathways: (1) dehydration to an enol followed by tautomerization, or (2) a pinacol-like rearrangement. In view of the following experiment, which pathway is suggested?

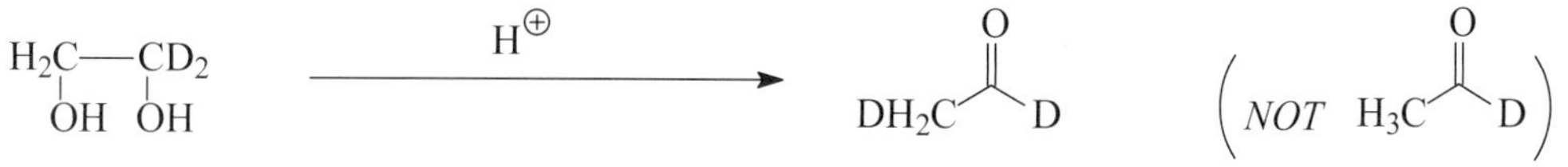

$$H_2C(OH)-CD_2(OH) \xrightarrow{H^{\oplus}} DH_2C-C(=O)-D \quad (NOT\ H_3C-C(=O)-D)$$

15. Cyclohexene glycol in the presence of acid forms cyclohexanone. Similar to problem 14, two pathways are possible: dehydration/tautomerization vs. a pinacol-like rearrangement:

[mechanism scheme showing cyclohexane-1,2-diol → (+H⊕, −H₂O) → carbocation intermediate → (~H:⁻, −H⊕) → cyclohexanone; alternative path via −H₂O to enol → taut → cyclohexanone]

cyclohexene glycol

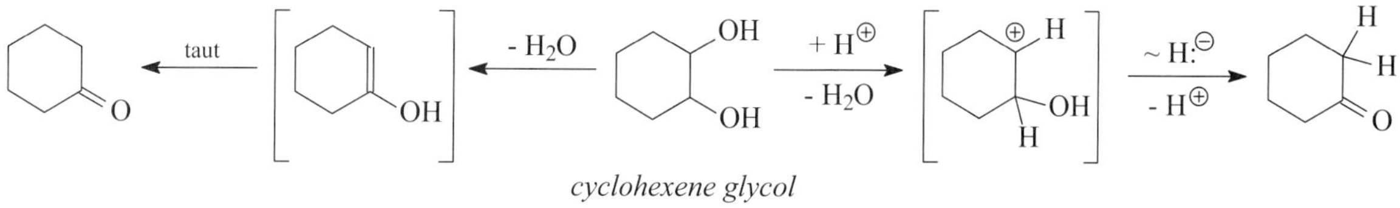

Synthesis of deuterium-labeled glycol **A**, when treated with acid, yields **B**:

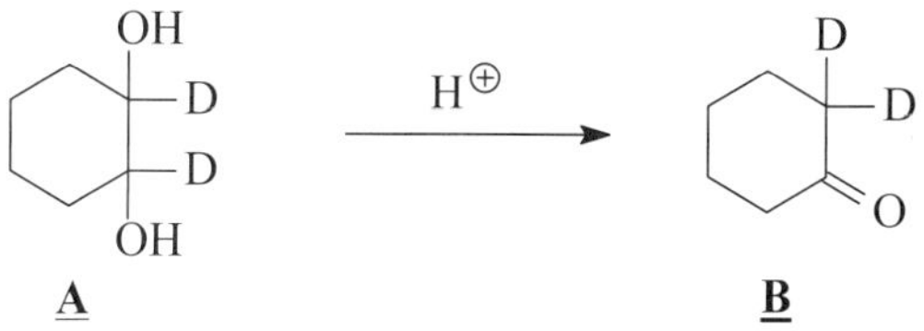

[Structure A: cyclohexane ring with OH, OH on adjacent carbons, each bearing a D] $\xrightarrow{H^{\oplus}}$ [Structure B: cyclohexanone with two D's on the α-carbon]

A **B**

a. Which pathway is consistent with this observation?

b. Suggest a preparation of **A** from cyclohexene.

128 • Chapter 13 Alcohols

16.  [cyclohexanone with ★ on carbonyl C and α-OH] $\xrightarrow{H^{\oplus}}$ [cyclohexanone with ★ on α-C bearing OH and adjacent C=O] (★ = ^{13}C)

17. 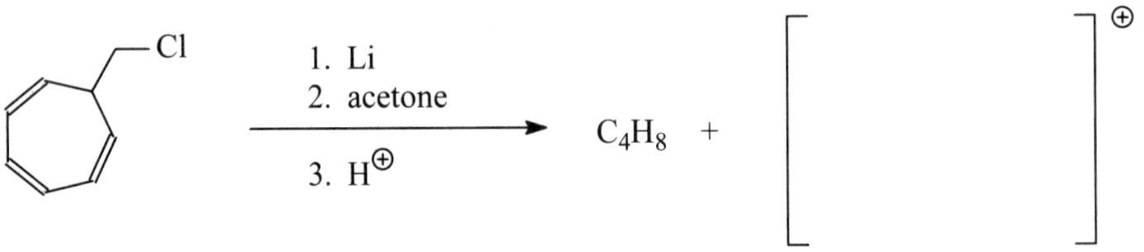 [cycloheptatrienyl-CH₂Cl] $\xrightarrow[\text{3. H}^{\oplus}]{\text{1. Li; 2. acetone}}$ C₄H₈ + [ ]$^{\oplus}$

(^{1}H NMR shows only a singlet at δ 8.2)

18. 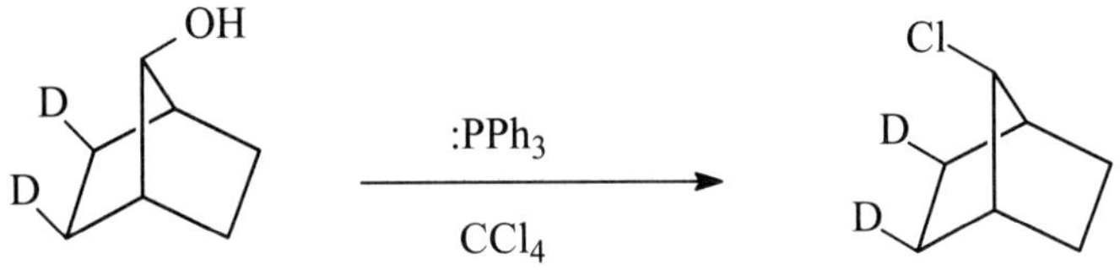 [norbornane with OH and two D] $\xrightarrow[\text{CCl}_4]{:PPh_3}$ [norbornane with Cl and two D]

19. 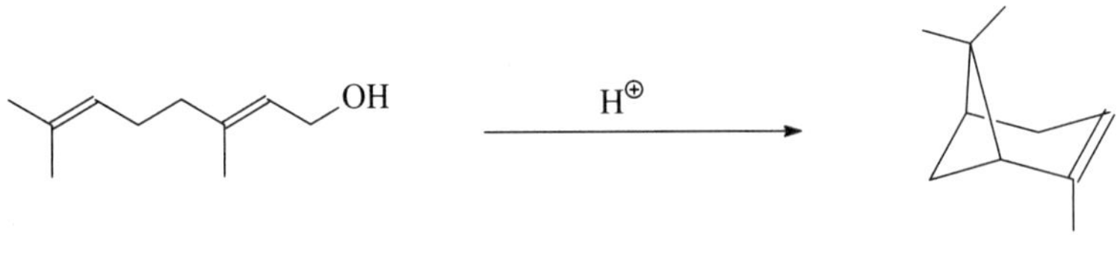 [geraniol] $\xrightarrow{H^{\oplus}}$ [α-pinene structure]

α-pinene (a constituent in oil of turpentine - interestingly, the dextrorotatory form is found in North American oils and the levorotatory form in European oils)

13.3 Mechanisms

CHAPTER 14
ETHERS

14.1 Reactions
Draw the structural formula of the major organic product(s). Show stereochemistry where appropriate.

1. 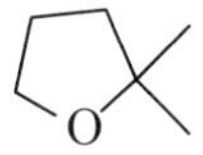 [2,2-dimethyltetrahydrofuran]
   1. HBr (1 equiv)
   2. TsCl
   3. KOAc, 18-crown-6

2. benzyl phenyl ether → (XS) HI

3. phenyl mercaptan + Me$_3$S$^\oplus$ I$^\ominus$ →

4. 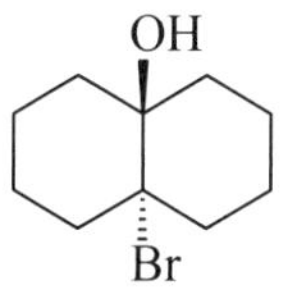 [decalin with OH (wedge) and Br (dash)] KOH

5. 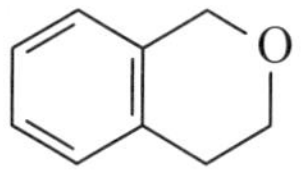 [isochroman]
   1. HF
   2. PCC

6. 2-isopropyloxirane → NaCN / MeOH

7. PhLi
   1. styrene epoxide
   2. H$_2$SO$_4$

8. 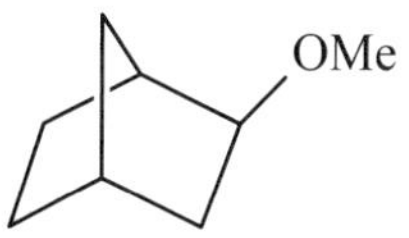 [norbornane with OMe]
   1. HI (1 equiv)
   2. CrO$_3$, H$^\oplus$
   3. NaBD$_4$
   4. H$^\oplus$

9.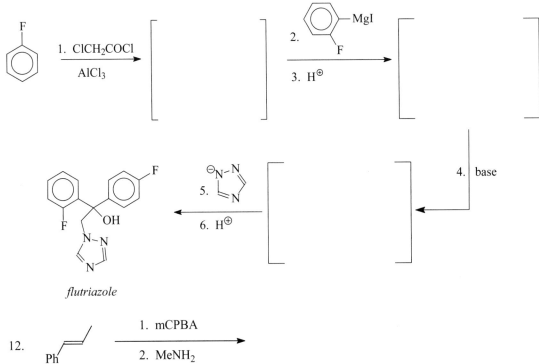

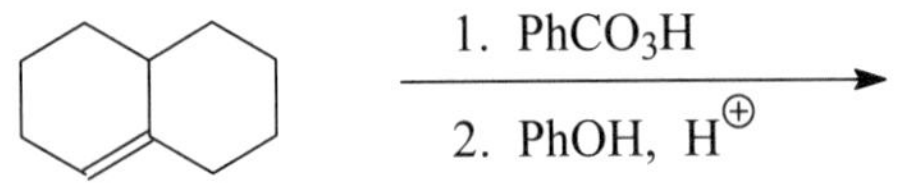

10.

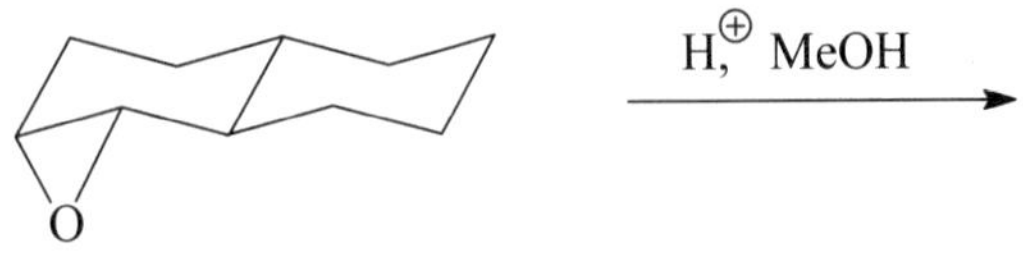

11. The fungicide *flutriazole* can be synthesized by the following scheme:

12.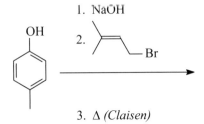

ephedrine (bronchodilator)

13. The *Claisen rearrangement* of allyl phenyl ethers:

14.1 Reactions

14. The Claisen rearrangement can be generalized to include *allyl vinyl ethers*:

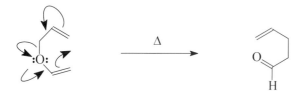

Draw the expected Claisen rearrangement product for each of the following:

a.

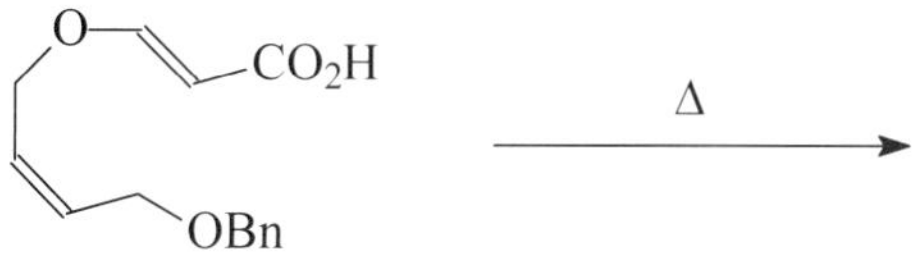

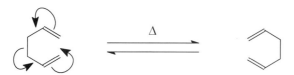

b. A stage in the biosynthesis of aromatic amino acids (draw the structure of prephenic acid and give a mechanism for its conversion to phenylpyruvic acid):

15. Mechanistically similar to the Claisen rearrangement is the *Cope rearrangement*:

This specific example became known as the "degenerate Cope," a moniker that did not particularly please its discoverer, Prof. A. Cope! Of course, the degeneracy can be removed:

14.1 Reactions

16. Going back to problem 14.1, 13, if the *ortho* positions are blocked the initial Claisen rearrangement product may be followed by a Cope rearrangement. Fill in the brackets.

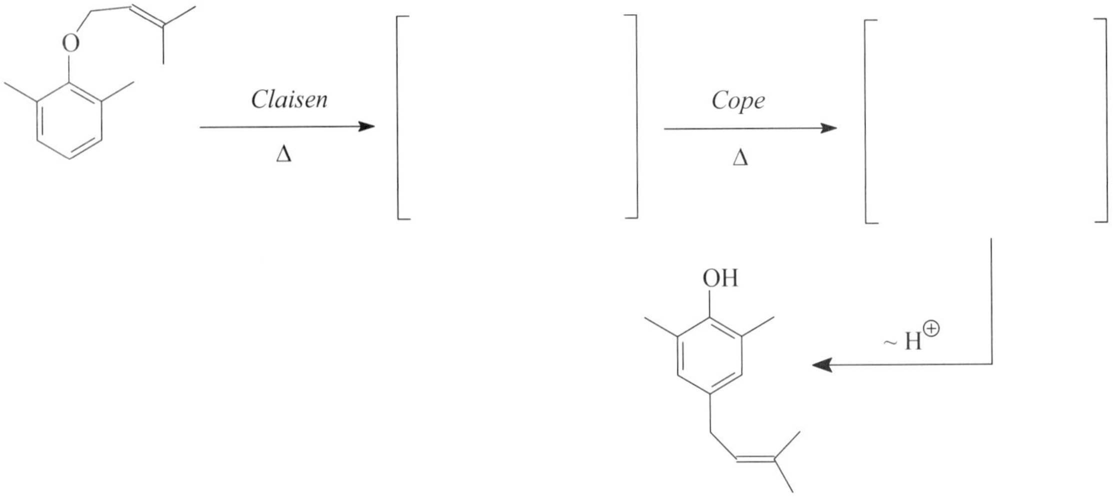

17. A slight variation of problem 14.1, 15 is the *oxy-Cope rearrangement*:

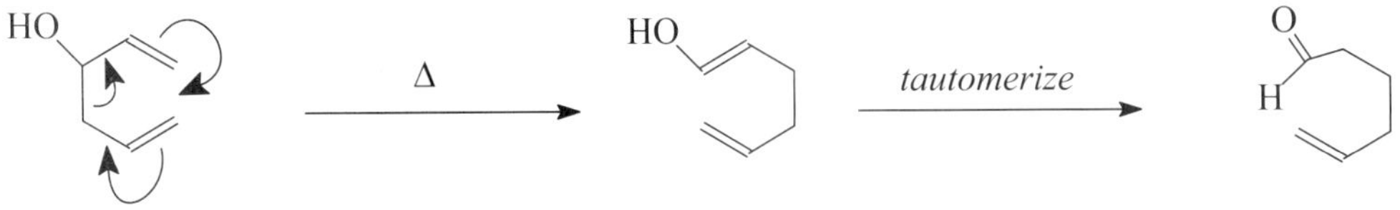

Predict the oxy-Cope product for the reaction below:

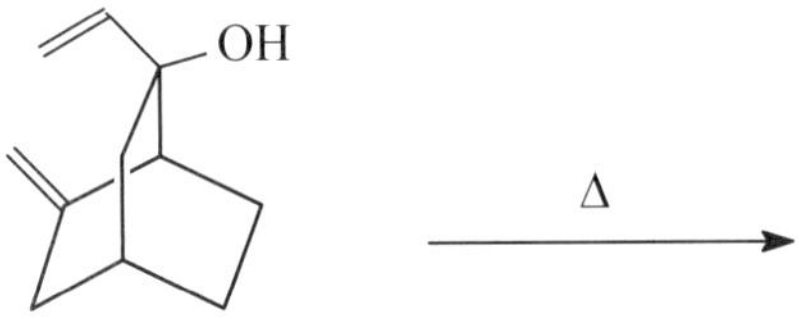

18. 2 cysteine $\xrightarrow{[O]}$ cystine [crystallization in kidneys can lead to one type of calculi (stone)]

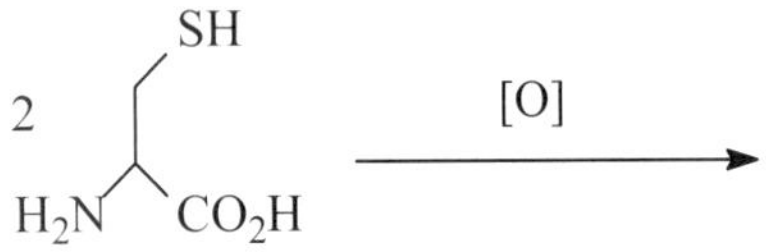

19. dithiothreitol $\xrightarrow{[O]}$ $C_4H_8O_2S_2$

14.1 Reactions

14.2 Syntheses
Supply a reagent or sequence of reagents that will effect the following conversions.

1. 3-methylpentane ⟶ 3-methoxy-3-methylpentane

2. cyclohexene ⟶ *trans*-cyclohexene glycol

3. 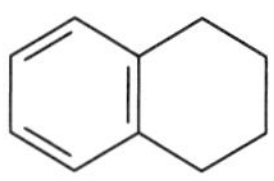 [tetralin] ⟶ [1-(2-tosyloxyethyl)tetralin]

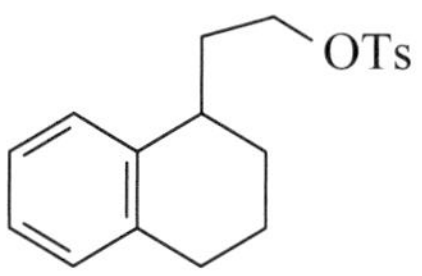

4. propylene ⟶ (allyl-S)₂ *diallyldisulfide* (found in garlic)

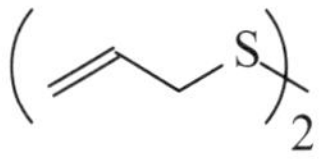

5. cyclohexane ⟶ cyclohexylacetaldehyde *via an epoxide*

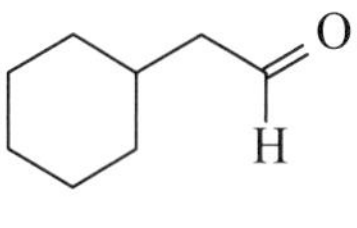

6. cyclohexene oxide ⟶ cyclohexane

134 • Chapter 14 Ethers

7.  [structure: methacryloyl proline with CO₂H] → [structure: HS-CH₂-CH(CH₃)-C(O)-N-proline-CO₂H]

captopril (antihypertensive)

8. 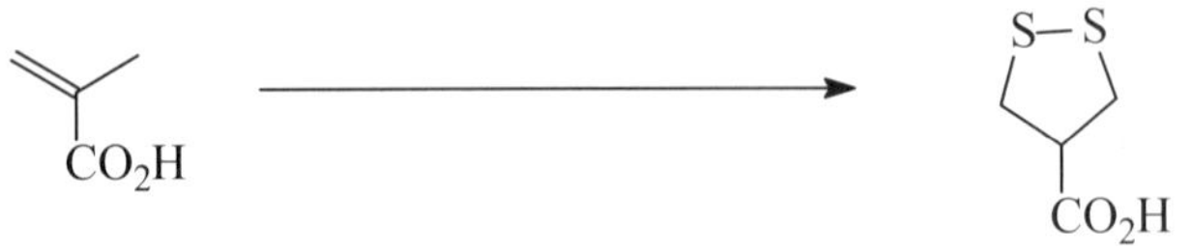 [structure: methacrylic acid, CH₂=C(CH₃)-CO₂H] → [structure: 1,2-dithiolane-4-carboxylic acid]

asparagusic acid (isolable from asparagus)

14.3 Mechanisms

Outline a detailed mechanism for each of the following. No other reagents than those given are necessary. Use arrows to explain the flow of electrons and show all intermediates.

1. styrene epoxide $\xrightarrow{\text{BF}_3, \text{Et}_2\text{O}}$ PhCH₂CHO

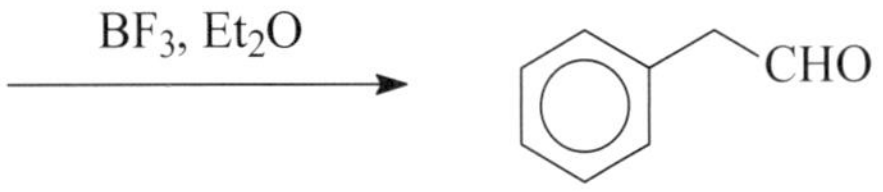

2. methyloxirane $\xrightarrow[\text{2. H}^\oplus]{\text{1. LDA}}$ allyl alcohol

3.  [structure: farnesyl epoxide] $\xrightarrow{\text{H}^\oplus}$ [structure: bicyclic decalin alcohol]

14.3 Mechanisms

4. 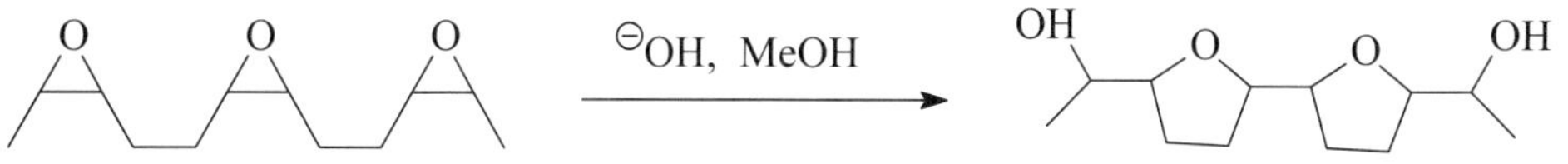 [structure: triepoxide] → ⁻OH, MeOH → [structure: bis-tetrahydrofuran diol]

5. Complex ladder polyether natural products, so named for their rung-like structure, are the active toxins found in harmful algal blooms known as *red tides*, which cause devastating ecological damage. *Brevetoxin B* is an example.

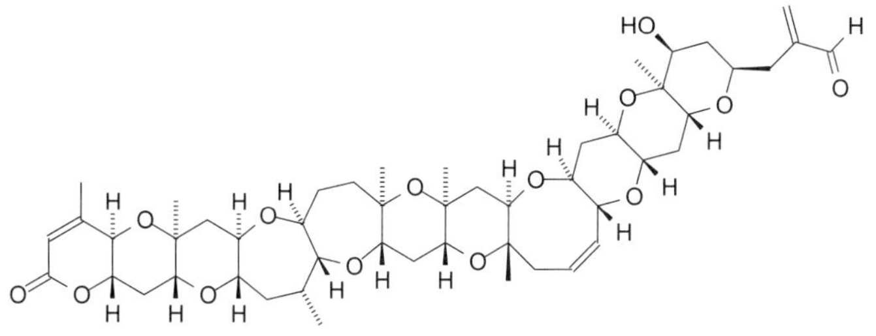

brevetoxin B

Twenty years ago Nakanishi (*Columbia*) proposed such products arise biosynthetically from an elaborate cascade of epoxide ring-opening reactions that zip up the polyether structure. The following reaction, discovered by Jamison (*MIT*) in 2007, supports this hypothesis.

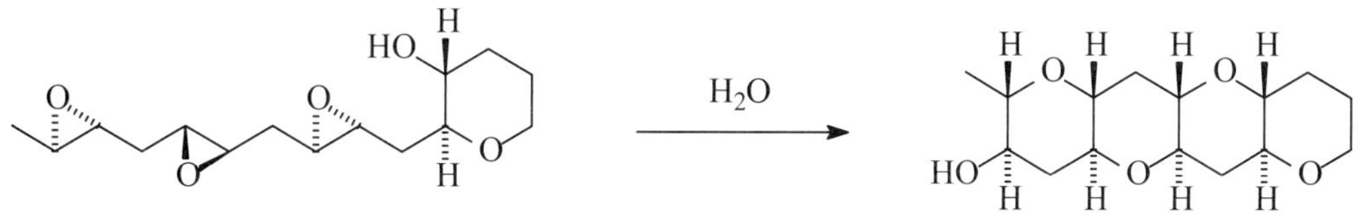

[reaction: hydroxy-triepoxide + H₂O → tris-tetrahydropyran polyether]

6. The biosynthesis of steroids involves an absolutely gorgeous (!) polycyclization reaction of *squalene epoxide*, followed by two sequential 1,2-hydride shifts and two 1,2-methide shifts to form *lanosterol* (lanosterol is then converted to *cholesterol*, the precursor to most other steroid hormones):

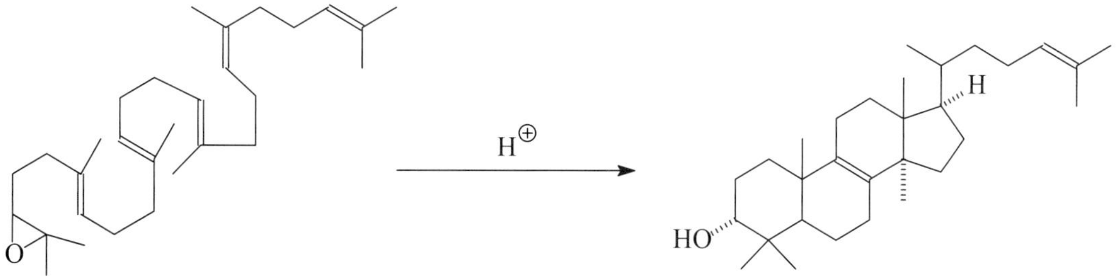

squalene epoxide → H⁺ → *lanosterol*

14.3 Mechanisms

7. Biochemical hydroxylation of aromatic compounds proceeds *via* arene oxides, which subsequently undergo ring opening to form phenols:

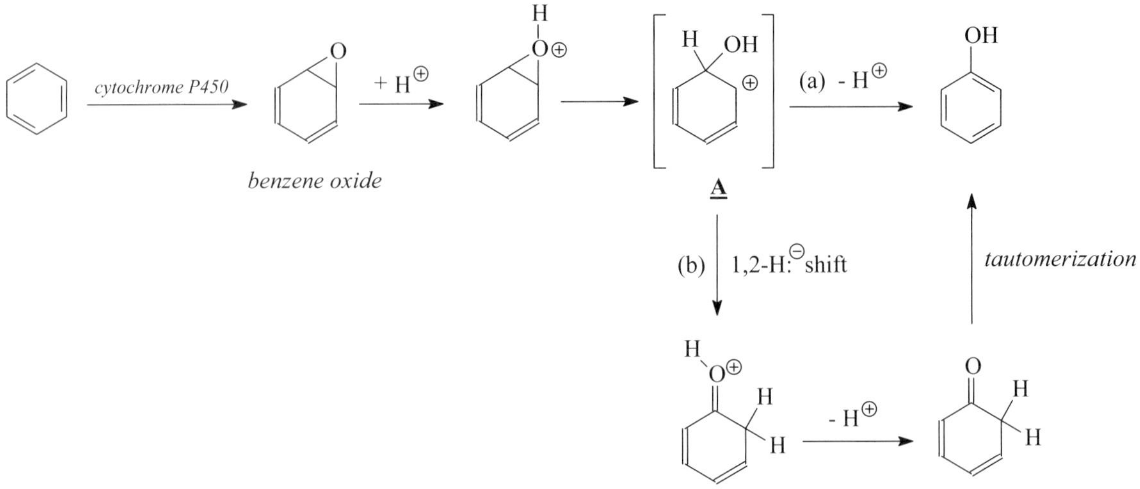

Phenol could be formed from intermediate **A** simply by an E1-like loss of a proton (*path a*) or, alternatively, by a pinacol-like rearrangement followed by tautomerization (*path b*). Support for *path b* was provided by chemists at the NIH who observed the following conversion:

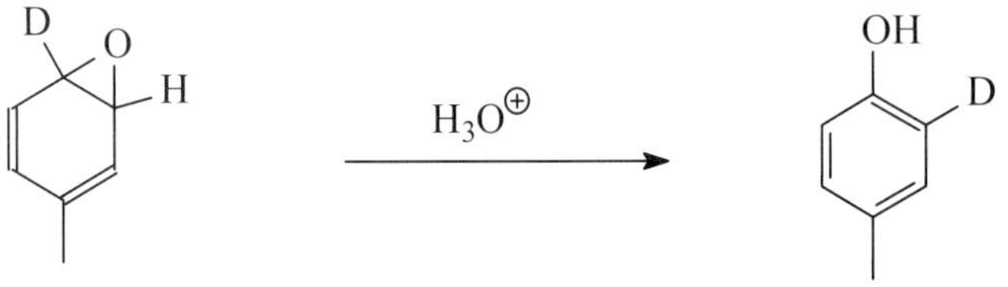

Explain. Account for the role of the methyl substituent. (This rearrangement of an arene oxide has now become known as the *NIH shift*!)

8.

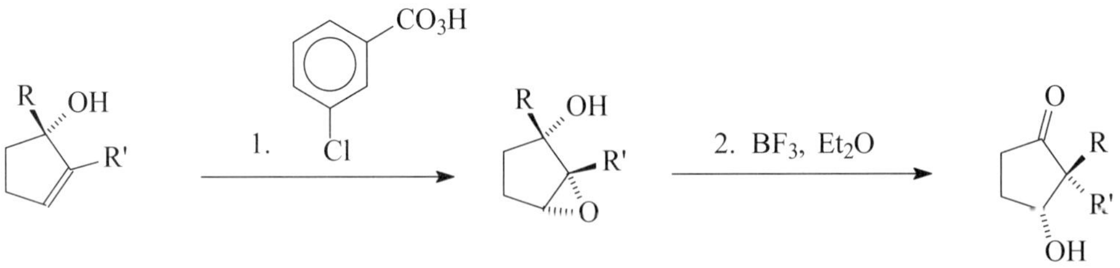

Step 2 illustrates a semi-pinacol type rearrangement. Propose a mechanism for that step.

14.3 Mechanisms

9. 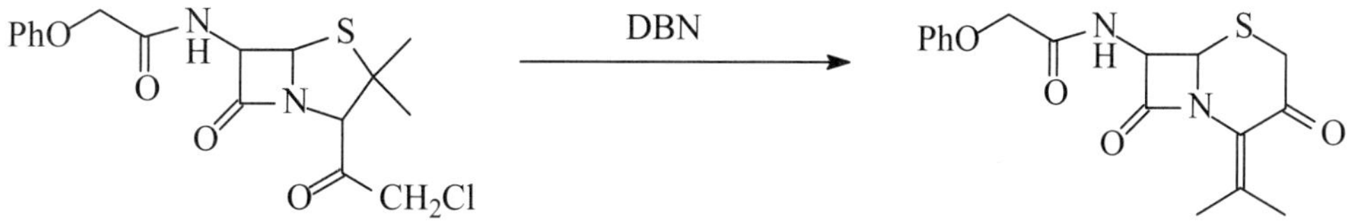 [Structure: PhO-CH2-C(=O)-NH- attached to β-lactam fused with thiazolidine bearing gem-dimethyl, with N-CH(C(=O)CH2Cl)] →(DBN)→ [Structure: PhO-CH2-C(=O)-NH- attached to β-lactam fused with 6-membered ring containing S, C=O, and exocyclic =C(CH3)2]

Note: DBN (1,5-diazabicyclo[4.3.0]non-5-ene) is a sterically hindered nitrogen base that favors elimination over substitution:

DBN + H–A ⟶ [DBN-H]⁺ A⁻

14.3 Mechanisms

CHAPTER 15
ALDEHYDES AND KETONES

15.1 Reactions
Draw the structural formula of the major organic product(s). Show stereochemistry where appropriate.

1. 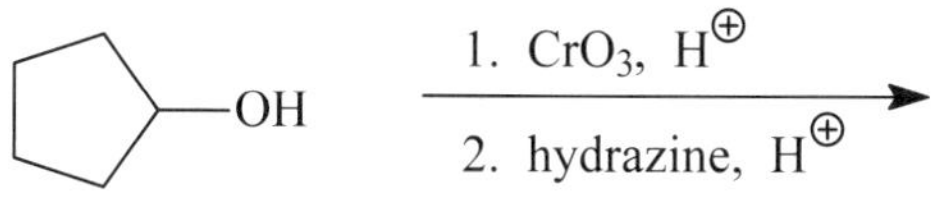 cyclopentanol —OH
   1. CrO_3, $H^⊕$
   2. hydrazine, $H^⊕$

2. PhOCH$_2$Br
   1. Ph$_3$P
   2. MeLi
   3. methyl ethyl ketone

   a *vinyl ether* (see 15.1, 12, 13 and 15.3, 3, 33 for examples of their reactivity)

3.  [1,3-dioxane with Ph substituent] $\xrightarrow{H_3O^⊕}$

4. 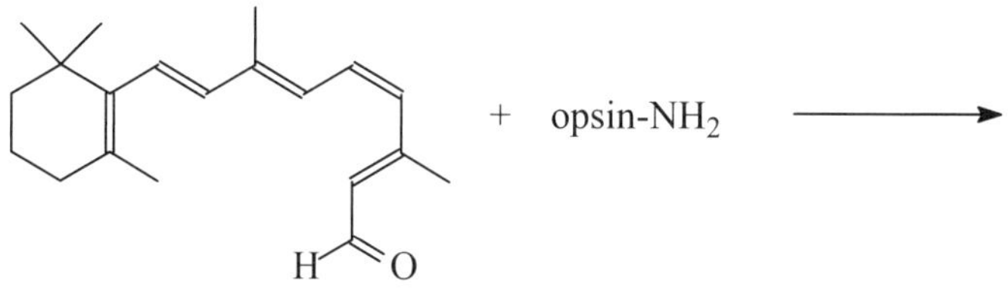 [11-cis-retinal structure] + opsin-NH$_2$ ⟶

   11-cis-retinal   (*a protein*)   *rhodopsin*

5. 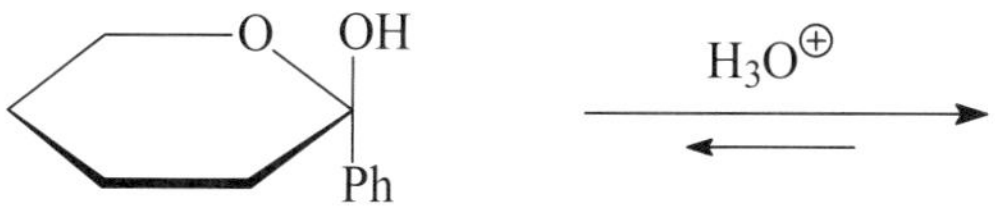 [tetrahydropyran with OH and Ph] $\xrightleftharpoons{H_3O^⊕}$

6. 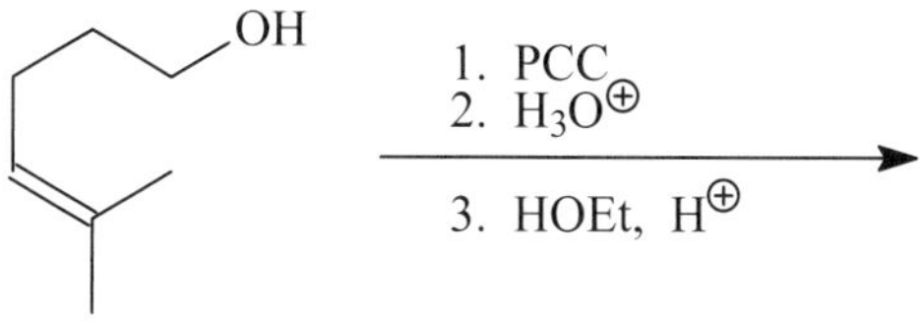 [alkene with OH chain]
   1. PCC
   2. $H_3O^⊕$
   3. HOEt, $H^⊕$

7. 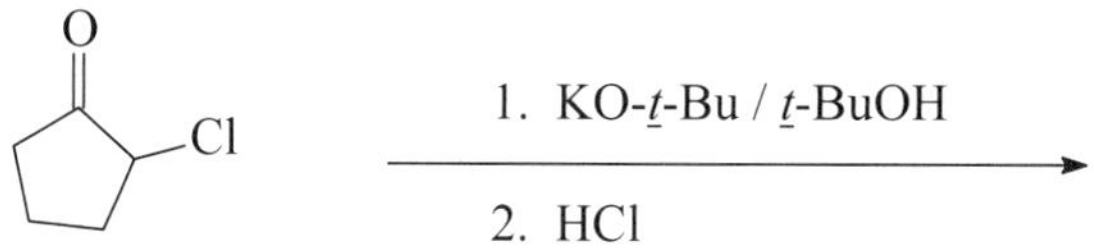 [2-chlorocyclopentanone]
   1. KO-*t*-Bu / *t*-BuOH
   2. HCl

15.1 Reactions

140 • Chapter 15 Aldehydes and Ketones

8. methyl *n*-propyl ketone $\xrightarrow{\text{1. NaBD}_4 \quad \text{2. H}^\oplus \quad \text{3. H}_2\text{SO}_4 \text{ (E1)}}$

9. vitamin D $\xrightarrow{\text{1. KMnO}_4 \quad \text{2. semicarbazide} \quad \text{3. H}_2/\text{Pt (XS)}}$

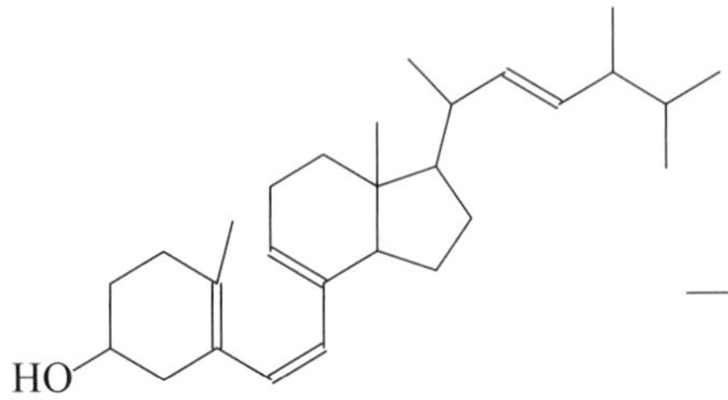

10. (methyl 3-methyl-5-oxo... ester with CHO group) $\xrightarrow{\text{1. ethylene glycol, H}^\oplus \quad \text{2. DIBAH, -78°} \quad \text{3. Ph}_3\text{P=CMe}_2 \quad \text{4. H}_3\text{O}^\oplus}$ citronellal

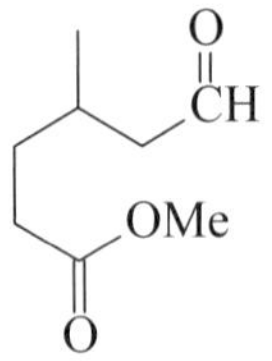

11. *p*-nitrobenzaldehyde $\xrightarrow{\text{1. Ph}_3\overset{\oplus}{\text{P}}\text{-CH}_2\text{-CH=CH-(1,3-dioxolane)} \,\ominus \quad \text{2. H}_3\text{O}^\oplus}$ a fluorescent "spy dust" ingredient

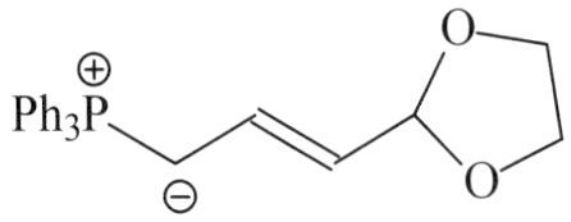

12. cyclohexanone $\xrightarrow{\text{1. Ph}_3\overset{\oplus}{\text{P}}\text{-}\overset{\ominus}{\text{C}}\text{HOCH}_3 \quad \text{2. H}_3\text{O}^\oplus}$ an aldehyde

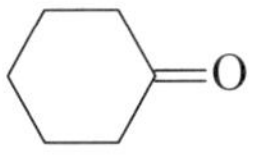

13. Using the above reaction (12) as a model, how could you prepare pentanal from butanal?

15.1 Reactions

14. 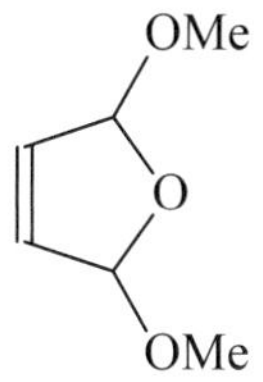 [2,5-dimethoxy-2,5-dihydrofuran]
 1. CH$_2$I$_2$ / Zn (Cu)
 2. H$_3$O$^\oplus$
 3. Ph$_3$P=CHC=CH$_2$

15. cyclopropanecarbaldehyde hydrazone
 1. H$_3$O$^\oplus$
 2. EtMgI
 3. H$^\oplus$

16. acetophenone diphenyl ketal
 1. H$_3$O$^\oplus$
 2. H$_2$NOH

17. 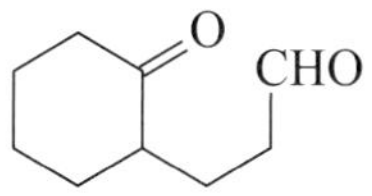 [2-(2-oxocyclohexyl)ethyl-CHO] $\xrightarrow{H_3O^\oplus}$ *a heterocycle*

18. 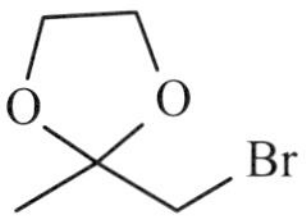 [2-methyl-2-(bromomethyl)-1,3-dioxolane]
 1. Ph$_3$P
 2. *n*-BuLi
 3. butanal
 4. H$_3$O$^\oplus$

19. 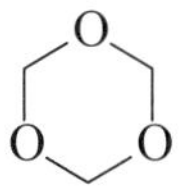 [1,3,5-trioxane] $\xrightarrow{H_3O^\oplus}$

20. *t*-butylacetylene
 1. H$_3$O$^\oplus$, Hg^{2+}
 2. hydrazine, $^\ominus$OH

15.1 Reactions

142 • Chapter 15 Aldehydes and Ketones

21. 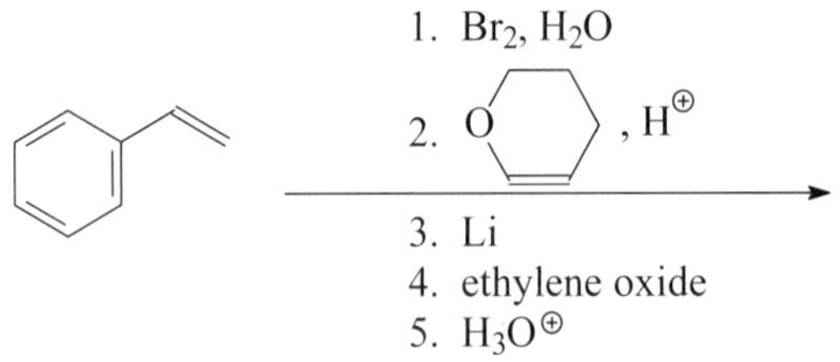 styrene
$$\xrightarrow{\begin{array}{l}1.\ Br_2,\ H_2O\\ 2.\ \text{(dihydropyran)},\ H^\oplus\\ 3.\ Li\\ 4.\ \text{ethylene oxide}\\ 5.\ H_3O^\oplus\end{array}}$$

22. 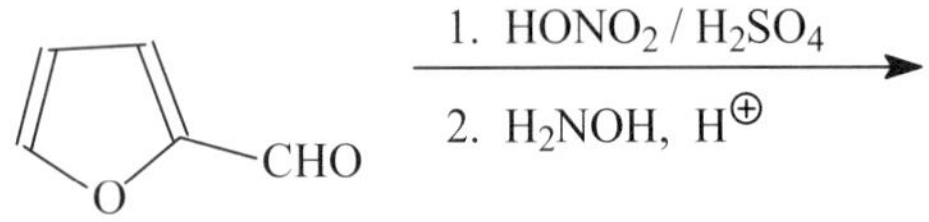 furan-2-carbaldehyde $\xrightarrow{\begin{array}{l}1.\ HONO_2\,/\,H_2SO_4\\ 2.\ H_2NOH,\ H^\oplus\end{array}}$

nitrofuroxime (used in treating urinary tract infections)

23. 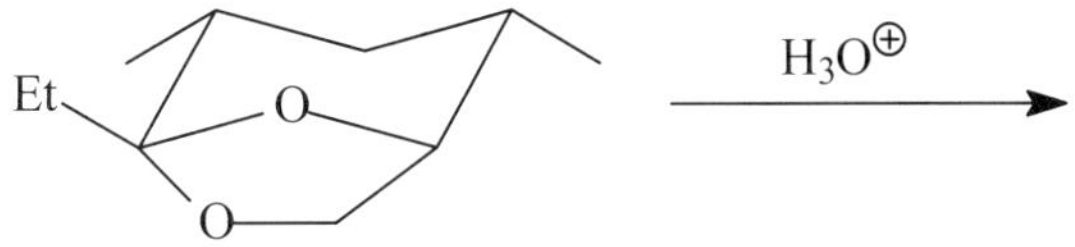 Et–(bicyclic acetal) $\xrightarrow{H_3O^\oplus}$

multistriatin
(European elm bark beetle pheremone)

24. 2-oxopropanal $\xrightarrow[\text{(intra-Cannizzaro)}]{^\ominus OH\,/\,ROH}$

25. 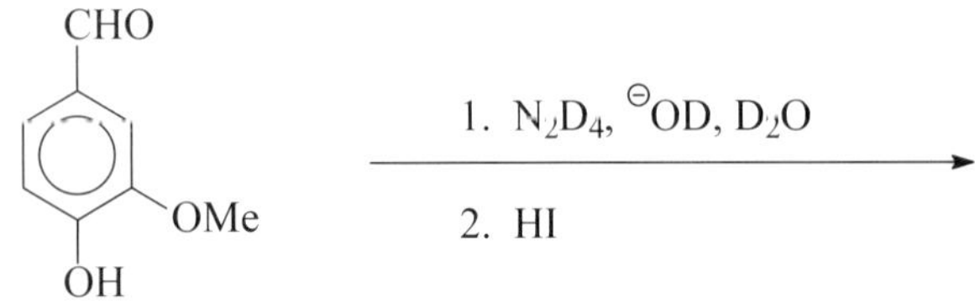 vanillin (3-methoxy-4-hydroxybenzaldehyde) $\xrightarrow{\begin{array}{l}1.\ N_2D_4,\ ^\ominus OD,\ D_2O\\ 2.\ HI\end{array}}$

26. 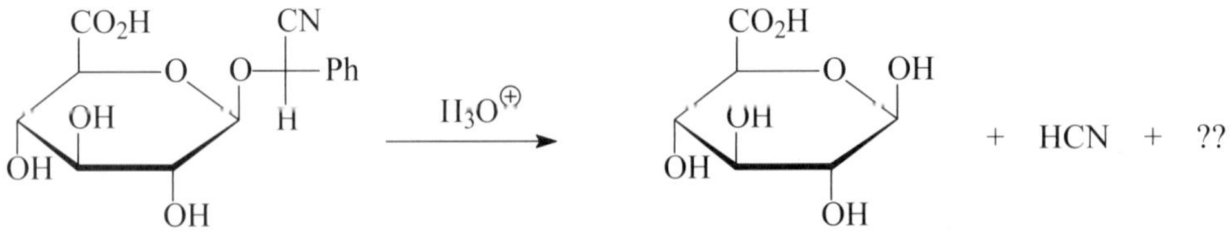 (sugar with CO₂H, CN, OPh, OH groups) $\xrightarrow{H_3O^\oplus}$ (sugar with CO₂H, OH groups) + HCN + ??

This reaction, with the release of the very toxic HCN, provides a defense mechanism for millipedes.

15.1 Reactions

27. 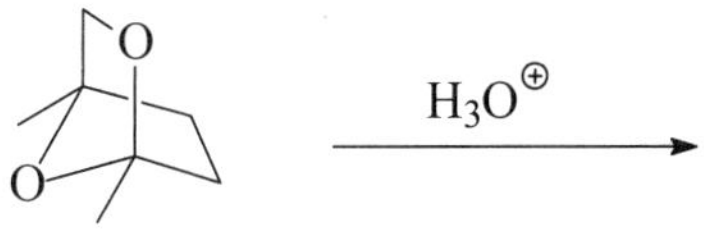 *frontalin* (insect pheremone) — H$_3$O$^⊕$ →

28. 3-oxobutanal — MeOH, H$^⊕$ →

$C_6H_{12}O_3$ IR: 1715 cm^{-1}
^{1}H NMR: δ 2.2 (s, 3H), 2.8 (d, 2H), 3.4 (s, 6H), 4.9 (t, 1H)

29.  H$_3$O$^⊕$ →

30. 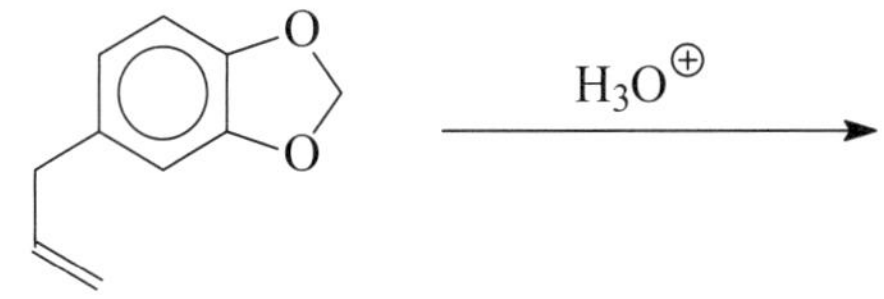 *safrole* (odor of sassafras) — H$_3$O$^⊕$ →

31. 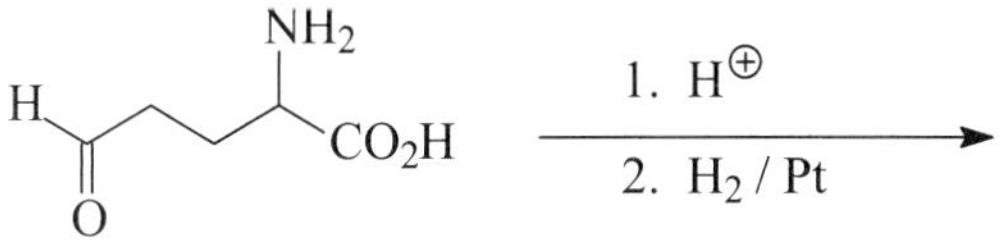 1. H$^⊕$ 2. H$_2$/Pt → *proline* (an amino acid)

32. 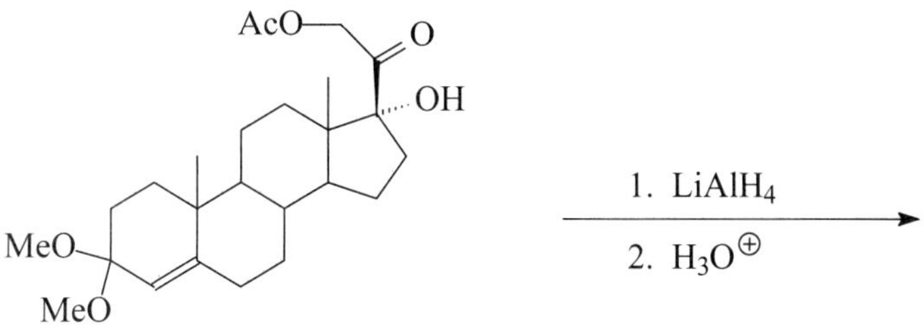 *cortisone acetate dimethyl ketal* — 1. LiAlH$_4$ 2. H$_3$O$^⊕$ →

15.1 Reactions

144 • Chapter 15 Aldehydes and Ketones

33.  OHC–CHO $\xrightarrow{\ominus OH}$

34. 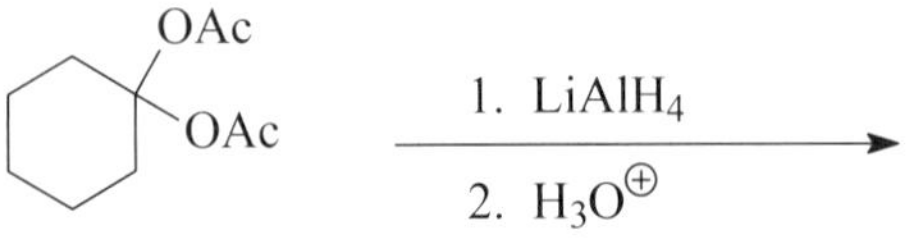 1,1-diacetoxycyclohexane $\xrightarrow{\text{1. LiAlH}_4 \quad \text{2. H}_3O^\oplus}$

35. 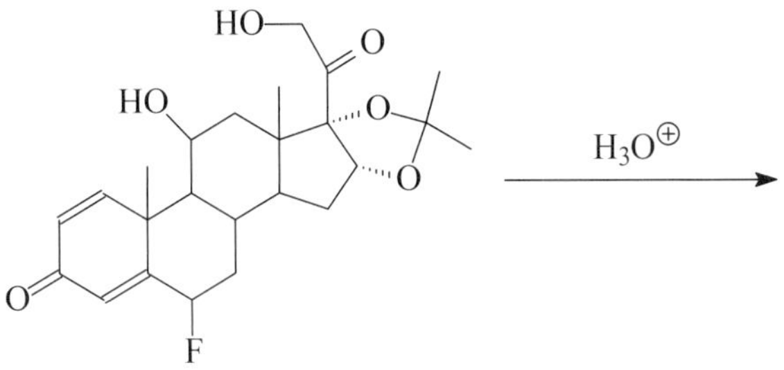 flunisolide (anti-inflammatory in allergy medication) $\xrightarrow{H_3O^\oplus}$

36. 1,2-cycloheptanedione $\xrightarrow{\text{(XS) hydroxylamine}}$

heptoxime (used in quantitative determination of Ni)

37. 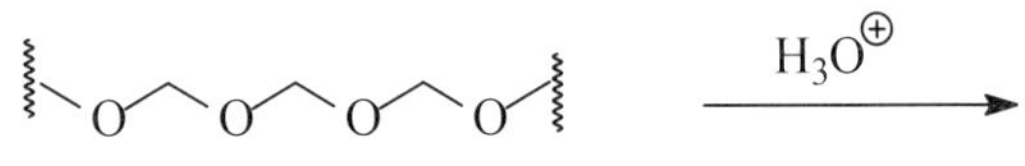 paraformaldehyde $\xrightarrow{H_3O^\oplus}$

38. Chain degradation of a hexose:

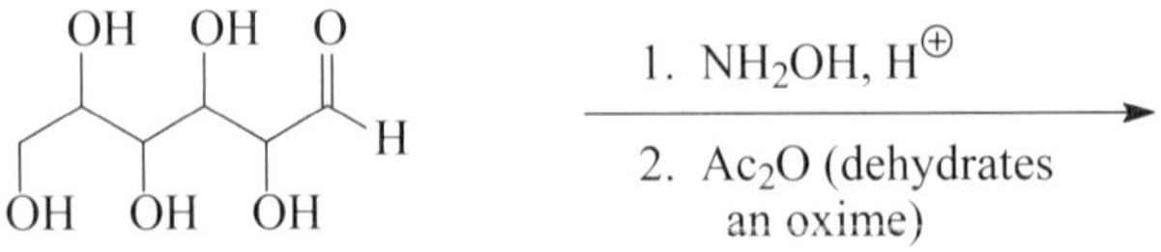

$\xrightarrow{\text{1. NH}_2\text{OH, H}^\oplus \quad \text{2. Ac}_2\text{O (dehydrates an oxime)}}$

15.1 Reactions

Problems • 145

39. 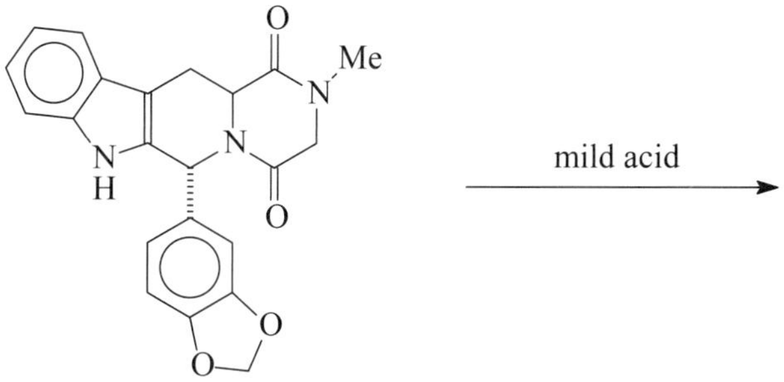 *tadalafil* (Cialis™) —mild acid→

40. 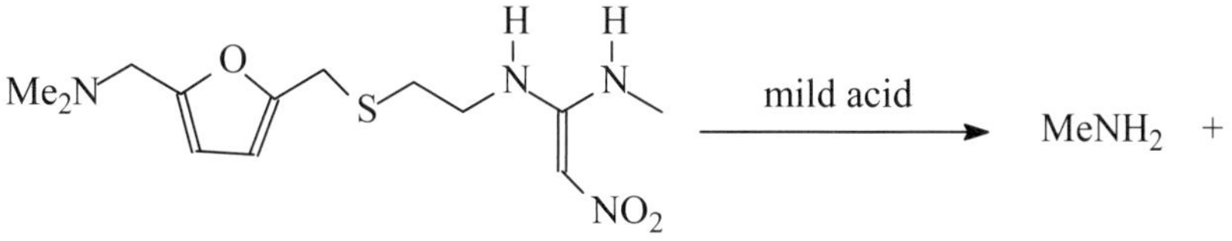 *ranitidine* (Zantac™ - antiulcerative) —mild acid→ MeNH$_2$ +

41. 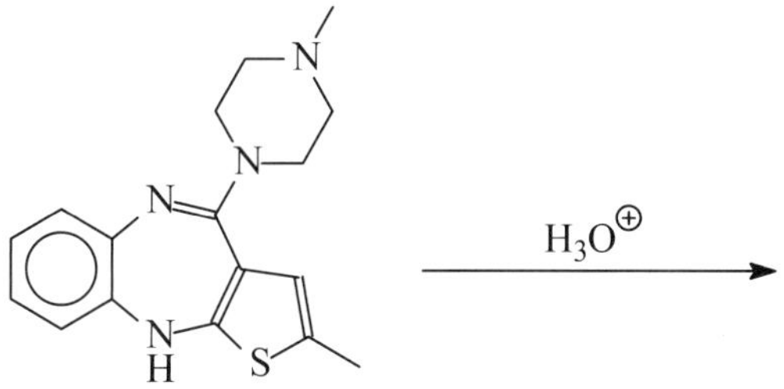 *olanzapine* (Zyprexa™ - antipsychotic) —H$_3$O$^\oplus$→

42. 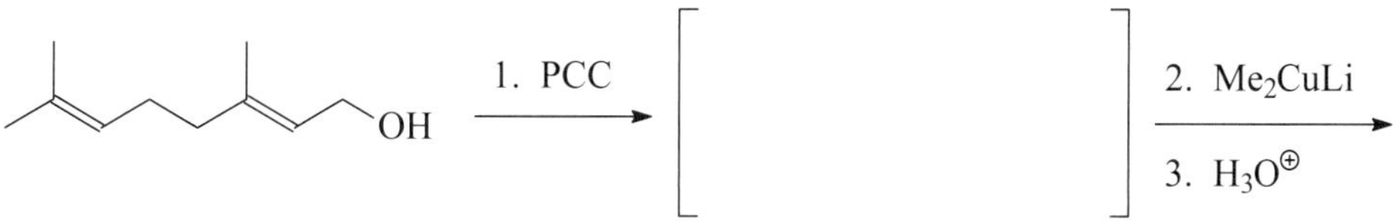 —1. PCC→ [*citral*] —2. Me$_2$CuLi / 3. H$_3$O$^\oplus$→

43. 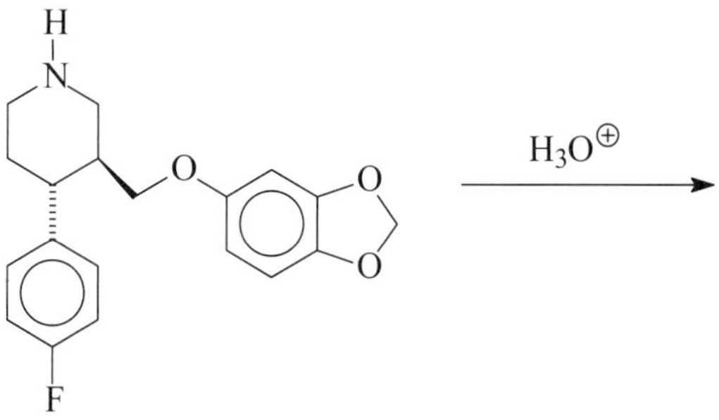 *paroxetine* (Paxil™ - antidepressant) —H$_3$O$^\oplus$→

15.1 Reactions

146 • Chapter 15 Aldehydes and Ketones

44. Ammonia is produced in the mitochondria primarily by the oxidation of *glutamate* to produce an imine, which is subsequently hydrolyzed:

$$^\ominus O_2C\text{-CH}_2\text{-CH}_2\text{-CH(NH}_2)\text{-CO}_2^\ominus \xrightarrow{[O]} \left[\quad \right] \xrightarrow{H_3O^\oplus} \left[\quad \right] + \overset{\oplus}{N}H_4$$

glutamate → *α-ketoglutarate*

45. 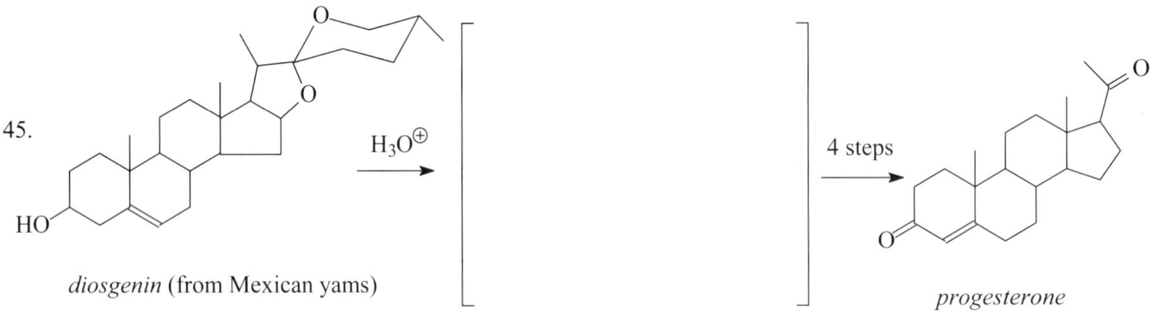 *diosgenin* (from Mexican yams) $\xrightarrow{H_3O^\oplus}$ [ ] $\xrightarrow{\text{4 steps}}$ *progesterone*

46. 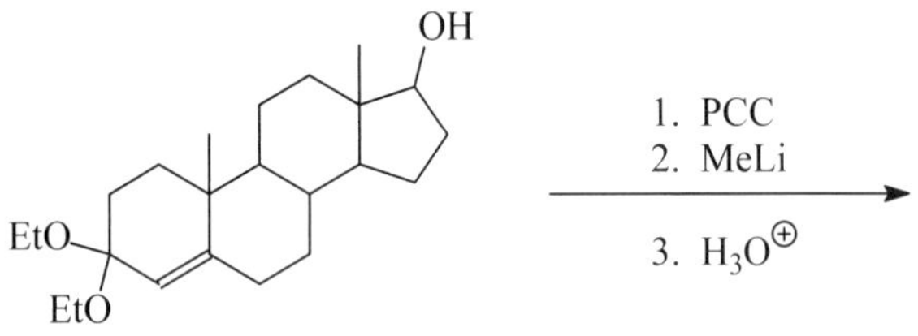 *testosterone diethyl ketal* $\xrightarrow{\text{1. PCC} \\ \text{2. MeLi} \\ \text{3. } H_3O^\oplus}$ *17-methyltestosterone* (an anabolic steroid)

47. A step in Woodward's (*Harvard*) synthesis of strychnine:

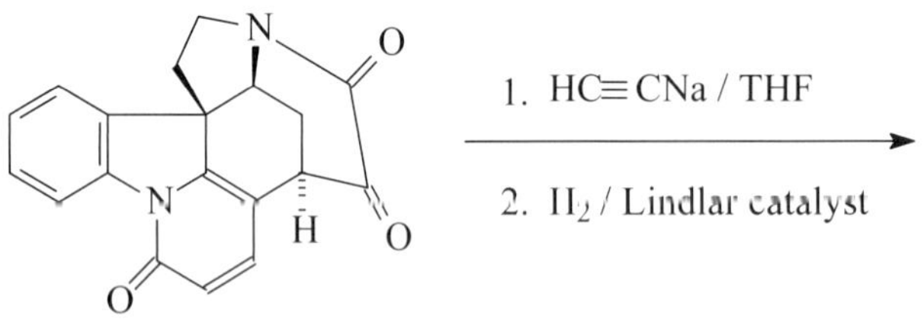

dehydrostrychninone $\xrightarrow{\text{1. HC≡CNa / THF} \\ \text{2. H}_2\text{/ Lindlar catalyst}}$

15.1 Reactions

48. Aldehyde protons are non-acidic. However, if the aldehyde is converted to a 1,3-dithiane (the sulfur analog of an acetal), the proton can then be quantitatively removed by NaNH$_2$ or organolithiuim bases. The resultant anion (Corey-Seebach reagent) readily undergoes S$_N$2 or carbonyl addition reactions. Subsequent hydrolysis of the product unmasks the starting carbonyl.

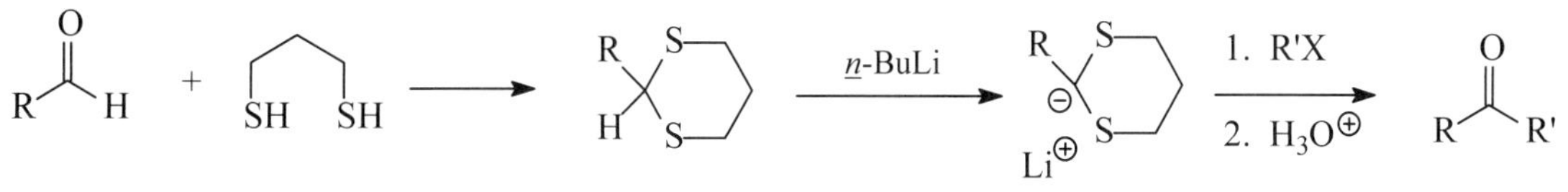

$$R-CHO + HS(CH_2)_3SH \longrightarrow \text{a dithiane} \xrightarrow{n\text{-BuLi}} \text{Corey-Seebach reagent} \xrightarrow[2.\ H_3O^\oplus]{1.\ R'X} R-CO-R'$$

Predict the products of the following reactions:

a. benzaldehyde
 1. HS(CH$_2$)$_3$SH, H$^\oplus$
 2. MeLi
 3. EtI
 4. H$_3$O$^\oplus$

b.  (1,3-dithiane)
 1. n-BuLi
 2. n-decyl bromide
 3. H$_3$O$^\oplus$

c. acetaldehyde
 1.  HS(CH$_2$)$_3$SH, H$^\oplus$
 2. NaNH$_2$
 3. cyclohexanone
 4. H$_3$O$^\oplus$

49. The amino acid *serine* can undergo a retro-aldol-like reaction (*see* CARBONYL CONDENSATION REACTIONS) to form glycine and formaldehyde; in cells this reaction is catalyzed by a derivative of pyridoxine (vitamin B6):

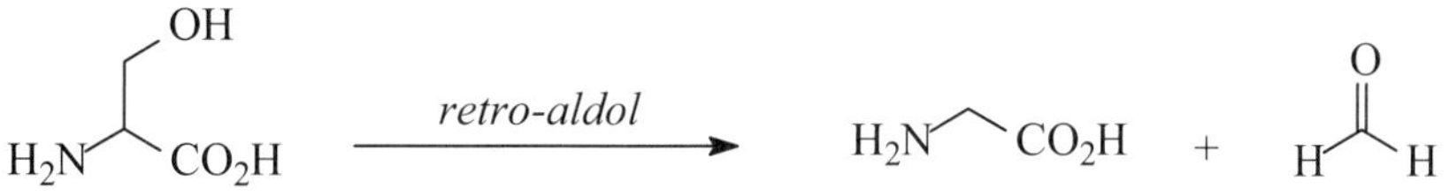

serine $\xrightarrow{\text{retro-aldol}}$ glycine + HCHO

(*cont. on next page*)

15.1 Reactions

Catabolic reactions that produce formaldehyde, as above, generally occur in the presence of another vitamin derivative, tetrahydrofolic acid (FH$_4$). The later detoxifies formaldehyde by reacting with it to produce **A**. On the other hand, many *anabolic* reactions require formaldehyde as a building block (*e.g.*, biosyntheses of the nucleoside bases). In those instances **A** undergoes hydrolysis to yield FH$_4$ and formaldehyde *in situ*. Draw the structure of FH$_4$.

FH$_4$ + CH$_2$O → **A** $\xrightarrow{H_3O^{\oplus}}$ FH$_4$ + CH$_2$O

catabolism   *anabolism*

[Note: Unlike us, bacteria can synthesize FH$_4$ *de novo* from precursors such as *p*-aminobenzoic acid (PABA). Sulfa drugs are effective competitive inhibitors to enzymes that utilize PABA, thus destroying the ability of the bacteria to synthesize FH$_4$.]

50. [structure] $\xrightarrow[-H_2O]{H^{\oplus}}$ Xanax^{TM} (anxiolytic)

15.1 Reactions

15.2 Syntheses
Supply a reagent or sequence of reagents that will effect the following conversions.

1. *cis*-2-butene → (HO, CO₂H substituted compound)

   cis-2-butene → (sec-butylamine, NH₂)

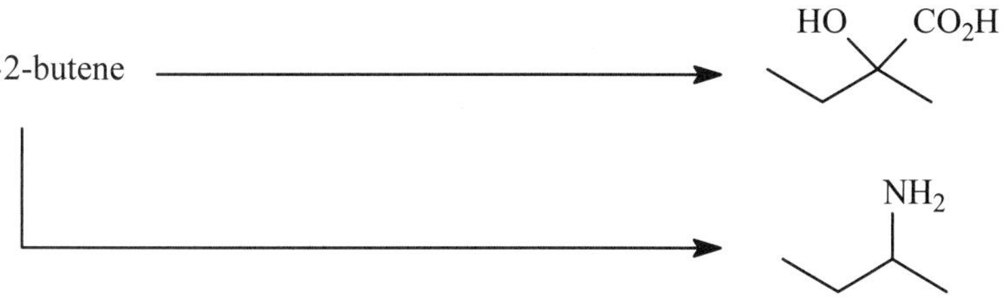

2. (methoxy norbornane with H) → (OH, D norbornane)

   → (norbornane)

3. cyclopentanone → cyclopentane-CO₂H

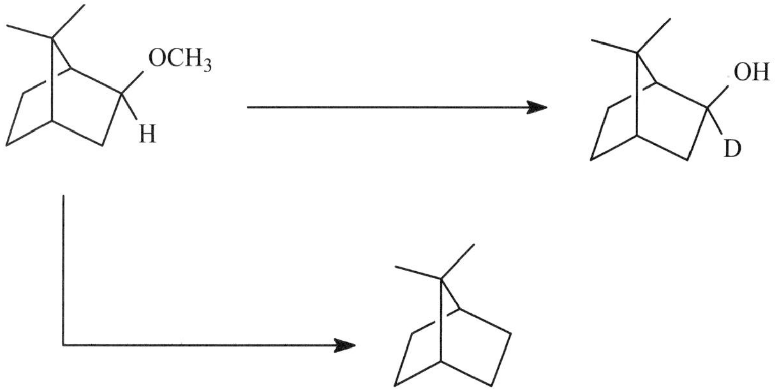

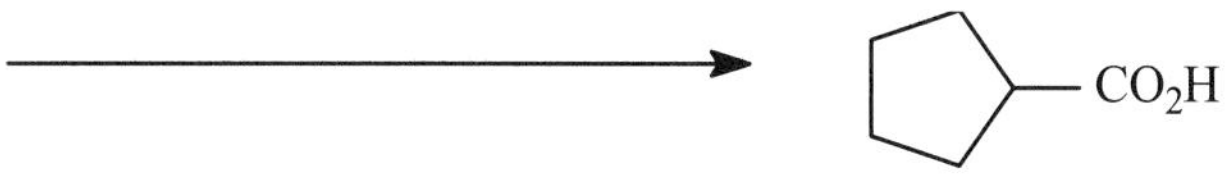

4. (camphor-like ketone) → (Cl, D substituted bicyclic)

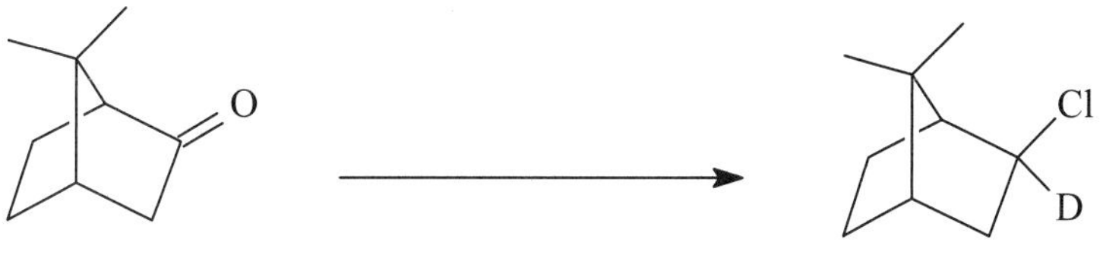

5. benzaldehyde → Ph-C(=O)-CO₂H

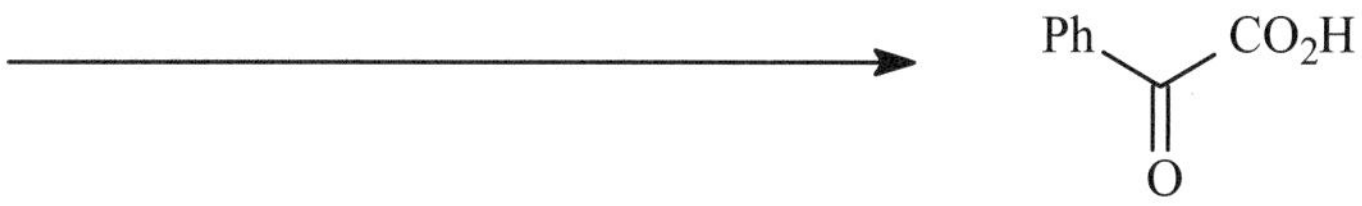

150 • Chapter 15 Aldehydes and Ketones

6. 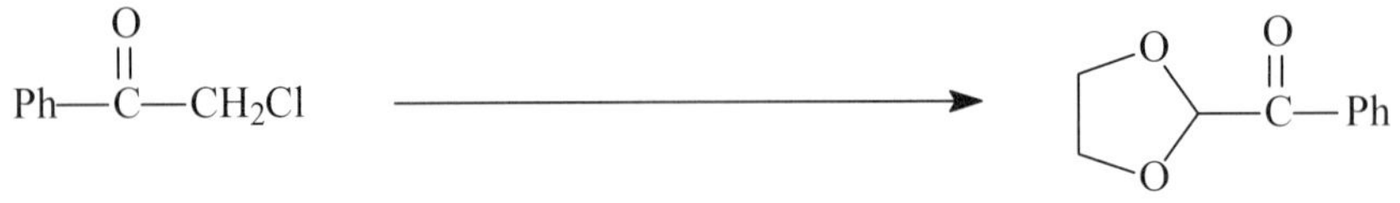 Ph—C(=O)—CH$_2$Cl ⟶ [1,3-dioxolan-2-yl]—C(=O)—Ph

7. 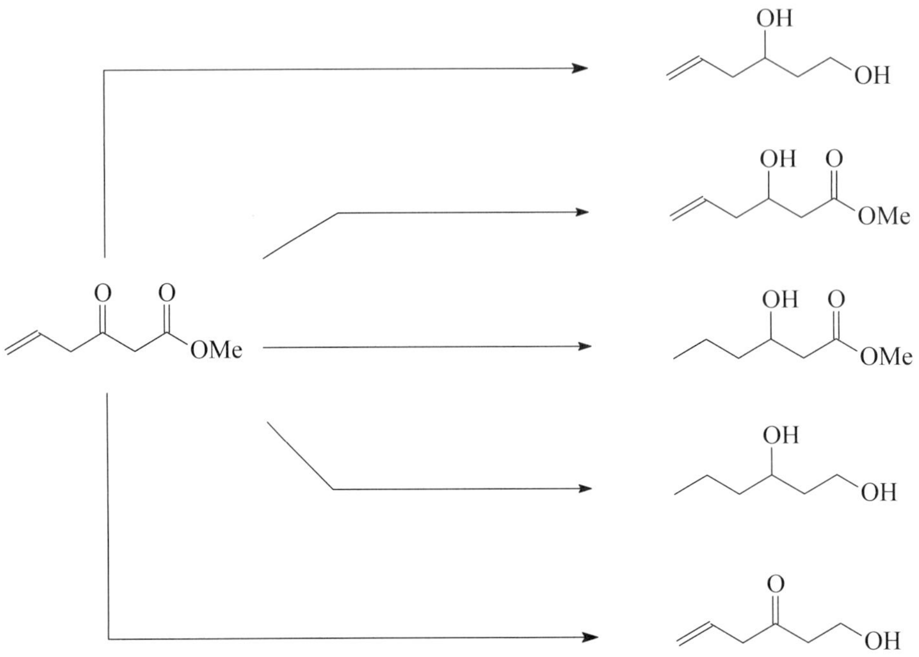 CH$_2$=CHCH$_2$—C(=O)—CH$_2$—C(=O)—OMe ⟶
   - CH$_2$=CHCH$_2$CH(OH)CH$_2$CH$_2$OH
   - CH$_2$=CHCH$_2$CH(OH)CH$_2$C(=O)OMe
   - CH$_3$CH$_2$CH$_2$CH(OH)CH$_2$C(=O)OMe
   - CH$_3$CH$_2$CH$_2$CH(OH)CH$_2$CH$_2$OH
   - CH$_2$=CHCH$_2$C(=O)CH$_2$CH$_2$OH

8.  methylenecyclohexane ⟶ *via a Wittig* ⟶ cyclohexyl-CH=CH-CH=CH$_2$

9. MVK (methyl vinyl ketone) ⟶ 2-methoxy-2-methyloxetane

15.2 Syntheses

10. 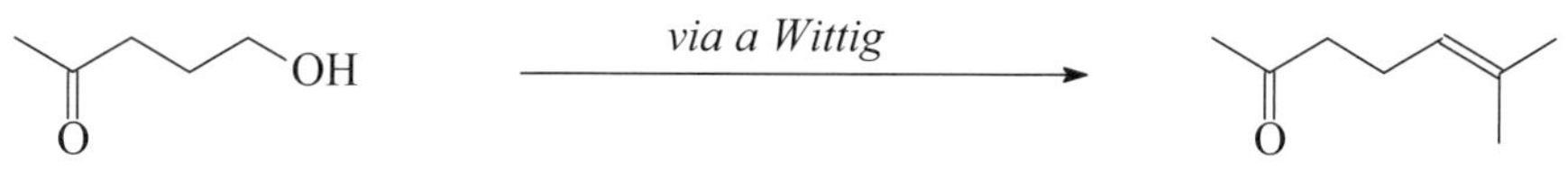 [structure: 5-hydroxypentan-2-one] —via a Wittig→ [structure: 6-methylhept-5-en-2-one]

11. 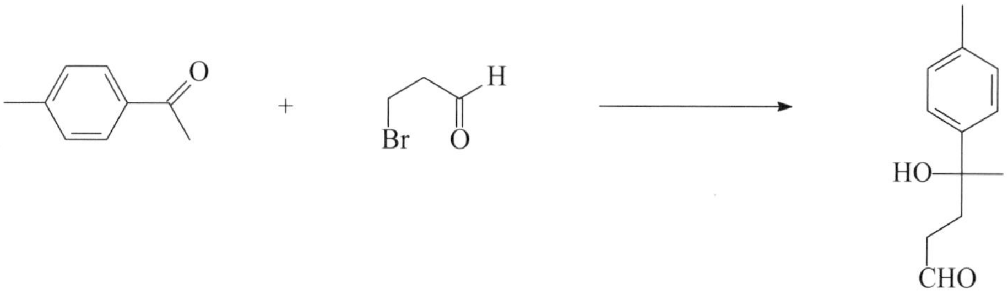 [4-methylacetophenone] + [3-bromopropanal] ⟶ [product: 4-methylphenyl group attached to C(OH)(CH3) with CH2CH2CHO chain]

12. Hydrazones can be deprotonated by strong bases to give carbanions that act as nucleophiles, *e.g.*,

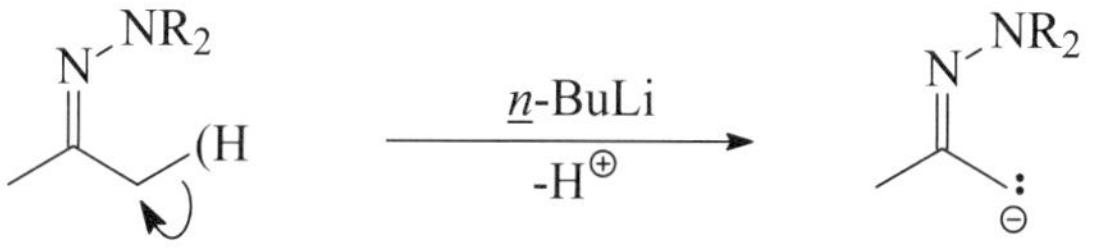

[CH3-C(=NNR2)-CH2(H)] —n-BuLi, -H⊕→ [CH3-C(=NNR2)-CH2⊖]

How could this observation be used to form [structure: CH3-CO-CH2-CH(OH)-Ph] from acetone and benzaldehyde?

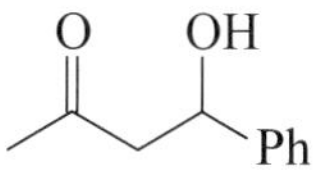

13.  [cyclohex-3-ene-1-carbaldehyde] ⟶ [4-oxo-cyclohexane with CHO and adjacent diketone: 3,4-dioxocyclohexane-1-carbaldehyde]

14.  [cyclohexene-1-carbaldehyde] ⟶ [1,2-dimethylcyclohexane]

15.2 Syntheses

15. [structure of 17-hydroxy steroid] ⟶ [structure with 17-OH and 17-C≡CH]

major component in OCPs

16. Ph-CO-C6H4-R , benzaldehyde ⟶ [tamoxifen structure with Ph, Et, Ph substituents on alkene]

(R = -OCH₂CH₂NMe₂)

tamoxifen (Nolvadex™ - antiestrogen)

17. 1-butene ⟶ [2-pentanone]

18. [cyclohexane with two CO₂Me groups] ⟶ [cyclohexane with two vinyl groups]

19. [N-substituted tetrahydropyridine with methyl and CH(CH₃)CH₂OH] ⟶ [chrysomelidial structure]

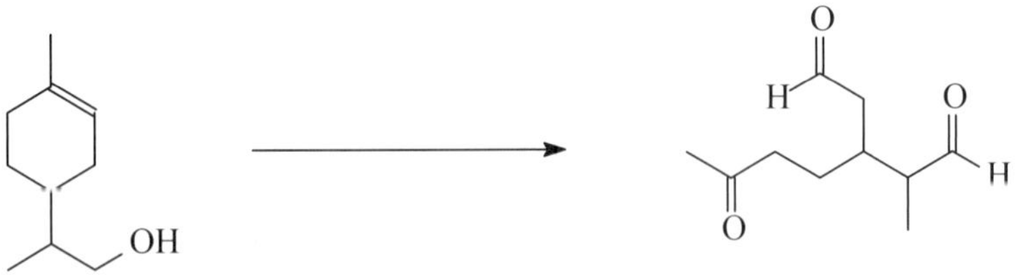

chrysomelidial (secreted by larvae of some beetles in self-defense)

15.2 Syntheses

20. The hotness of chili peppers can be quantified by determining their *Scoville heat units* (SHUs). An SHU is the amount of dilution needed before the chili is undetectable. The hottest, according to the *Guinness Book of World Records*, is the *bhut jolokia* from India, firing up at around 1,041,427 SHU, *i.e.*, a drop of extract needs about a million drops of water! (Jalapeño and Tabasco range a mild 5,000 – 25,000 and 100,000 – 200,000, respectively, on the SHU scale.) The active ingredient is *capsaicin*. Formulate a synthesis of the carboxylic acid moiety from 6-bromo-1-hexanol.

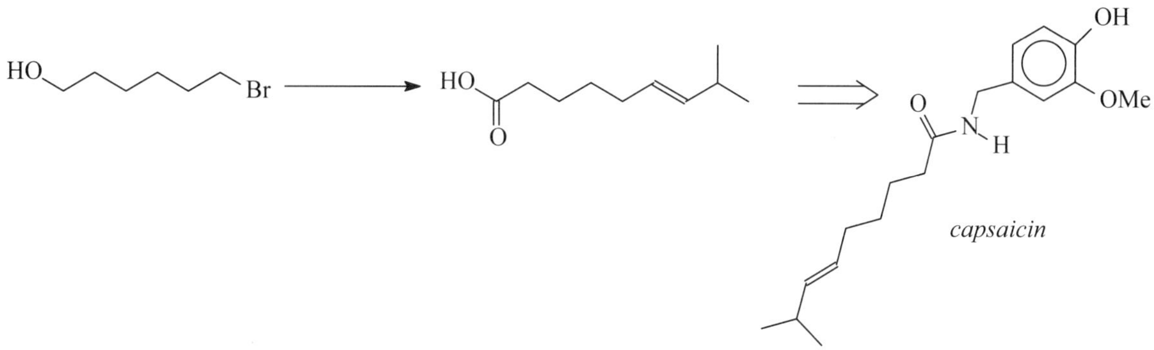

capsaicin

21. Similar to benzyl carbon-oxygen single bonds, carbon-sulfur single bonds readily undergo *hydrogenolysis*. This observation provides a more gentle reduction of aldehyde or ketone carbonyls than the highly alkaline Wolff-Kishner or acidic Clemmensen reductions. Complete the following illustration of this approach:

a *dithiane* (see 15.1, 48)
(or *thioketal*)

22. 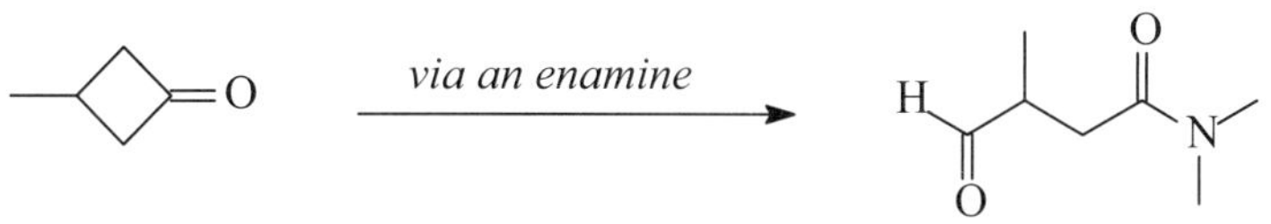   ⎯⎯ *via an enamine* ⎯⎯→

15.2 Syntheses

15.3 Mechanisms

Outline a detailed mechanism for each of the following. No other reagents than those given are necessary. Use arrows to explain the flow of electrons and show all intermediates.

1. 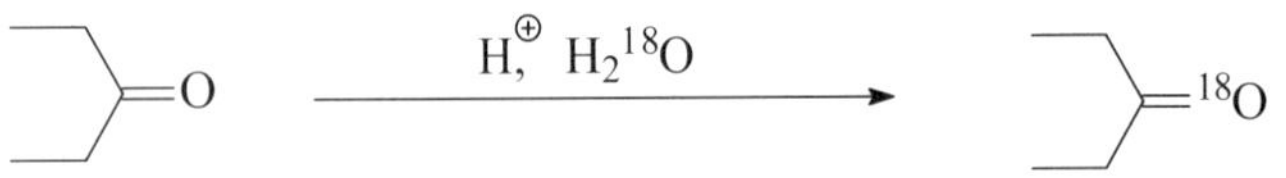 (ketone) + $H^{\oplus}$, $H_2{}^{18}O$ → (ketone with ^{18}O)

2. (cyclic imine, tetrahydropyridine) + $H_3O^{\oplus}$ → H_2N—(CH$_2$)$_3$—CHO

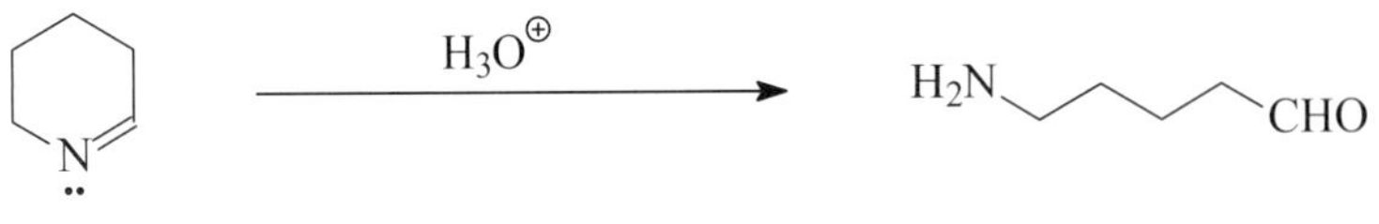

3. *Vinyl ethers*, unlike ordinary ethers, hydrolyze rapidly in water with just a trace of added acid. Draw the products and mechanism for

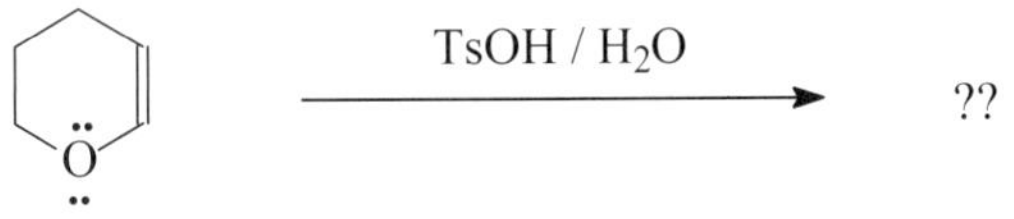

   (dihydropyran) + TsOH / H$_2$O → ??

4. 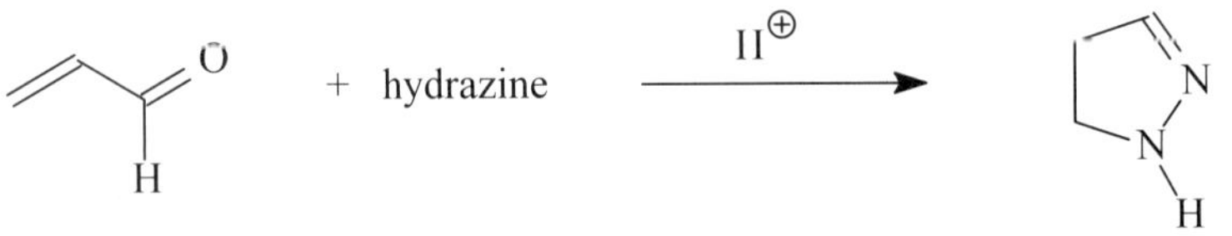 (acrolein, CH$_2$=CH–CHO) + hydrazine $\xrightarrow{H^{\oplus}}$ (pyrazoline)

5. 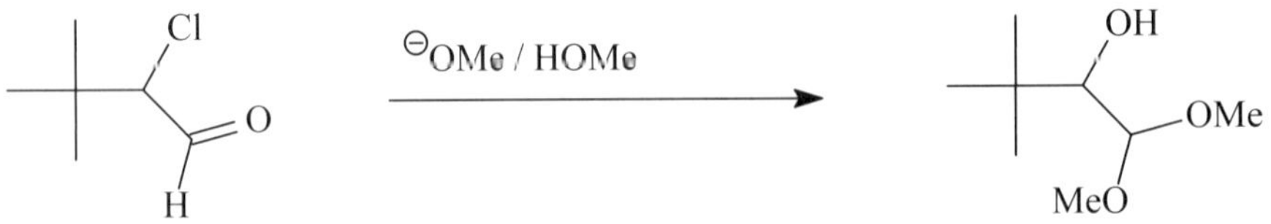 (2-chloro-3,3-dimethylbutanal with Cl) + $^{\ominus}$OMe / HOMe → (product with OH and two OMe groups)

6. [reaction: R-CO-CH2-CO-R + H2NOH, H⁺ → 3,5-disubstituted isoxazole]

7. [reaction: spirocyclic dioxolane with pendant -CH2CH2OH + H⁺ → rearranged product with HOCH2CH2-O- and new tetrahydrofuran ring fused to cyclopentane]

8. *Fugu*, a fish, is a Japanese delicacy. Unfortunately it produces a very toxic substance, tetrodotoxin (an adult fugu contains enough to kill 30 people), in organs that must be removed by a licensed chef. To become a fugu chef requires training for years with a master and culminates in a battery of state-administered exams, including eating a fugu prepared by oneself ! Though the risk of fugu poisoning is practically nil, *if prepared by a master*, a handful of diners succumb to fugu each year; perhaps that is why Japan's Imperial Family is forbidden from tasting one of their country's choicest dishes. Deduce the structure and outline the mechanism of the carboxylic acid produced when tetrodotoxin is treated with aqueous acid.

tetrodotoxin + $H_3O^⊕$ →

9. E. J. Corey (*Harvard*) found that sulfur ylids, similar to the Wittig reagent, can be prepared as follows:

[reaction: Me2S=O + CH3I → (1. S_N2, 2. n-BuLi) → Me2-S(⊕)-C̈H2(⊖) with O]

When treated with cycloheptanone a 70% yield of **A** is obtained. Explain, showing clearly how the intermediate betaine's behavior to form an epoxide differs from that of a typical Wittig intermediate.

A [spiro epoxide of cycloheptane]

15.3 Mechanisms

156 • Chapter 15 Aldehydes and Ketones

10.  HCl →

11. 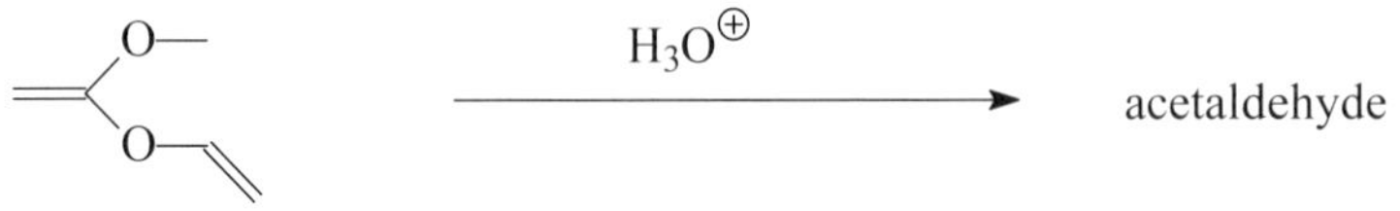 (structure) H₃O⁺ → acetaldehyde

12. 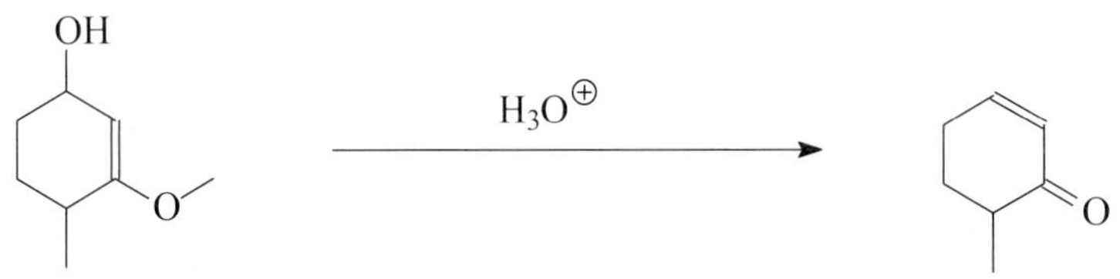 (structure) H₃O⁺ → (structure)

13. cyclohexanone — diazomethane → cycloheptanone

14. 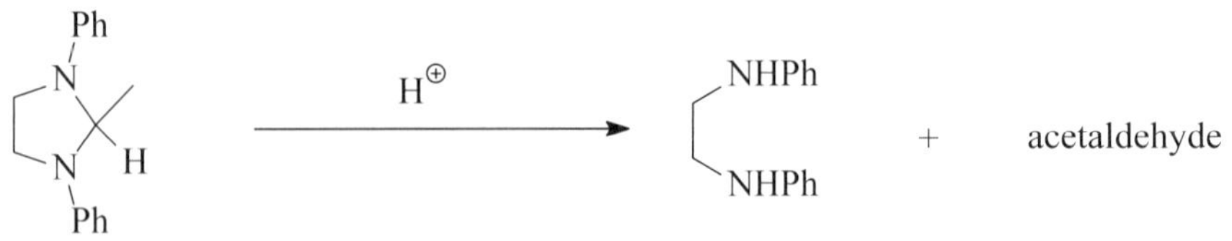 (structure) H⁺ → (structure) + acetaldehyde

15.3 Mechanisms

15. The *Vedejs olefin inversion reaction* readily converts *cis-to-trans* or *trans-to-cis* stereoisomers:

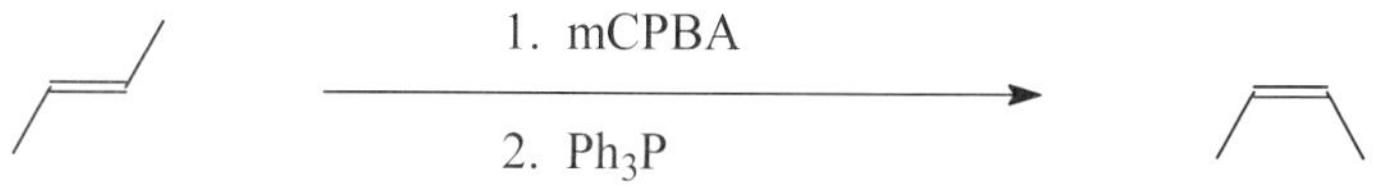

$$\text{trans-2-butene} \xrightarrow[\text{2. Ph}_3\text{P}]{\text{1. mCPBA}} \text{cis-2-butene} \quad \text{(Hint: think Wittig-like)}$$

16. propylene epoxide $\xrightarrow{:\text{P(OMe)}_3}$ propylene

17. [furanose tetraol] + NH$_2$R $\xrightarrow{\text{H}^\oplus}$ [furanose with NHR at anomeric position]

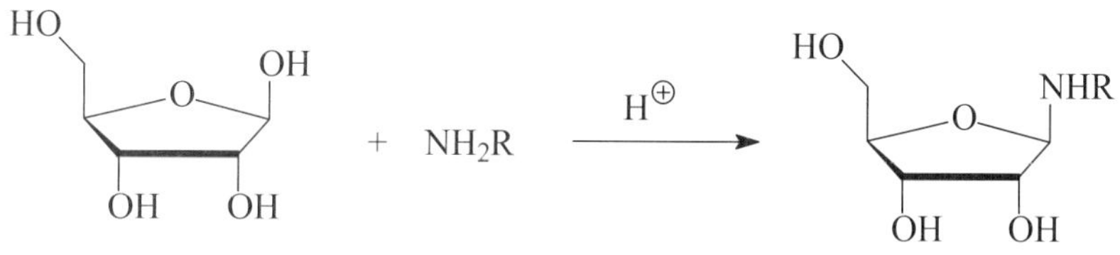

18. 2 phenol + acetone $\xrightarrow{\text{H}^\oplus}$ C$_{15}$H$_{16}$O$_2$

    bisphenol A (a starting material in the synthesis of Lexan™)

19. [cis-2-butenal] $\xrightarrow{\underline{n}\text{-PrNH}_2}$ [2-ethyl-1-propyl-2H-azete]

15.3 Mechanisms

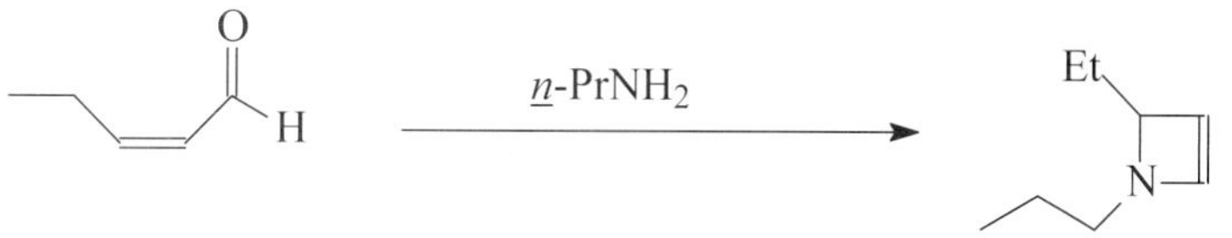

158 • Chapter 15 Aldehydes and Ketones

20. Another protecting group for alcohols (in addition to TMS or vinyl ether derivatives) is MOM (methoxymethyl). MOM is stable to base, but can be cleaved upon treatment with mild acid. The following sequence illustrates its use:

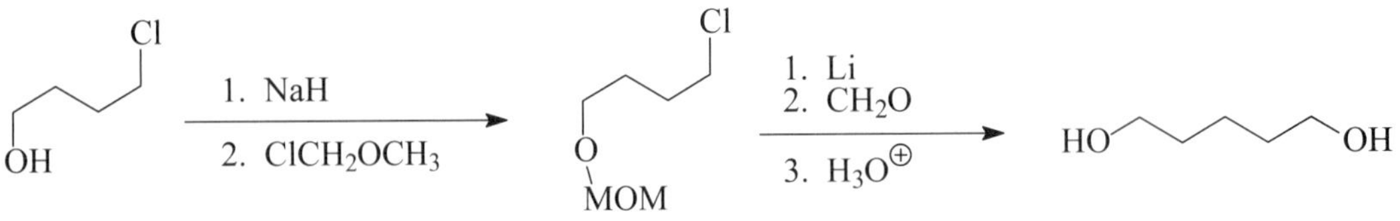

a. Draw the structure of the MOM derivative and explain its mechanism of formation.

b. Outline the mechanism of the last step. What other two organic products are formed from the MOM group?

21.

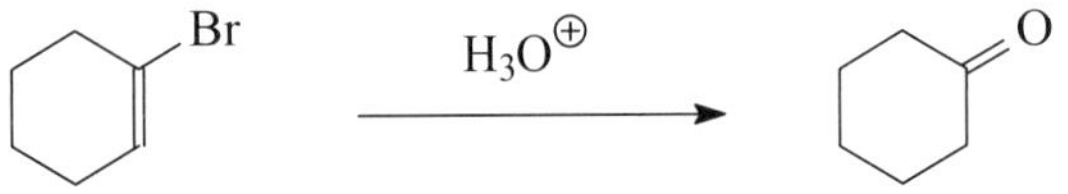

22.

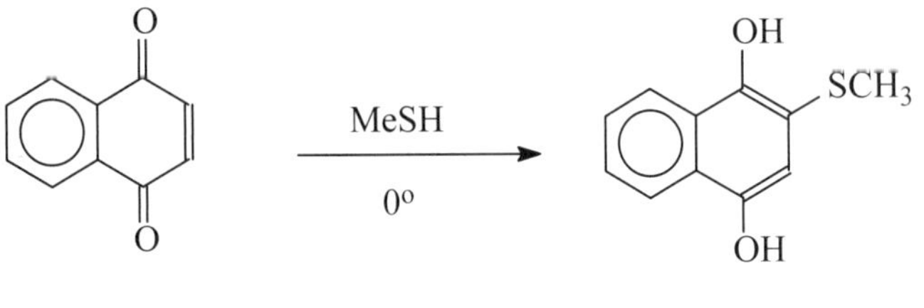

23.

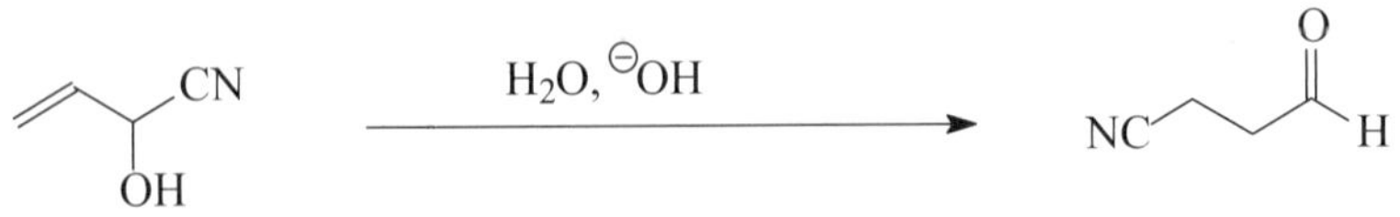

15.3 Mechanisms

24. 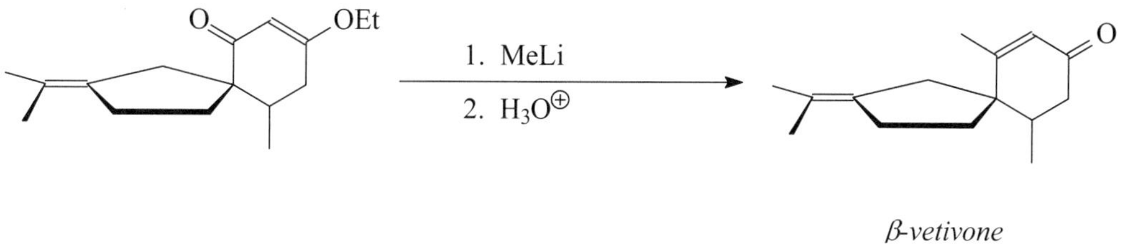 [reaction scheme: enone with OEt group, reagents 1. MeLi, 2. H₃O⁺, yields β-vetivone]

25. 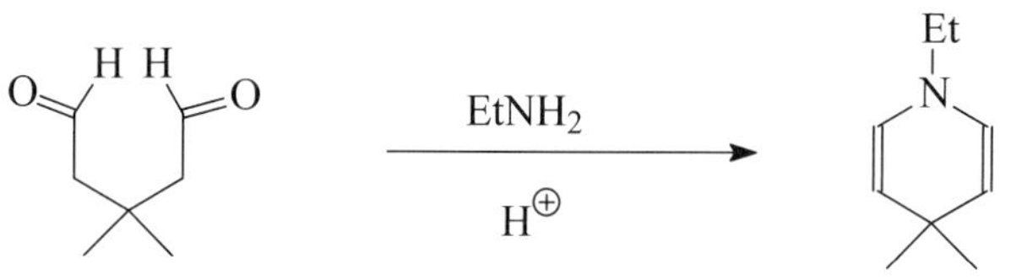 [reaction scheme: diketone + EtNH₂, H⁺ → N-ethyl dihydropyridinone]

26. Outline the mechanism for steps 2 and 3.

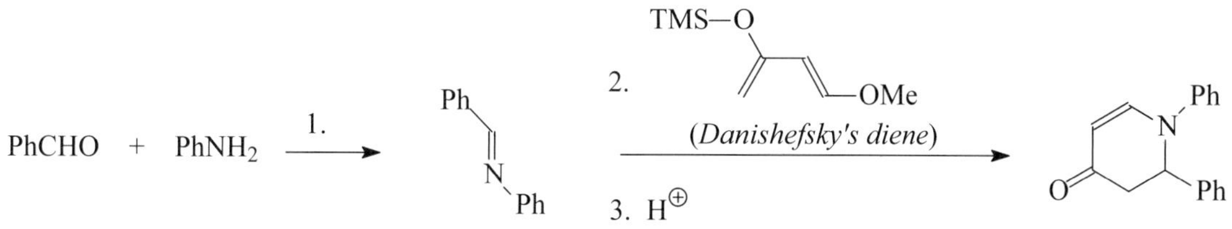

PhCHO + PhNH₂ —1.→ PhCH=N-Ph —2. TMS-O-CH=CH-CH=CH-OMe (Danishefsky's diene); 3. H⁺→ N-phenyl-2-phenyl-dihydropyridinone

27. Aromatic aldehydes cannot be prepared by direct Friedel-Crafts acylation (formyl chloride is unstable). One alternative is the Gatterman-Koch reaction (12.4, 10). Other options include the following two reactions:

a. the *Reimer-Tiemann reaction*

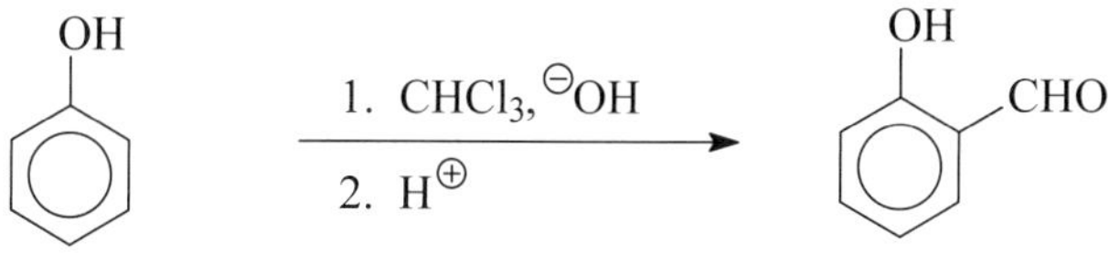

[phenol + 1. CHCl₃, ⁻OH; 2. H⁺ → salicylaldehyde (o-hydroxybenzaldehyde)]

15.3 Mechanisms

b. the *Vilsmeier reaction* (outline a mechanism for *both* steps)

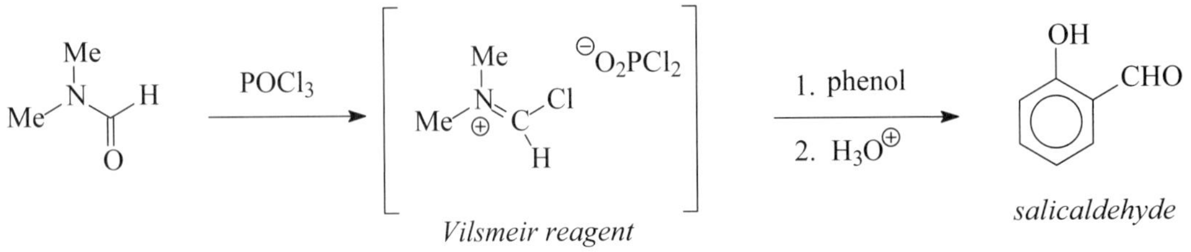

[Reaction scheme: Me₂N-CHO + POCl₃ → [Vilsmeir reagent: Me₂N⁺=CHCl, ⁻O₂PCl₂] →(1. phenol, 2. H₃O⁺) salicaldehyde (2-hydroxybenzaldehyde)]

28. Another approach to enhancing the acidity of an aldehyde proton (see 15.1, 48 – Corey-Seebach reaction) is illustrated by the *benzoin condensation reaction*:

2 PhCHO →(⁻CN) Ph-C(=O)-CH(OH)-Ph (benzoin)

29. [Reaction: 2-(piperidinyl)methyl-3-hydroxybenzene + formaldehyde, H⁺ → fused tricyclic amine with OH]

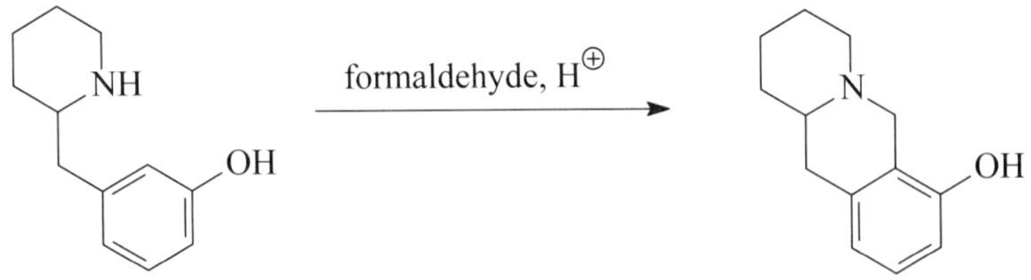

30. β-D-ribose ⇌(H⁺) α-D-ribose

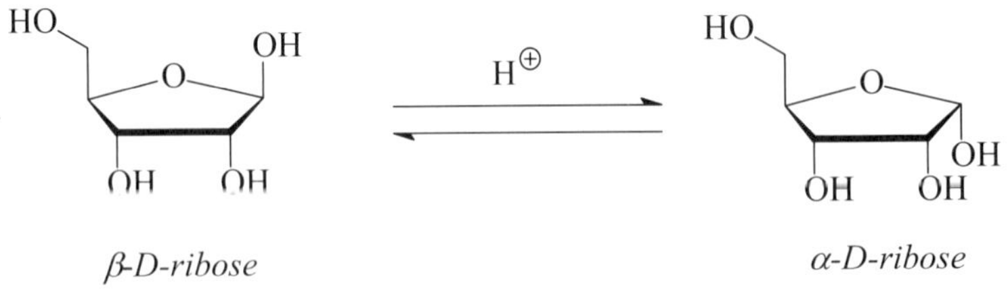

(Carbohydrate chemists call this process "mutarotation" and refer to the two epimeric diastereomers as "anomers.")

15.3 Mechanisms

31. The final step in the *urea cycle*:

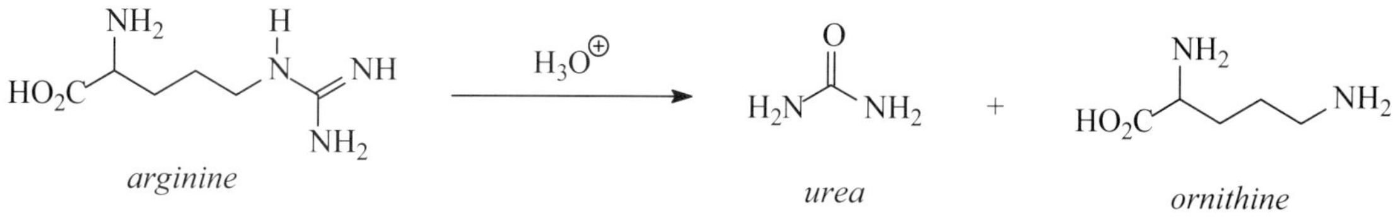

[Structure: arginine] → (H$_3$O$^+$) → urea (H$_2$N-CO-NH$_2$) + ornithine

32. *Fluorescamine* reacts with amines to give a highly fluorescent derivative. As little as a nanogram of an amino acid, for example, can be detected by this method. (Warning: do not attempt this one alone!)

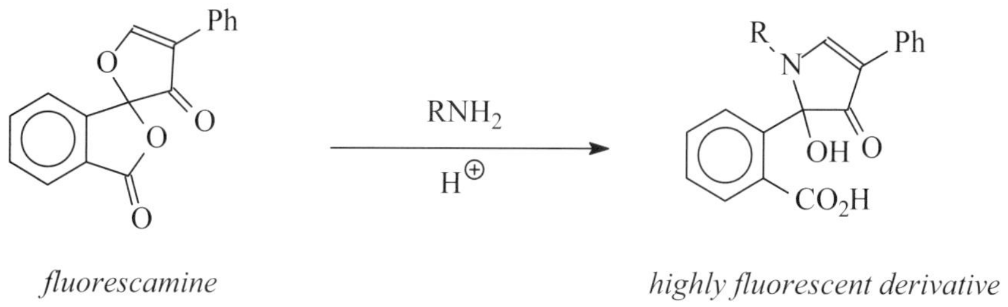

fluorescamine + RNH$_2$, H$^+$ → highly fluorescent derivative

33. Unlike other types of phospholipids, *plasmalogens* undergo hydrolysis to produce not only fatty acids but also fatty *aldehydes*. Explain the formation of the latter.

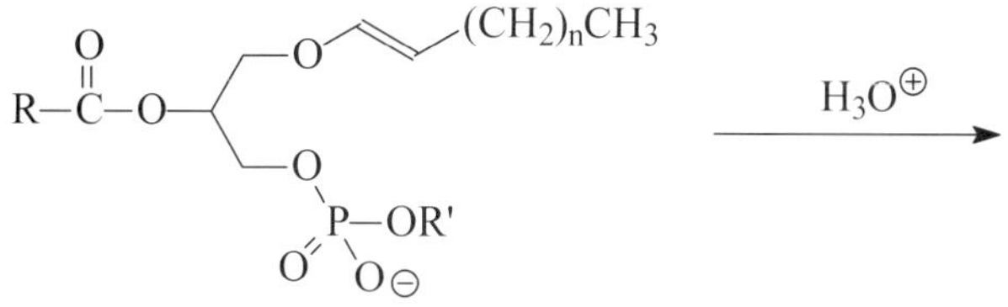

a *plasmalogen* (platelet activating factor) + H$_3$O$^+$ →

34. Although ketones are generally not reactive with most oxidizing agents, they are readily oxidized to esters when treated with peracids (*Baeyer-Villager reaction*).

Ph-CO-Ph + RCO$_3$H → Ph-CO-O-Ph

15.3 Mechanisms

35. Many historians of chemistry credit the discovery of molecular rearrangements to the *benzilic acid rearrangement*:

Ph–C(=O)–C(=O)–Ph (benzil) → KOH, EtOH → Ph–C(OH)(Ph)–CO₂⁻ (CB of benzilic acid)

Discovered by Liebig in 1838, it is a rare example of a rearrangement under alkaline conditions (most require acidic environments). Because of (1) disagreements over atomic weights at the time (the "conventional" weights for carbon and oxygen were thought to be 6 and 8!), and (2) the (erroneous as we now know) dogma propagated by Kekulé that carbon skeletal rearrangements could not occur in the course of chemical reactions, many wrong structures for benzilic acid were proposed -- until Baeyer finally got it right nearly forty(!) years later in 1877.

a. Propose a mechanism for the benzilic acid rearrangement.

b. Baeyer observed a benzilic acid-type rearrangement when phenanthrenequinone is treated with base. Draw the expected product.

phenanthrenequinone → 1. NaOH 2. H⁺

c. Another more modern benzilic acid-type rearrangement:

Ph–C(=O)–C(=O)–Ph → 1. (o-tolyl)–MgX 2. H⁺ → (o-tolyl)–C(=O)–C(Ph)(Ph)–OH

15.3 Mechanisms

Problems 36 – 40 illustrate the *dienone – phenol-type rearrangements*.

36. 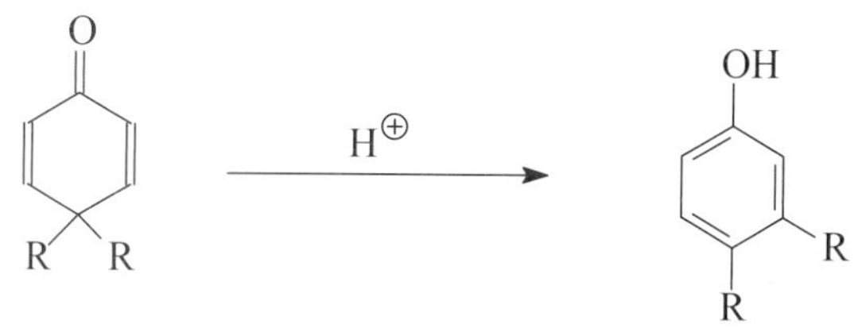 [cyclohexadienone with gem-R,R at C4] $\xrightarrow{H^\oplus}$ [phenol with OH, and two R groups at adjacent positions]

37.  [4-hydroxyphenyl-CH$_2$CH$_2$-OTs] $\xrightarrow{\text{base}}$ [ ] $\xrightarrow{\text{acid}}$ [benzocyclobutene with HO substituent]

38. 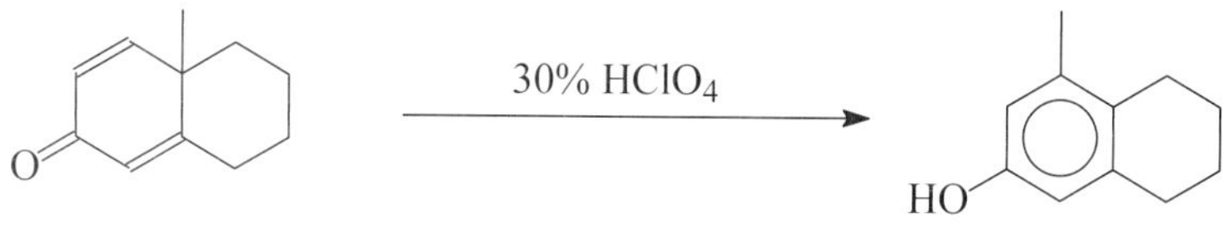 [4a-methyl-octahydronaphthalenone dienone] $\xrightarrow{30\% \text{ HClO}_4}$ [methyl-tetrahydronaphthalenol, HO meta to methyl]

However (!),

39. 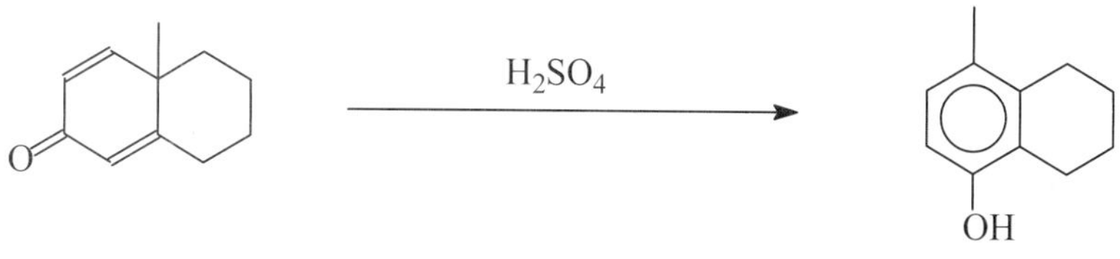 [4a-methyl dienone isomer] $\xrightarrow{H_2SO_4}$ [methyl-tetrahydronaphthalenol with OH peri to methyl]

15.3 Mechanisms

164 • Chapter 15 Aldehydes and Ketones

And, lastly, a steroid dienone – phenol rearrangement:

40. 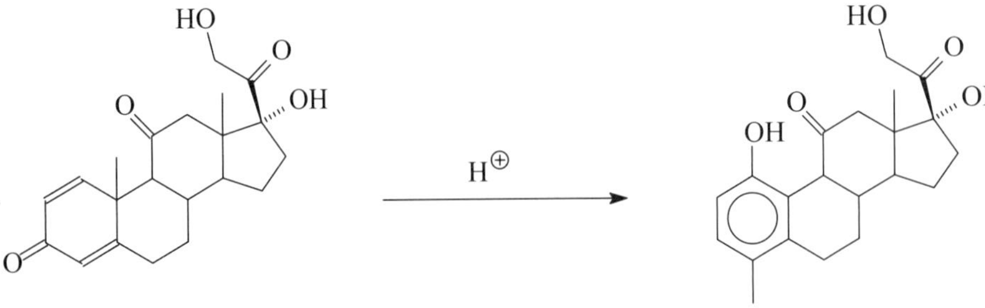 prednisone (anti-inflammatory) $\xrightarrow{H^+}$

Problems 41 and 42 illustrate the *Favorskii-type rearrangement*.

41. 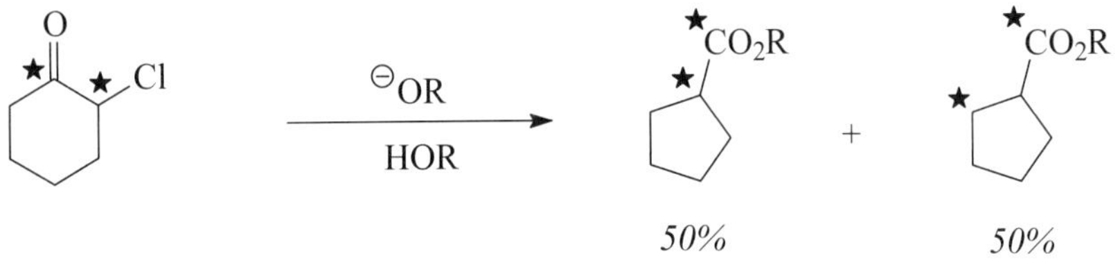 [2-chlorocyclohexanone with ^{13}C tags] $\xrightarrow[\text{HOR}]{^\ominus OR}$ [cyclopentane carboxylate ester, 50%] + [cyclopentane carboxylate ester, 50%] ($\star = {}^{13}C$ tag)

42. 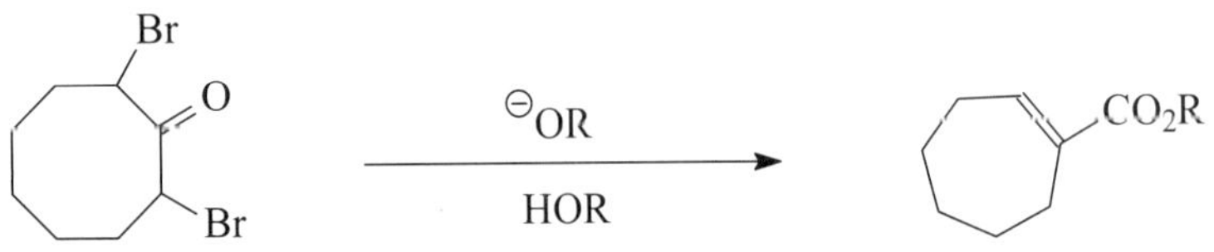 [2,8-dibromocyclooctanone] $\xrightarrow[\text{HOR}]{^\ominus OR}$ [cycloheptene carboxylate ester, CO_2R]

43. 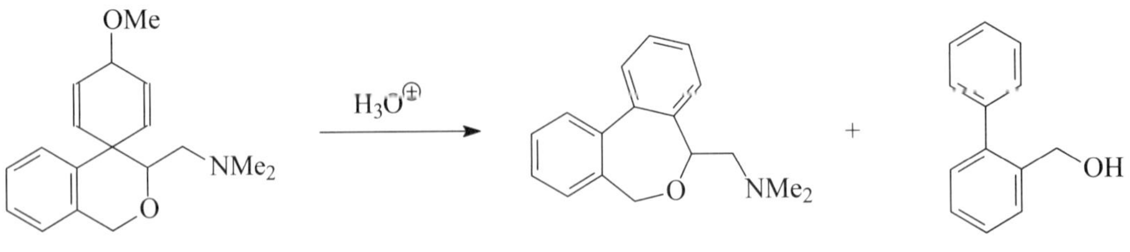 [spiro isochroman with OMe dienone and CH_2NMe_2] $\xrightarrow{H_3O^+}$ [dibenzoxepine with CH_2NMe_2] + [biphenyl-2-methanol, OH]

15.3 Mechanisms

44. 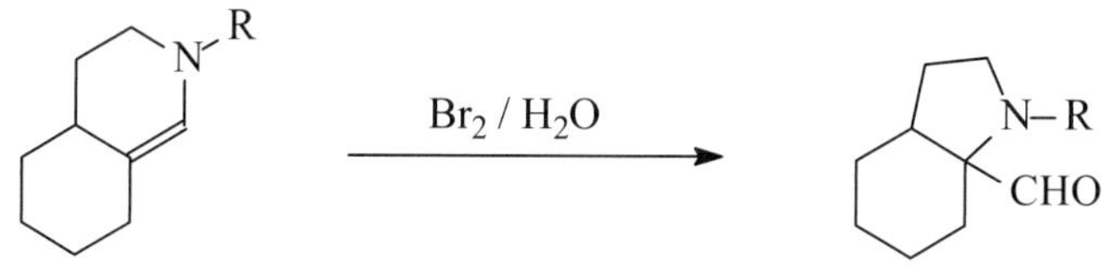 [structure] → Br₂ / H₂O → [structure with CHO]

45. Sheehan's (*MIT*) classic total synthesis of penicillin V involved a condensation step between the following reactants. Formulate a mechanism. (Note: the product is simply a nitrogen – sulfur analog of an acetal.)

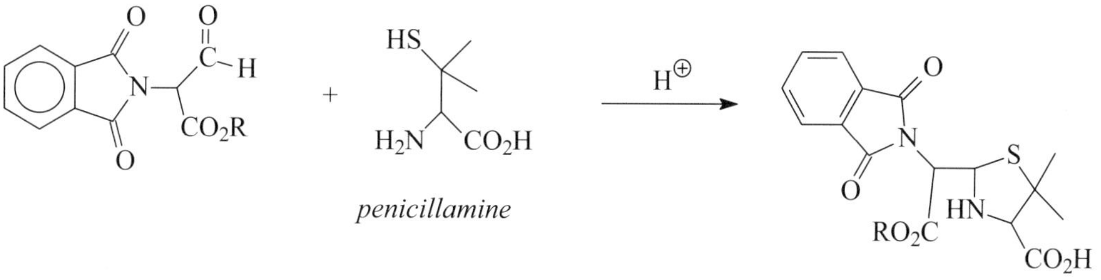

penicillamine

46. Woodward (*Harvard*) envisioned the biosynthesis of *strychnine* as beginning with a condensation of derivatives of the amino acids tryptophan (trp) and phenylalanine (phe). Sketch a likely sequence of events.

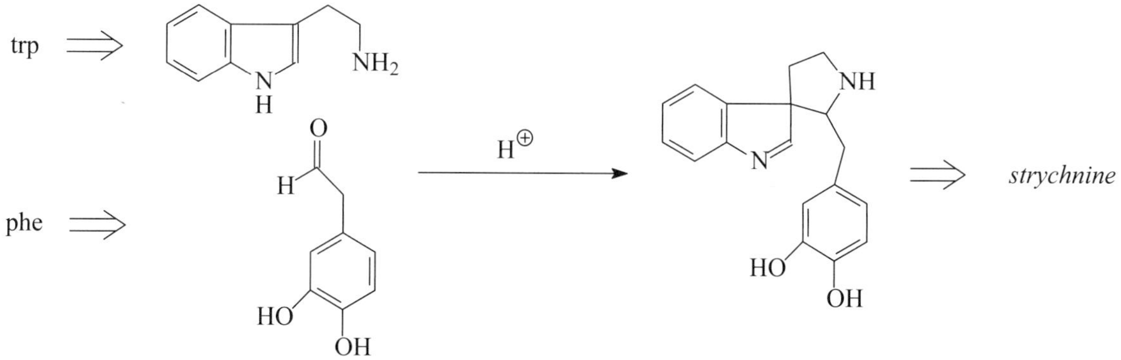

trp ⇒

phe ⇒

⇒ *strychnine*

15.3 Mechanisms

47. Thioketones, in the presence of aqueous acid, form hydrates *via* an intermediate ketone:

48. Many aldehydes autooxidize in air. For example, a white powder (benzoic acid) may often be seen around the cap of a bottle of previously opened benzaldehyde (liquid). Such autooxidation is thought to proceed by the addition of O_2 to a molecule of benzaldehyde *via* a free radical process to form perbenzoic acid. The perbenzoic acid then reacts with a second molecule of benzaldehyde to form two molecules of benzoic acid. Outline a mechanism for the second step. Hint: recall the Baeyer-Villager oxidation of ketones to esters (15.3, 34).

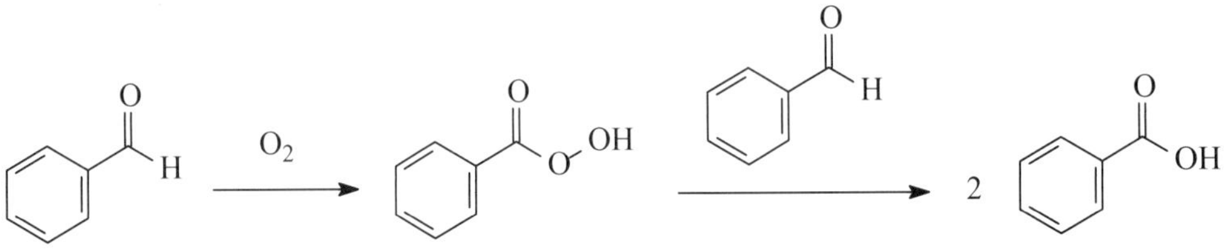

49. An impressive biomimetic conversion in Johnson's (*Stanford*) total synthesis of *progesterone* (see 19.3, 16 for the final stage):

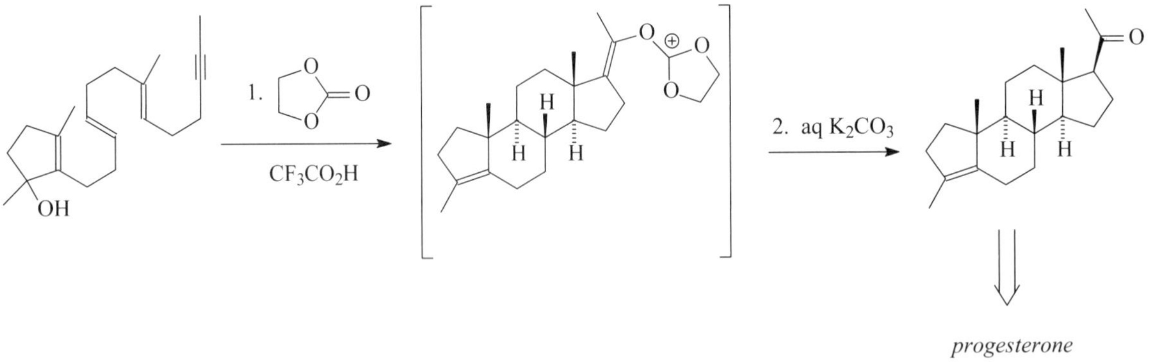

15.3 Mechanisms

CHAPTER 16
CARBOXYLIC ACIDS

16.1 Reactions
Draw the structural formula of the major organic product(s). Show stereochemistry where appropriate.

1. 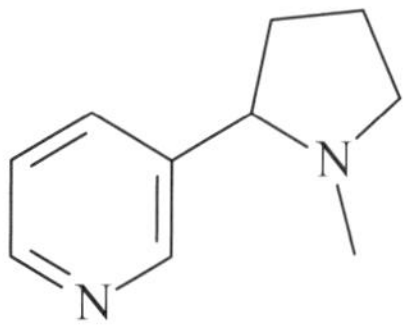 nicotine + KMnO$_4$, H$^+$ → niacin

2. phenylacetic acid
   1. NaOH
   2. Me$_3$O$^+$ BF$_4^-$

3. 3-oxobutanoic acid
   1. NaBH$_4$
   2. H$^+$

4. 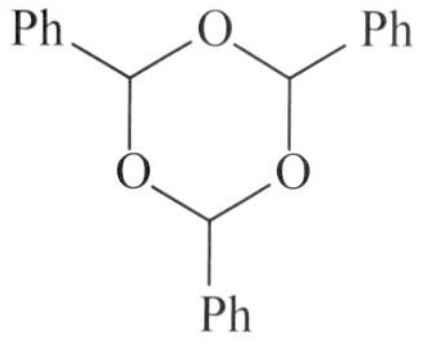 [cyclic triacetal with three Ph groups]
   1. H$_3$O$^+$
   2. CrO$_3$, H$^+$

5. γ-bromobutyric acid
   1. NaOH (1 equiv)
   2. Δ
   3. LiAlH$_4$
   4. H$_3$O$^+$

6. 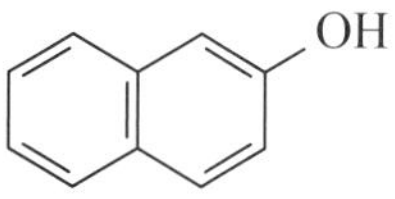 2-naphthol
   1. KOH
   2. acrylic acid (propenoic acid)
   3. BH$_3$
   4. H$_3$O$^+$

7. benzyl chloride
   1. NaCN
   2. PhMgCl
   3. H$_3$O$^+$

8. 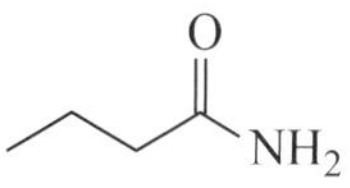 butanamide
   1. SOCl$_2$
   2. DIBAH, -78°
   3. H$_3$O$^+$

168 • Chapter 16 Carboxylic Acids

9. 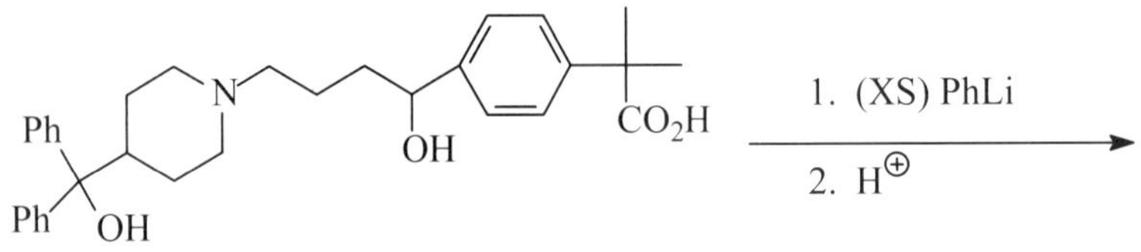 fexofenadine (Allegra™ - antihistaminic)

1. (XS) PhLi
2. H$^\oplus$

10. 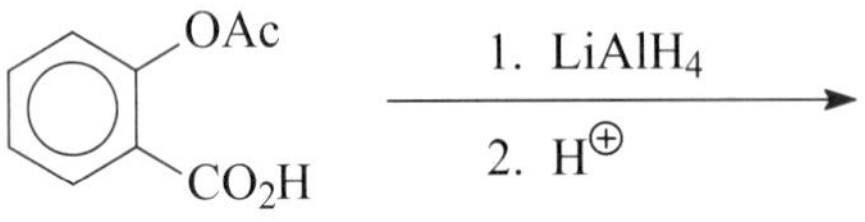 aspirin

1. LiAlH$_4$
2. H$^\oplus$

11. 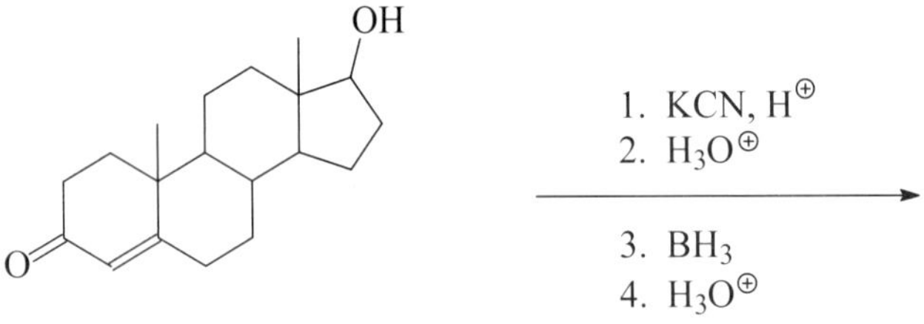 testosterone

1. KCN, H$^\oplus$
2. H$_3$O$^\oplus$
3. BH$_3$
4. H$_3$O$^\oplus$

12.

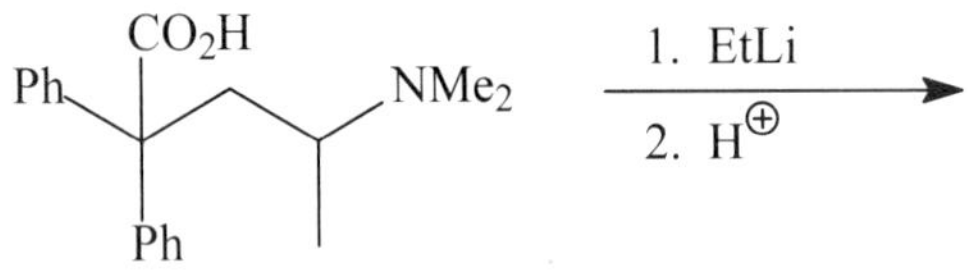

1. EtLi
2. H$^\oplus$

methadone

13.

1. H$_3$O$^\oplus$
2. PCC

[ ] ⟶ CO$_2$ +

14. A reaction in the biosynthesis of the amino acid *leucine*:

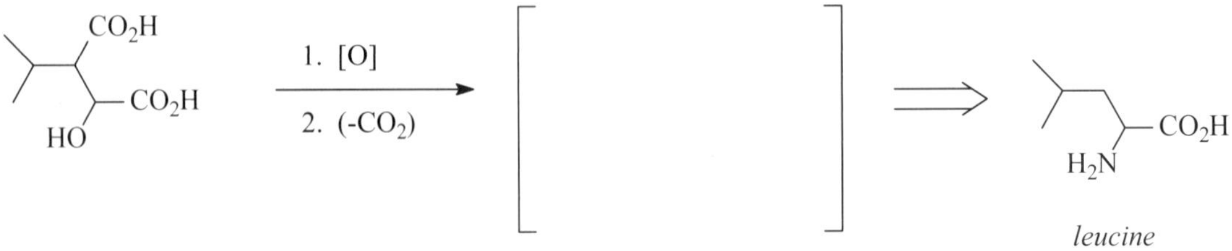

1. [O]
2. (-CO$_2$)

[ ] ⟹ leucine

16.1 Reactions

15. The alkaloid *cocaine*, isolable from coca leaves, can be converted to *tropinone*, a precursor to the antispasmodic *atropine* (see 20.3, 12). Deduce the structure of tropinone.

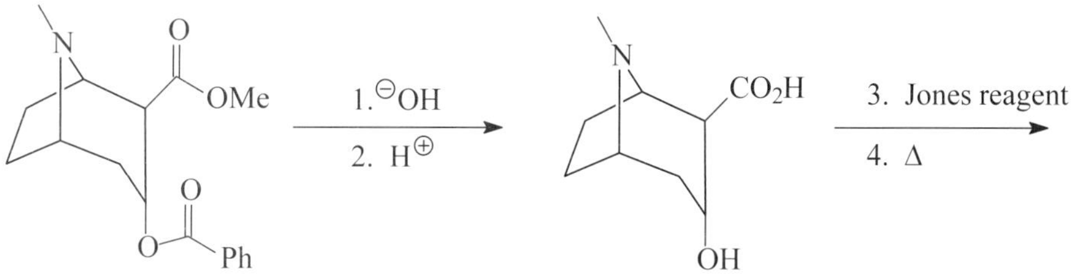

cocaine → (1. $^{\ominus}$OH; 2. H$^{\oplus}$) → *ecgonine* → (3. Jones reagent; 4. Δ) → *tropinone*

16. salicylic acid (2-hydroxybenzoic acid) → 1. NaOH (2 equiv); 2. MeI (1 equiv); 3. H$^{\oplus}$

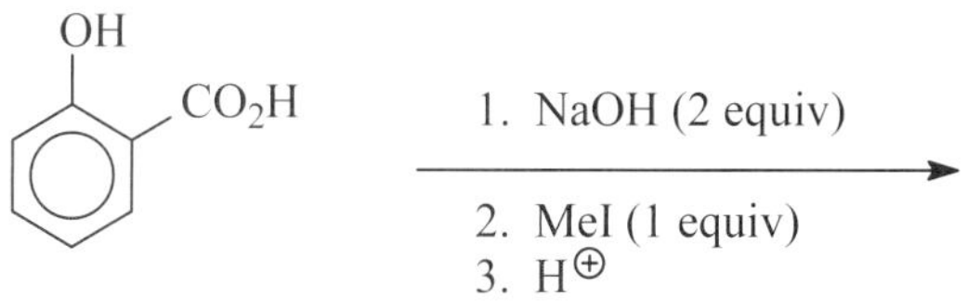

17. Chemical structures for medicinals that contain acid-base components are routinely drawn incorrectly in prescription information supplied by drug companies. For example, *sumatriptan succinate*, an active ingredient of Treximet^{TM} (prescribed for migraines) is drawn as shown below. Draw its *correct* structure.

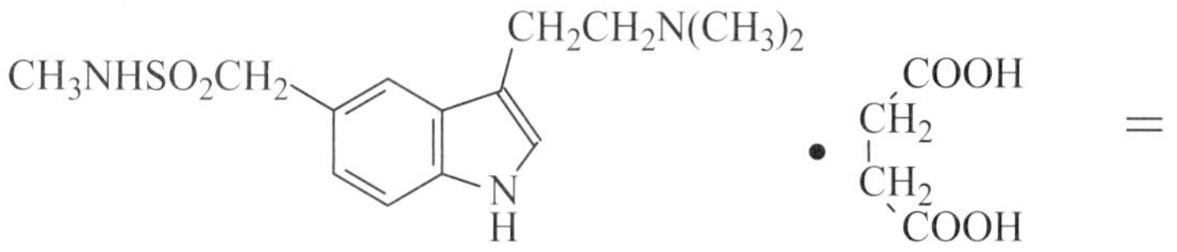

16.2 Syntheses

Supply a reagent or sequence of reagents that will effect the following conversions.

1. benzoic acid ⟶ PhCH$_2$CO$_2$H

2. propylene ⟶ pentanedioic acid

Chapter 16 Carboxylic Acids

3.  CH₃-CH=CH-CO-OH $\longrightarrow$ acetone + CO_2

4. *o*-chloroacetophenone $\longrightarrow$ 2-acetylbenzoic acid (aryl with COCH₃ and CO_2H ortho)

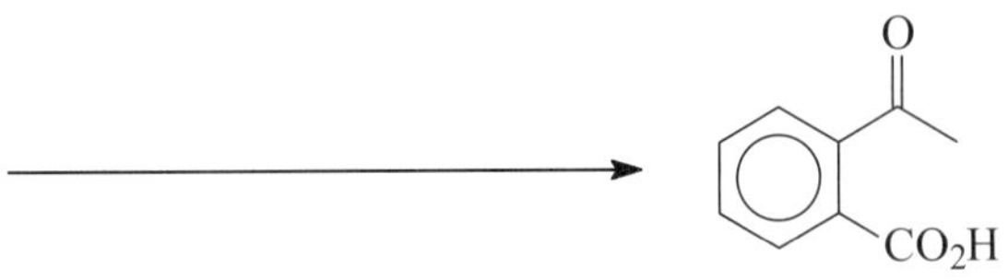

5.  styrene $\longrightarrow$ *ibuprofen* (Motrin™ - antipyretic)

6. 1-butanol $\longrightarrow$ *sodium valproate* (used in the treatment of epilepsy) [2-propylpentanoate, CO_2Na]

7. 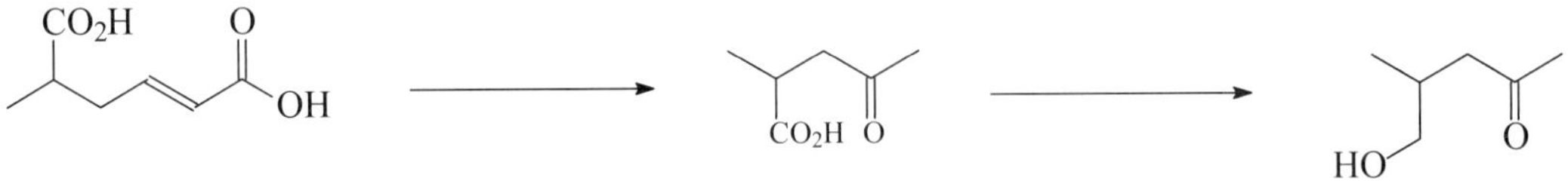 (CO₂H)CH(CH₃)CH₂CH=CHCO-OH $\longrightarrow$ (CH₃)CH(CO₂H)CH₂C(O)CH₃ $\longrightarrow$ (CH₃)CH(CH₂OH)CH₂C(O)CH₃

8. benzyl bromide $\xrightarrow{\textit{via a nitrile}}$ PhCH₂CHO

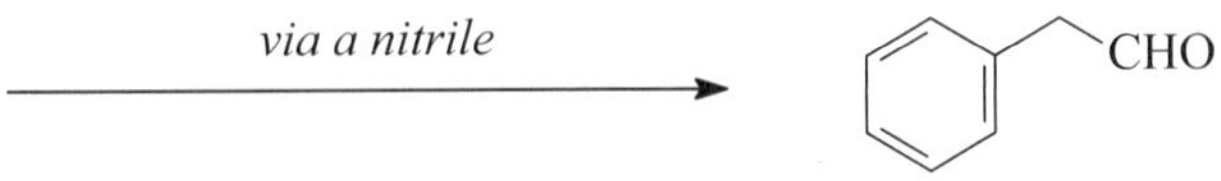

16.2 Syntheses

9.

10. 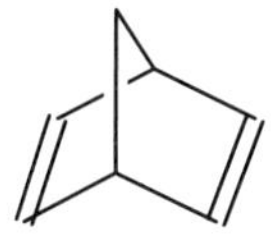 pentanedioic acid

11. RCO₂H → RCH₂C(O)R'

    (11. $RCO_2H \longrightarrow RCH_2\underset{\underset{}{}}{\overset{O}{\text{C}}}R'$)

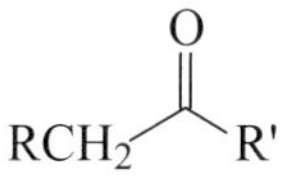

12. 3-oxobutanoic acid → CH₃C(O)CH₂CH(OH)Ph

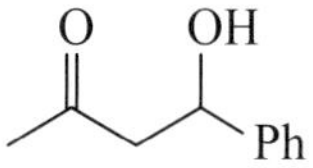

13. ethanol → butanedioic acid (*succinic acid*)

14.  benzene → benzene-1,3-dicarboxylic acid (CO₂H at 1,3 positions)

16.2 Syntheses

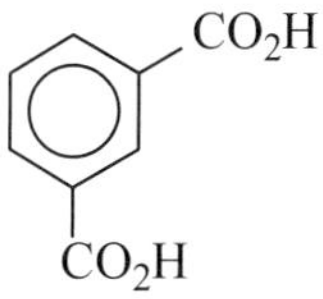

15. acetylene ⟶ hexanoic acid

16.  [structure: 2-(2-hydroxyphenyl)ethanol] ⟶ [structure: 2-(2-ethoxyethyl)phenol]

16.3 Mechanisms

Outline a detailed mechanism for each of the following. No other reagents than those given are necessary. Use arrows to explain the flow of electrons and show all intermediates.

1. 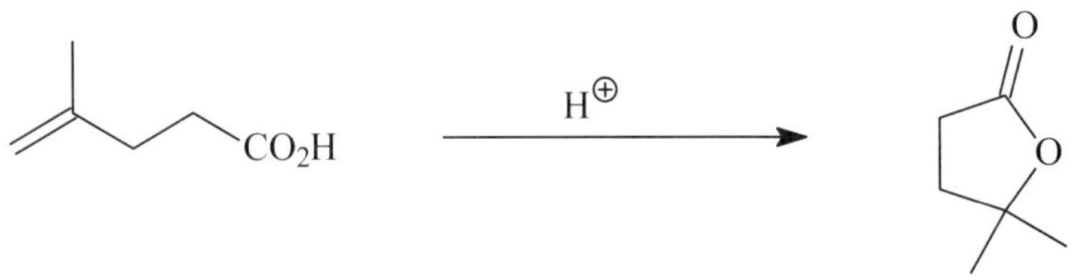 [4-methyl-4-pentenoic acid] —H⁺→ [γ,γ-dimethyl-γ-butyrolactone]

2. 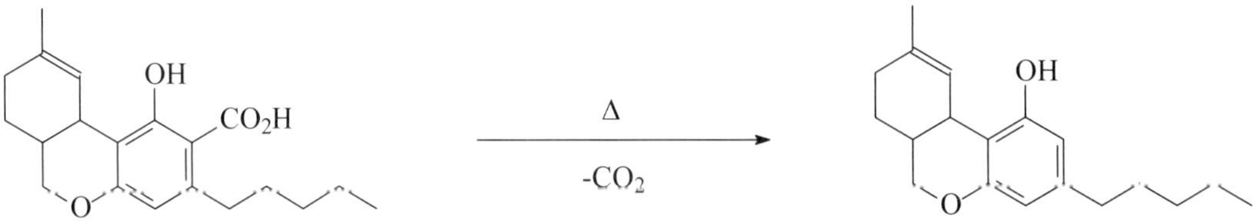 tetrahydrocannabinolic acid —Δ, −CO$_2$→ THC

3. 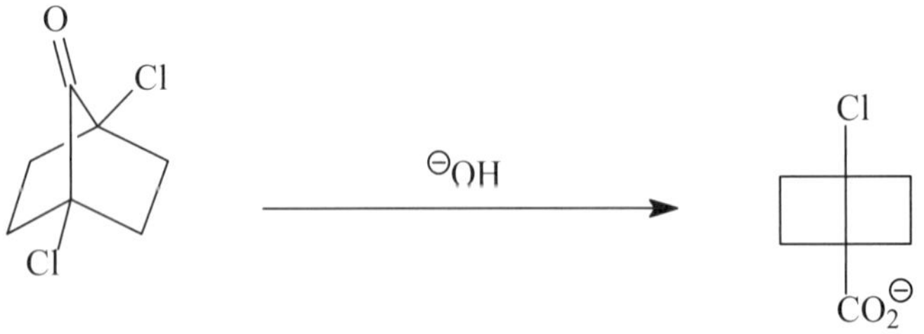 [1,4-dichlorobicyclic ketone] —⁻OH→ [1-chloro-3-chloro-cyclobutane carboxylate]

16.3 Mechanisms

4. Isobutylene and carbon monoxide, in the presence of acid, give dimethylpropionic acid. Explain.

5. The carboxyl group may be protected by allowing it to react with 2-amino-2-methylpropanol to form an oxazoline derivative. Outline the mechanism. (Acid hydrolysis of the oxazoline regenerates the carboxylic acid.)

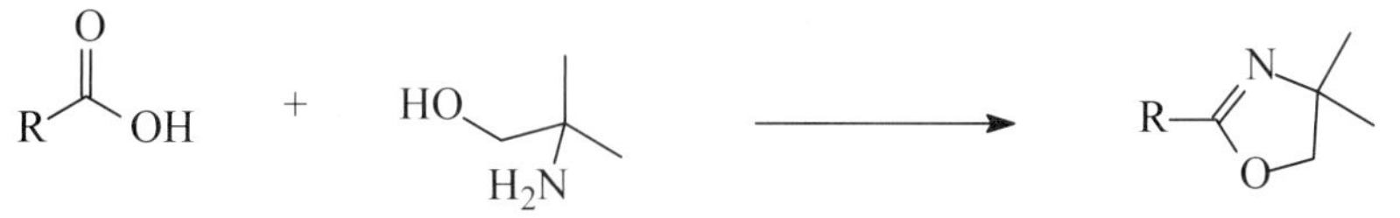

R-C(=O)-OH + HO-CH2-C(CH3)2-NH2 ⟶ R-(oxazoline)

(an *oxazoline*)

6. 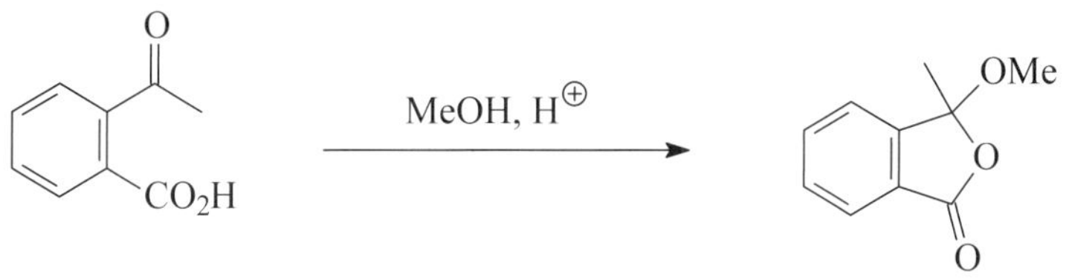 [o-acetylbenzoic acid] + MeOH, H⊕ ⟶ [methoxy phthalide product]

7. The aldehyde flavorings formed in the roasting of cocoa beans is caused by the *Strecker degradation* of amino acids:

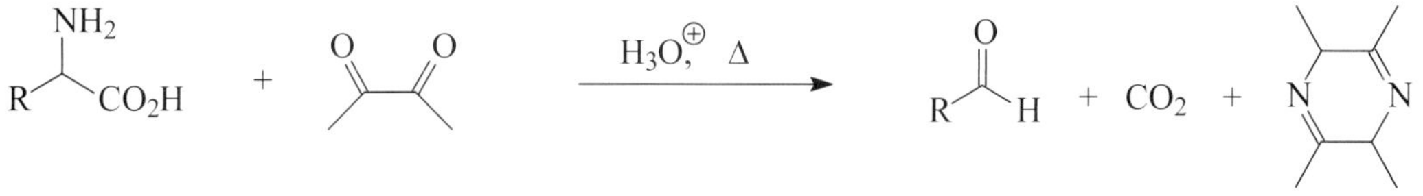

R-CH(NH2)-CO2H + [diketone] →(H3O⊕, Δ) R-CHO + CO2 + [pyrazine]

8. Strecker also developed a synthesis of amino acids:

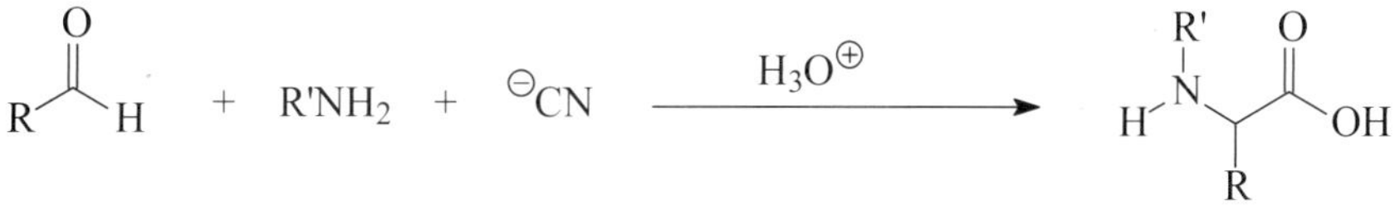

R-CHO + R'NH2 + ⊖CN →(H3O⊕) R'-NH-CHR-CO2H

16.3 Mechanisms

174 • Chapter 16 Carboxylic Acids

9. The biosynthesis of the amino acid phenylalanine involves an acid-catalyzed decarboxylation of *prephenic acid*:

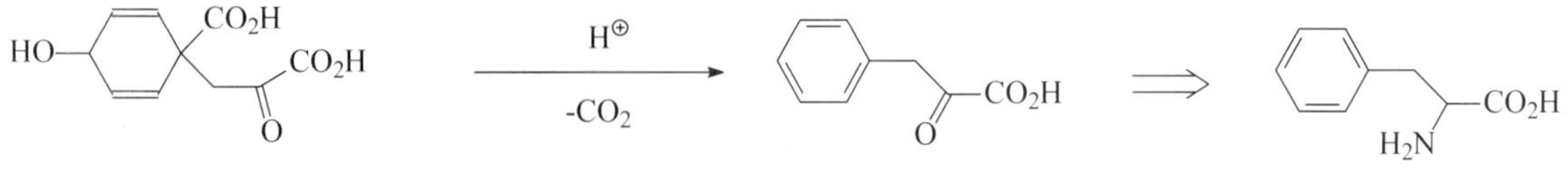

prephenic acid → (H⊕, -CO_2) → phenylacetic acid derivative ⟹ phenylalanine

10. *Ninhydrin* reacts with amino acids to give a blue dye which can be colorimetrically assayed. Sketch the intermediates.

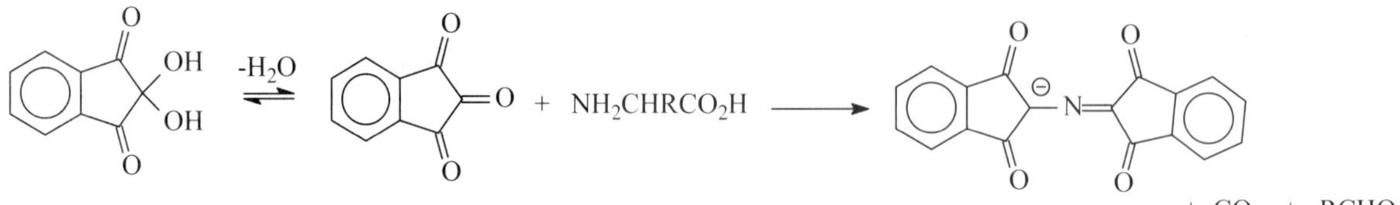

ninhydrin ⇌ (-H_2O) triketone + NH_2CHRCO_2H → a blue dye + CO_2 + RCHO

11. The vitamin *niacin* is used to form nicotinamide adenosine dinucleotide, which readily shuttles between its oxidized (NAD^+) and reduced (NADH) forms. The latter serves as a cellular equivalent to $NaBH_4$. The essential portions of the structures are shown below. Outline a mechanism for the cellular conversion of pyruvate to lactate. (Note: like $NaBH_4$, NADH cannot reduce carboxylic acid carbonyls.)

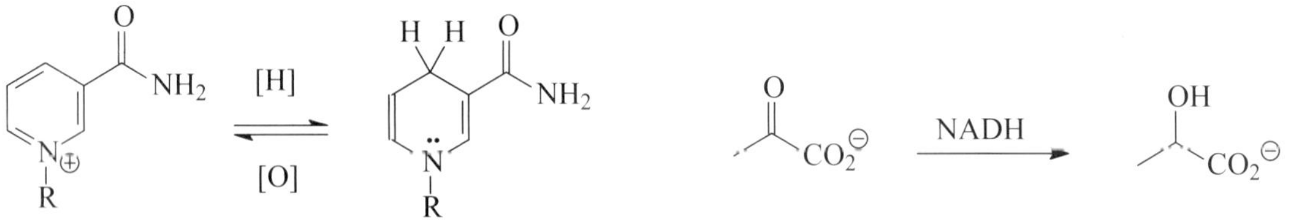

NAD^+ ⇌ ([H]/[O]) NADH

pyruvate → (NADH) → lactate

12. The cellular biosynthesis of glucose (*gluconeogenesis*) begins with the conversion of oxaloacetate (OAA) to phosphoenolpyruvate (PEP!) via a decarboxylation-phosphorylation pathway. Provide arrows.

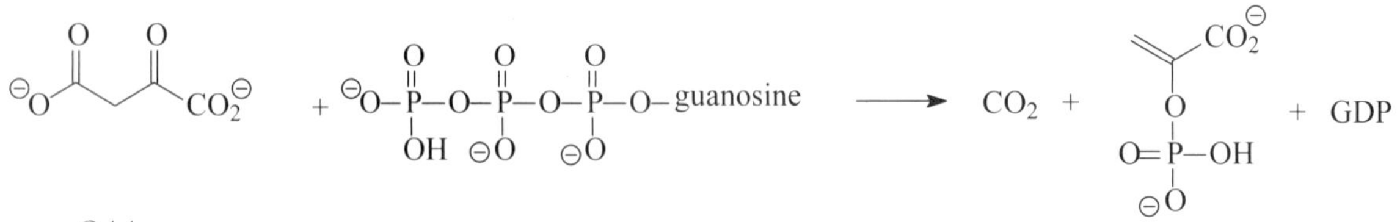

OAA + GTP → CO_2 + PEP + GDP

16.3 Mechanisms

13. Unlike β-ketocarboxylic acids, α-ketocarboxylic acids do NOT undergo mild thermal decarboxylation. However, the enzyme pyruvate decarboxylase (PDC) gently converts pyruvate to acetaldehyde at 37°. The key is provided by an essential cofactor, a derivative of *vitamin B1* (thiamine). The activity of thiamine resides in the *thiazolium* ring, shown below. A mechanistic clue was offered by Breslow's (*Columbia*) discovery that H_a rapidly undergoes exchange with deuterium when thiamine is dissolved in D_2O, suggesting that H_a is relatively acidic. Propose a mechanism for thiamine-assisted decarboxylation of α-ketocarboxylic acids. (Hint: begin with the conjugate base of thiamine, then consider how the thiazolium nitrogen can serve as an 'electron sink' to accept the electrons from decarboxylation.)

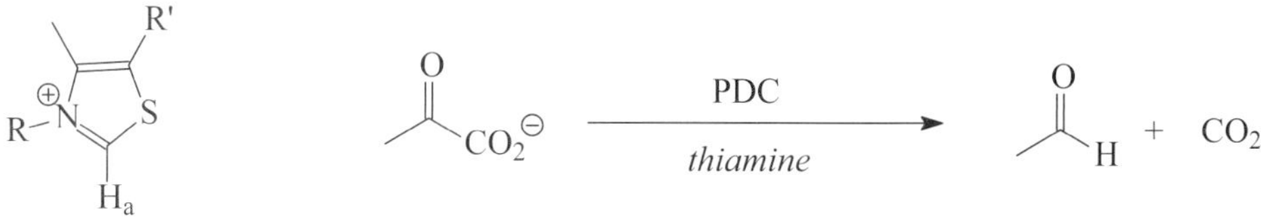

thiazolium ring

pyruvate

14. Another biochemical approach to decarboxylation:
Vitamin B6 (pyridoxine) is a precursor to the coenzyme PLP (pyridoxal phosphate), a catalyst for many reactions, such as decarboxylations, that involve amino acids. Outline a mechanism. (Hint: form an imine from PLP and the amino acid, then consider the role of the pyridinium nitrogen as an 'electron sink.')

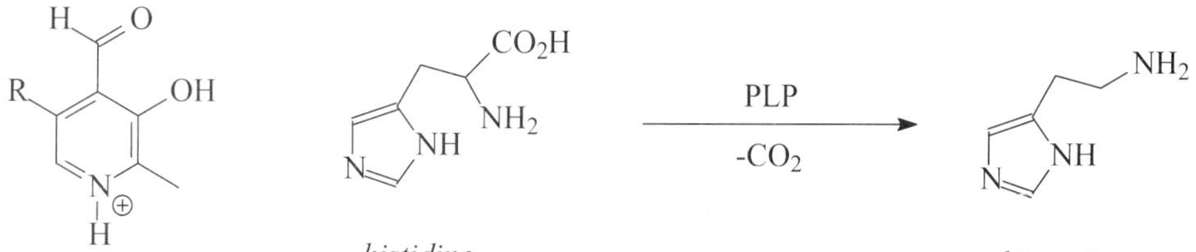

PLP

histidine

histamine

15. Determination of the molecular mass of acetic acid in a nonpolar solvent, *e.g.*, hexane, yields a value of 120. Explain.

16.3 Mechanisms

CHAPTER 17
CARBOXYLIC ACID DERIVATIVES

17.1 Reactions
Draw the structural formula of the major organic product(s). Show stereochemistry where appropriate.

1. 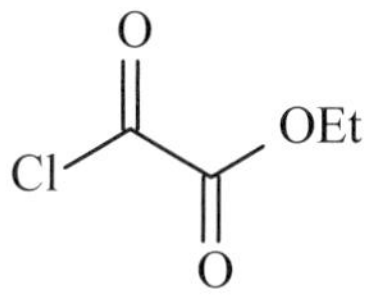 Cl-C(=O)-C(=O)-OEt → Et$_2$NH (1 equiv) / pyridine

2. butyric anhydride + methylamine ⟶

3. cyclopropyl cyclohexanecarboxylate → EtOH (XS), H$^⊕$

4. propane-1,3-diol → phosgene

5. oxalic acid
   1. PCl$_3$
   2. LiAlH(O-*t*-Bu)$_3$
   3. H$^⊕$

6. N,N-diisopropylpropionamide → H$_3$O$^⊕$

7. phenyl hexanoate
   1. isopropyl magnesium bromide
   2. H$_3$O$^⊕$

8. 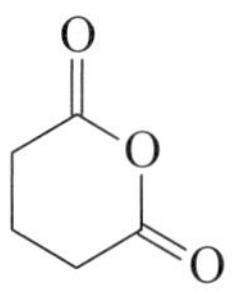 (glutaric anhydride) → MeOH, H$^⊕$

9. 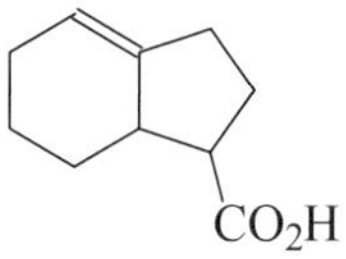 (bicyclic alkene with CO$_2$H)
   1. LiAlH$_4$
   2. H$^⊕$
   3. Ac$_2$O

17.1 Reactions

178 • Chapter 17 Carboxylic Acid Derivatives

10. ε-caprolactone $\xrightarrow{\text{saponification}}$

11. penicillin G $\xrightarrow{H_2O,\ ^{\ominus}OH}$

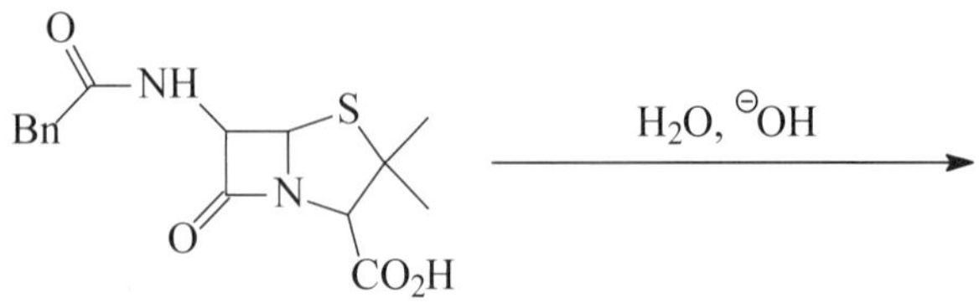

12.

$$\underset{\text{Et}}{\overset{\text{Et}}{>}}C(CO_2R)_2 \ +\ \text{urea} \longrightarrow$$

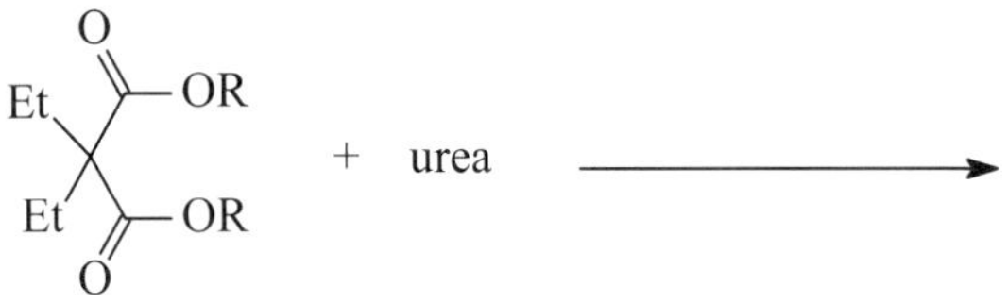

Veronal™ (a barbiturate, sedative)

13. cocaine $\xrightarrow{\begin{array}{l}1.\ ^{\ominus}OH\\ 2.\ H^{\oplus}\\ 3.\ \text{Jones reagent}\\ 4.\ \Delta\end{array}}$

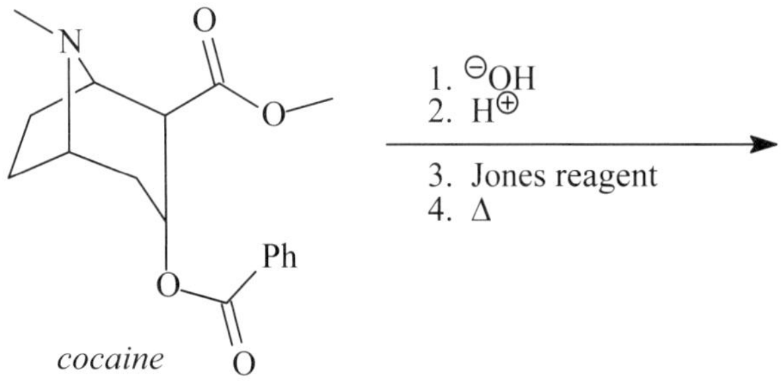

14. cholyl coenzyme A (a rare example of cis-fused A-B rings in steroids) $\xrightarrow{NH_2CH_2CO_2H\ (\textit{glycine})}$ glycocholate (a major bile salt)

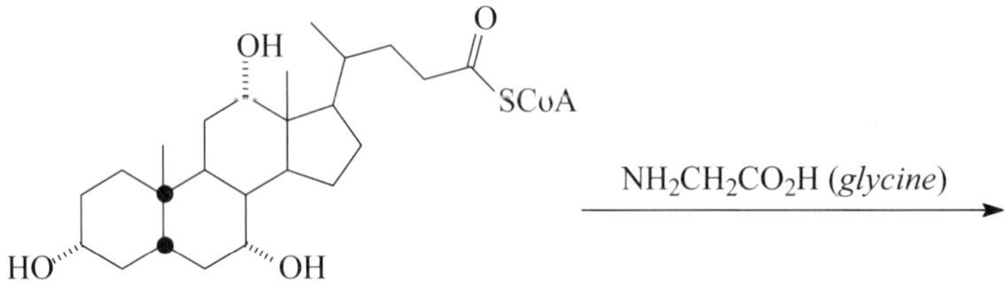

15. strychnine $\xrightarrow{\begin{array}{l}1.\ H_3O^{\oplus}\\ 2.\ CH_2N_2\end{array}}$

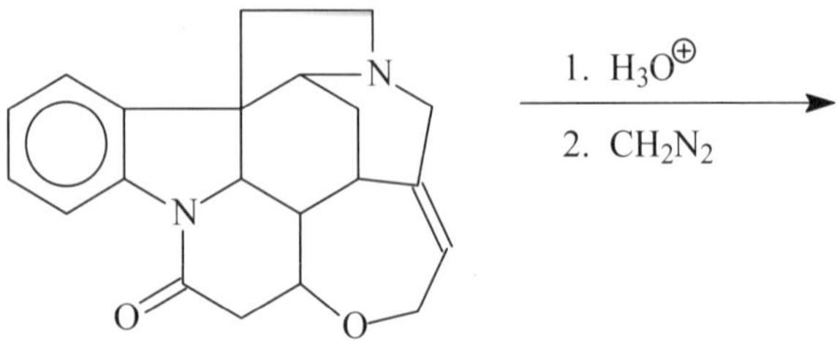

17.1 Reactions

16.

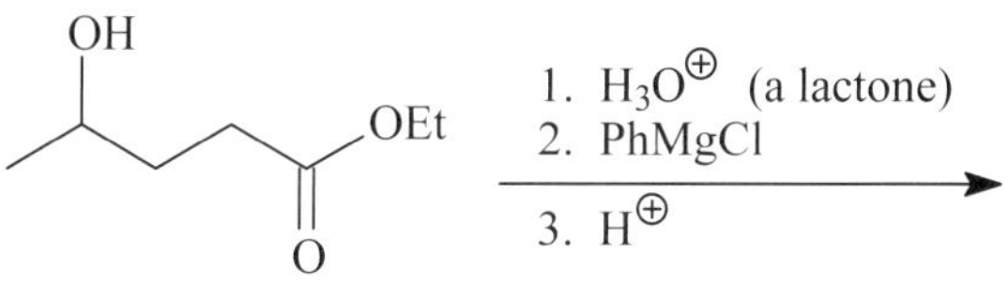

$\underset{\text{OH}}{\text{CH}_3\text{CH(OH)CH}_2\text{CH}_2\text{CO}_2\text{Et}}$ $\xrightarrow[\text{3. H}^\oplus]{\text{1. H}_3\text{O}^\oplus \text{ (a lactone)} \\ \text{2. PhMgCl}}$

17.

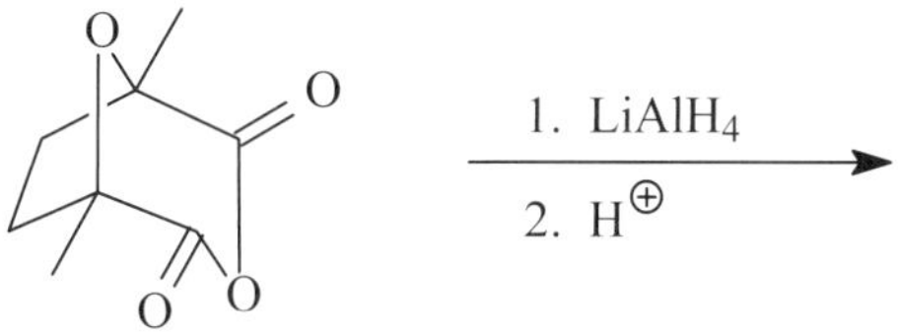

Spanish fly $\xrightarrow[\text{2. H}^\oplus]{\text{1. LiAlH}_4}$

18.

$\text{CH}_3\text{NH-CH}_2\text{CH}_2\text{-NHCH}_3$ $\xrightarrow[\text{3. H}^\oplus]{\text{1. phosgene} \\ \text{2. LiAlH}_4}$

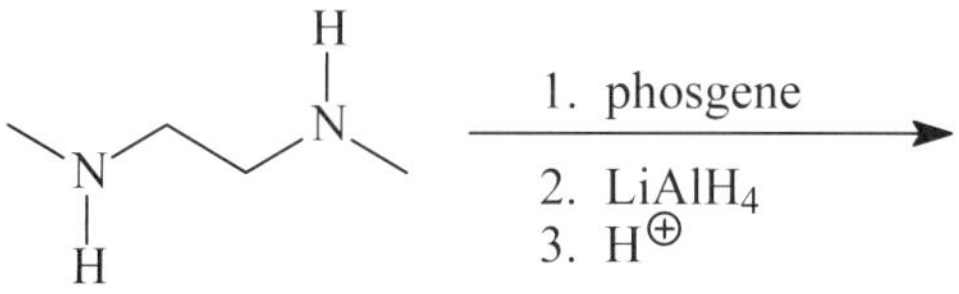

19.

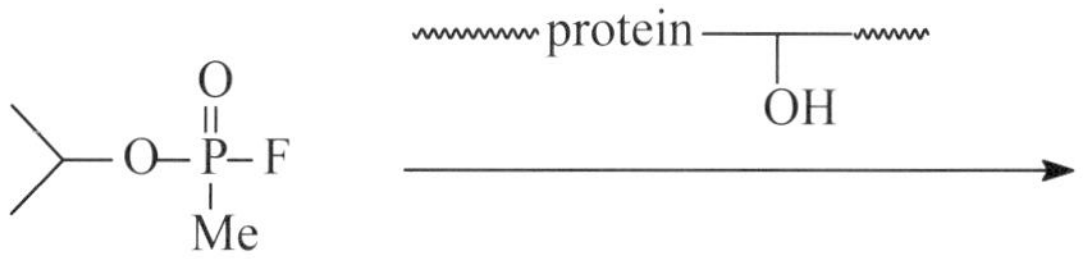

Sarin™ (a cholinesterase inhibitor) $\xrightarrow{\text{protein—OH}}$

20.

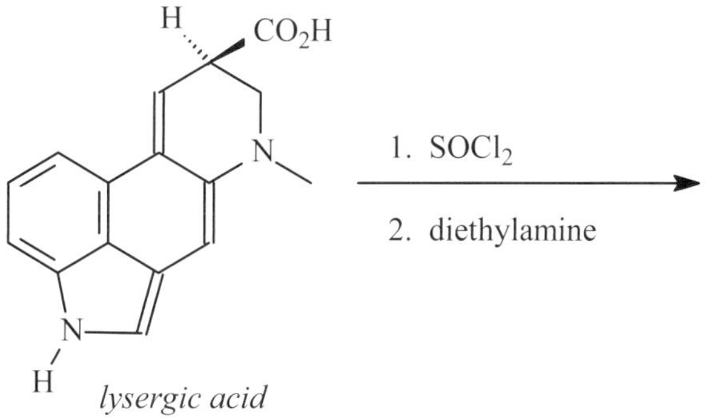

lysergic acid $\xrightarrow[\text{2. diethylamine}]{\text{1. SOCl}_2}$ *LSD*

21.

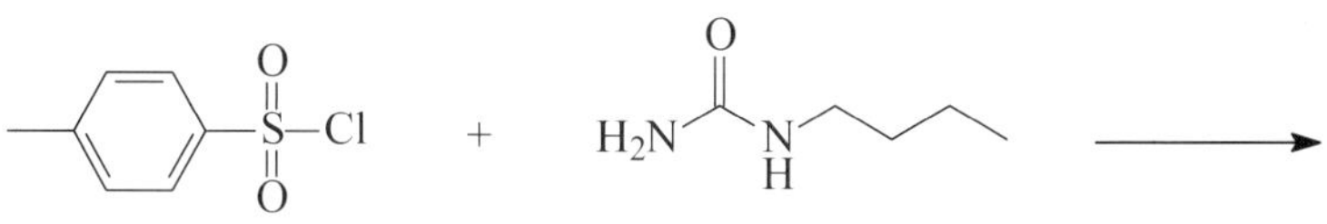

$\text{p-CH}_3\text{-C}_6\text{H}_4\text{-SO}_2\text{Cl}$ + $\text{H}_2\text{N-C(=O)-NH-C}_4\text{H}_9}$ $\longrightarrow$ Orinase™ (for diabetes)

22.

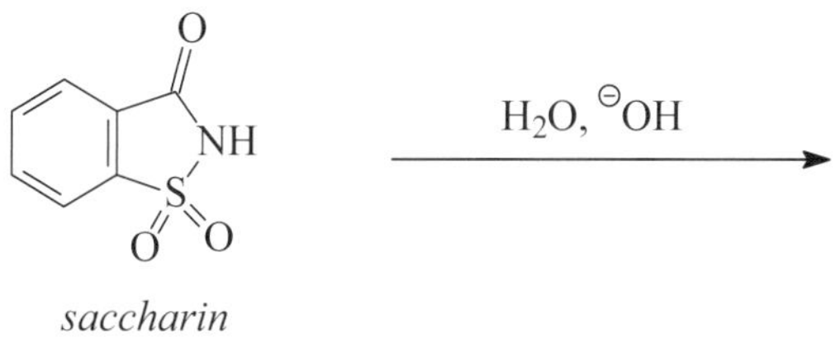

saccharin $\xrightarrow{\text{H}_2\text{O, }^\ominus\text{OH}}$

17.1 Reactions

180 • Chapter 17 Carboxylic Acid Derivatives

23. p-hydroxyaniline $\xrightarrow{\text{acetic anhydride}}$ acetaminophen (Tylenol™ - antipyretic)
    (1 equiv)

24. acetic acid $\xrightarrow[\text{(-H}_2\text{O)}]{\text{Al}_2\text{O}_3, \Delta}$ H$_2$C=C=O $\xrightarrow{\text{aniline}}$

    ketene

25. $\left[\begin{array}{c} \text{MeO-C(=O)-C}_6\text{H}_4\text{-C(=O)-OMe} \end{array} \right]_n$ + [ethylene glycol]$_n$ $\xrightarrow{\text{H}^{\oplus}}$ Dacron™

    dimethyl phthalate

26. Lexan™, a high-molecular weight "polycarbonate," is manufactured by mixing *bisphenol A* (see 15.3, 18) with phosgene (COCl$_2$) in the presence of pyridine. Draw a partial structure for Lexan™.

    HO—C$_6$H$_4$—C(CH$_3$)$_2$—C$_6$H$_4$—OH

    bisphenol A

27. Me—N=C=O + [1-naphthol with OH] $\longrightarrow$

    methyl isocyanate

    (active ingredient in the insecticide Sevin™)

28. [cyclohexanone ethylene ketal with Cl substituent] $\xrightarrow{\begin{array}{l}\text{1. Li} \\ \text{2. CuI} \\ \text{3. benzoyl chloride} \\ \text{4. H}_3\text{O}^{\oplus}\end{array}}$

29. [structure: F$_3$C-C$_6$H$_4$-O-CH(Ph)-CH$_2$-CH$_2$-NHMe] $\xrightarrow{\begin{array}{l}\text{1. propionic anhydride} \\ \text{2. LiAlH}_4 \\ \text{3. H}^{\oplus}\end{array}}$

    fluoxetine (Prozac™ - antidepressant)

17.1 Reactions

30. 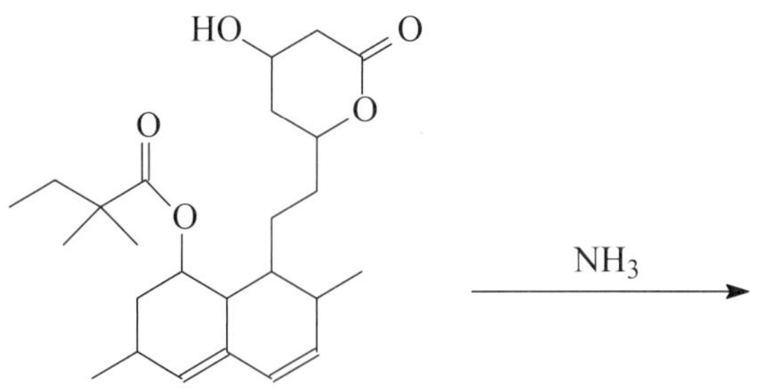 **simvastatin** (Zocor™ - antilipemic) $\xrightarrow{\text{NH}_3}$

31. 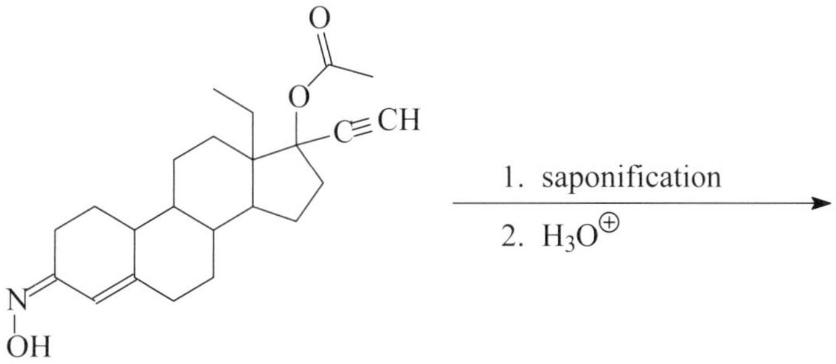 **Ortho TriCyclin™** (OCPs) $\xrightarrow[\text{2. H}_3\text{O}^\oplus]{\text{1. saponification}}$

32. 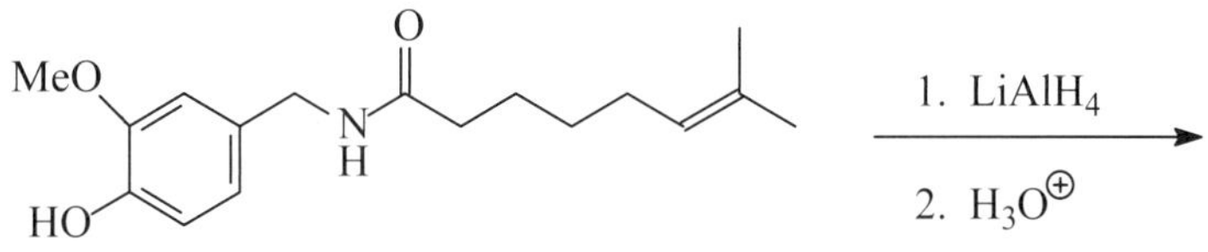 **capsaicin** (active agent in cayenne pepper) $\xrightarrow[\text{2. H}_3\text{O}^\oplus]{\text{1. LiAlH}_4}$

33. 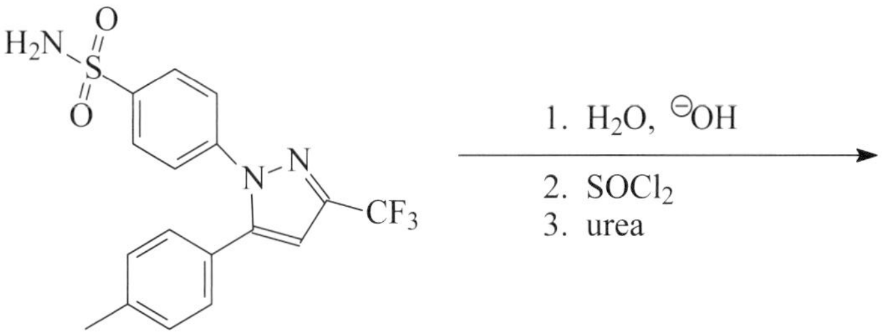 **celecoxib** (Celebrex™ - anti-inflammatory) $\xrightarrow[\substack{\text{2. SOCl}_2 \\ \text{3. urea}}]{\text{1. H}_2\text{O, }^\ominus\text{OH}}$

34. 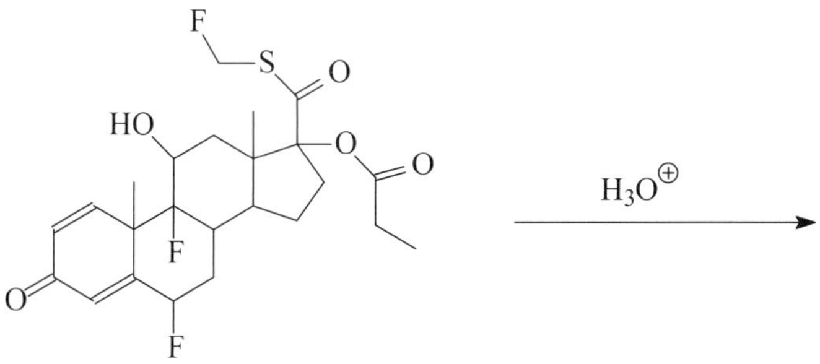 **fluticasone propionate** (Flonase™ - anti-inflammatory) $\xrightarrow{\text{H}_3\text{O}^\oplus}$

17.1 Reactions

182 • Chapter 17 Carboxylic Acid Derivatives

35. 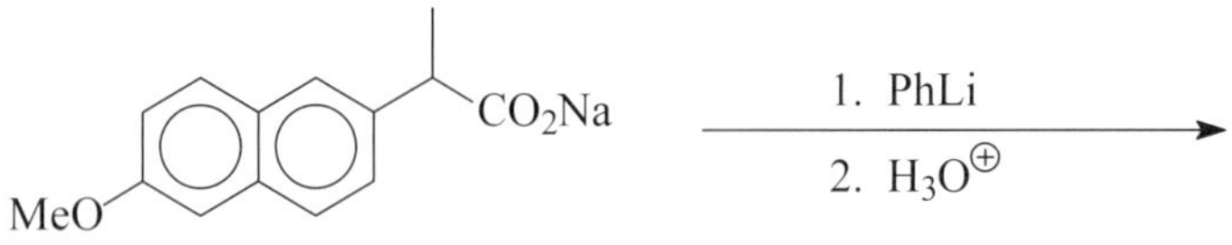 [structure of naproxen sodium] $\xrightarrow{\text{1. PhLi} \quad \text{2. H}_3\text{O}^\oplus}$

naproxen sodium (Aleve™ - anti-inflammatory)

36. 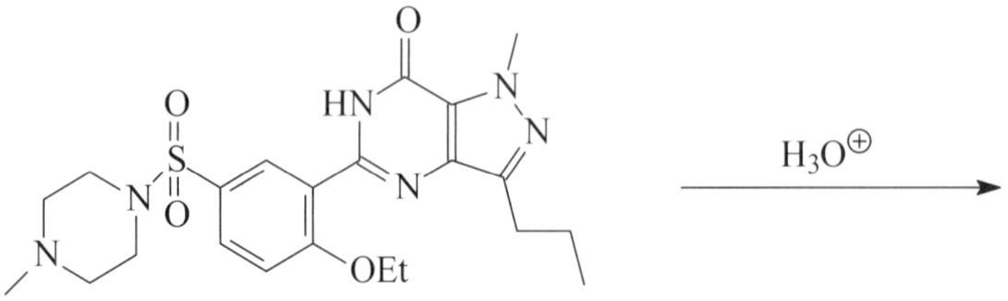 [structure of sildenafil] $\xrightarrow{\text{H}_3\text{O}^\oplus}$

sildenafil (Viagra™ - treatment of ED)

37. 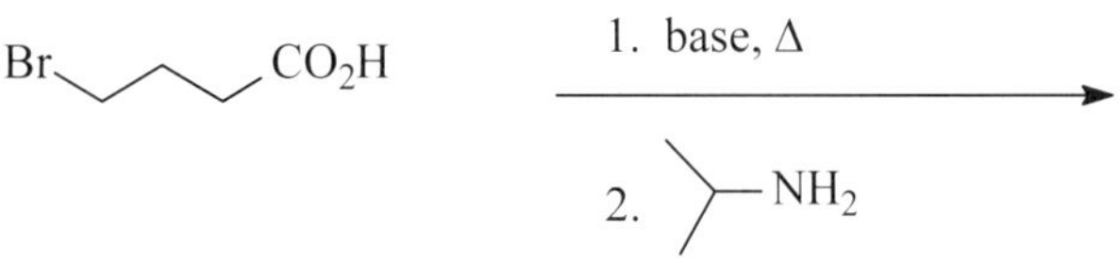 Br~~~CO₂H $\xrightarrow{\text{1. base, } \Delta \quad \text{2. } i\text{-Pr-NH}_2}$

38. n  [ε-caprolactam structure] $\xrightarrow{{}^\ominus\text{OH}}$ *Nylon 6* (a polyamide)

ε-caprolactam

39. 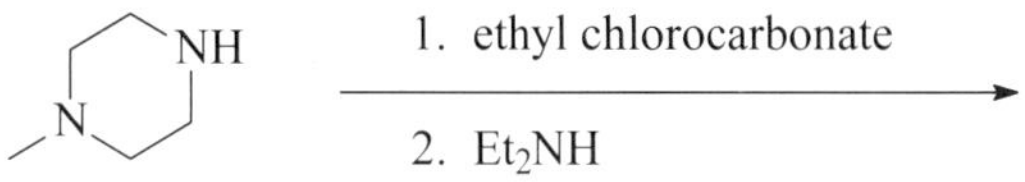 [N-methylpiperazine] $\xrightarrow{\text{1. ethyl chlorocarbonate} \quad \text{2. Et}_2\text{NH}}$

diethylcarbamazine (anthelmintic)

40. 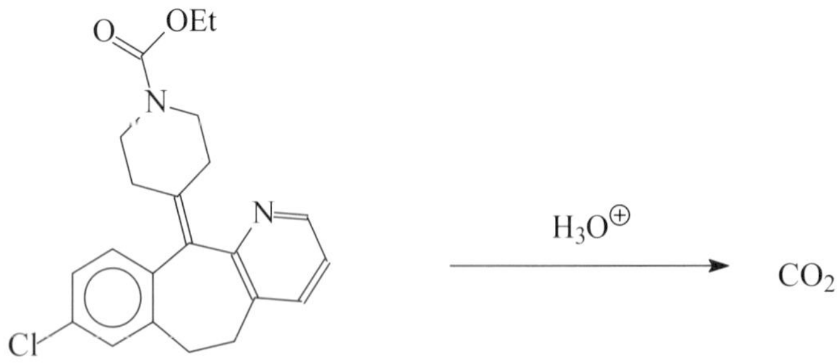 [structure of loratadine] $\xrightarrow{\text{H}_3\text{O}^\oplus}$ CO_2

loratadine (Claritin™ - antihistaminic)

17.1 Reactions

Problems • 183

41. 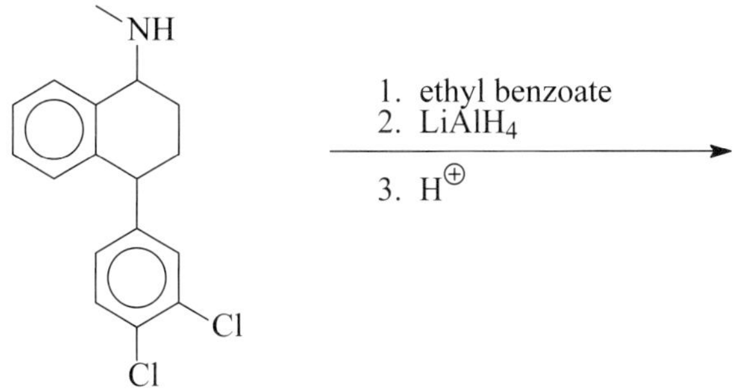 [structure]
  1. ethyl benzoate
  2. LiAlH₄
  3. H⁺

  sertraline (Zoloft™ - antidepressant)

42. 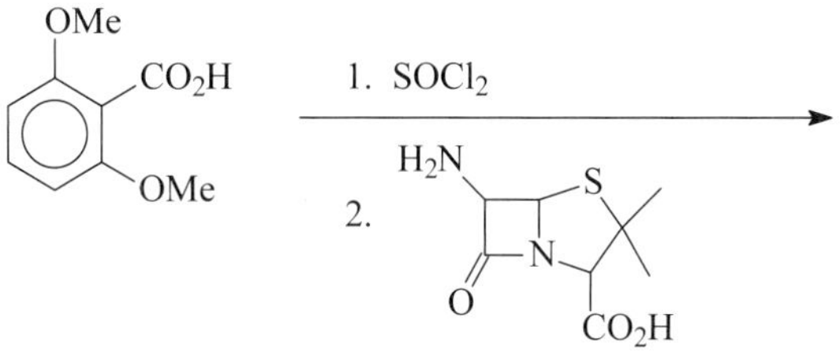 [structure]
  1. SOCl₂
  2. [structure]

  methicillin [an estimated 90,000 people in the US fall ill each year from MRSA (methicillin resistant *Staphylococcus aureus*)]

43. 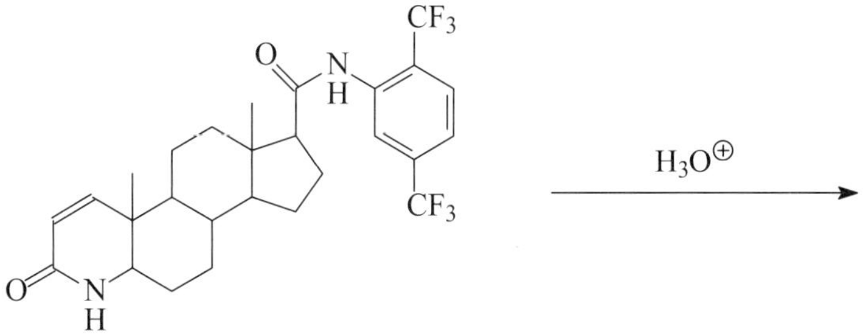 [structure] H₃O⁺

  dutasteride (Avodart™ - treatment of BPH)

44. 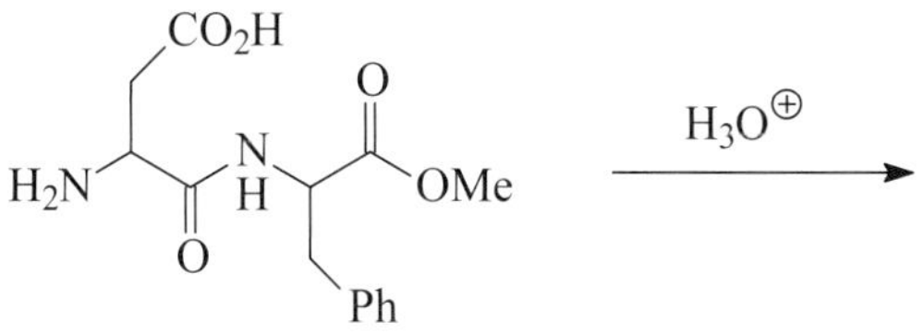 [structure] H₃O⁺

  aspartame

45. 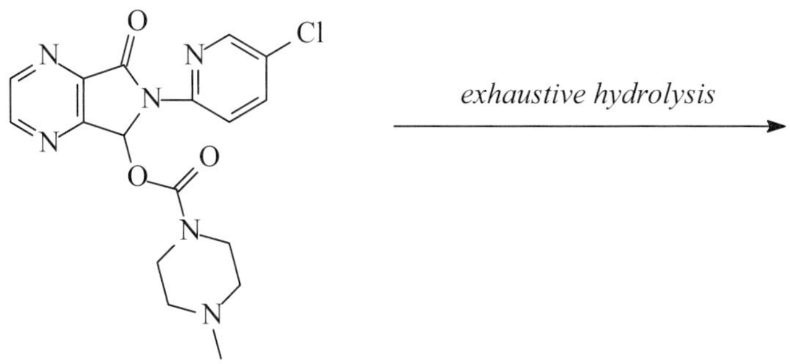 [structure] *exhaustive hydrolysis*

  zopiclone (Lunesta™ - sedative, hypnotic)

17.1 Reactions

184 • Chapter 17 Carboxylic Acid Derivatives

46.

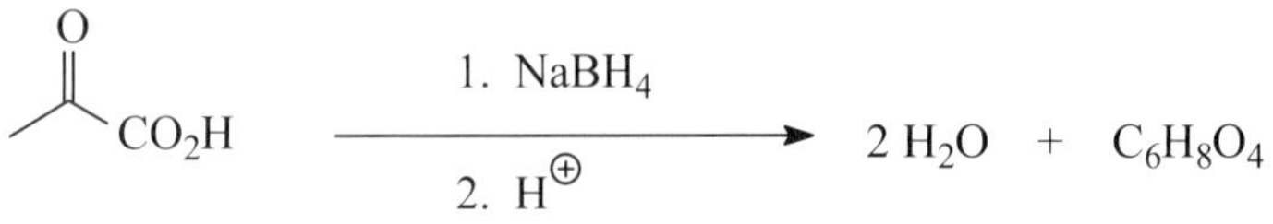

pyruvic acid (CH₃-CO-CO₂H) $\xrightarrow[\text{2. H}^\oplus]{\text{1. NaBH}_4}$ 2 H₂O + C₆H₈O₄

47. Chain *elongation* of a tetrose sugar:

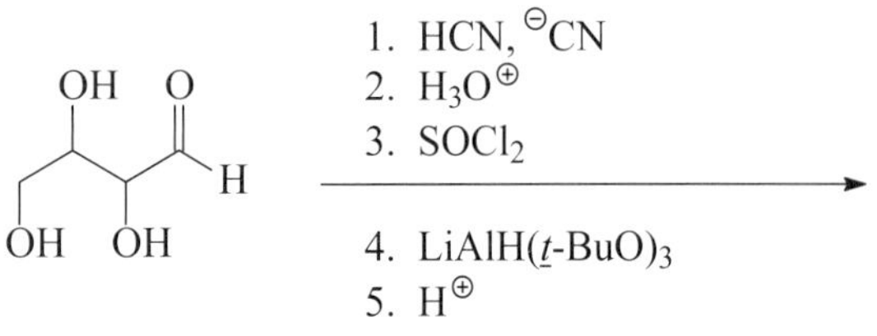

HOCH₂-CH(OH)-CH(OH)-CHO $\xrightarrow[\begin{array}{l}\text{4. LiAlH(}t\text{-BuO)}_3\\ \text{5. H}^\oplus\end{array}]{\begin{array}{l}\text{1. HCN, }^\ominus\text{CN}\\ \text{2. H}_3\text{O}^\oplus\\ \text{3. SOCl}_2\end{array}}$

(see chain *degradation* of a sugar, 15.1, 38)

48. Consider the reaction of amino acid **A** with amino acid **B**. Four possible products are possible: **A-A**, **B-B**, **A-B**, and **B-A**, if simply **A** and **B** are heated together. A more rational synthesis of, for example, **A-B** is to first treat **A** with *t*-butyl chlorocarbonate (**C**), which has the effect of eliminating (blocking) the nucleophilicity of the nitrogen in **A**. The blocked species is termed a *t*-BOC amino acid (*t*-butoxy-carbonyl).

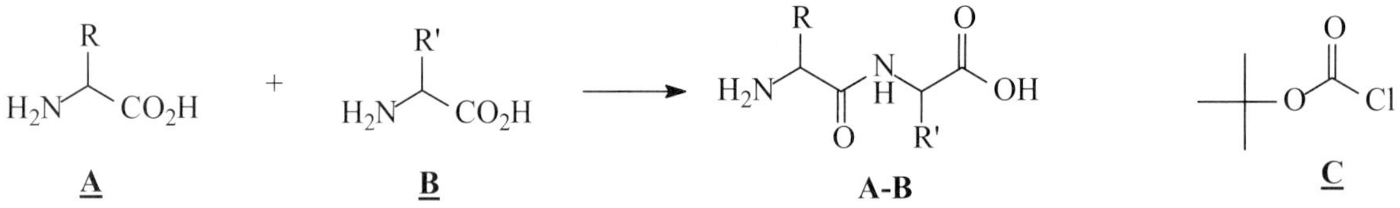

H₂N-CHR-CO₂H + H₂N-CHR'-CO₂H ⟶ H₂N-CHR-C(O)-NH-CHR'-CO₂H   (t-Bu)O-C(O)-Cl

A   **B**   **A-B**   **C**

a. Draw the product of the reaction of **A** with **C**.

b. The *t*-BOC-**A** is then condensed with **B** to yield a derivative of **A-B**. **A-B** is formed by treating that derivative with mild acid. Show the mechanism of removing the blocking group to form **A-B**. (Hint: CO₂ and isobutylene are by-products.)

17.1 Reactions

49. In order for most thioesters to traverse the inner mitochondrial membrane they must first undergo a transesterification with carnitine:

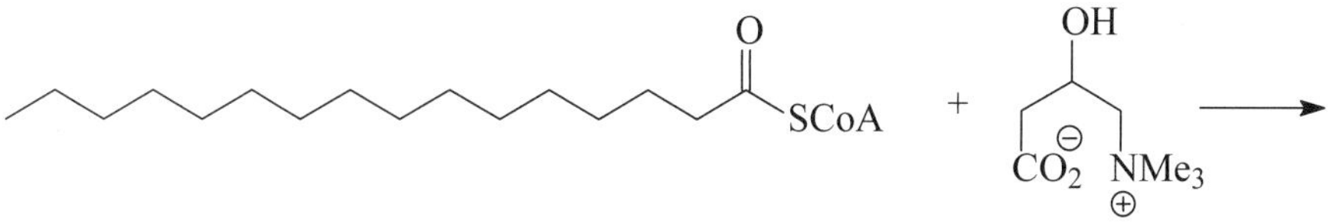

palmitoyl coenzyme A + *carnitine* ⟶

50. $CH_3(CH_2)_{14}CO_2H$ + *lanosterol* $\xrightarrow{H^{\oplus}}$ *lanolin* (a biochemical wax)

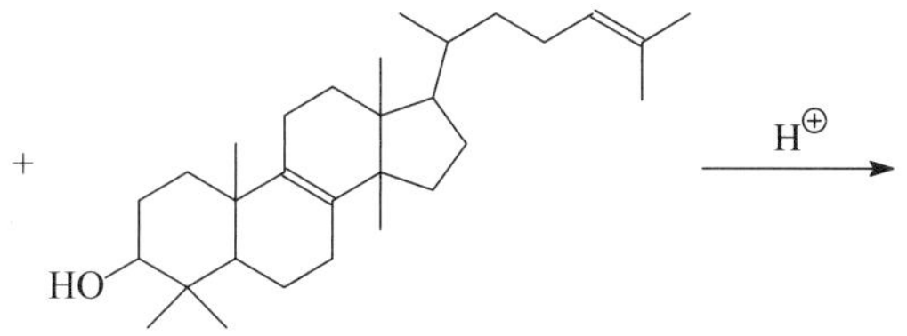

palmitic acid

51. The catabolism of pyrimidines, *e.g.*, *thymine*, proceeds by the following reactions:

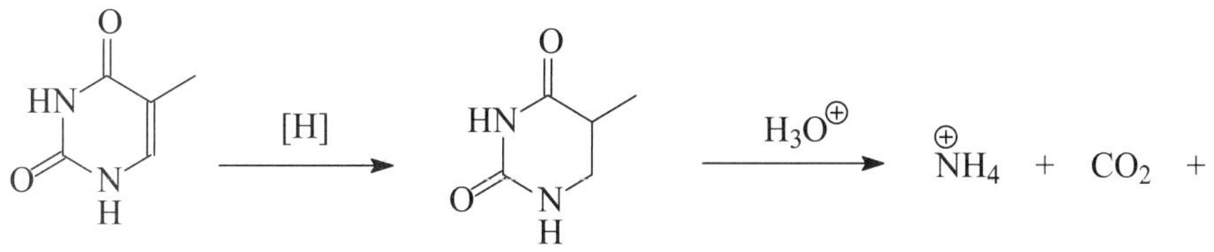

thymine $\xrightarrow{[H]}$ (dihydrothymine) $\xrightarrow{H_3O^{\oplus}}$ $NH_4^{\oplus}$ + CO_2 +

52. Conversion of carboxylic acids to acid halides does NOT occur in cells because the latter generally are too reactive in aqueous media. However, they are often activated to *mixed anhydrides* by a phosphoryation process:

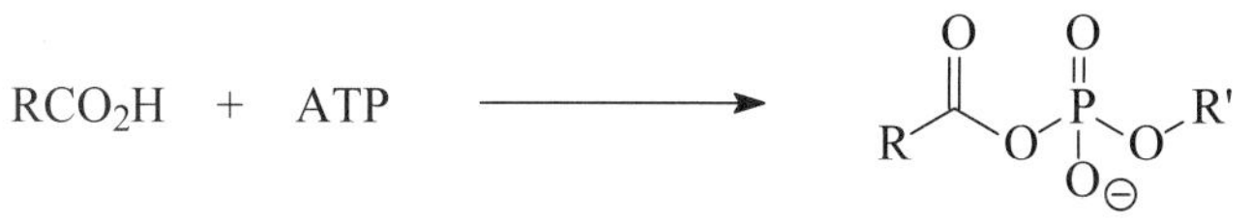

RCO_2H + ATP ⟶ *a mixed anhydride*

Biochemists characterize the mixed anhydride as having a "high acylating potential" – nucleophilic acyl substitution at the carbonyl carbon readily transfers the acyl moiety to the attacking nucleophile. Predict the products in the following metabolic reactions. (*Coenzyme A*, a derivative of the vitamin *pantothenic acid*, simply behaves as a thiol and is abbreviated HSCoA.)

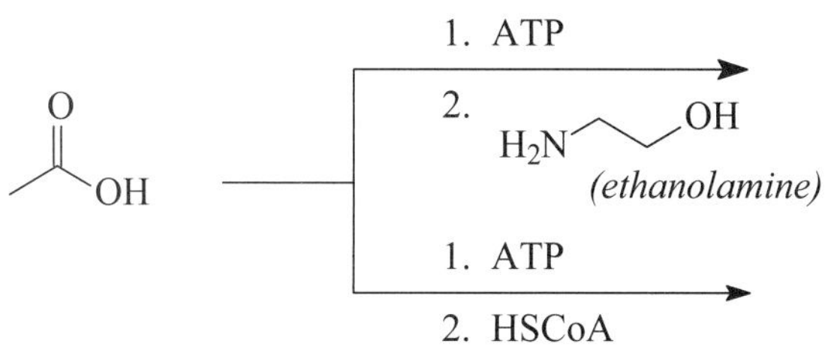

CH₃CO₂H
1. ATP
2. H_2N⌒OH (ethanolamine) ⟶ *acetyl choline*

1. ATP
2. HSCoA ⟶ *acetyl coenzyme A*

17.1 Reactions

53.  [structure of tuliposide: CH₂=C(CH₂CH₂OH)–C(=O)–O–glucose] →(H⊕) + glucose

tuliposide
(found in tulip bulbs)

tulipaline (a γ-lactone)
(produced when bulbs are damaged - a fungicide)

54. 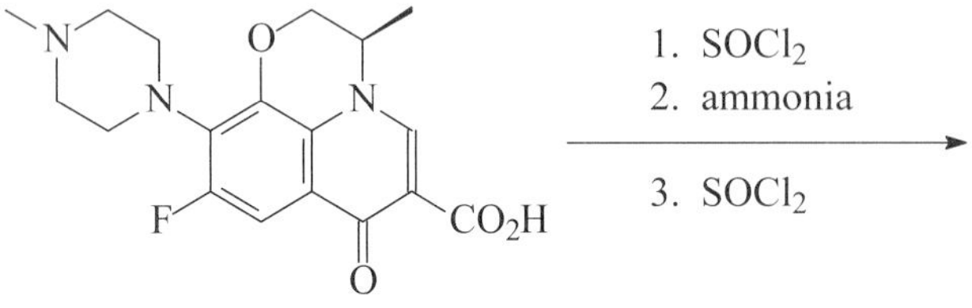 Levaquin™ (antibacterial) →
1. SOCl₂
2. ammonia
3. SOCl₂

55. 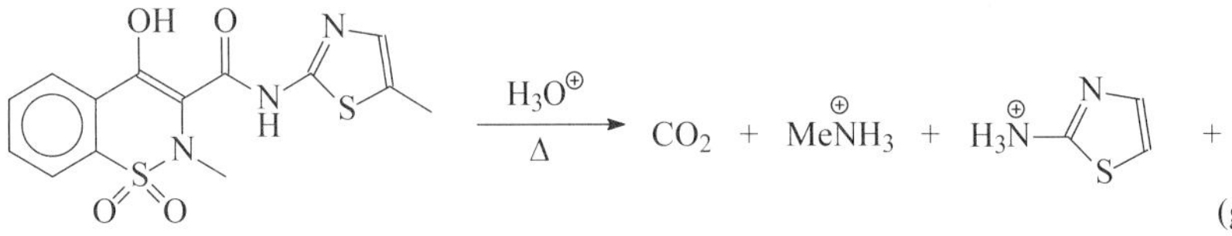 *meloxicam* (Metacam™ - anti-inflammatory) →(H₃O⊕, Δ) CO_2 + MeNH₃⁺ + H₃N⁺–(thiazole) + ??

(gives a positive Tollens' test)

17.2 Syntheses

Supply a reagent or sequence of reagents that will effect the following conversions.

1. *sec*-butyl acetate ⟶ methyl 2-methylbutanoate

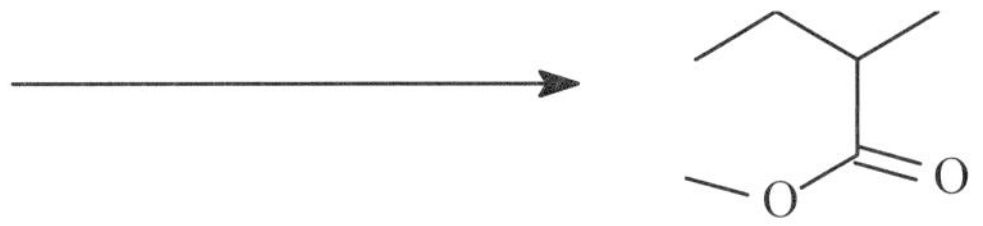

2. R–C(=O)–NH₂ ⟶ R–C(=O)–H

   R–C(=O)–NH₂ ⟶ R–C(=O)–CH₃

17.2 Syntheses

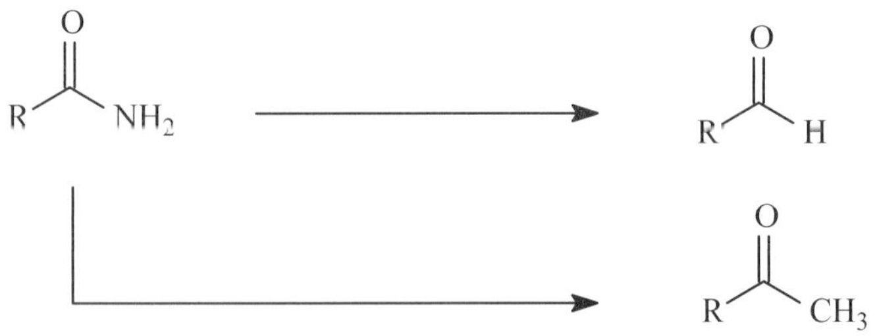

3.

4.  [phenylpropene] ⟶ [indanone]

5. 1,3-cyclopentadiene, acetylenedicarboxylic acid ⟶ [norbornene-fused anhydride]

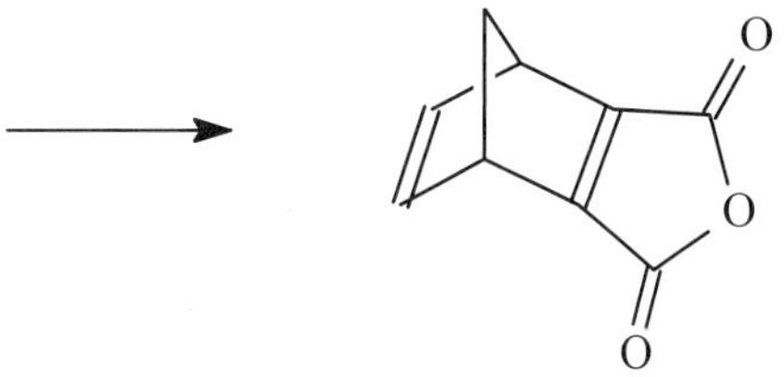

6.  OHC-CH₂CH₂CH₂-CO₂H ⟶ [2-hydroxytetrahydropyran]

7. benzamide ⟶ Ph-[1,3-dioxolane]

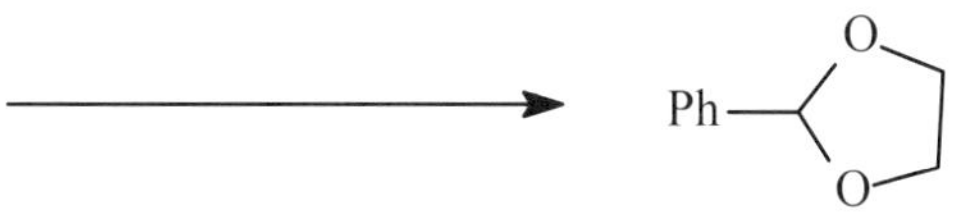

8. butanal ⟶ 2-pentanone

17.2 Syntheses

9. Following is an outline for the synthesis of *diazepam* (Valium™). Supply the appropriate reagents for each step.

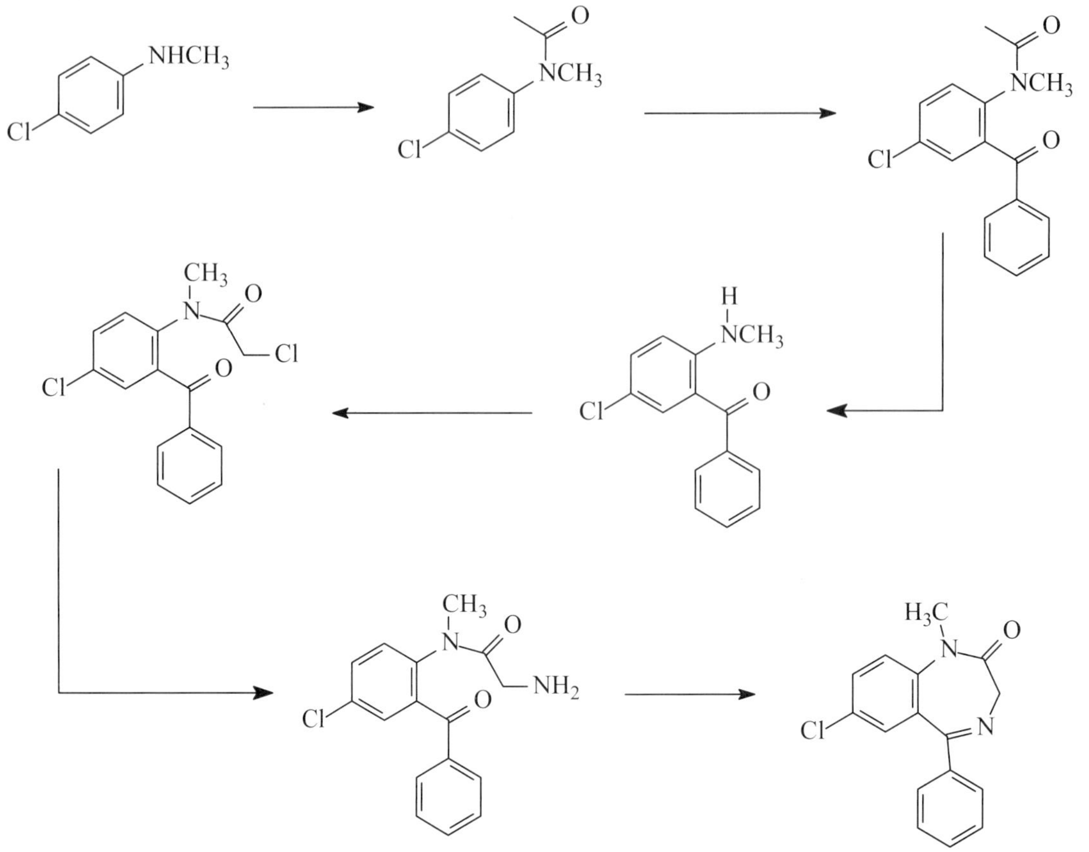

10. 3-oxohexanedioic acid ⟶ (γ-butyrolactone with methyl group)

11. methyl benzoate ⟶ methyl phenylacetate

17.2 Syntheses

12. Following is an outline for the synthesis of *fluoxetine* (Prozac™). Supply reagents for each step.

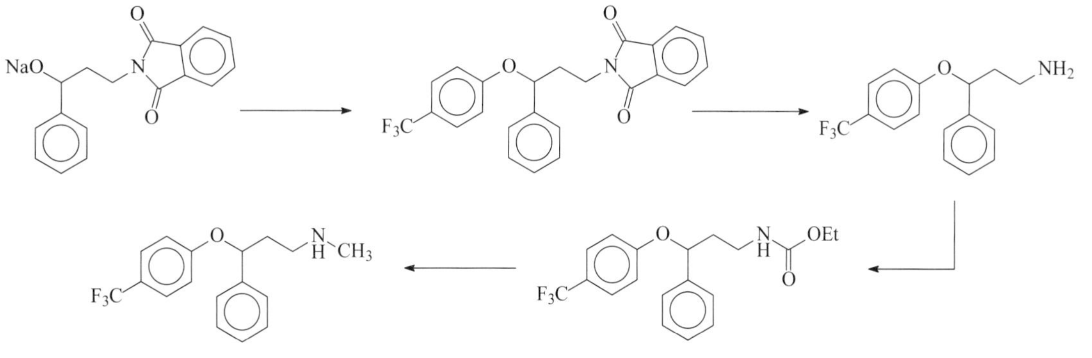

Prozac™ (antidepressant)

13.

Demerol™ (analgesic)

14. Following is an alternative synthesis of Prozac™ (see 17.2, 12). The reagent for step 5 is indicated; supply reagents for all the other steps. Outline a mechanism for step 5.

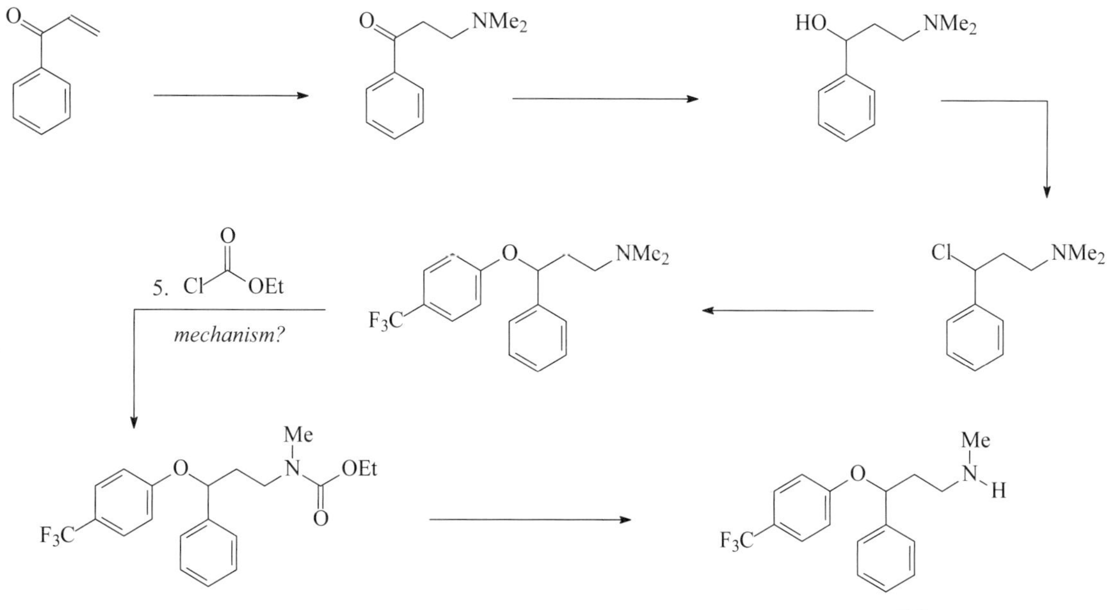

5. Cl—C(=O)—OEt

mechanism?

Prozac™

mechanism:

17.2 Syntheses

190 • Chapter 17 Carboxylic Acid Derivatives

15. 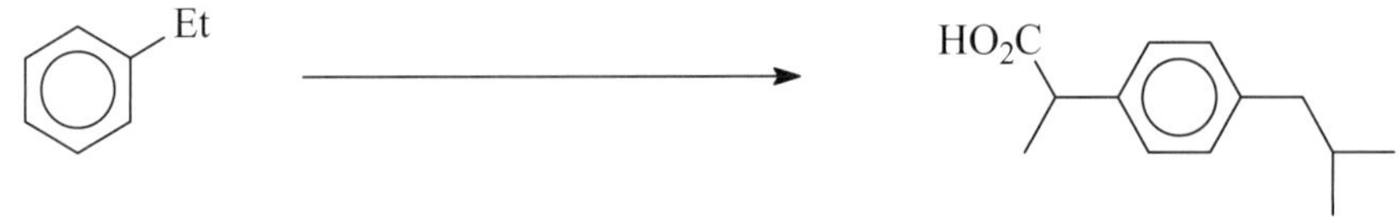 ethylbenzene ⟶ ibuprofen

16. 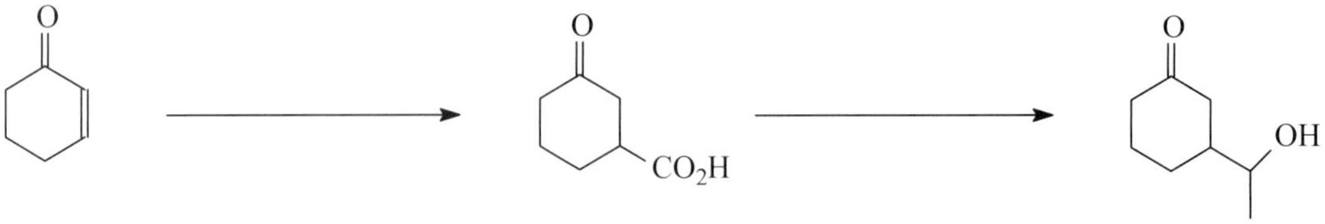 cyclohex-2-enone ⟶ 3-oxocyclohexanecarboxylic acid ⟶ 3-(1-hydroxyethyl)cyclohexanone

17.  cyclohex-3-enone ⟶ 2-methyl-2-hydroxytetrahydrofuran

18. 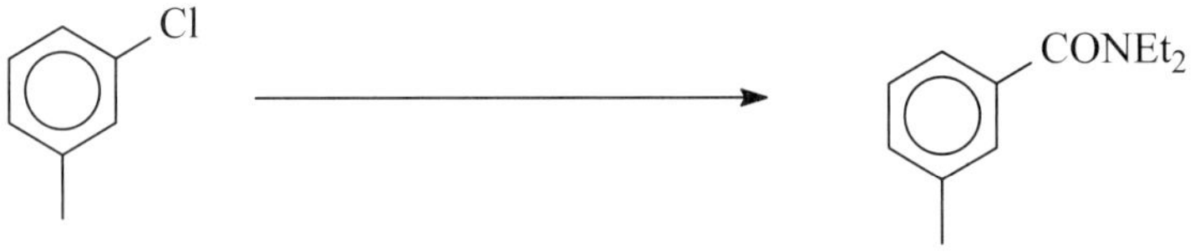 3-chlorotoluene ⟶ *DEET* (N,N-diethyl-*m*-toluamide - insect repellent)

19. *Melatonin* mediates circadian rhythm, the 24-hour sleep-wake cycle. Because its biosynthesis is inhibited by light, it is produced in the brain when the eye is not receiving light. Outline a synthesis from the neurotransmitter *serotonin*.

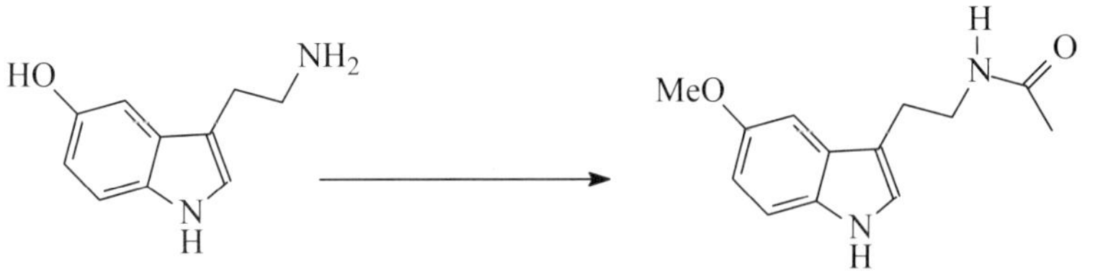

serotonin (5-HT) ⟶ *melatonin* 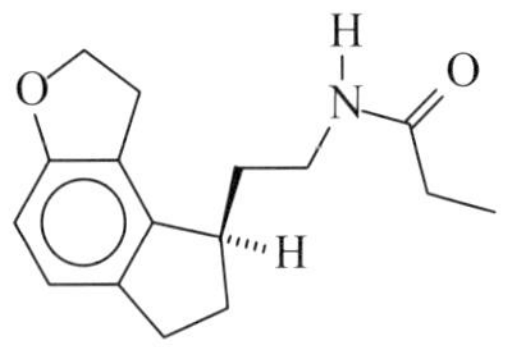 | Roserem™

Insomnia affects one in every eight people. Roserem™, a selective melatonin receptor agonist, is an example of several drugs approved to treat short- and long-term insomnia.

17.2 Syntheses

20. The two monomers (**B** and **C**) for the synthesis of *Nylon 66* can be prepared from a sugar derivative **A**. Supply the necessary reagents.

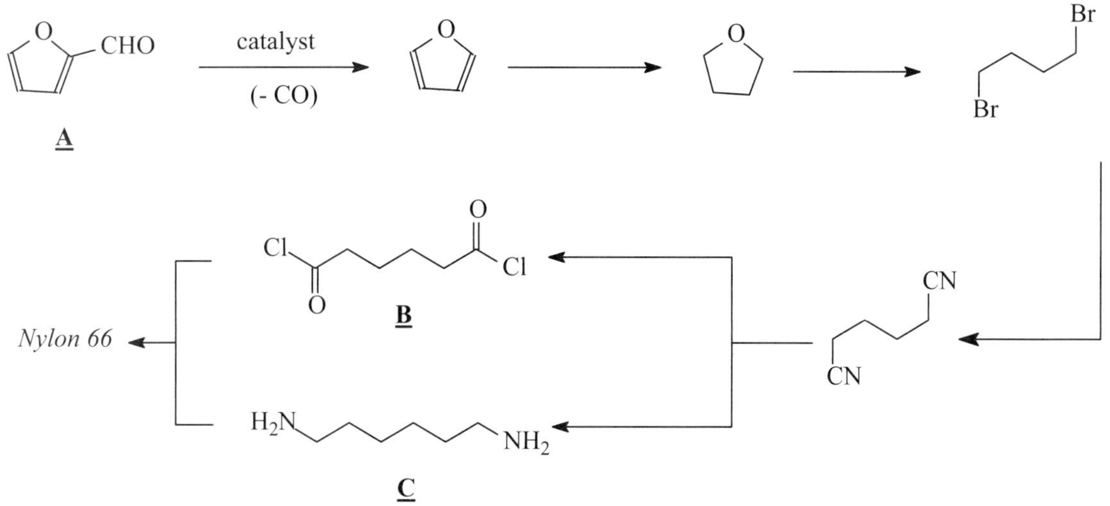

21. Name the following polymer and devise a synthesis for it. Remember, CH$_2$=CHOH is not an appropriate starting monomer. Why?

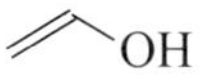

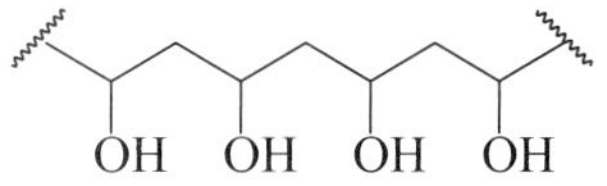

22. Some members of the morphine family of opium alkaloids…

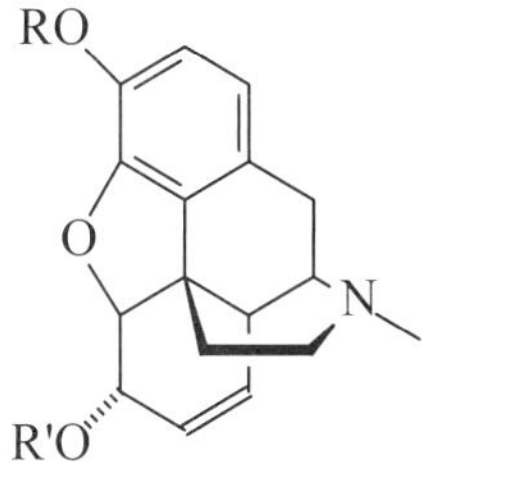

R, R' = H (*morphine*)

R = Me, R' = H (*codeine*)

R, R' = Ac (*heroin*)

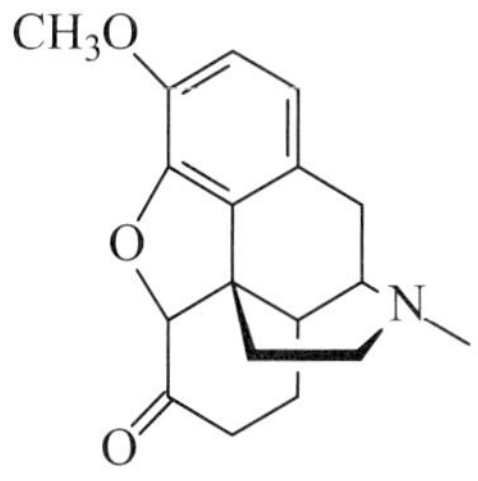

hydrocodone (a component of Vicodin™)

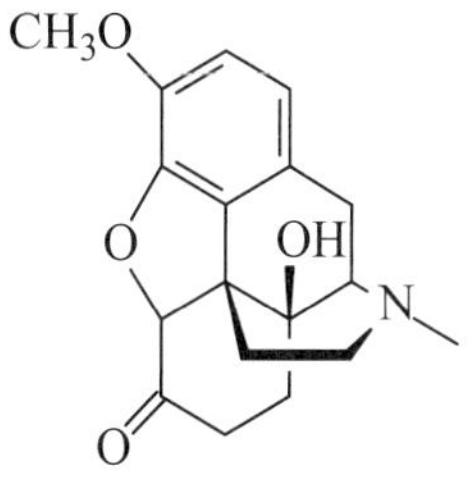

oxycodone (HCl salt = OxyContin™, a component of Percoset™)

How can the following conversions be accomplished?

a. morphine ⟶ codeine

17.2 Syntheses

192 • Chapter 17 Carboxylic Acid Derivatives

b. morphine ⟶ heroin

c. codeine ⟶ hydrocodone

d. In aqueous solution *codeinone* exists in dynamic equilibrium with its β,γ-unsaturated isomer, *neopinone*, hydration of which yields *oxycodone*. Write a mechanism for the equilibration.

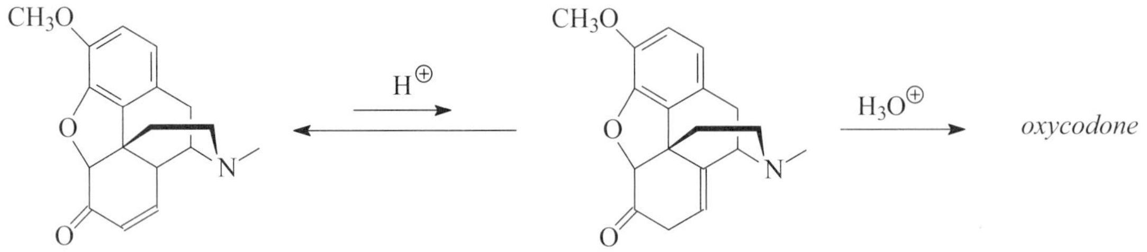

codeinone ⇌ (H⁺) *neopinone* ⟶ (H_3O^+) *oxycodone*

23.

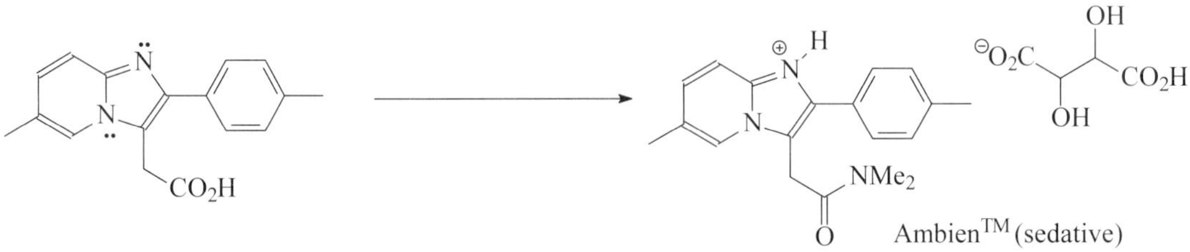

Ambien™ (sedative)

17.2 Syntheses

17.3 Mechanisms

Outline a detailed mechanism for each of the following. No other reagents than those given are necessary. Use arrows to explain the flow of electrons and show all intermediates.

1. γ-butyrolactone $\xrightarrow{H_2{}^*O,\ H^{\oplus}}$ (show location of the labeled oxygen)

2. 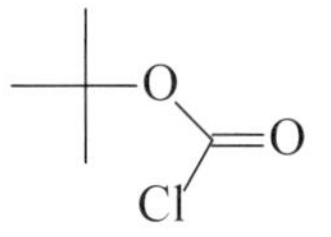 [pivaloyl chloride structure] $\xrightarrow[\text{2. } H^{\oplus}]{\text{1. (XS) RMgX}}$ *t*-BuOH + ??

3. Lactic acid (α-hydroxypropanoic acid) forms a cyclic compound, $C_6H_8O_4$. Formulate a structure for this compound. Why does lactic acid not form a simple α-lactone?

4. 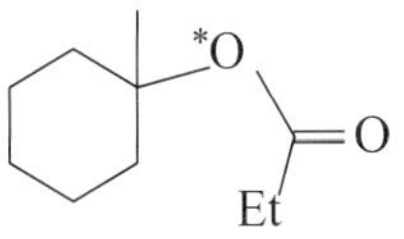 [cyclohexyl propanoate with labeled *O] $\xrightarrow{TsOH}$ an alkene + ?? (show location of the labeled oxygen)

5. ethyl 5-oxohexanoate $\xrightarrow{\text{PhMgCl (1 equiv)}}$ [δ-lactone with Ph and Me at C-6]

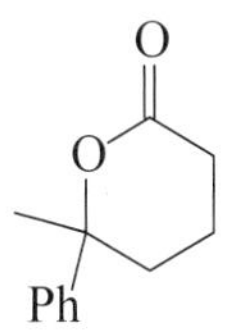

194 • Chapter 17 Carboxylic Acid Derivatives

6. Phenylisothiocyanate (**A**, *PITC, Edman reagent*) can be used to sequence proteins, *i.e.*, to determine the order of amino acids (primary structure). For example, treatment of dipeptide **B** with **A** in the presence of acid yields **C** (a phenylthiohydantoin, or PTH, derivative of the amino acid). Characterization of **C** identifies the first (from the N-terminal end) amino acid, in this case alanine.

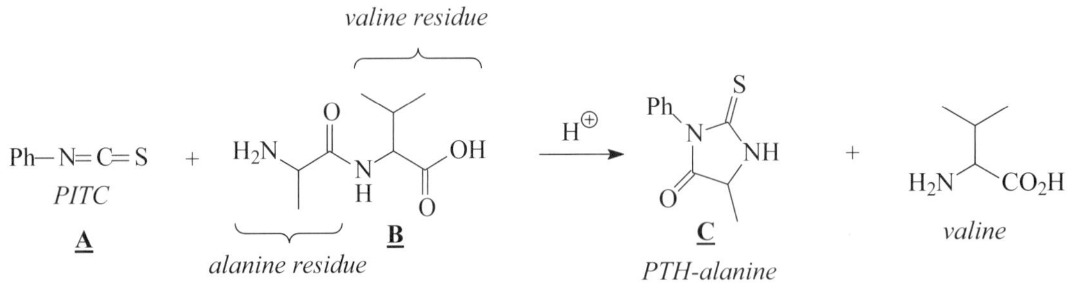

Ph—N=C=S + [dipeptide B: alanine residue–valine residue] $\xrightarrow{H^\oplus}$ **C** (PTH-alanine) + H$_2$N–CH(iPr)–CO$_2$H (valine)

PITC **A**   **B**

7. [isobutyramide] $\xrightarrow{\text{1. Me}_3\text{O}^\oplus \text{BF}_4^\ominus}_{\text{2. H}_3\text{O}^\oplus}$ [methyl isobutyrate, OMe ester]

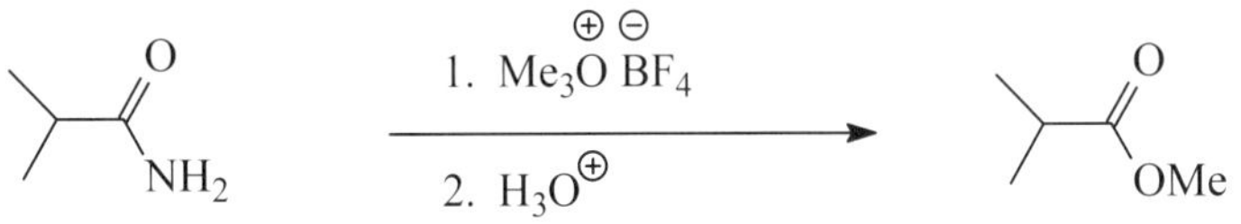

8. The *Swern oxidation*:

a. "activation" of DMSO step:

Me–S(O$^\ominus$)–Me (DMSO, S$^\oplus$) + ClC(O)C(O)Cl (oxalyl chloride) $\xrightarrow[\text{– CO}]{\text{– CO}_2}$ [*a chlorosulfonium salt*]

b. oxidation step:

[chlorosulfonium salt] + R–CH(OH)–R $\xrightarrow{\text{NR}_3}$ R–C(O)–R

17.3 Mechanisms

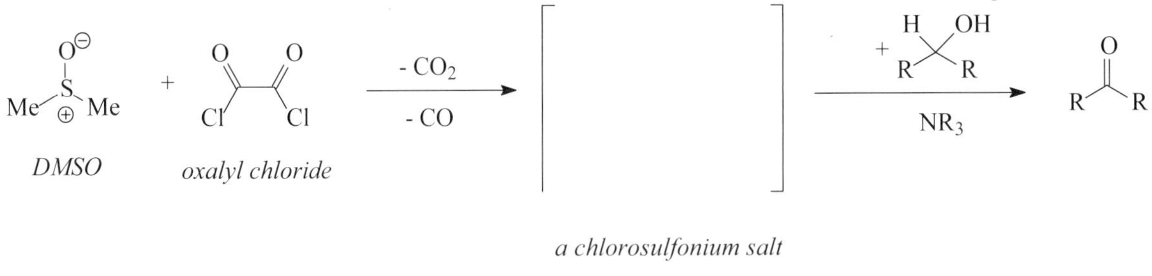

9. Similar to the Swern is the *Corey-Kim oxidation*:

[Reaction scheme: dimethyl sulfide + N-chlorosuccinimide → [a chlorosulfonium salt] → with R₂CH(OH) and NR₃ → R-CO-R]

10. The biosynthesis of pyrimidine bases, *e.g.*, *uracil*, begins with the formation of dihydoorotic acid. Formulate a mechanism.

[Reaction scheme: carbamoyl phosphate + aspartic acid → dihydroorotic acid ⇒ uracil]

11. The antimalarial *mefloquine* can be synthesized from substituted 4-quinolones by the following sequence of reactions. Outline a mechanism for step 1 and draw the structures in brackets.

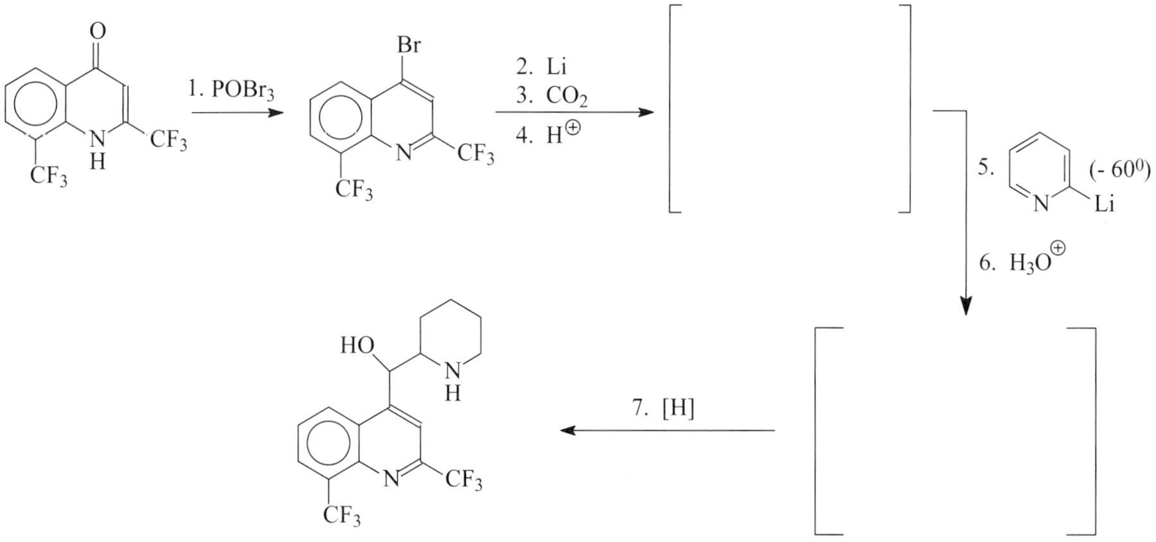

[Reaction scheme showing the synthesis of mefloquine from a 4-quinolone via steps: 1. POBr₃; 2. Li; 3. CO₂; 4. H⊕; 5. 2-lithiopyridine (−60°); 6. H₃O⊕; 7. [H]]

17.3 Mechanisms

12. Acid halides react with diazomethane to give diazomethyl ketones, which, like diazomethane, decompose to give carbenes.

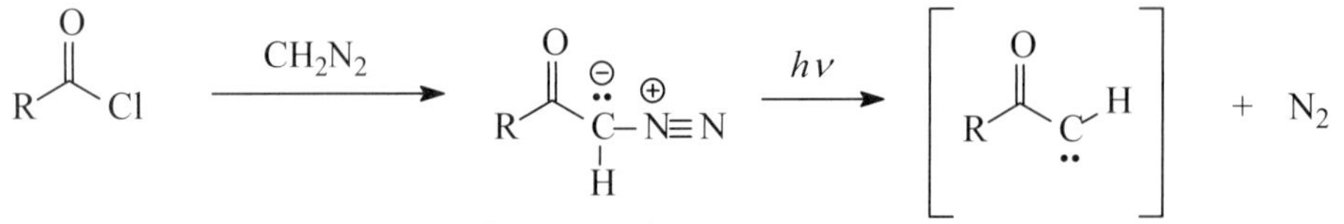

$$R-C(=O)-Cl \xrightarrow{CH_2N_2} R-C(=O)-C^{(-)}H-N^{(+)}\equiv N \xrightarrow{h\nu} [R-C(=O)-C-H] + N_2$$

a diazomethyl ketone

a. Formulate a mechanism.

b. This reaction was used in the synthesis of *twistane*. Draw the structures in brackets.

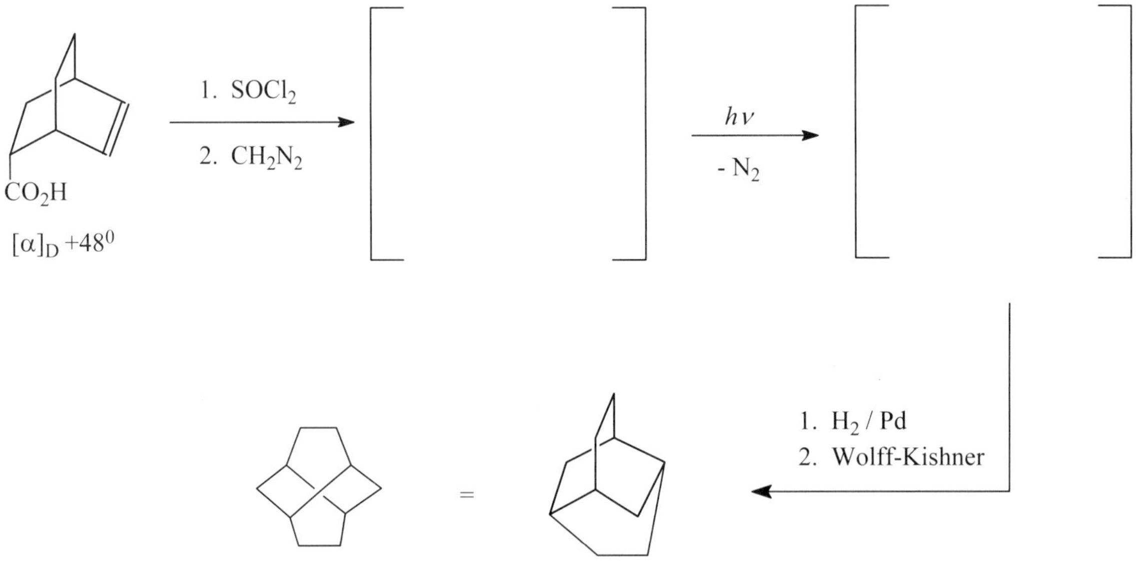

[structure with CO$_2$H, $[\alpha]_D$ +48°] $\xrightarrow[\text{2. CH}_2\text{N}_2]{\text{1. SOCl}_2}$ [ ] $\xrightarrow[-N_2]{h\nu}$ [ ]

$\xleftarrow[\text{2. Wolff-Kishner}]{\text{1. H}_2/\text{Pd}}$

twistane
$[\alpha]_D$ +434°

13. Cyanogen bromide (CNBr) specifically cleaves certain peptide (amide) bonds to yield a lactone:

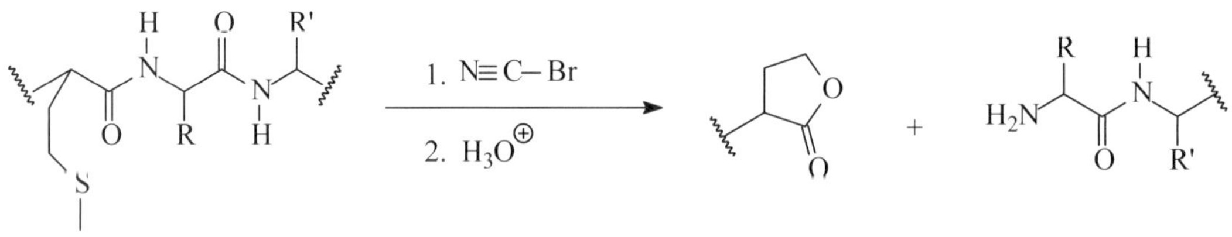

$\xrightarrow[\text{2. H}_3\text{O}^{\oplus}]{\text{1. N}\equiv\text{C-Br}}$

17.3 Mechanisms

14. Draw the structure in brackets and give a mechanism for the conversion of **A** to strychnone.

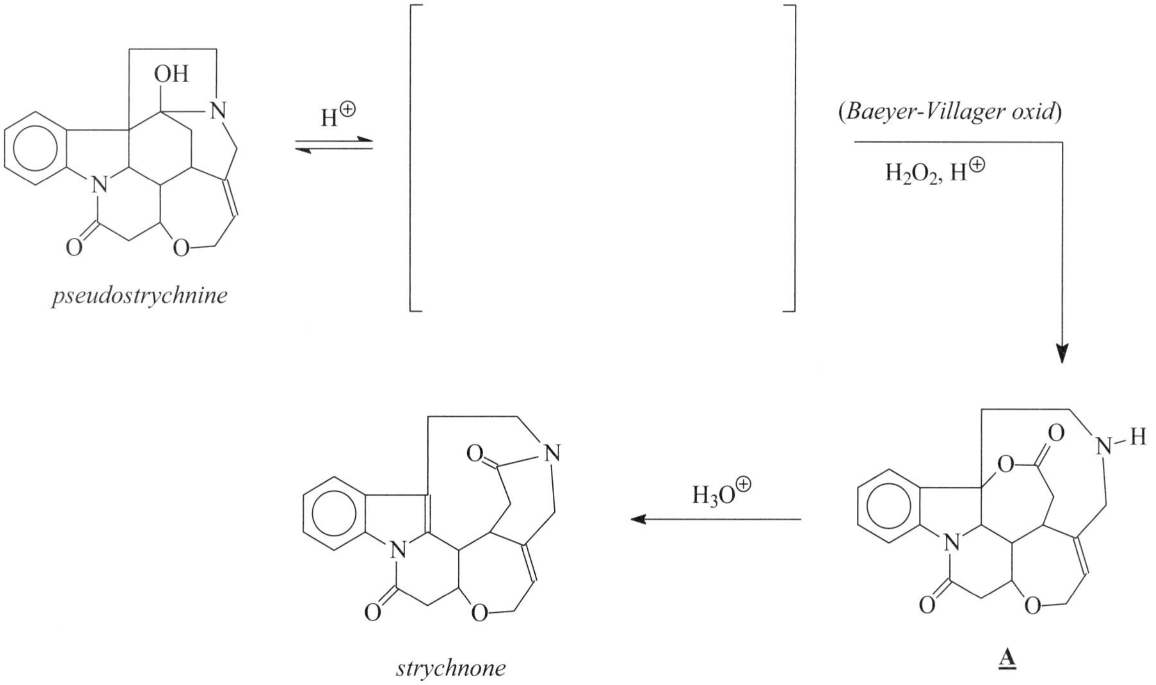

pseudostrychnine ⇌ (H⊕) [ ] → (Baeyer-Villager oxid) H₂O₂, H⊕ → **A** → (H₃O⊕) *strychnone*

15. A step in Woodward's (*Harvard*) classic total synthesis of strychnine:

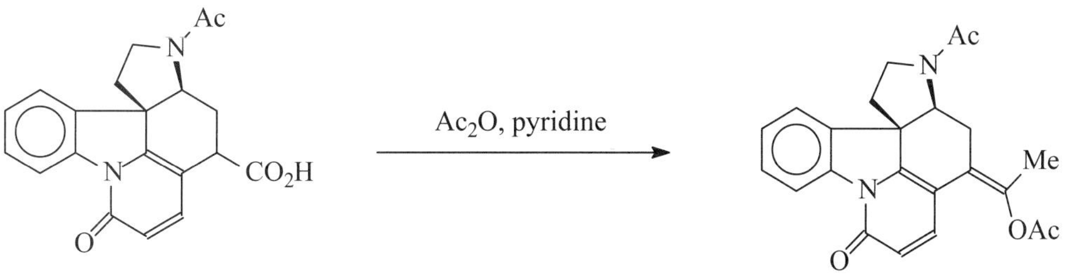

Ac₂O, pyridine

17.3 Mechanisms

16. The final step in Sheehan's (*MIT*) total synthesis of *penicillin V* involved the formation of a strained β-lactam. To accomplish this he employed a new reagent, *dicyclohexylcarbodiimide* (DCC), first reported from his lab two years earlier to smoothly form amides from an aqueous mixture of a carboxylic acid and an amine at room temperature. [That important advance in the state of the art for forming amide bonds was subsequently utilized by Merrifield (*Rockefeller*) in his solid-phase approach to synthesizing proteins by linking amino acids together through amide (peptide) bonds.] Propose a mechanism for the lactamization reaction.

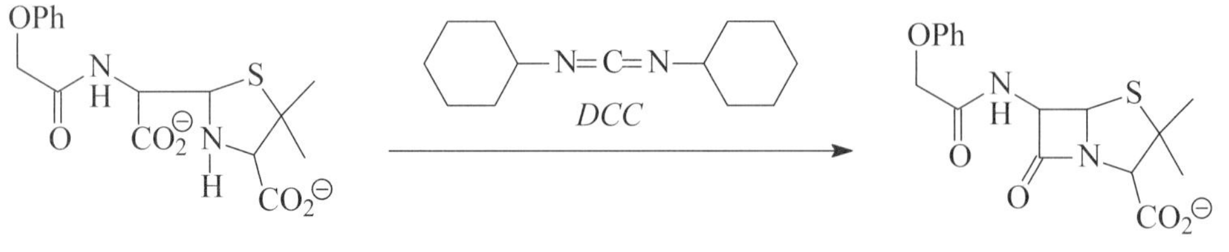

salt of penicillin V

17.

(a bisphosphonic acid)

Fosamax™ (alendronate sodium, bone resorption inhibitor)

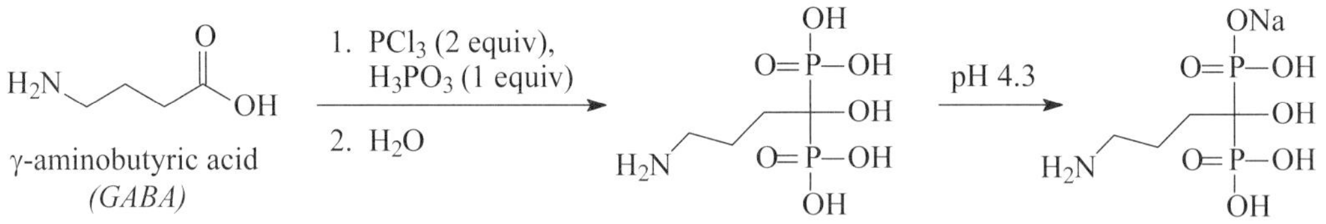

Hint: phosphorous acid exists in two tautomeric forms; use the nucleophilic form to attack the product of the reaction of GABA with PCl_3. This one is rather challenging.

18. Tertiary alcohols are weakly nucleophilic because of steric hindrance near the hydroxyl group and, therefore, do not readily undergo Fischer esterification. One approach to form acetate esters of such alcohols is to allow them to react with isopropenyl acetate in the presence of an acid catalyst. Hint: the actual acetylation step involves an S_N1-like reaction of the alcohol with an acylium ion.

19. In contrast to phenyl acetate, the conjugate base of aspirin (acetylsalicylic acid) readily undergoes hydrolysis in water, suggesting kinetic enhancement by the latter's carboxylate moiety. Consider two possible pathways and outline a mechanism for each.

a. The carboxylate anion acts as a *nucleophile*, attacking the acetate ester to form a mixed anhydride, which is subsequently hydrolyzed by water:

b. The carboxylate anion acts as a *base*, removing a proton from water to form hydroxide, which subsequently attacks the ester:

c. Experimental evidence indicates that when the reaction is conducted in the presence of ^{18}O-labeled water, no label is found in the salicylic acid product. Which pathway is supported by this experiment?

20.

17.3 Mechanisms

CHAPTER 18
CARBONYL α-SUBSTITUTION REACTION AND ENOLATES

18.1 Reactions
Draw the structural formula of the major organic product(s). Show stereochemistry where appropriate.

1. 2,2-dimethylcyclopentanone $\xrightarrow{\text{1. LDA} \quad \text{2. } \underline{n}\text{-PrBr}}$

2. methyl 3-oxopentanoate $\xrightarrow[\text{3. } H_3O^{\oplus}, \Delta]{\text{1. } ^{\ominus}OMe, \text{ MeOH} \quad \text{2. benzoyl bromide}}$

3. cyclohexanone $\xrightarrow[\text{3. EtI} \quad \text{4. } H_3O^{\oplus}, \Delta]{\text{1. } EtOCO_2Et, \text{ LDA} \quad \text{2. } ^{\ominus}OEt}$

4. diethyl malonate $\xrightarrow[\text{3. } H_3O^{\oplus}, \Delta]{\text{1. base} \quad \text{2. benzoic anhydride}}$

5. 2-chloro-3-cyano-5-nitrobenzene + $NC-CH_2-CO_2Et$ $\xrightarrow[\text{2. } H_3O^{\oplus}, \Delta]{\text{1. } ^{\ominus}OEt}$

6. *trans*-4-methylcyclohexyl tosylate + $NaCH(CO_2R)_2 \longrightarrow$

7. (E)-3-pentene-2-one $\xrightarrow{\begin{array}{l}\text{1. } H_3O^{\oplus} \\ \text{2. } CrO_3, H^{\oplus} \\ \text{3. NaH (2 equiv)} \\ \text{4. benzyl chloride (1 equiv)} \\ \text{5. } H^{\oplus}\end{array}}$

18.1 Reactions

8. diethyl malonate
   1. ⁻OEt
   2. isobutylene epoxide
   3. H_3O^+, Δ

9. ethyl acetoacetate
   1. (XS) NaOEt
   2. $Br(CH_2)_4Br$
   3. H_3O^+, Δ

   ($C_7H_{12}O$)

10. 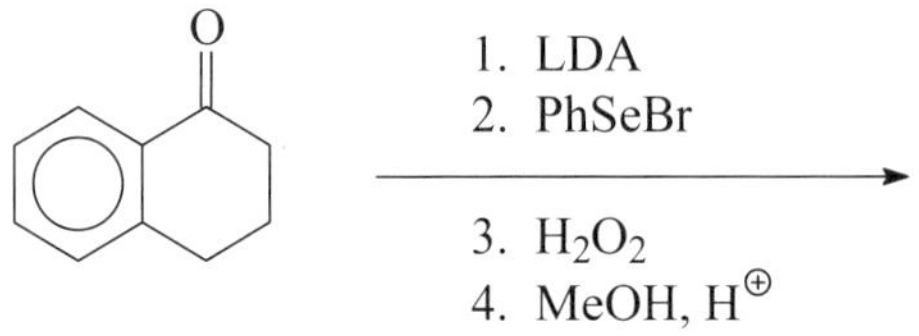 α-tetralone
    1. LDA
    2. PhSeBr
    3. H_2O_2
    4. MeOH, H^+

11. 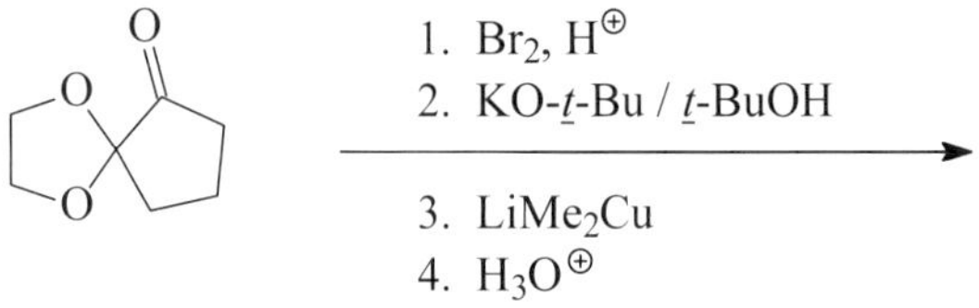 (1,4-dioxaspiro[4.4]nonan-6-one)
    1. Br_2, H^+
    2. KO-t-Bu / t-BuOH
    3. $LiMe_2Cu$
    4. H_3O^+

12. t-butyl methyl ketone
    1. Br_2, H^+
    2. $(CN)_2CH$:⁻
    3. H_3O^+, Δ

13. diethyl malonate
    1. ⁻OEt, HOEt
    2. butanoyl chloride
    3. H_3O^+, Δ

    → $2\ CO_2$ +

14. 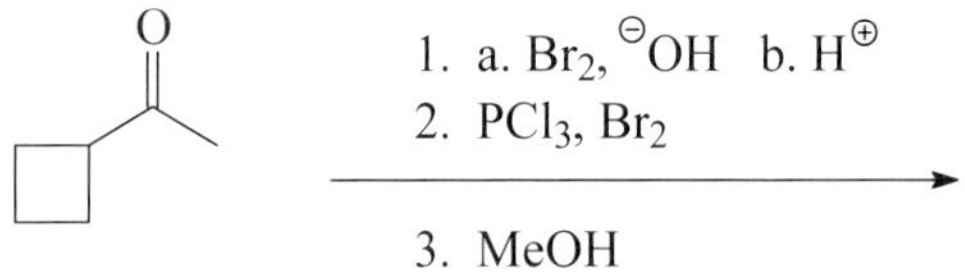 cyclobutyl methyl ketone
    1. a. Br_2, ⁻OH   b. H^+
    2. PCl_3, Br_2
    3. MeOH

15. 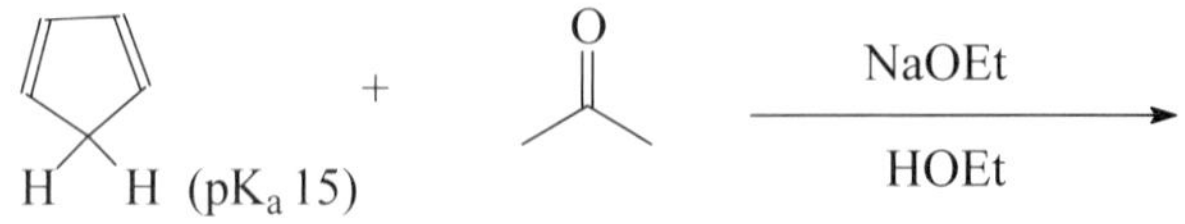 cyclopentadiene (H H, pK_a 15) + acetone

    NaOEt / HOEt

    (C_8H_{10})

18.1 Reactions

16. 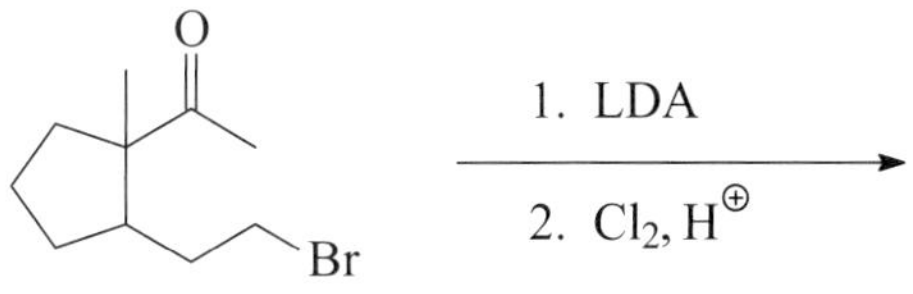 [structure: 1-acetyl-1-methyl-2-(2-bromoethyl)cyclopentane] → 1. LDA 2. Cl$_2$, H$^+$

17. 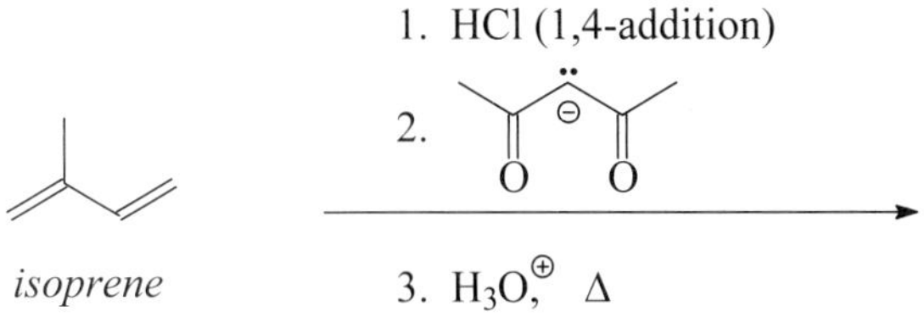 isoprene →
   1. HCl (1,4-addition)
   2. [acetylacetonate anion: CH(COCH$_3$)$_2^-$]
   3. H$_3$O$^+$, Δ

18. 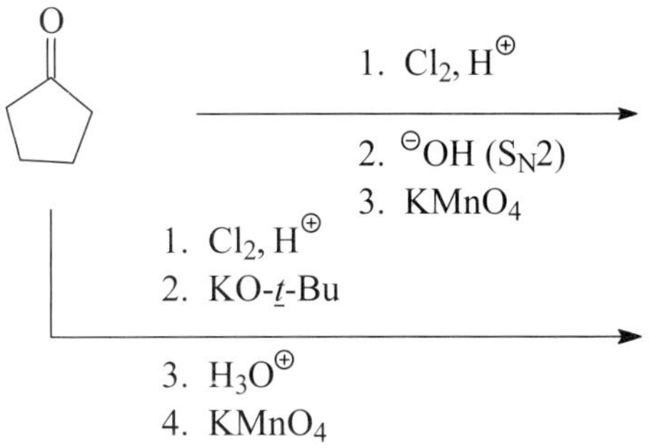 cyclopentanone →

   (top pathway)
   1. Cl$_2$, H$^+$
   2. $^-$OH (S$_N$2)
   3. KMnO$_4$

   (bottom pathway)
   1. Cl$_2$, H$^+$
   2. KO-t-Bu
   3. H$_3$O$^+$
   4. KMnO$_4$

19. 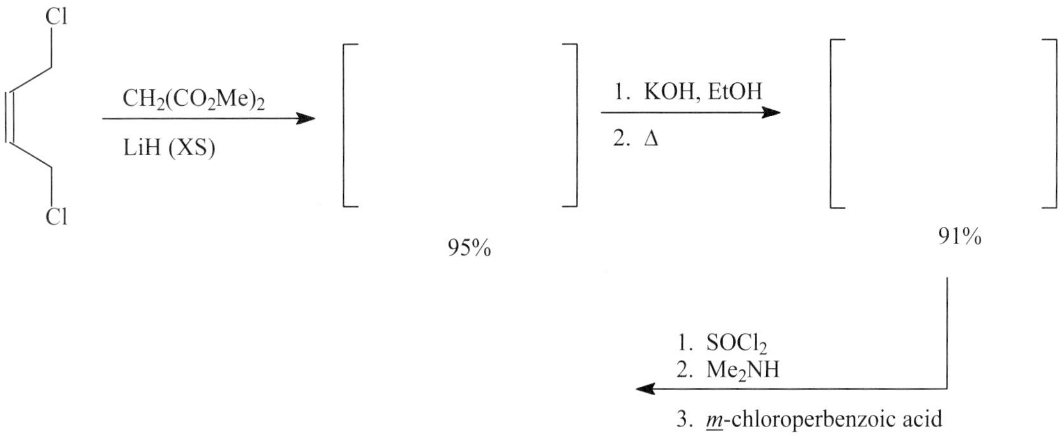 (Z)-1,4-dichloro-2-butene $\xrightarrow{\text{CH}_2(\text{CO}_2\text{Me})_2,\ \text{LiH (XS)}}$ [ ] 95% $\xrightarrow[\text{2. }\Delta]{\text{1. KOH, EtOH}}$ [ ] 91%

   $\xleftarrow{\text{1. SOCl}_2;\ \text{2. Me}_2\text{NH};\ \text{3. }m\text{-chloroperbenzoic acid}}$

20. [4-hydroxycoumarin / chromane-2,4-dione] $\xrightarrow{\text{1. base;\ 2. benzalacetone (PhCH=CHCOCH}_3\text{)}}$ [ ] $\xrightarrow{tautomerize}$ coumadin (Warfarin™ - an anticoagulant)

18.1 Reactions

18.2 Syntheses
Supply a reagent or sequence of reagents that will effect the following conversions.

1. butyric acid $\longrightarrow$ ethylpropanedioic acid

2. dimethyl malonate $\longrightarrow$ δ-valerolactone
   (valeric acid is a common name for pentanoic acid)

3. dimethyl malonate $\longrightarrow$ butanedial

4. dimethyl malonate $\longrightarrow$ seconal (a sedative)

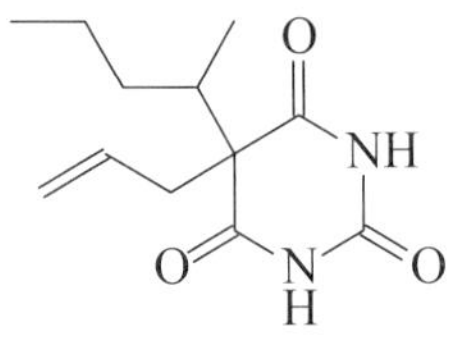

5. dimethyl malonate $\longrightarrow$ 2-benzylbutanoic acid

6. dimethyl malonate + styrene $\longrightarrow$ Ph-CH=CH-CH$_2$-COOH

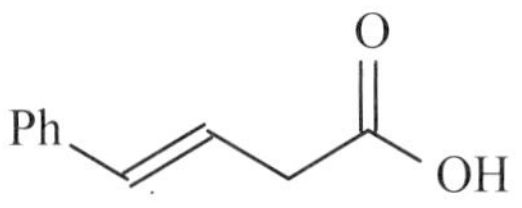

7. dimethyl malonate + methyl acrylate $\longrightarrow$ Br-(CH$_2$)$_5$-Br
   (acrylic acid is propenoic acid)

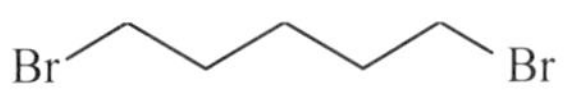

8. ethyl acetoacetate $\longrightarrow$ s-Bu-CONH$_2$

9. ethyl acetoacetate ⟶ CH₃COCH₂COCOOH

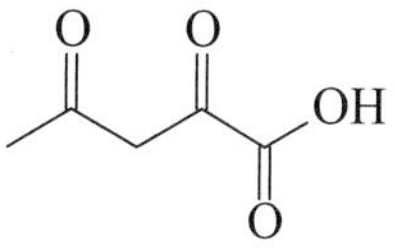

10. ethyl acetoacetate ⟶ 3-acetyltetrahydro-2H-pyran-2-one

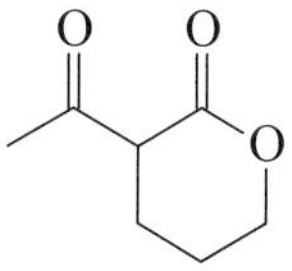

11. ethyl acetoacetate ⟶ 2,5-hexanedione (heptane-2,6-dione) ⟶ 2,6-dichloroheptane

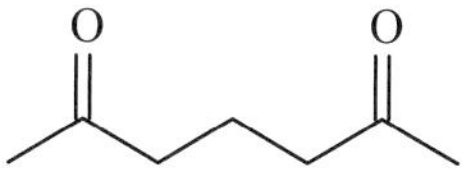

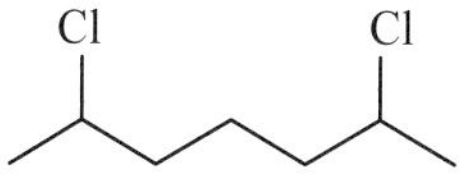

12. ethyl acetoacetate ⟶ *s*-butyl methyl ketone ⟶ 2-methylbutanoate

13. cyclopentanone ⟶ 3-acetoxycyclopentanone (OAc)

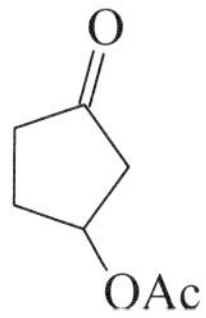

14. cyclopentanol ⟶ 2-oxocyclopentanecarboxylic acid

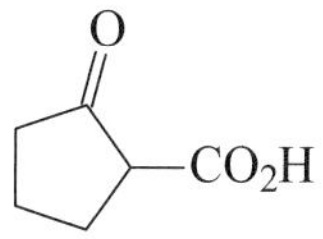

15. 2,4-pentanedione ⟶ 3-(1-methylcyclohexyl)-2,4-pentanedione

18.2 Syntheses

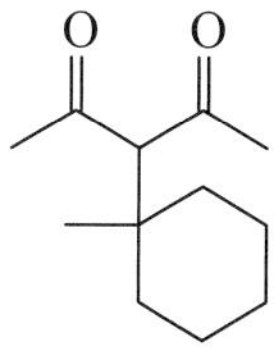

16. methyl acetoacetate ⟶ 2,4-pentanedione

17. 3-pentanone —*via an organoselenium cmpd*→ ethyl vinyl ketone ⟶ [4-ethyl-1,3-dioxane structure]

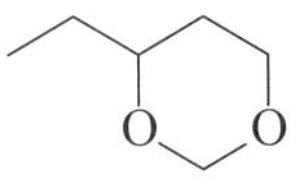

18. chlorobenzene 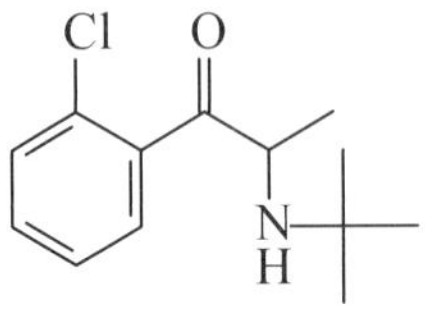 ⟶ Wellbutrin™ (an antidepressant)

19. 1-naphthol , acrylic acid ⟶ *propranolol* (a β-adrenergic blocker, developed by Sir James Black, recipient of '88 Nobel Prize in medicine; greatest breakthrough in pharmaceuticals for heart illness since discovery of *digitalis* approximately 200 years ago)

20. dimethyl malonate ⟶ *sodium pentothal* (used to induce pre-surgical anesthesia in combination with sedatives)

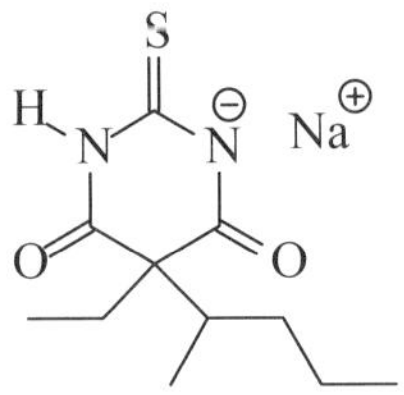

21. PhCH₂CN ⟶ [4-cyano-4-phenyl-1-methylpiperidine] ⟶ *meperidine* (an analgesic)

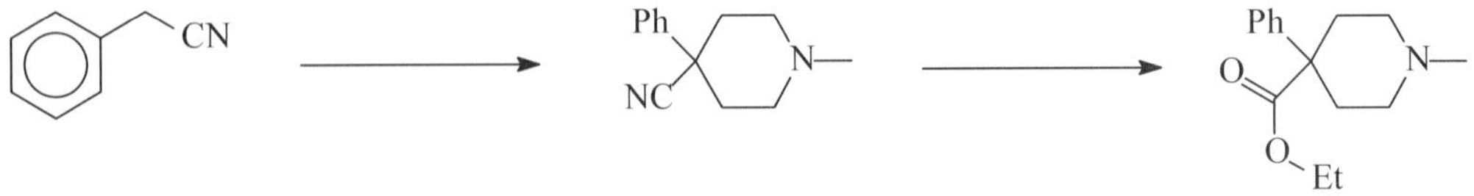

18.2 Syntheses

18.3 Mechanisms

Outline a detailed mechanism for each of the following. No other reagents than those given are necessary. Use arrows to explain the flow of electrons and show all intermediates.

1. [structure: 1,3-dihydroxyacetone] $\xrightarrow{H^{\oplus}}$ [structure: glyceraldehyde]

   (a central reaction in glycolysis catalyzed by the enzyme *TIM*, triose isomerase)

2. ethyl acetoacetate $\xrightarrow[\text{2. propylene oxide}]{\text{1. EtO}^{\ominus}\text{, HOEt}}$ [structure: 3-acetyl-5-methyl-γ-butyrolactone]

3. 1,3-Diphenyl-1,3-propanedione gives a positive iodoform test even though it is not a methyl ketone. In addition to CHI$_3$, two equivalents of benzoate are formed. Explain.

4. [bicyclic ketone with methyl substituent] $\xrightarrow[\text{racemization}]{H^{\oplus}}$ [bicyclic ketone with methyl substituent, inverted]

5. [bicyclic ketone with acetate] $\xrightarrow[\text{MeOH}]{^{\ominus}\text{OMe}}$ [cycloheptane-1,4-dione]

208 • Chapter 18 Carbonyl α-Substitution Reactions and Enolates

6. 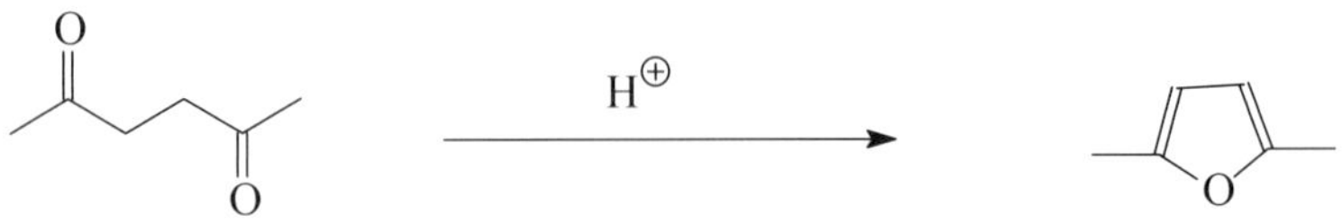 [hexane-2,5-dione] + H⁺ ⟶ [2,5-dimethylfuran]

7. The vitamin *biotin* is necessary for many metabolic carboxylation reactions. It reacts initially with CO_2 to form unstable **A**, which then "donates" CO_2 to a substrate. Outline the mechanism for carboxylation of pyruvic acid to oxaloacetic acid (OAA).

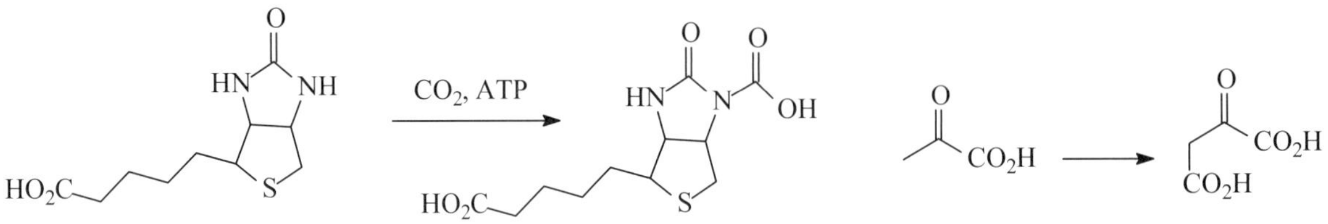

biotin —CO₂, ATP→ **A**   pyruvic acid ⟶ OAA

8. 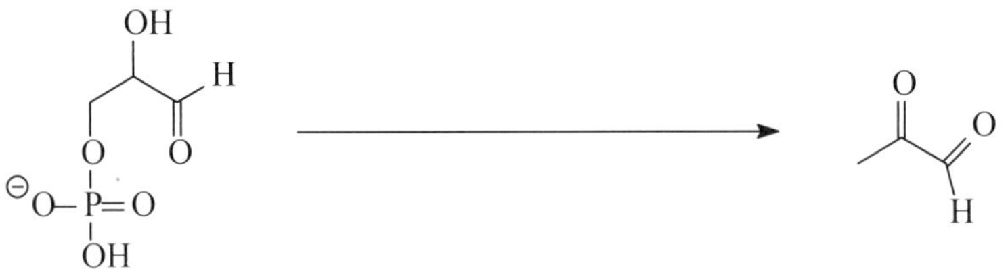 [glyceraldehyde-3-phosphate] ⟶ [methylglyoxal / pyruvaldehyde]

9. 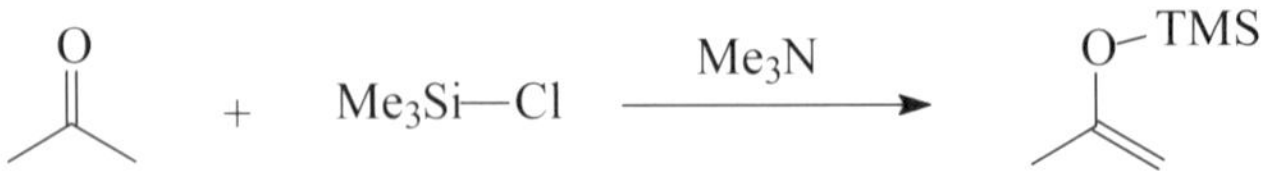 [acetone] + Me₃Si—Cl —Me₃N→ [TMS enol ether of acetone]

18.3 Mechanisms

CHAPTER 19
CARBONYL CONDENSATION REACTIONS

19.1 Reactions
Draw the structural formula of the major organic product(s). Show stereochemistry where appropriate.

1. benzaldehyde + acetophenone
   1. $^{\ominus}$OH, (-H$_2$O)
   2. NaCH(CO$_2$R)$_2$
   3. H$_3$O$^{\oplus}$

2. isobutyraldehyde $\xrightarrow{H^{\oplus}}$

3. (cyclohexane-1,3-dione) + (t-Bu)CHO $\xrightarrow{^{\ominus}OEt,\ EtOH}$

4. (cyclopentane with -CH$_2$CO$_2$Me and -CH$_2$COCH$_3$ substituents) $\xrightarrow{^{\ominus}OMe,\ MeOH}$

5. dimethyl heptanedioate $\xrightarrow[\text{2. }H_3O^{\oplus}\ \Delta]{\text{1. }^{\ominus}OMe}$ CO$_2$ +

6. (2,3-dimethylcyclopent-2-enone) $\xrightarrow[\text{(conj. add'n + retro-aldol)}]{^{\ominus}OH,\ ROH}$

7. (2,5-dioxo compound with cis alkene) $\xrightarrow{^{\ominus}OH}$

   cis-jasmone (a perfume)

8. propionaldehyde $\xrightarrow[\text{2. }H_3O^{\oplus}\ \Delta\ (Darzen's\ cond.)]{\text{1. }BrCH_2CO_2Me,\ t\text{-BuO}^{\ominus}}$

9. 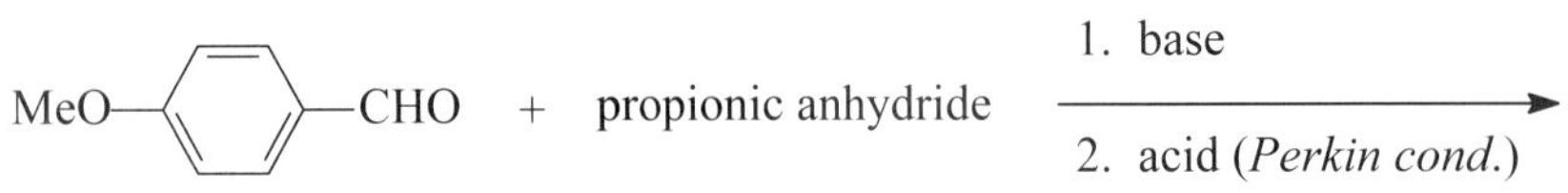 MeO-C6H4-CHO + propionic anhydride $\xrightarrow{\text{1. base} \atop \text{2. acid (Perkin cond.)}}$

10. 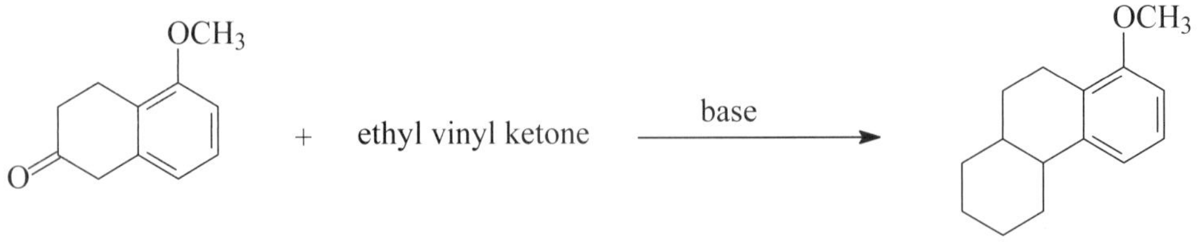 [5-methoxy-2-tetralone] + ethyl vinyl ketone $\xrightarrow{\text{base}}$ [tricyclic product] (complete)

11. benzaldehyde + CH$_3$NO$_2$ $\xrightarrow{^{\ominus}\text{OH}}$

12. 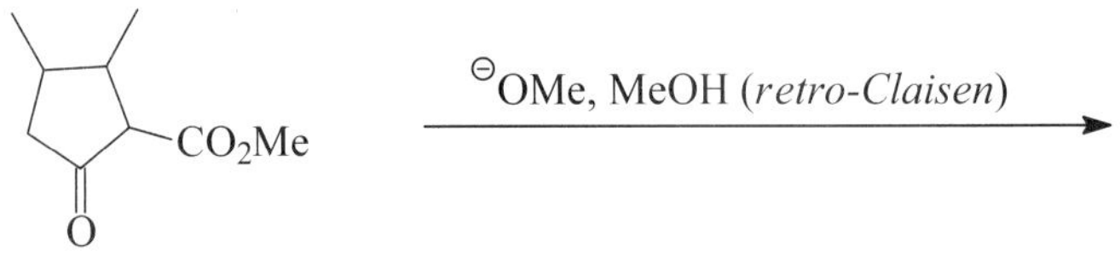 [2,3-dimethyl-5-oxocyclopentane carboxylic acid methyl ester] $\xrightarrow{^{\ominus}\text{OMe, MeOH (retro-Claisen)}}$

13. cyclopentanone $\xrightarrow{\text{1. NaOEt / EtOH (-H}_2\text{O)} \atop \text{2. NH}_2\text{NHPh}}$

14. 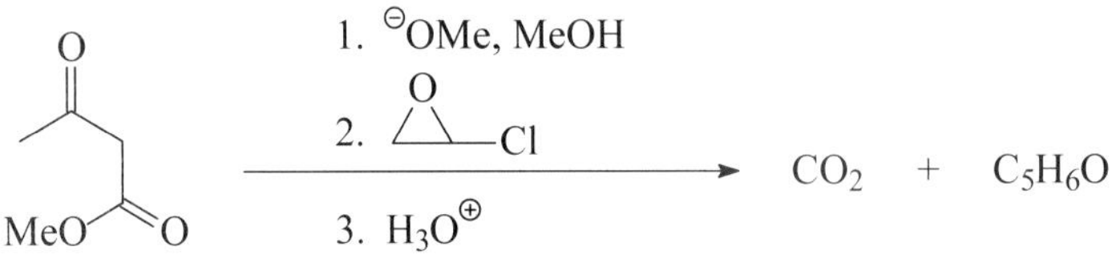 MeO-CO-CH$_2$-CO-CH$_3$ $\xrightarrow{\text{1. }^{\ominus}\text{OMe, MeOH} \atop \text{2. epoxide-Cl} \atop \text{3. H}_3\text{O}^{\oplus}}$ CO$_2$ + C$_5$H$_6$O

15. acetone $\xrightarrow[\text{(aldol)}]{^{\ominus}\text{OH}}$ C$_6$H$_{10}$O $\xrightarrow{\text{1. CH}_2\text{(CO}_2\text{Me)}_2, ^{\ominus}\text{OMe} \atop \text{2. H}_3\text{O}^{\oplus}, \Delta}$ C$_8$H$_{12}$O$_2$

19.1 Reactions

16. 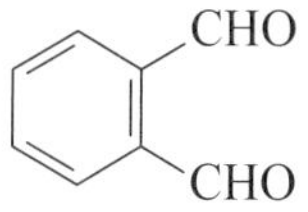 benzene-1,2-dicarbaldehyde + 3-pentanone $\xrightarrow{\ominus OH\ (-H_2O)}$

17. 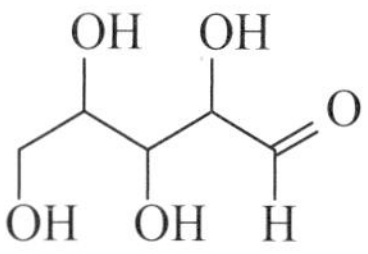 (2,3,4,5-tetrahydroxypentanal) $\xrightarrow{retro\text{-}aldol}$

18. 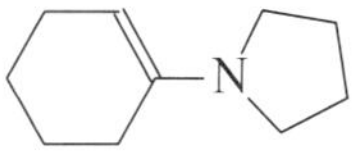 1-(cyclohex-1-en-1-yl)pyrrolidine $\xrightarrow[\text{2. }H_3O^\oplus]{\text{1. methyl vinyl ketone}}_{\text{3. NaOH }(aldol)}$

19. acetaldehyde + (XS) formaldehyde $\xrightarrow{\ominus OH}$

$C(CH_2OH)_4$ -- *pentaerythritol*

20. 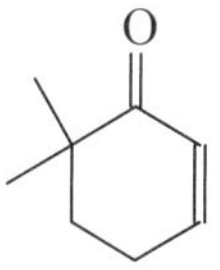 6,6-dimethylcyclohex-2-en-1-one $\xrightarrow{PhCHO,\ H^\oplus}$

21. 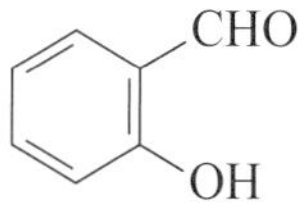 salicylaldehyde + $CH_2(CO_2Et)_2$ $\xrightarrow[\text{2. }H^\oplus]{\text{1. base}}$

$C_{12}H_{10}O_4$

22. 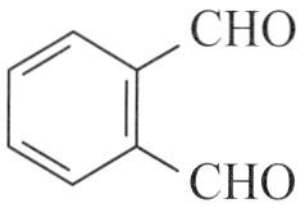 benzene-1,2-dicarbaldehyde + 1,4-cyclohexanedione $\xrightarrow{KOH}$

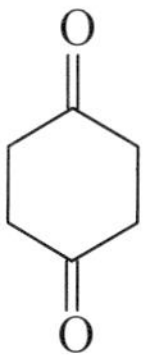

(a pentacyclic dione)

19.1 Reactions

212 • Chapter 19 Carbonyl Condensation Reactions

23. A reaction in the biosynthesis of the amino acid *leucine*:

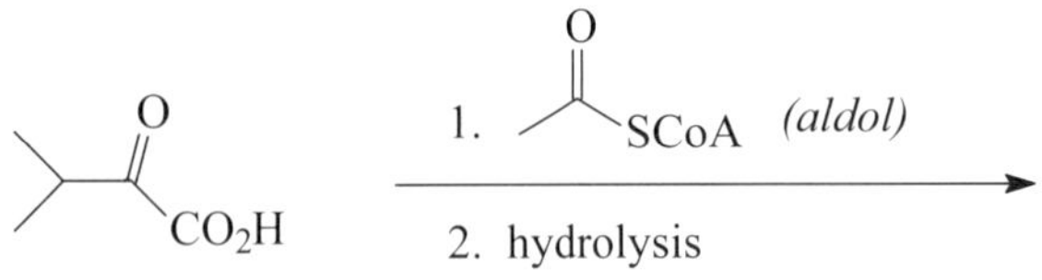

(CH₃)₂C(CO₂H)–C(=O)– 　1. CH₃C(=O)SCoA (aldol)　2. hydrolysis →

24. *Trans-resveratrol*, isolable from red wine, has been implicated as a cardioprotective and can be synthesized as follows:

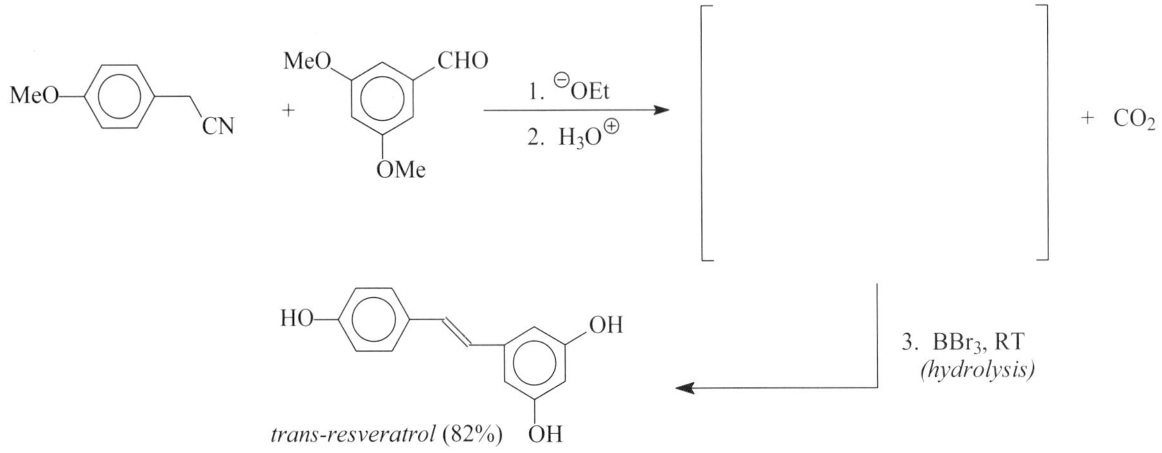

MeO–C₆H₄–CH₂CN + (3,5-dimethoxy)C₆H₃–CHO → 1. ⁻OEt 2. H₃O⁺ → [intermediate] + CO₂

3. BBr₃, RT (*hydrolysis*) →

trans-resveratrol (82%) (HO–C₆H₄–CH=CH–C₆H₃(OH)₂)

25.

[bicyclic ketone with methyl group and C=C] 　1. H₃O⁺　2. KO-*t*-Bu (*retro-aldol*) →

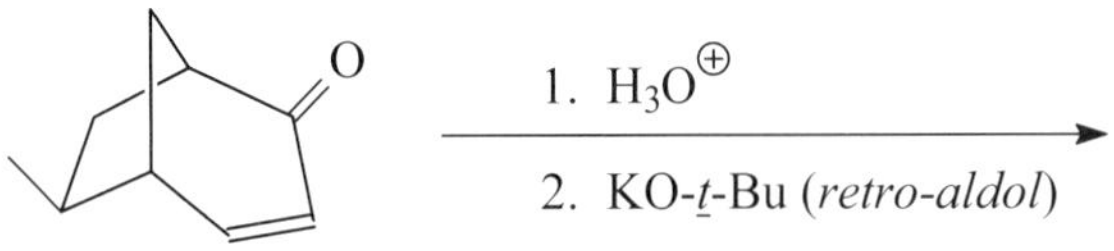

26. Forward and retro-aldol-like reactions that occur in plants:

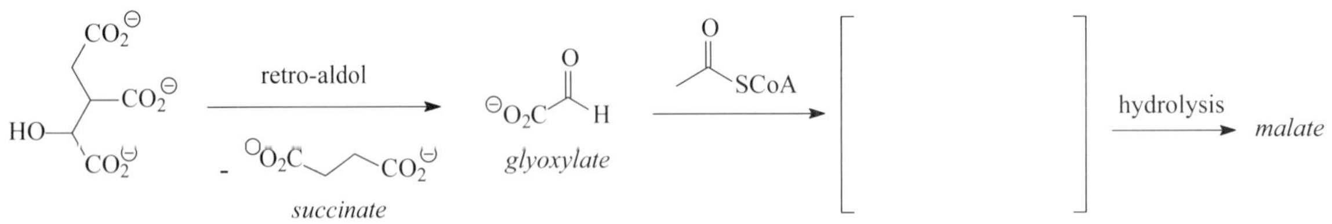

[HOCH(CO₂⁻)–C(CO₂⁻)–CH₂CO₂⁻] —retro-aldol→ ⁻O₂C–CHO (*glyoxylate*) + ⁻O₂C–CH₂CH₂–CO₂⁻ (*succinate*)

glyoxylate + CH₃C(=O)SCoA → [intermediate] —hydrolysis→ malate

19.1 Reactions

27. CH₃-CO-C≡CH + 1,3-cyclohexanedione $\xrightarrow[\text{(Robinson annulation)}]{^{\ominus}\text{OEt / EtOH}}$

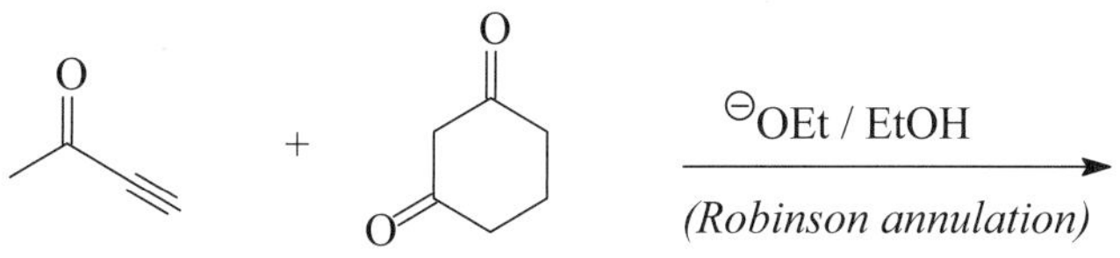

28. The biosynthesis of glucose involves *aldolase*, an enzyme that catalyzes both forward and retro-aldol reactions. The forward process illustrates a mixed aldol wherein the enzyme initially binds with **A**, promoting its tautomerization and subsequent reaction with **B** to form a ketohexose:

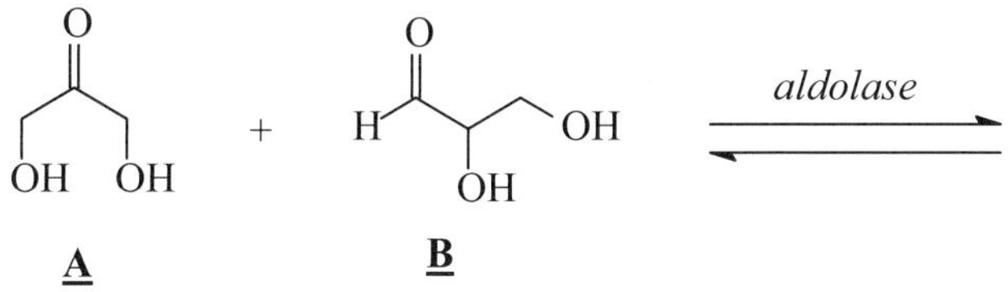

A: HOCH₂-CO-CH₂OH (dihydroxyacetone)
B: OHC-CH(OH)-CH₂OH (glyceraldehyde)

$$\mathbf{A} + \mathbf{B} \xrightleftharpoons{aldolase}$$

29. The *Krebs Cycle* begins with an aldol-like condensation of a thioester (acetyl coenzyme A) with oxaloacetate, followed by hydrolysis:

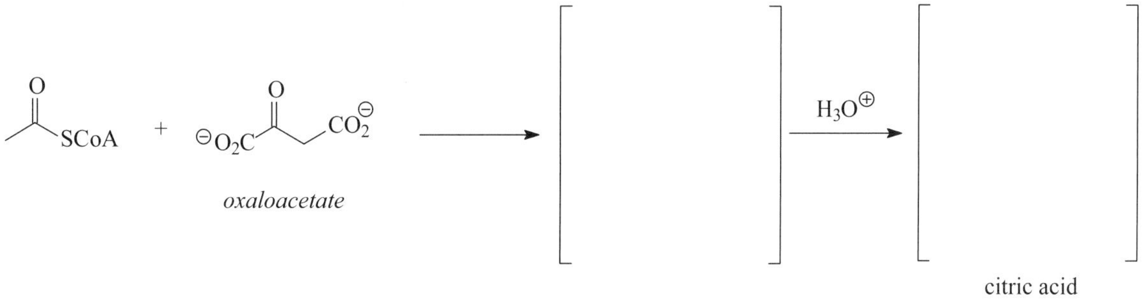

CH₃-CO-SCoA + $^{\ominus}$O₂C-CO-CH₂-CO₂$^{\ominus}$ (oxaloacetate) → [ ] $\xrightarrow{H_3O^{\oplus}}$ [citric acid]

30. The following sequence illustrates how fatty acids are catabolized to acetyl coenzyme A, a process known as β-oxidation. Fill in the brackets.

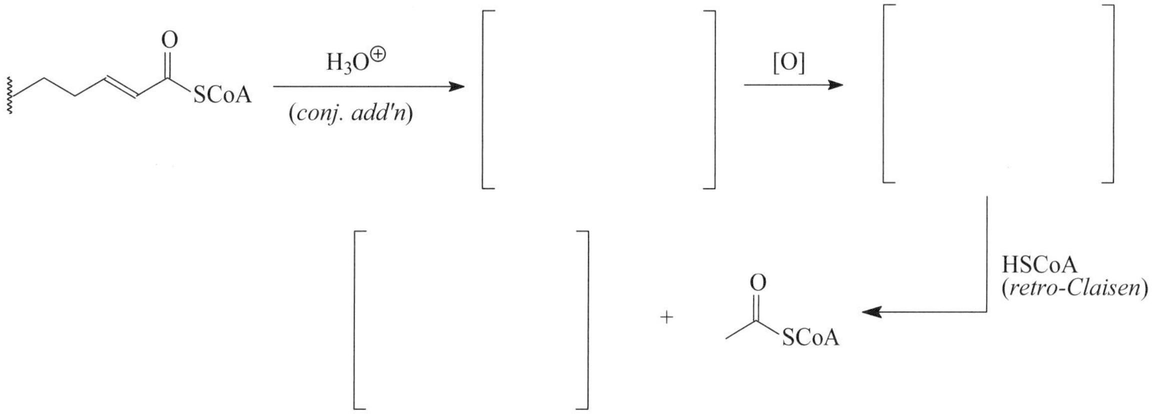

~~~CH₂-CH=CH-CO-SCoA $\xrightarrow[\text{(conj. add'n)}]{H_3O^{\oplus}}$ [ ] $\xrightarrow{[O]}$ [ ]

[ ] + CH₃-CO-SCoA $\xleftarrow[\text{(retro-Claisen)}]{\text{HSCoA}}$

19.1 Reactions

214 • Chapter 19 Carbonyl Condensation Reactions

31. Excessive accumulation of acetyl CoA can lead to metabolic *ketosis* by the following pathway:

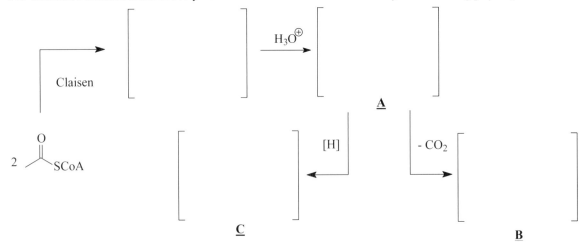

[**A**, **B**, and **C** (unfortunately!) are referred to as "*ketone bodies*;" accumulation of acids **A** and **C** lowers blood pH (*acidosis*).]

---

32. A *cortisone story…*

Cortisone is one of 43 steroids found in adrenal cortical glands. It was first isolated by Kendall (*Mayo Clinic*) in 1934 (extraction of ~ 1,000 lbs of beef adrenal glands yielded only 85 – 200 mg of cortisone). One of the earliest total syntheses of cortisone was published by Sarett (*Merck*) in 1952. The following reactions illustrate his strategy.

a. The initial sequence of reactions formed the A-B-C rings. Draw the missing structures.

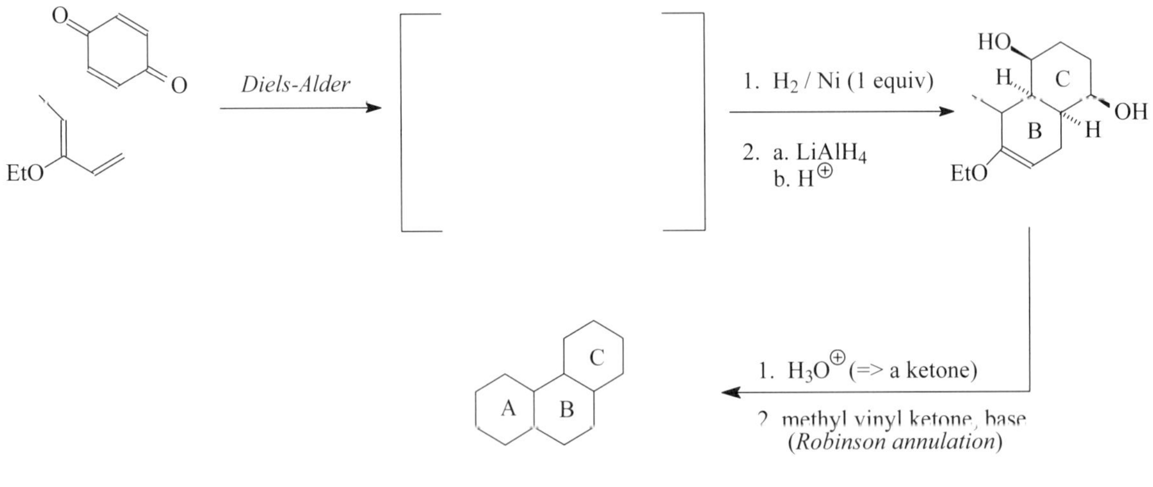

19.1 Reactions

b. Construction of the D ring began as follows. Fill in the bracketed structure and outline the mechanism for step 3.

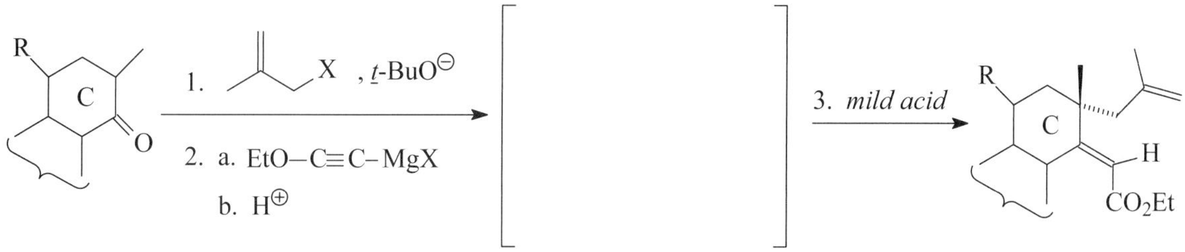

c. Subsequent selective reduction followed by tosylation produced the indicated structure, which was then treated with the sequence of reagents shown. Draw the product of step 2 and give the mechanism for step 3.

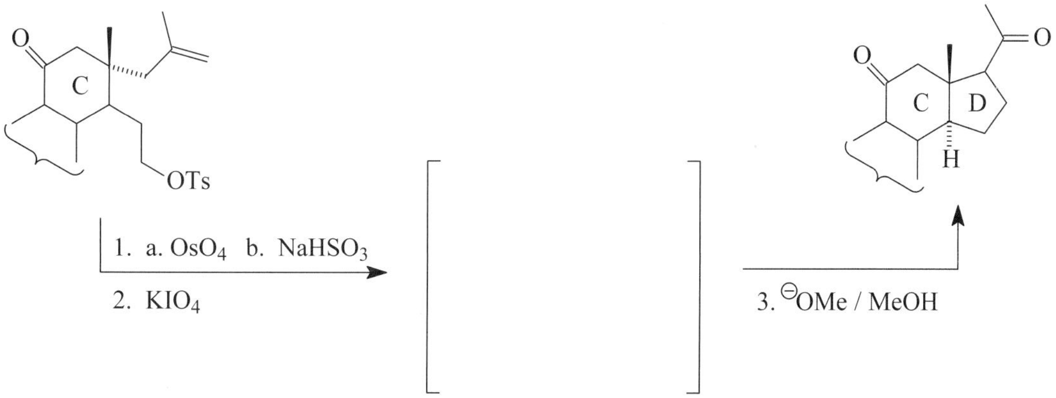

d. The above product was then subjected to the following steps. Draw structures for the critical intermediates in steps 2-4.

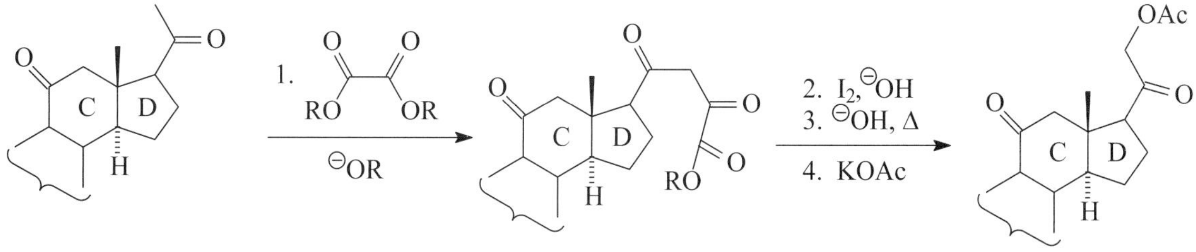

19.1 Reactions

216 • **Chapter 19** Carbonyl Condensation Reactions

e. The final four steps yielded *cortisone*. Deduce the structure of cortisone acetate.

$$\text{steroid with OAc} \xrightarrow[\text{2. } (-H_2O)]{\text{1. HCN}} [\quad] \xrightarrow[\text{(- HCN)}]{\text{3. KMnO}_4, {}^{\ominus}\text{OH}} [\quad \text{cortisone acetate} \quad] \xrightarrow{\text{4. H}_3\text{O}^{\oplus}} \text{cortisone}$$

33.

$$\text{diester} \xrightarrow[\text{2. H}_3\text{O}^{\oplus}, \Delta]{\text{1. }{}^{\ominus}\text{OMe, MeOH}}$$

34.

$$\text{2-nitrobenzaldehyde} \xrightarrow[\text{NaOMe / MeOH}]{\text{CH}_3\text{NO}_2} [\quad \text{C}_8\text{H}_7\text{N}_2\text{O}_5{}^{\ominus} \text{ Na}^{\oplus} \quad] \Rightarrow \text{indigo blue}$$

indigo blue (probably oldest known coloring agent - used to dye bluejeans)

35. A step involving an intra-Michael reaction in Corey's (*Harvard*) synthesis of *longifolene*, a component of Indian turpentine oil:

$$\text{precursor} \xrightarrow{\text{Et}_3\text{N}} [\quad] \Rightarrow \text{longifolene}$$

19.1 Reactions

19.2 Syntheses
Supply a reagent or sequence of reagents that will effect the following conversions.

1. ? + ? $\xrightarrow{\text{via an aldol}}$ 1,3-diphenyl-1-propanol

2. cyclohexanone $\xrightarrow{\text{via an enamine}}$ [2-(3-oxobutyl)cyclohexanone]

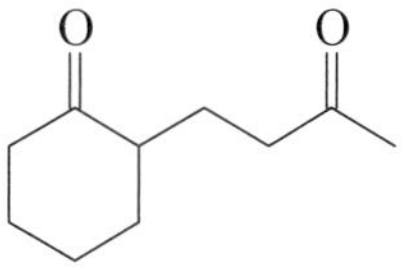

3. ? $\xrightarrow{\text{via an aldol}}$ [2-hydroxycyclohexane-1-carbaldehyde]

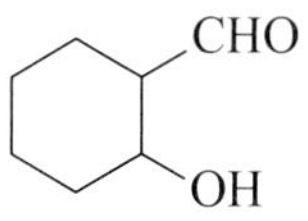

4. ? $\xrightarrow{\text{via an aldol}}$ [1-(2-methylcyclopent-1-en-1-yl)ethan-1-one]

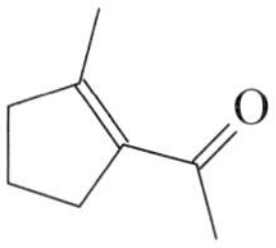

5. diethyl ketone $\longrightarrow$ [3-ethyl-4-methylheptane]

6. acetone $\longrightarrow$ [5-methylhex-4-en-3-one]

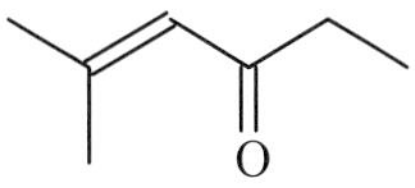

7. 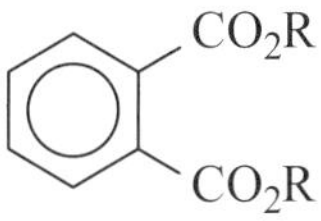 [dialkyl phthalate] $\longrightarrow$ [2-alkoxycarbonyl-1,3-indandione]

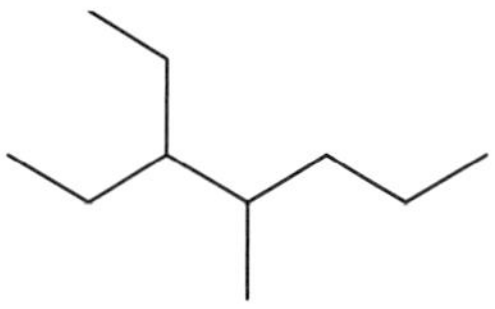

19.2 Syntheses

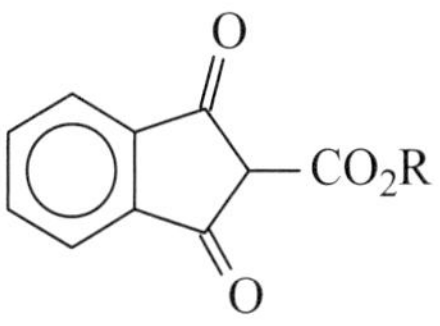

8. cyclohexanone ⟶ 4-benzyl-1,3-cyclohexadione

9. 1-pentene ⟶ [structure: CH$_2$=CH-CH$_2$-CH(CHO)-CH$_2$-CH$_2$-CH$_3$]

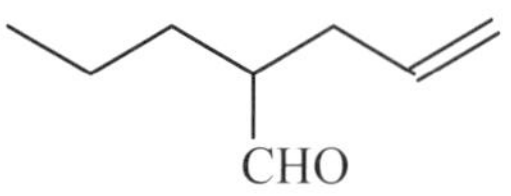

10. phenylacetaldehyde ⟶ PhCH(CH$_2$OH)$_2$

11. ? + ? *via a Robinson annulation* ⟶ [bicyclic dione structure]

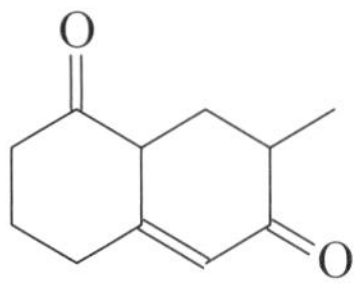

12. 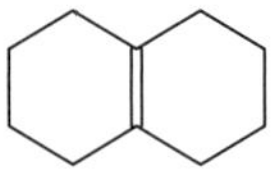 [decalin structure] ⟶ [bicyclic ketone structure]

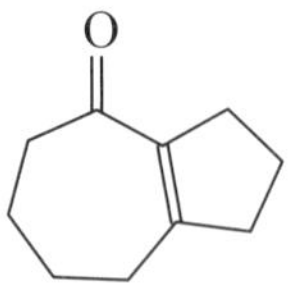

13. cyclohexanone, acetone *via a Wittig** ⟶ [cyclohexylidene acetone]   *Why not *via* an aldol?

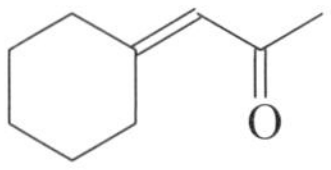

14.  [1,3-cyclopentanedione] *via a Robinson annulation* ⟶ [bicyclic enone dione with methyl]

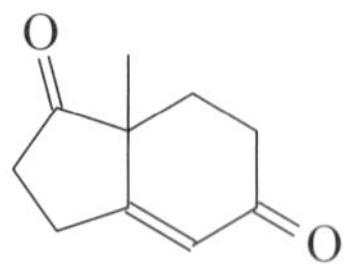

19.2 Syntheses

15.

16. acetone ⟶ [structure: 3,5,5-trimethylcyclohex-2-enone]

19.3 Mechanisms
Outline a detailed mechanism for each of the following. No other reagents than those given are necessary. Use arrows to explain the flow of electrons and show all intermediates.

1. [1-(2-hydroxy-2-methylcyclobutyl)ethanone] —H⁺→ [3-methylcyclohex-2-enone]

2. [cyclopentanone] + CH₂N₂ ⟶ [cyclohexanone] + N₂

3. [2,2-dimethyl-1,3-cyclohexanedione] —NaOEt / EtOH→ [ethyl 4,4-dimethyl-5-oxohexanoate]

4. 2 methyl acrylate $\xrightarrow[\text{2. }^{\ominus}\text{OMe}]{\text{1. CH}_3\text{NH}_2}$ [4-oxo-1-methylpiperidine-3-carboxylic acid methyl ester]

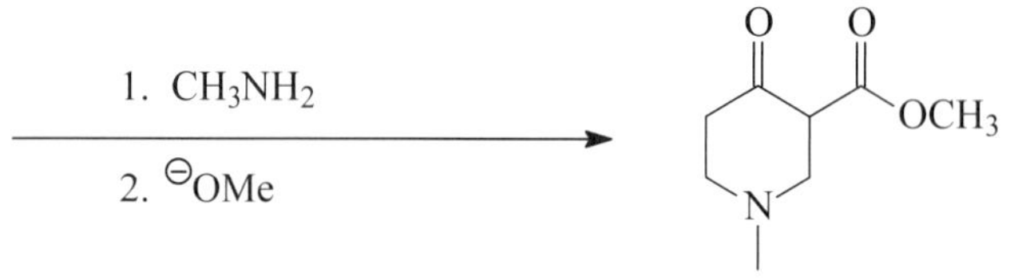

5. *p*-chlorobenzaldehyde + CHBr$_3$ $\xrightarrow{^{\ominus}\text{OH}}$ Cl-C$_6$H$_4$-CH(OH)-CO$_2^{\ominus}$

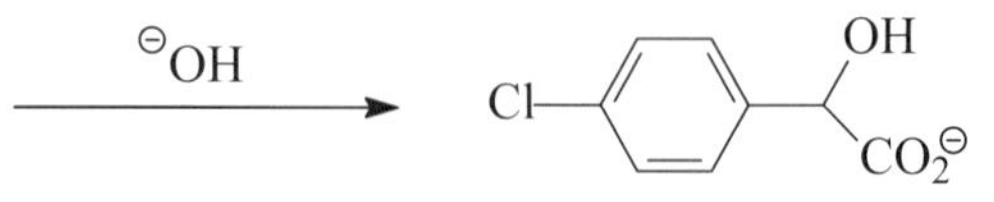

6. 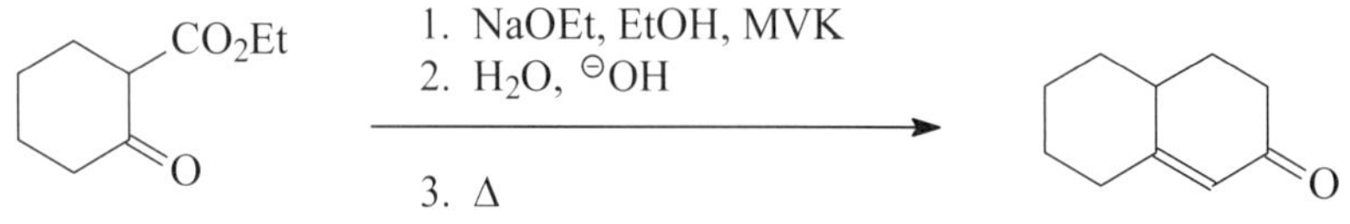 [ethyl 2-oxocyclohexanecarboxylate] $\xrightarrow[\text{3. }\Delta]{\text{1. NaOEt, EtOH, MVK} \atop \text{2. H}_2\text{O, }^{\ominus}\text{OH}}$ [octalone]

7. 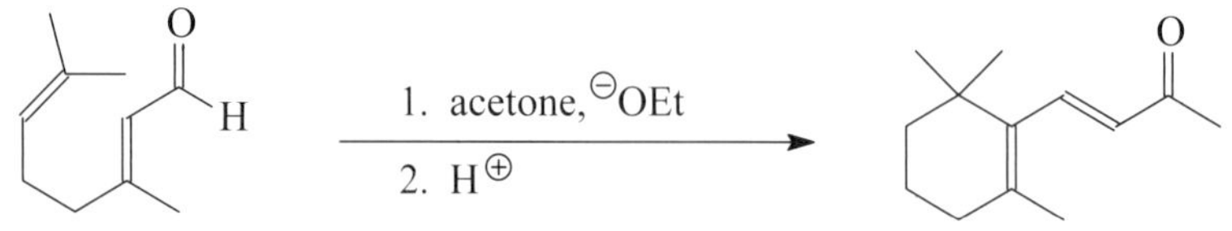 *geranial* $\xrightarrow[\text{2. H}^{\oplus}]{\text{1. acetone, }^{\ominus}\text{OEt}}$ [β-ionone]

(*a step in the commercial synthesis of vitamin A*)

8. 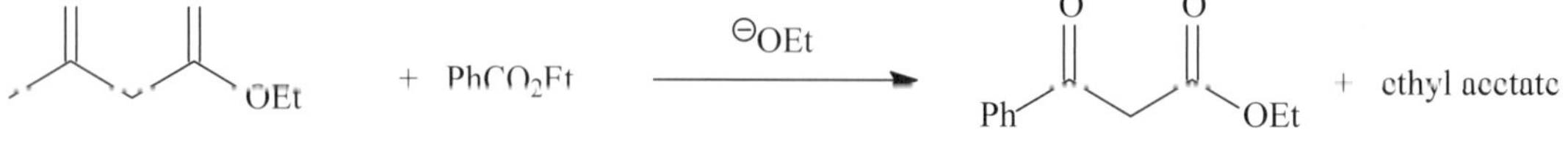 [ethyl acetoacetate] + PhCO$_2$Et $\xrightarrow{^{\ominus}\text{OEt}}$ PhCOCH$_2$CO$_2$Et + ethyl acetate

19.3 Mechanisms

9. 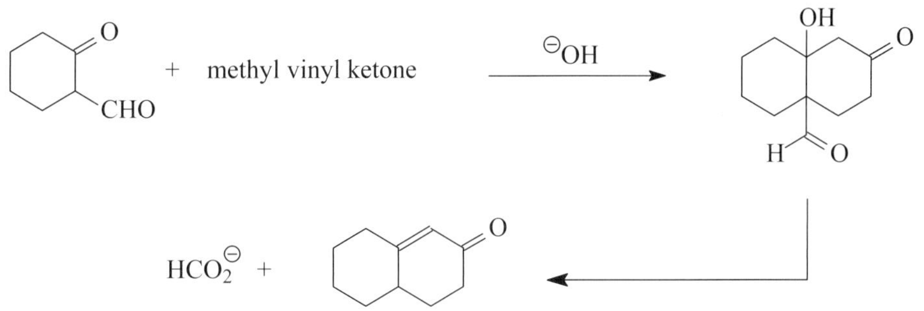 [cyclohexanone with CHO substituent] + methyl vinyl ketone $\xrightarrow{\ominus OH}$ [decalone intermediate with OH and CHO]

$HCO_2^{\ominus}$ + [octahydronaphthalenone] ←

10. The anaerobic breakdown of glucose (glycolysis) involves the following isomerization and retro-aldol:

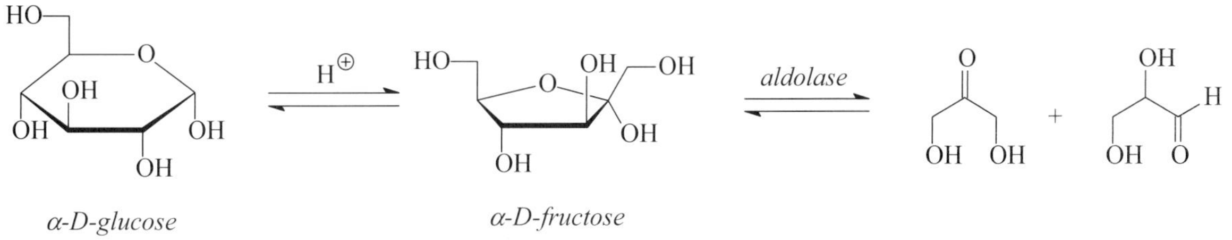

α-D-glucose $\xrightleftharpoons{H^{\oplus}}$ α-D-fructose $\xrightleftharpoons{aldolase}$ [dihydroxyacetone] + [glyceraldehyde]

11. 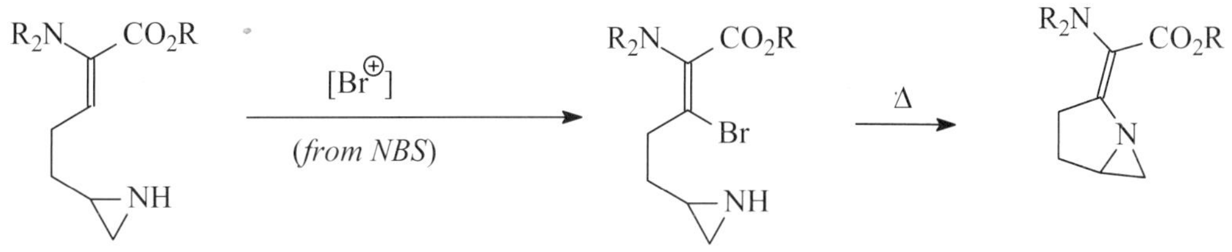 [enamine ester with aziridine] $\xrightarrow[\text{(from NBS)}]{[Br^{\oplus}]}$ [α-bromo intermediate] $\xrightarrow{\Delta}$ [bicyclic product]

12. The following illustrates the *Stobbe reaction*. Hint: a key intermediate is a γ-lactone.

[diethyl succinate] + $Ph_2C=O$ $\xrightarrow{\text{base}}$ [Ph₂C=C(CO₂R)CH₂CO₂⁻]

19.3 Mechanisms

13. 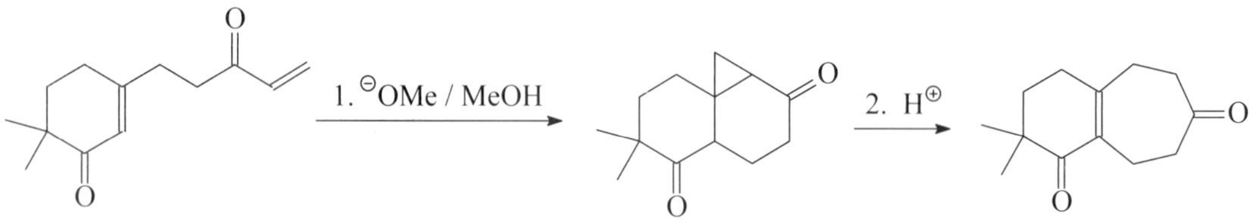 [reaction scheme]

14. 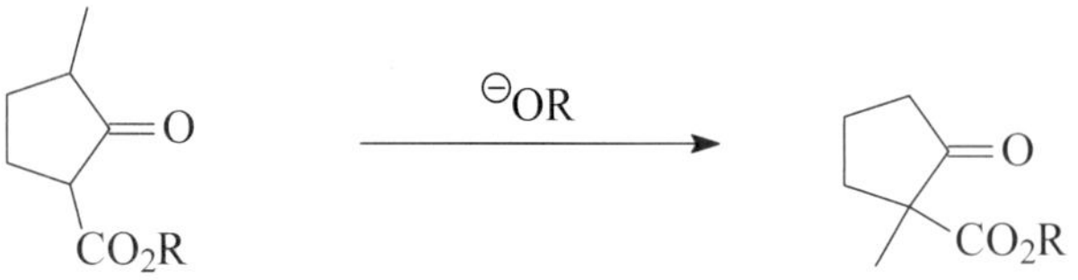 [reaction scheme]

15. 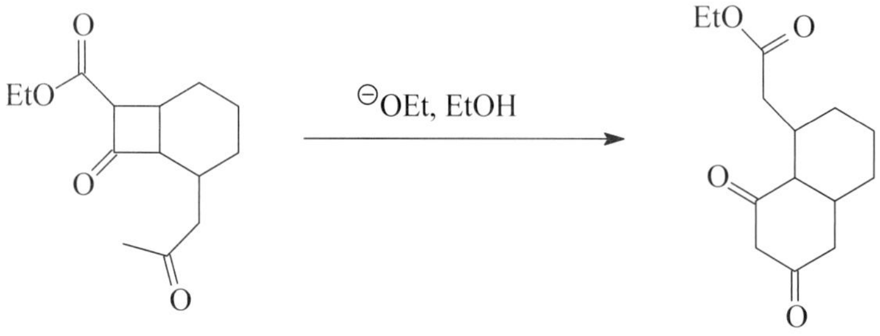 [reaction scheme]

16. The final stage of Johnson's (*Stanford*) historic total synthesis of *progesterone* (give a mechanism for step 2):

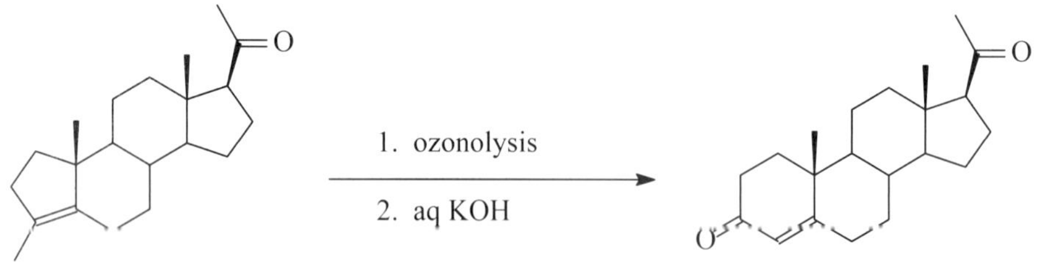

[reaction scheme] → *progesterone*

1. ozonolysis
2. aq KOH

19.3 Mechanisms

23. The biosynthesis of porphyrin rings (*e.g.*, *heme*) begins with an annulation reaction that involves an aldol reaction and imine formation in the dimerization of δ-aminolevulinic acid (ALA) to form *porphobilinogen*.

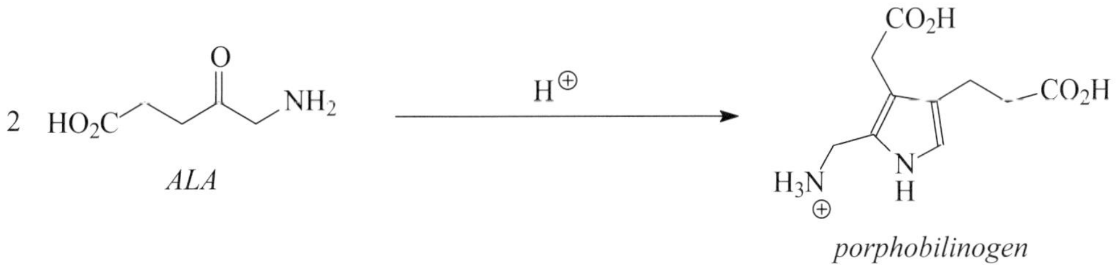

2 HO₂C−CH₂−CH₂−C(O)−CH₂−NH₂ (ALA) $\xrightarrow{H^\oplus}$ porphobilinogen

24. Several steps in Sheehan's (*MIT*) total synthesis of *penicillin V* are shown below.

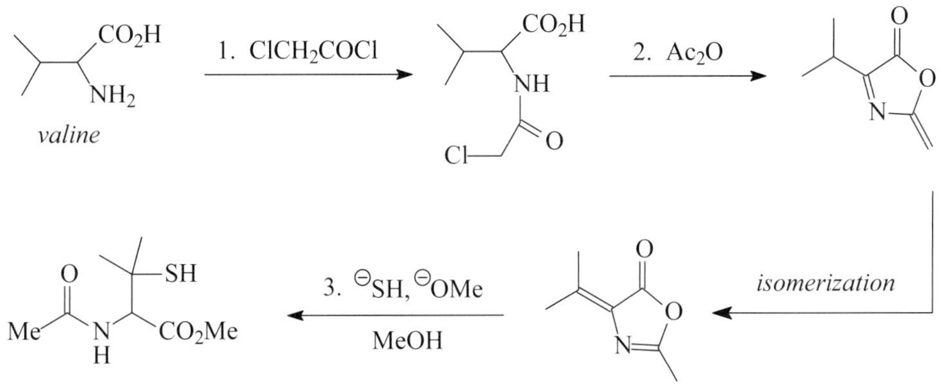

valine (iPr-CH(NH₂)-CO₂H) →[1. ClCH₂COCl]→ iPr-CH(NHCOCH₂Cl)-CO₂H →[2. Ac₂O]→ (oxazolone with =CH₂) →[isomerization]→ (oxazolone with =C(CH₃)₂, 2-methyl) →[3. ⁻SH, ⁻OMe / MeOH]→ MeC(O)NH-C(CH₃)₂(SH)-CH(...)-CO₂Me

a. Propose a mechanism for step 2.

b. Propose a mechanism for step 3.

25. The biosynthesis of fatty acids begins with a Claisen-like reaction:

CH₃C(O)SR + ⁻O₂C-CH₂-C(O)SR → CO₂ + CH₃C(O)CH₂C(O)SR

19.3 Mechanisms

26. The $C_\alpha - C_\beta$ bond in β-hydroxyketones is easily cleaved via a retro-aldol reaction; the carbonyl – C_α bond is unreactive. In α-hydroxyketones, however, the $C_\alpha - C_\beta$ bond is unreactive; but, <u>in the presence of *thiamine*</u>, the carbonyl – C_α bond can be cleaved (2).

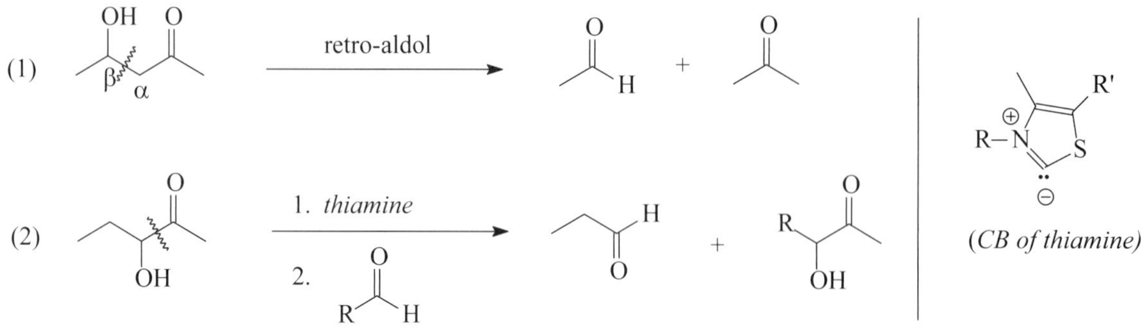

Recalling the mechanism for *thiamine*-assisted decarboxylation of α-ketocarboxylic acids (problem 16.3, 13), formulate a mechanism for reaction (2).

27. The biosynthesis of cholesterol begins with the formation of HMG-CoA (3-<u>h</u>ydroxy-3-<u>m</u>ethyl<u>g</u>lutaryl coenzyme A):

a. Formulate a mechanism.

19.3 Mechanisms

b. HMG-CoA is subsequently reduced to *mevalonate* by an enzyme, HMG-CoA reductase. Because this reaction is the major control (rate-limiting) step, considerable research has been devoted toward developing a class of medicines that inhibits the action of this enzyme, notably the statins [*e.g., atorvastatin* (Lipitor™)].

HMG-CoA $\xrightarrow{\text{a reductase}}$ mevalonate $\xrightarrow{\text{ATP}}$ [phosphorylated product]

$$P = -\overset{\overset{O}{\|}}{\underset{OH}{P}}-O^{\ominus}$$

$$PP = -\overset{\overset{O}{\|}}{\underset{O^{\ominus}}{P}}-O-\overset{\overset{O}{\|}}{\underset{OH}{P}}-O^{\ominus}$$

Mevalonate then undergoes phosphorylation and decarboxylation to form *I-PP* (isopentenyl pyrophosphate) and *DMA-PP* (dimethylallyl pyrophosphate) – recall problem 9.4, 19a. Outline the mechanisms for decarboxylation to form *I-PP* and isomerization of the latter to form *DMA-PP*.

[structure] $\xrightarrow{-CO_2}$ I-PP $\rightleftharpoons$ DMA-PP

Lipitor™

19.3 Mechanisms

CHAPTER 20
AMINES

20.1 Reactions
Draw the structural formula of the major organic product(s). Show stereochemistry where appropriate.

1. phthalimide
   1. base
   2. ClCH(CO$_2$Et)$_2$
   3. base
   4. isopropyl chloride
   5. H$_3$O$^\oplus$

   → *alanine*

2. 2-methyl-2-hexylamine
   1. (XS) CH$_3$I
   2. Ag$_2$O, H$_2$O, Δ
   3. OsO$_4$
   4. NaHSO$_3$
   5. (XS) COCl$_2$
   6. NH$_3$

   → C$_9$H$_{18}$N$_2$O$_4$ - Miltown™

3. N-ethylcyclohexylamine
   1. (XS) MeI
   2. Ag$_2$O, H$_2$O
   3. Δ

4.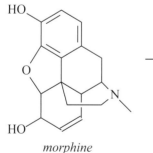

   1. (XS) MeI
   2. Ag$_2$O, H$_2$O
   3. Δ

   coniine (toxin in hemlock, killed Socrates)

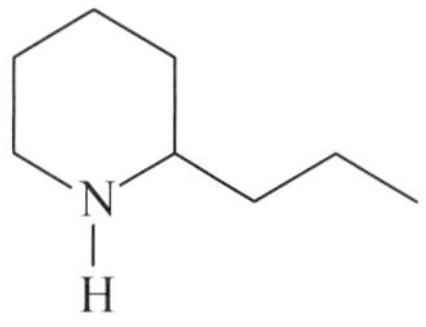

5. *morphine*
   1. (XS) CH$_3$I
   2. Ag$_2$O, H$_2$O, Δ

6. Ph$_2$N—C(Me)(H)—C(H)(Me)(Et)
   1. H$_2$O$_2$
   2. Δ

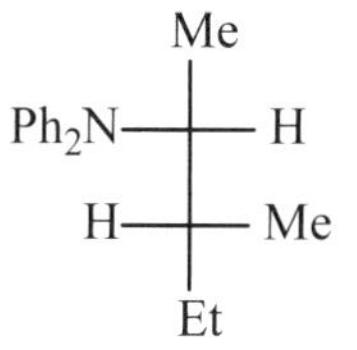

7. cyclopropanecarboxylic acid $\xrightarrow{\text{1. SOCl}_2 \text{ 2. NaN}_3 \text{ 3. }\Delta, \text{H}_2\text{O}}$

8. 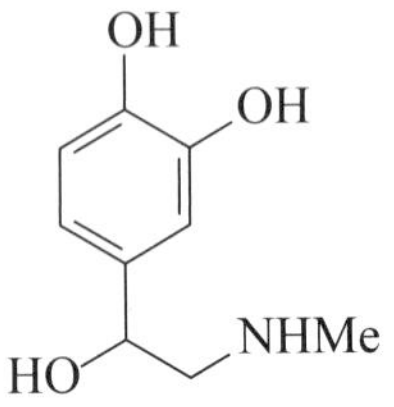 epinephrine $\xrightarrow{\text{1. (XS) MeI} \text{ 2. Ag}_2\text{O, H}_2\text{O, }\Delta}$

9. 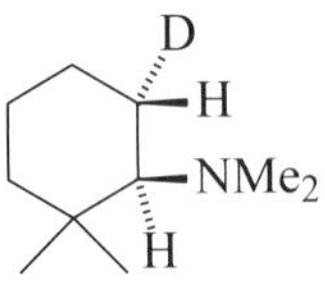 (cyclohexane with D, H, NMe$_2$, H substituents and gem-dimethyl)
   - Hofmann elimination
   - Cope elimination

10. 3-pentanone + dimethylamine + formaldehyde $\xrightarrow[\text{(Mannich rx)}]{\text{H}^\oplus}$

11. 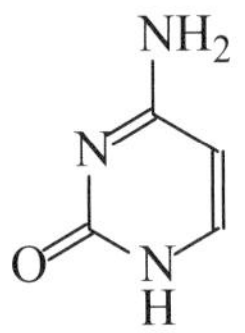 cytosine $\xrightarrow{\text{HONO, H}_2\text{O}}$ uracil

12. 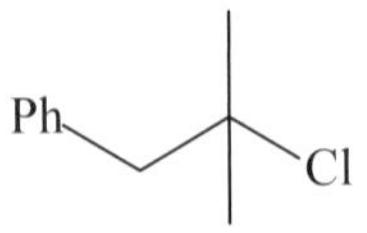 Ph-CH$_2$-C(CH$_3$)$_2$-Cl $\xrightarrow{\text{1. Li 2. CO}_2 \text{ 3. H}^\oplus \text{ 4. SOCl}_2 \text{ 5. NH}_3 \text{ 6. Br}_2, {}^\ominus\text{OH, H}_2\text{O}}$ phentermine (a diet drug)

20.1 Reactions

13.

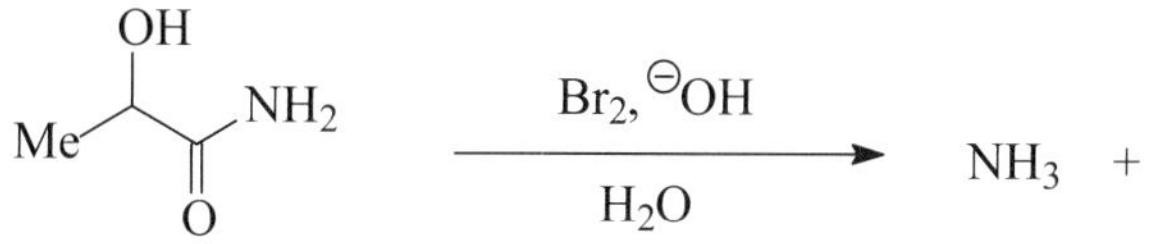

Me-CH(OH)-C(=O)-NH₂ →[Br₂, ⁻OH / H₂O] NH₃ +

14. Ph₂CHOH →
   1. NaNH₂
   2. ethylene oxide
   3. PBr₃
   4. Me₂NH

   diphenhydramine (Benadryl™ - antihistamine)

15. 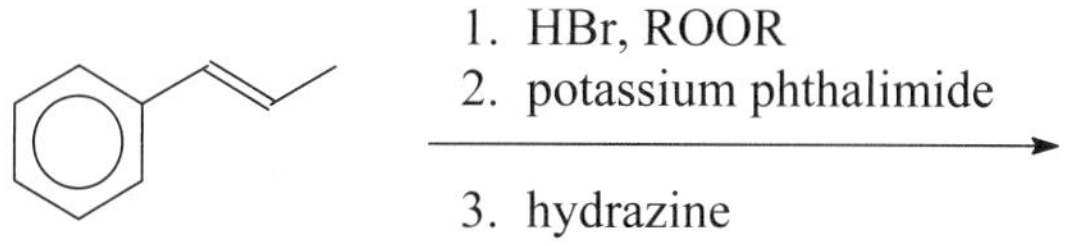 (trans-β-methylstyrene) →
   1. HBr, ROOR
   2. potassium phthalimide
   3. hydrazine

   amphetamine (CNS stimulant)

16. 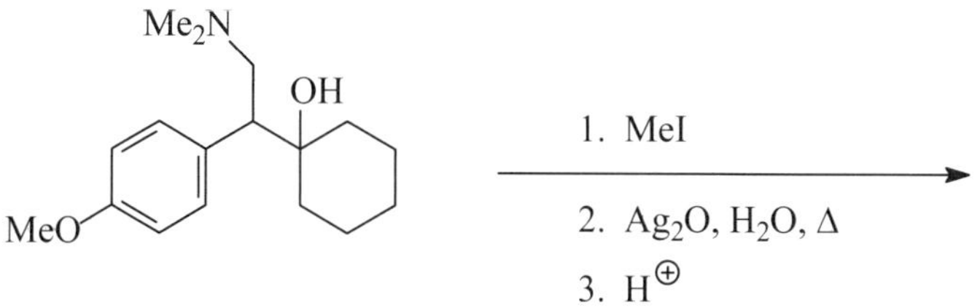 Effexor™ (antidepressant) →
   1. MeI
   2. Ag₂O, H₂O, Δ
   3. H⁺

   (gives a positive DNP test)

17. 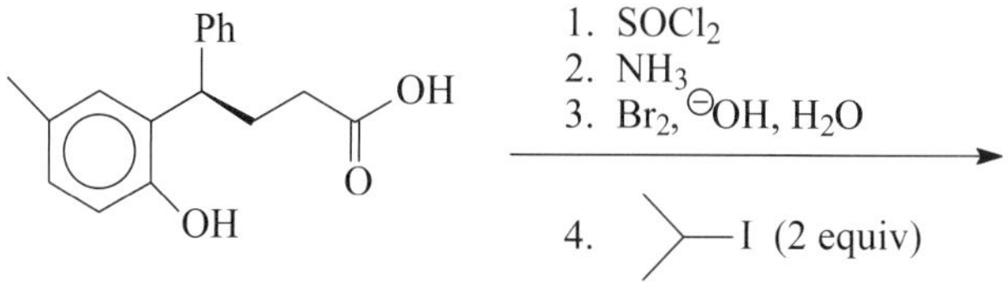 →
   1. SOCl₂
   2. NH₃
   3. Br₂, ⁻OH, H₂O
   4. (iPr)–I (2 equiv)

   Detrol™ (treatment of urinary incontinence)

18. 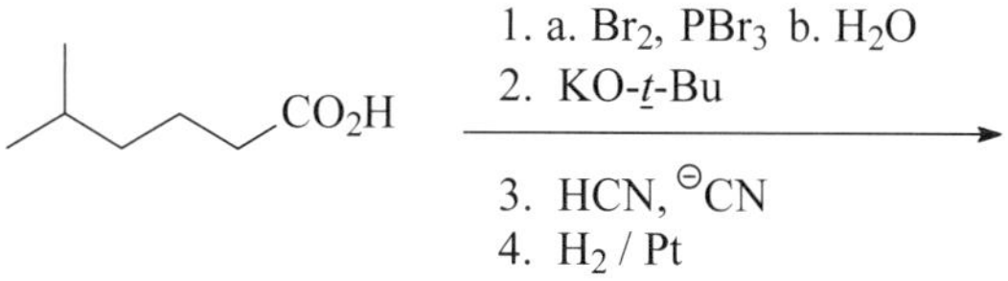 (4-methylpentyl-CO₂H) →
   1. a. Br₂, PBr₃ b. H₂O
   2. KO-*t*-Bu
   3. HCN, ⁻CN
   4. H₂ / Pt

   pregabalin (Lyrica™ - first treatment approved for fibromyalgia)

20.1 Reactions

19. *p*-toluidine (*p*-aminotoluene) $\xrightarrow{\begin{array}{l}1.\ Br_2\\ 2.\ NaNO_2,\ HCl\\ 3.\ KI\end{array}}$

20. benzonitrile $\xrightarrow{\begin{array}{l}1.\ Cl_2,\ FeCl_3\\ 2.\ NaNH_2\ /\ NH_3\\ 3.\ KNO_2,\ H^\oplus\\ 4.\ CuCN\\ 5.\ H_3O^\oplus\end{array}}$

21. 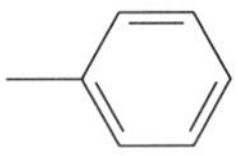 toluene $\xrightarrow{\begin{array}{l}1.\ KMnO_4,\ H^\oplus\\ 2.\ \text{fuming nitric acid}\\ 3.\ Fe,\ HCl\\ 4.\ NaNO_2,\ HCl\\ 5.\ HBF_4\end{array}}$

22. 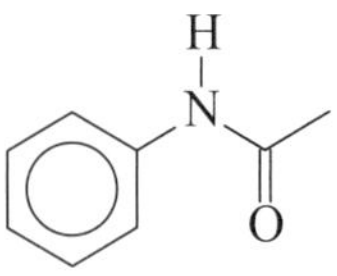 acetanilide $\xrightarrow{\begin{array}{l}1.\ Br_2,\ Fe\\ 2.\ Cl_2,\ Fe\\ 3.\ H_2O,\ ^\ominus OH\\ 4.\ I\text{-}Cl,\ Fe\\ 5.\ HONO\\ 6.\ H_3PO_2\end{array}}$

23. 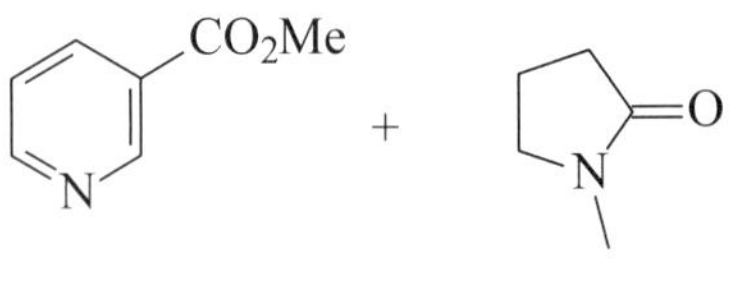 methyl nicotinate + N-methyl-2-pyrrolidone $\xrightarrow{\begin{array}{l}1.\ NaH\\ 2.\ H_3O^\oplus\\ 3.\ NaBH_4\\ 4.\ HBr\ (S_N1)\\ 5.\ \Delta\ (\text{-HBr})\end{array}}$ *nicotine*

24. 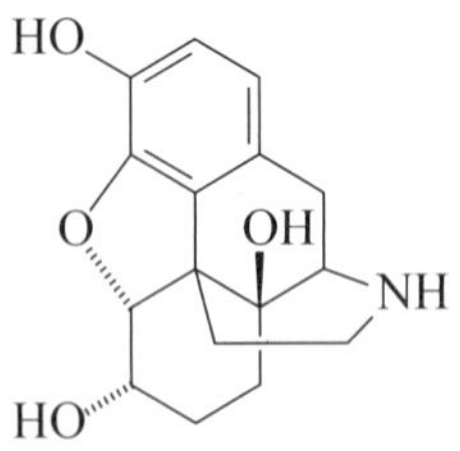 $\xrightarrow{\begin{array}{l}1.\ \text{cyclobutanecarbaldehyde, }NaBH_3CN,\ EtOH\\ 2.\ HCl\end{array}}$ Nubain™ (narcotic)

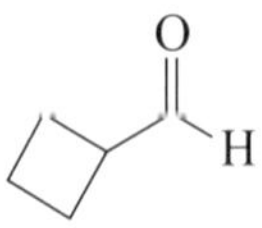

20.1 Reactions

20.2 Syntheses
Supply a reagent or sequence of reagents that will effect the following conversions.

1. methylcyclohexane $\longrightarrow$ 1-methyl-1-cyclohexylamine

2. cyclohexane $\longrightarrow$ cyclohexyl-CH$_2$NH$_2$

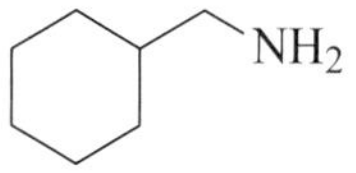

3. isopentane $\longrightarrow$ 3-methyl-1-butene (*via a Hofmann elimination*)

4. 3-methyl-1-butene $\longrightarrow$ isopropylamine

5. mescaline (from peyote cactus) [3,4,5-trimethoxyphenethylamine] $\longrightarrow$ 3,4,5-trimethoxybenzaldehyde $\longrightarrow$ 3,4,5-trimethoxy-N-phenylbenzylamine

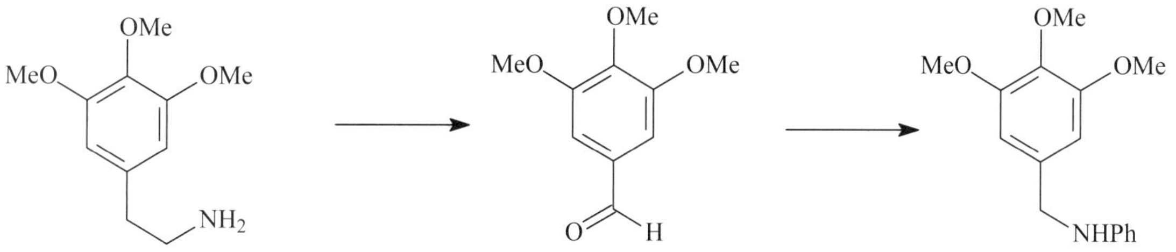

6. *p*-nitrotoluene $\longrightarrow$ *p*-nitrobenzylamine

7. *p*-nitrotoluene $\longrightarrow$ *p*-nitroaniline

8.  [1-methylcyclopentan-1-amine] ⟶ [1,4-dioxaspiro[4.4]nonane]

9. 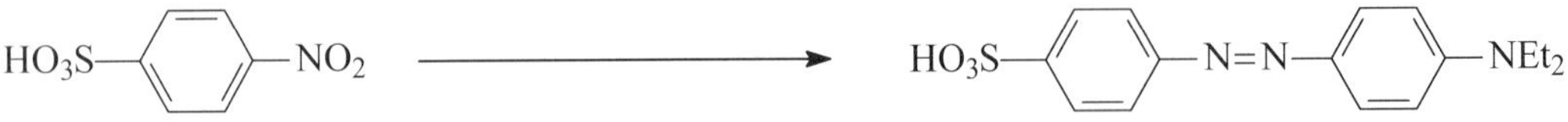 HO₃S—[C₆H₄]—NO₂ ⟶ HO₃S—[C₆H₄]—N=N—[C₆H₄]—NEt₂

   methyl orange

10. 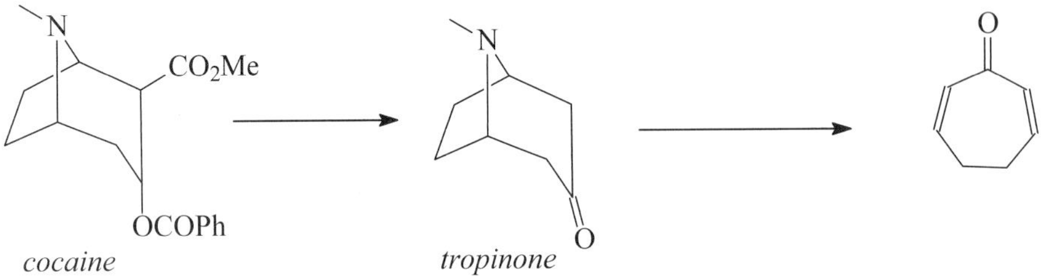 cocaine ⟶ tropinone ⟶ cyclohepta-2,6-dienone

11. toluene ⟶ 2,6-dichlorotoluene

12. *p*-nitroaniline ⟶ acetaminophen (Tylenol™)

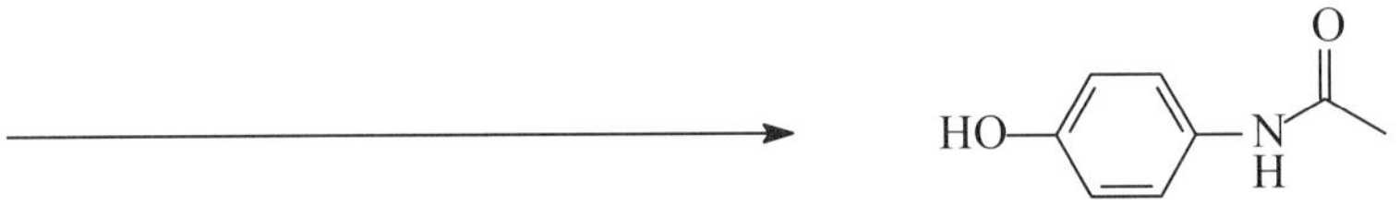

13. 1-nitro-2,6-dimethylbenzene, ethylene oxide, diethylamine ⟶ lidocaine

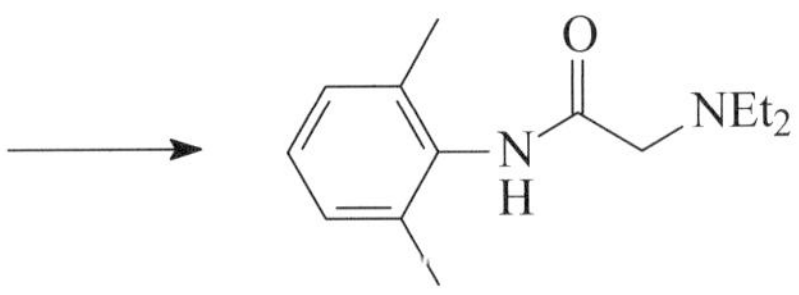

14. benzene ⟶ anisole

20.2 Syntheses

15.  [phenyl-NO₂] ⟶ [Ph-N=N-C₆H₄-NH₂]
butter yellow

16. benzyl methyl ketone ⟶ Ph-CH₂-CH(CH₃)-NH(CH₃)
methamphetamine

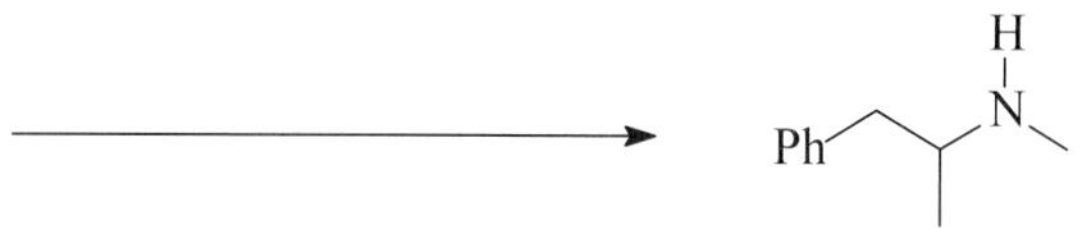

17. 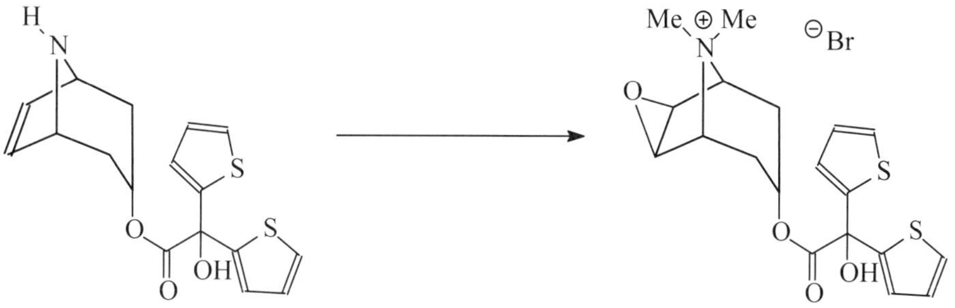 [tropane-like structure with NH, alkene, and diphenylthiophene glycolate ester] ⟶ [quaternary N,N-dimethyl ammonium with epoxide, Br⁻ counterion]
tiotropium bromide (Spireva™ - bronchodilator)

18. 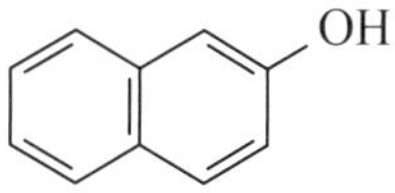 2-naphthol, benzene ⟶ [1-(4-nitrophenylazo)-2-naphthol]
para-red dye

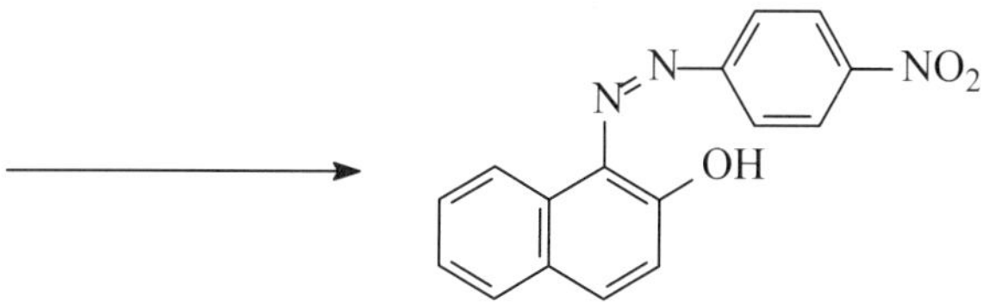

19. benzene ⟶ 1,3-dideuteriobenzene

20. 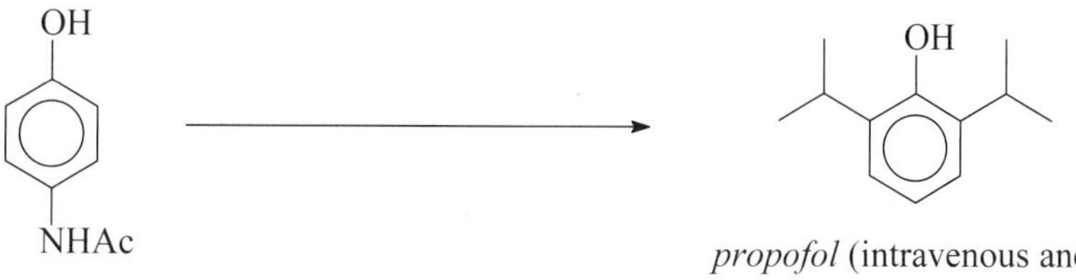 4-acetamidophenol (OH, NHAc) ⟶ 2,6-diisopropylphenol
propofol (intravenous anesthetic)

20.2 Syntheses

20.3 Mechanisms

Outline a detailed mechanism for each of the following. No other reagents than those given are necessary. Use arrows to explain the flow of electrons and show all intermediates.

1. 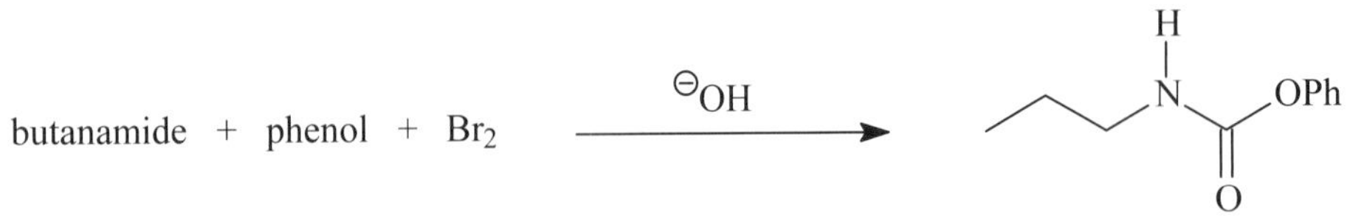 butanamide + phenol + Br_2 $\xrightarrow{\ominus OH}$ propyl-NH-C(=O)-OPh

2. phthalimide $\xrightarrow[\text{2. } H^{\oplus}]{\text{1. } Cl_2, H_2O, \ominus OH}$ anthranilic acid (2-aminobenzoic acid)

3. anthranilic acid $\xrightarrow[\text{3. 1,3-butadiene}]{\substack{\text{1. HONO} \\ \text{2. adjust to pH 8}}}$ 1,4-dihydronaphthalene

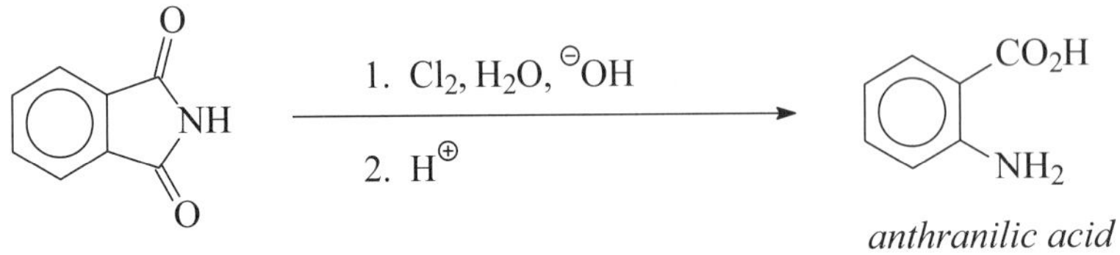

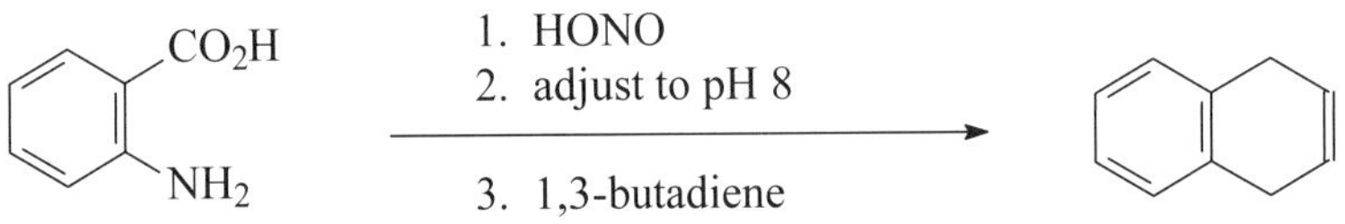

4. indanone $\xrightarrow[\text{3. } NaNO_2, HCl]{\substack{\text{1. HCN, } \ominus CN \\ \text{2. } H_2/Pt}}$ α-tetralone

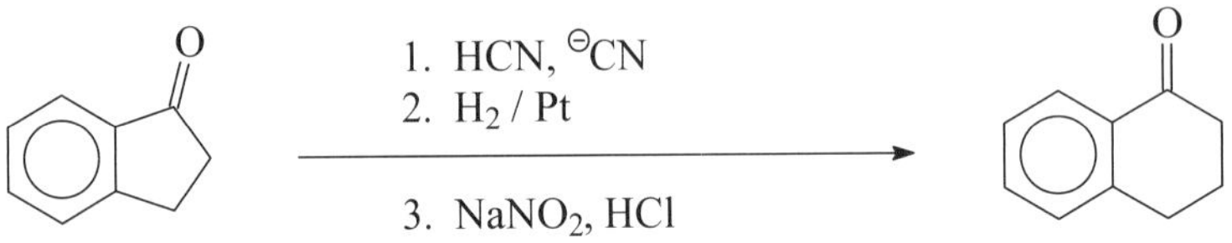

5. 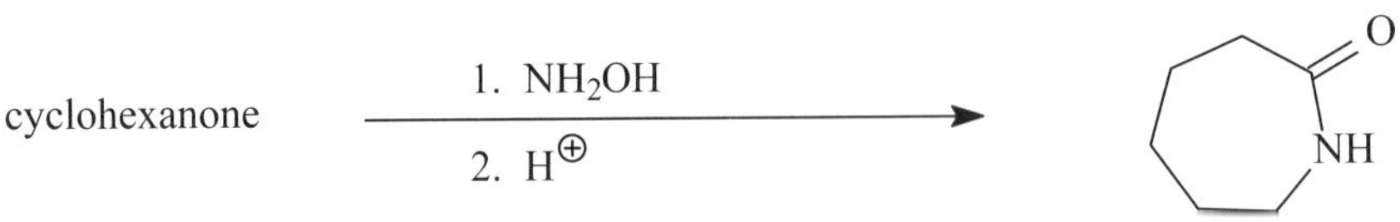 cyclohexanone $\xrightarrow[\text{2. } H^{\oplus}]{\text{1. } NH_2OH}$ ε-caprolactam

(This is an example of the *Beckmann rearrangement*, similar to the Hofmann and Curtius rearrangements.)

6. Sir Robert Robinson (*Oxford*) observed that *thebaine* (a dimethylated derivative of the alkaloid morphine) forms phenyldihydrothebaine when treated with phenylmagnesium halide. Formulate a mechanism and draw the Hofmann elimination product.

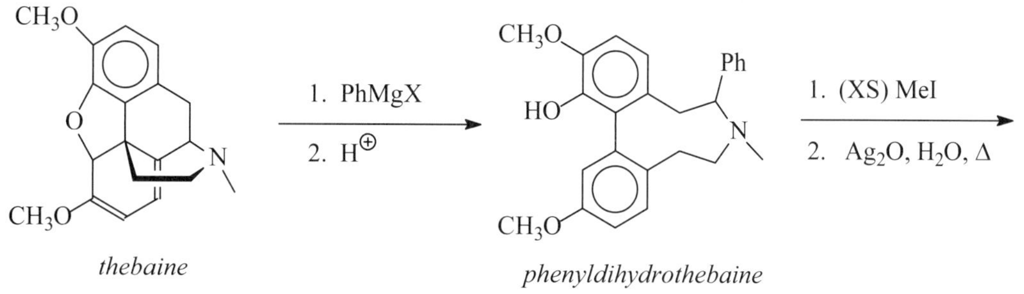

7. The *Curtius rearrangement* not only occurs with *acyl* halides but also *alkyl* azides. Draw the bracketed structure and deduce a mechanism for its formation.

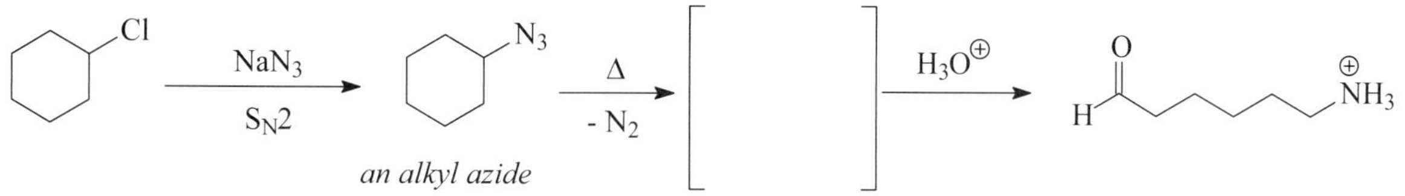

8. Hydrazoic acid (HN_3) undergoes addition to ketones to form a product that readily rearranges to an amide (*Schmidt reaction*):

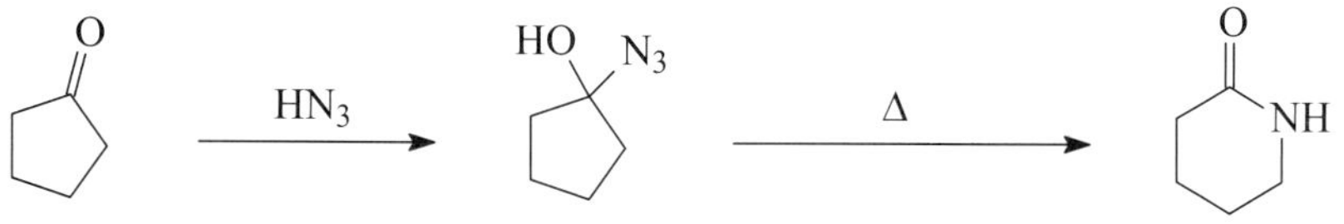

20.3 Mechanisms

9. The *Fischer indole synthesis* involves an isomerization known as a [3,3] sigmatropic rearrangement, shown by the arrows below:

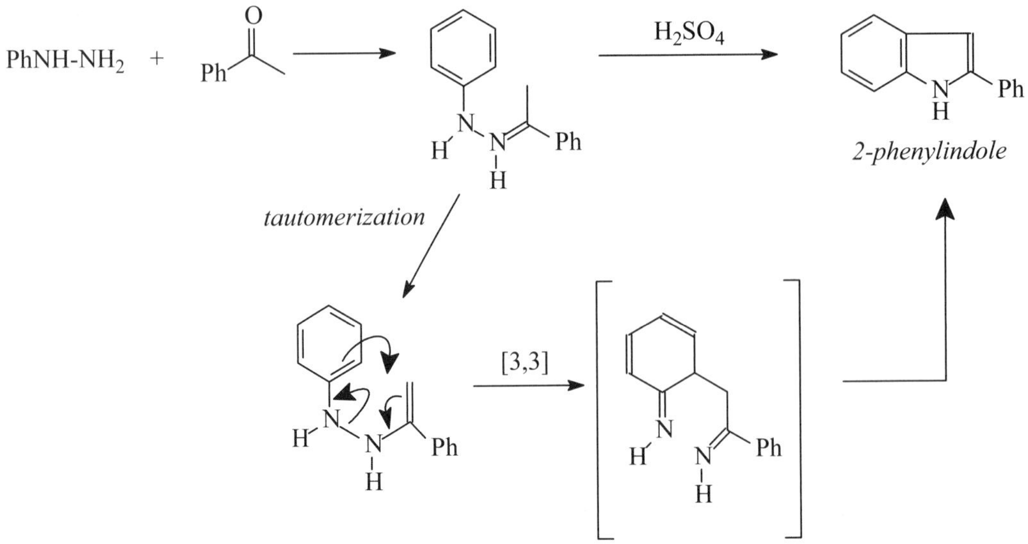

PhNH-NH₂ + PhC(O)CH₃ → [phenylhydrazone] —H₂SO₄→ 2-phenylindole

tautomerization → [ene-hydrazine] —[3,3]→ [bracketed intermediate] → 2-phenylindole

Outline a mechanism for conversion of the intermediate in brackets to the indole product.

10. 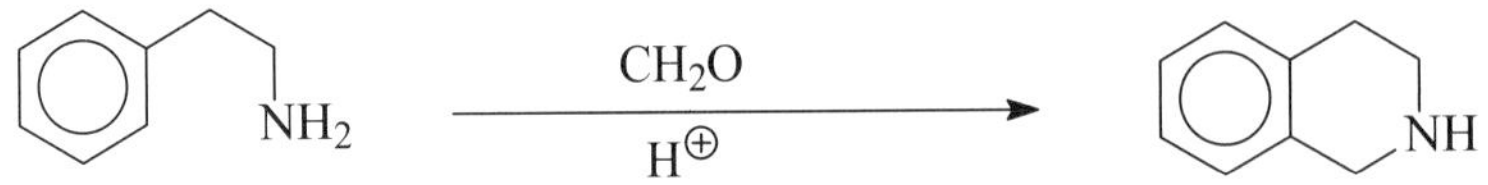 PhCH₂CH₂NH₂ —CH₂O, H⁺→ 1,2,3,4-tetrahydroisoquinoline

11. acetophenone + CH₂O + NH₃ —H⁺→ (PhCOCH₂CH₂)₃N

20.3 Mechanisms

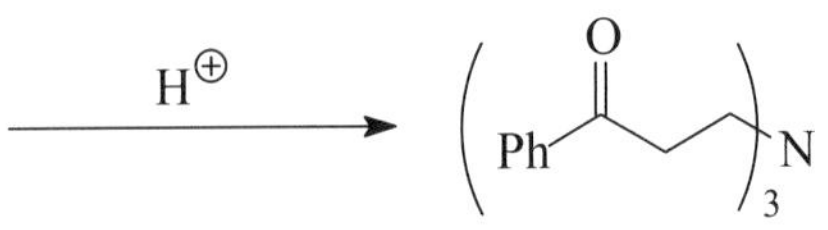

12. *Atropine*, an antidote to cholinesterase inhibitors (*e.g.*, nerve gases), can be easily synthesized from tropinone. The first total synthesis of tropinone required 17 steps. Years later Robinson (*Oxford*) accomplished its synthesis in a one-step, one-pot reaction (*Robinson-Schopf condensation*)! Sketch the critical intermediates in this synthesis.

13. The *Ritter reaction* offers a way to prepare amides (or, by subsequent hydrolysis, amines) from good precursors to carbocations:

a. [1-methylcyclohexene] + acetonitrile $\xrightarrow{H_2SO_4}$ [N-(1-methylcyclohexyl)acetamide]

b. [4-methoxybenzonitrile] + [t-butyl acetate] $\xrightarrow{H_2SO_4}$ [N-t-butyl-4-methoxybenzamide]

14. A convenient method of synthesizing pure *secondary* amines involves (1) treating the sulfonamide of a *primary* amine with hydroxide, followed by (2) an alkyl halide, then (3) hydrolysis. Outline such an approach to preparing N-methylaniline.

20.3 Mechanisms

15. The *Corey-Link reaction* (step 2) may be used to prepare α-amino acids:

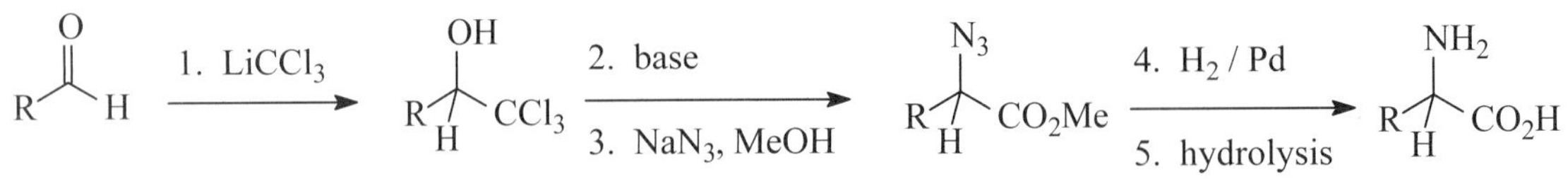

$$R-CHO \xrightarrow{\text{1. LiCCl}_3} R-CH(OH)-CCl_3 \xrightarrow[\text{3. NaN}_3,\text{ MeOH}]{\text{2. base}} R-CH(N_3)-CO_2Me \xrightarrow[\text{5. hydrolysis}]{\text{4. H}_2/\text{Pd}} R-CH(NH_2)-CO_2H$$

a. Outline a mechanism for steps 2 and 3.

b. Account for the product in a mechanistically similar reaction:

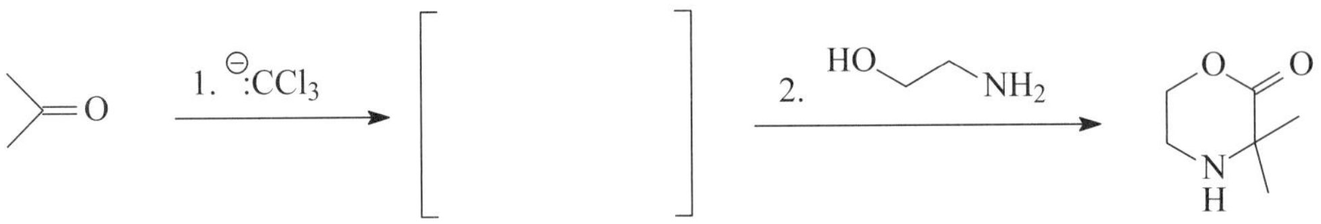

acetone $\xrightarrow{\text{1. }^{\ominus}:CCl_3}$ [ ] $\xrightarrow{\text{2. HOCH}_2CH_2NH_2}$ morpholin-2-one (3,3-dimethyl)

16. *Parkinson's disease* is associated with low levels of dopamine, a neurotransmitter. The enzyme monoamine oxidase (MAO) deaminates dopamine, thereby decreasing its concentration. One approach to treating Parkinson's utilizes (-)-Deprenyl™, a "suicide inhibitor" to MAO. The mechanism first involves oxidation of the drug by a flavin cofactor of MAO, followed by a conjugate addition reaction between the reduced flavin and oxidized drug to irreversibly "kill" any future normal activity by the MAO enzyme. Outline the mechanism for formation of the adduct.

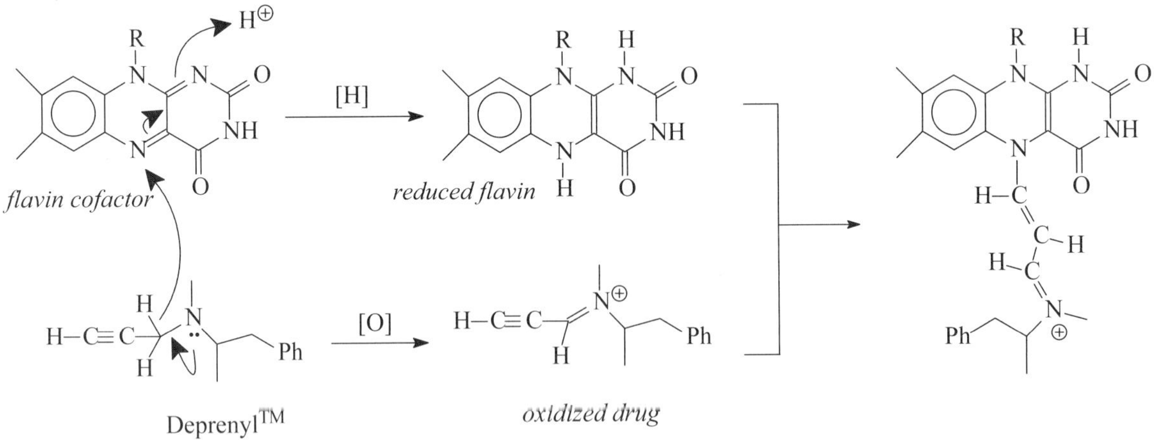

20.3 Mechanisms

SOLUTIONS TO PROBLEMS

CHAPTER 1
THE BASICS

1.1 Hybridization, formulas, physical properties

1. a. Seldane™: $C_{32}H_{42}NO_2$ Relenza™: $C_{12}H_{20}N_4O_7$

   b. [structures shown with labeled arrows a, b, c, d]

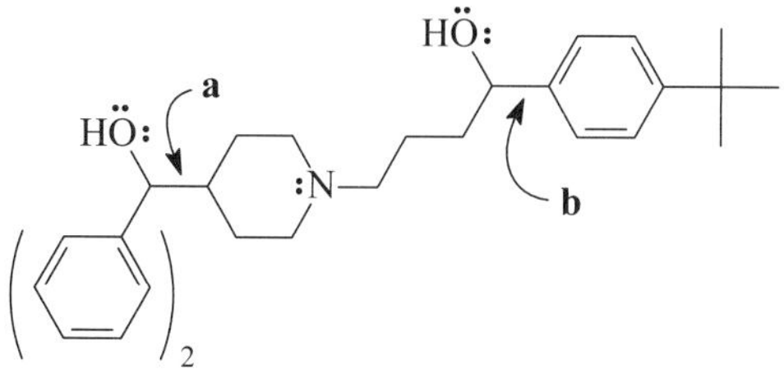

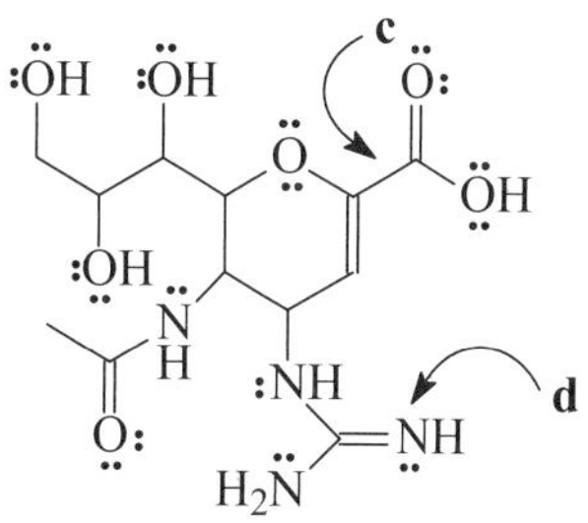

   c. **a**: $sp^3 - sp^3$; **b**: $sp^3 - sp^2$; **c**: $sp^2 - sp^2$

   d. Seldane™ oxygens: sp^3; nitrogen **d**: sp^2

2. a. :C≡C:⁻⁻   b. H–C≡O:⁺   c. :Ö=N=Ö:⁺   d. the *conjugate base* of :NH₂CH₃

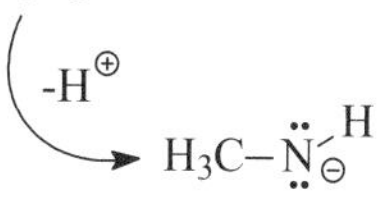

   $$\xrightarrow{-H^{\oplus}} H_3C-\ddot{N}^{\ominus}\diagdown^H$$

   e. [structure]   f. [structure]

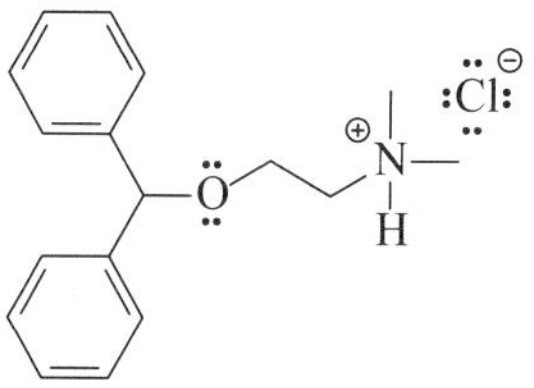

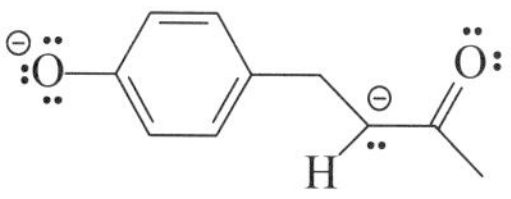

3. a. [orbital diagram] :C(H)(H) sp^2 => bond angle ~ 120°

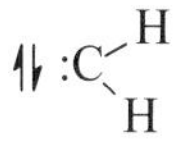

   b. H–[p orbital]–H = ↑·CH₂↑

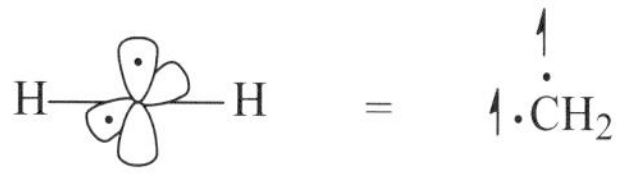

   a linear HCH bond angle implies sp hybridization; therefore, each lone electron lies in an un-hybridized p orbital with spins aligned (Hund's rule)

4. a. higher bp: [structure showing N–H⋯:N intermolecular H-bond, δ+ and δ-]

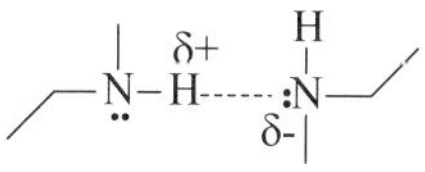

   this isomer is capable of *intermolecular* H-bonding, thereby increasing intermolecular attractive forces and raising its bp relative to the other amine

   b. lower mp: catechol

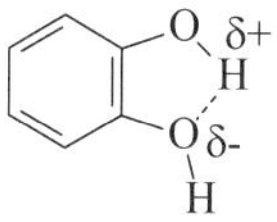

   [structure of catechol with intramolecular H-bond, δ+ and δ-]

   catechol, unlike hydroquinone, can undergo *intramolecular* H-bonding, which decreases intermolecular attractions and results in lowering its mp relative to hydroquinone

5. a. no   b. no   c. yes   d. yes   e. yes   f. no   g. yes   h. yes

6. a. $CHCl_3$ b. CH_3NO_2 c. SO_2

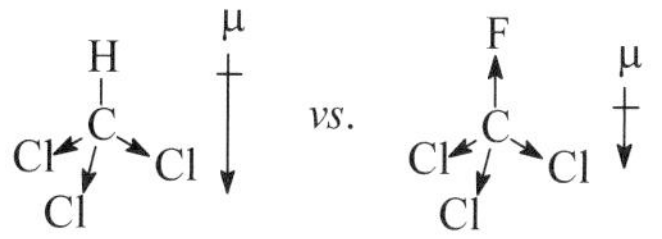

   [structures with dipole moment arrows]

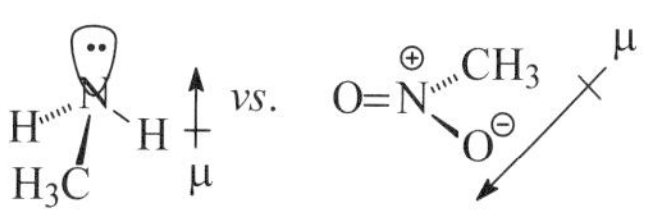

   permanent charge separation (> ε)

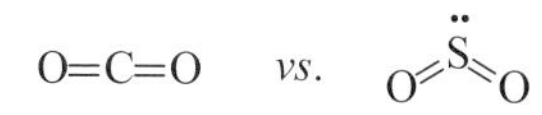

   linear, $\mu = 0$ bent, $\mu > 0$

7. a. *penicillin V*: $C_{16}H_{18}N_2O_4S$ *cimetidine*: $C_{10}H_{16}N_6S$

   b. arrow **a**: sp^2 **b**: sp^3 **c**: sp

c.

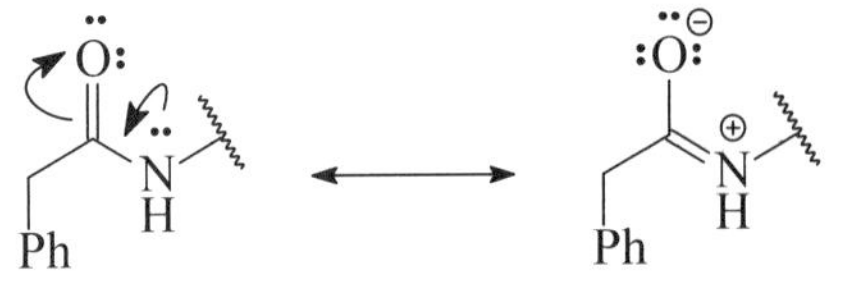

this resonance structure suggests *some* double bond character; electrons must be in a p orbital in order to resonate

d. lone pairs: *penicillin V*: 12;   *cimetidine*: 8.

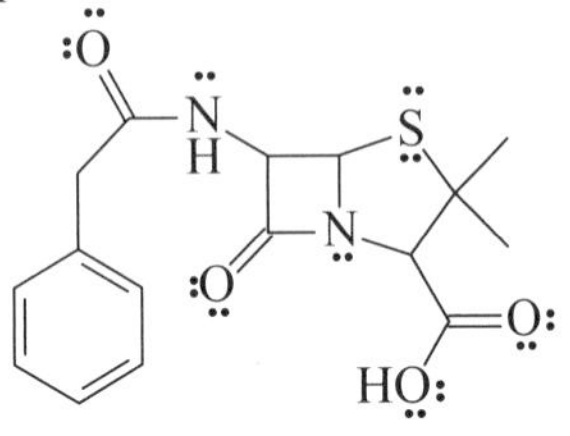

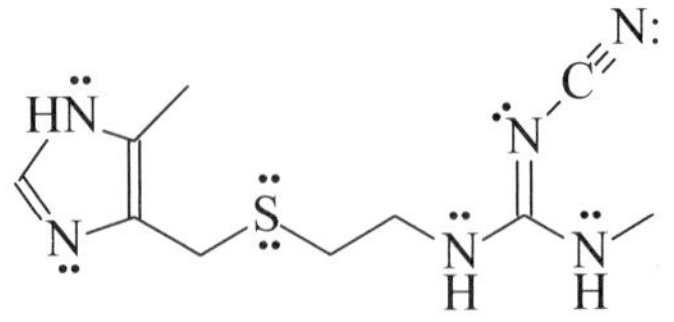

8. a. *sumatriptan*: $C_{14}H_{21}N_3O_2S$ *prostacyclin*: $C_{20}H_{32}O_5$

b. *Sumatriptan* contains 8 sp^2 and 6 sp^3 carbons; *prostacyclin* contains 5 sp^2 and 15 sp^3 carbons.

c. lone pairs: *sumatriptan*: 7;   *prostacyclin*: 10.

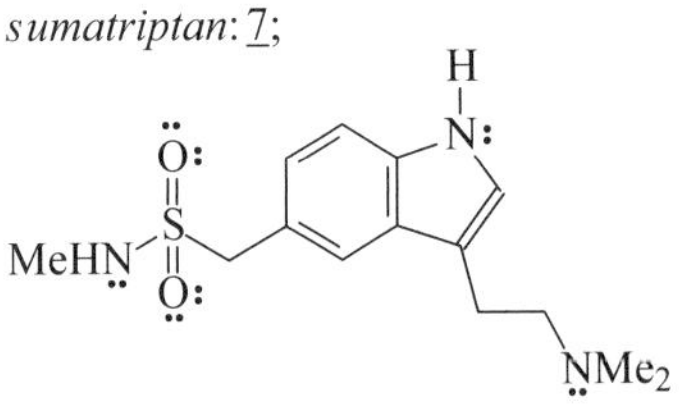

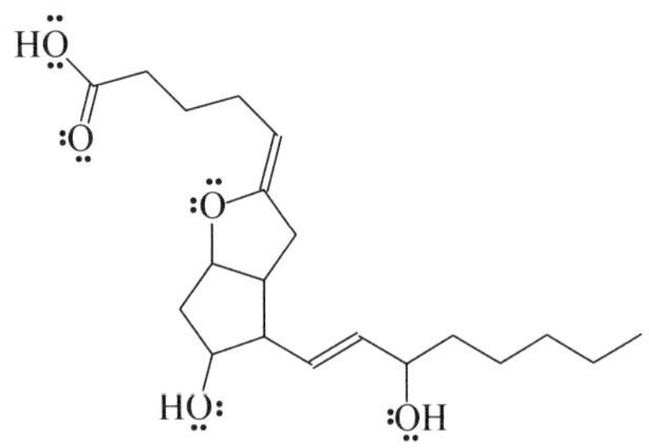

9. a. Rozerem™: $C_{16}H_{21}NO_2$ Chantix™: $C_{13}H_{13}N_3$ Ritalin™: $C_{14}H_{20}NO_2$

b. lone pairs: Rozerem™: 5;   Chantix™: 3; 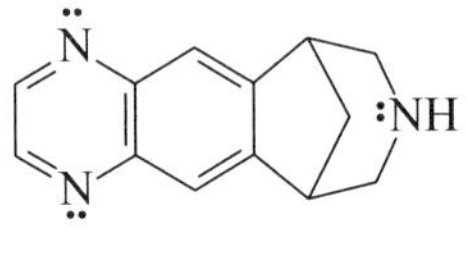   Ritalin™: 4.

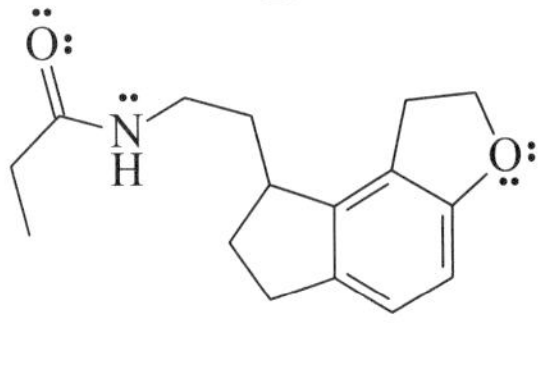

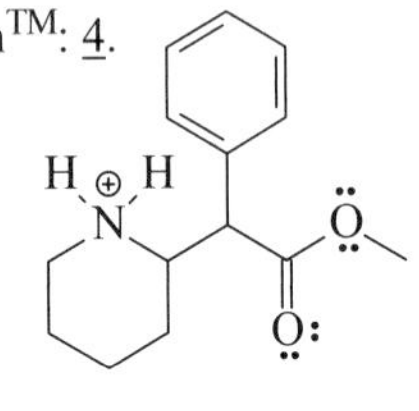

10. lone pairs: *theobromine*: 8; 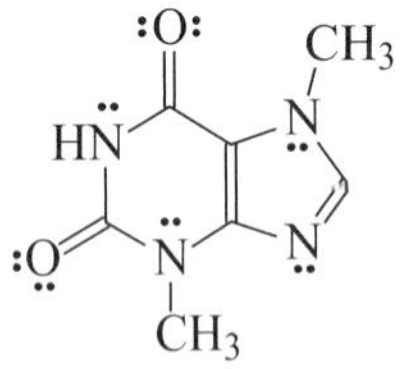   *melamine*: 6.

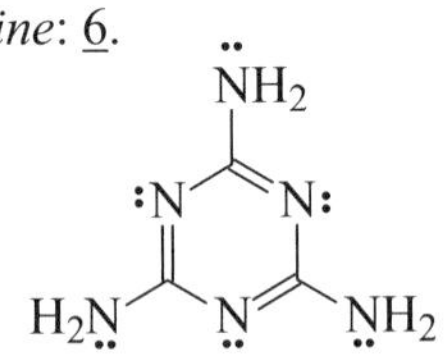

11. a. alkene, amide, amine, ester, ether   b. alkene, amine, arene, carboxylic acid, halide, ketone

c. alcohol, alkyne, arene.

1.1 Hybridization, formulas, physical properties

1.2 Acids and bases

1. strongest base in ammonia: $H_2\ddot{N}:^{\ominus}$ amide anion - the CB of ammonia

$H_2\ddot{N}-(H \curvearrowleft :H^{\ominus}) \longrightarrow H_2\ddot{N}:^{\ominus} + H_2$

2. stronger base: $(CH_3)_2NH$ nitrogen is more electron-releasing than oxygen

3. a. cyclohexanone + $AlCl_3$ → cyclohexanone-O-$AlCl_3$ adduct (O$^{\oplus}$, $AlCl_3^{\ominus}$)

b. $Ph_3P: \curvearrowright BF_3 \longrightarrow Ph_3\overset{\oplus}{P}-\overset{\ominus}{B}F_3$

c. morpholine N-methyl + BH_3 → N$^{\oplus}$-$BH_3^{\ominus}$ adduct

4. a. $CH_3-\ddot{O}:^{\ominus}$ $CH_3CH_2-\ddot{C}l: \longrightarrow CH_3-\ddot{O}-CH_2CH_3 + :\ddot{C}l:^{\ominus}$
    LB   LA

b. $H_2C=CH_2$ $BF_3 \longrightarrow \overset{\oplus}{C}H_2-CH_2-\overset{\ominus}{B}F_3$
    LB   LA

c. $H_3C-\ddot{O}-(H \curvearrowleft :CH_2-CH_3) \longrightarrow H_3C-\ddot{O}:^{\ominus} + H_3C-CH_3$
    LA   LB

(also Bronsted-Lowry acid-base, respectively)

d. $:\ddot{C}l-\ddot{C}l:$ $AlCl_3 \longrightarrow :\ddot{C}l:^{\oplus} + AlCl_4^{\ominus}$
    LB   LA

e. $CH_3-\ddot{N}=C=\ddot{S}:$ $:NH_3 \longrightarrow CH_3-\overset{\ominus}{\ddot{N}}-C\overset{\ddot{S}:}{\underset{NH_3^{\oplus}}{-}}$
    LA   LB

5. a. $C_{20}H_{28}O$ b. [structure: tertiary alkoxide with C≡C-H, $pK_a \sim 35$, H)—$\ddot{N}H_2$] $\rightleftharpoons$ [tertiary alcohol O-H, $pK_a \sim 16$] + $:\ddot{N}H_2^{\ominus}$ c. $K_{eq} \ll 1$
               WB   WA   SA   SB   ($\Delta pK_as > 4$)

6. a. [reaction showing ibuprofen carboxylate deprotonating acetaminophen phenol]

b. WB   WA   SA   SB ; pKa ~ 10 (phenol OH), pKa ~ 5 (COOH)

c. $K_{eq} \ll 1$ ($\Delta pK_a s > 4$)

7. lowest pK_a: **b**.   a. ~ 16   b. ~ 5   c. ~ 16   d. ~ 10   e. ~ 38   f. ~ 35

8. quantitative rx: **b**. $R-C\equiv C-H \;\; :NH_2^{\ominus} \xrightarrow{K_{eq} \gg 1} R-C\equiv C:^{\ominus} + :NH_3$

   pK_a 22   SB   WB pK_a ~ 35

   a. hydroxide ($K_{eq} \ll 1$)   c. H_2 (NR)   d. $EtO^{\ominus}$ ($K_{eq} \ll 1$)

9. [estradiol phenol OH pK_a ~ 16 / phenol OH deprotonated by $:CH_2NO_2^{\ominus}$, pK_a ~ 10; $K_{eq} \sim 1$ ($\Delta pK_a s \sim 0$); products: phenoxide + CH_3NO_2, pK_a ~ 10]

10. a. [pyridinium Cl^-]

    b. [pyridinium H deprotonated by carbonate $\rightarrow$ pyridine + $HCO_3^{\ominus}$]; pK_a ~ 5 SA ; pK_a ~ 10 WA

    c. $K_{eq} \gg 1$ ($\Delta pK_a s > 4$)

1.3 Resonance

1. $CH_3-\overset{\ominus}{\underset{..}{C}}=\underset{..}{\overset{..}{O}}:$ sp²   [cyclopentadienyl anion with H and lone pair, p*]   [benzyl anion $CH_2^{\ominus}$, p*]   [oxazoline-type ring with $\overset{..}{O}:^{\oplus}$ and $:N:$, sp²]

   * not sp³ per VSEPR; electrons may resonate if housed in a p orbital

1.3 Resonance

2. lower pK_a: H-O-CN

H–Ö–C≡N: $\xrightarrow{-H^{\oplus}}$ $^{\ominus}$:Ö–C≡N: ⟷ Ö=C=N:$^{\ominus}$

negative charge delocalization => more stable CB

H–C≡N: $\xrightarrow{-H^{\oplus}}$ $^{\ominus}$:C≡N: negative charge localized on C

3. a. <u>3</u>. HÖ–C(H)–$\overset{\oplus}{N}H_2$ ⟷ HÖ–C(H)=NH$_2$ ⟷ $\overset{\oplus}{H\ddot{O}}$=C(H)–$\ddot{N}H_2$

b. <u>1</u>. [diene with cation] localized charge — saturated, no p orbital

c. <u>5</u>. [five resonance structures of p-acetylphenoxide]

d. <u>5</u>. [five resonance structures of p-methoxybenzyl cation]

4. Me–NH(H)–C(=NH)–NH$_2$ ← most basic

$\xrightarrow{+H^{\oplus}}$

Me–$\overset{\oplus}{N}$H(H)–C(=NH)–NH$_2$ vs. Me–NH(H)–C(=NH)–$\overset{\oplus}{N}$H$_2$(H) vs. Me–NH(H)–C(=$\overset{\oplus}{N}$H(H))–NH$_2$

no important resonance contributions

4 resonance structures => a more stable CA

{ Me–$\overset{\oplus}{N}$H(H)–C(NH–H)–NH$_2$ ⟷ Me–NH(H)–$\overset{\oplus}{C}$(NH–H)–NH$_2$ ⟷ Me–NH(H)–C(NH–H)=$\overset{\oplus}{N}$H$_2$ }

5. a. $^{\oplus}$:Ö–Ö–Ö:$^{\ominus}$ ⟷ :Ö=$\overset{\oplus}{O}$–Ö:$^{\ominus}$

oxygen 'octetted,' closer charge separation

b. [decahydroquinoline-like cation] ⟷ [iminium with C=C]

carbon 'octetted,' additional π bond

1.3 Resonance

248 • Chapter 1 The Basics

c. [structure showing CH₂⁻−C≡N: ↔ CH₂=C=N:⁻]
negative charge is borne by more electronegative atom

d. [cyclohexenyl cation with :ÖH ↔ cyclohexene with ⁺ÖH]
carbon 'octetted,' additional π bond

6. a. **2**. [pyrrolium resonance structures]

b. **3**. [oxepinyl cation three resonance structures]

c. **1**. [cyclohexenyl-CH₂⁺ structure]

d. **3**. [cyclohexenone enolate three resonance structures]

7. [pentane-2,4-dione A —$-H_a^+$→ three resonance structures of enolate]

A

CB of **A** is stabilized by charge delocalization over three nuclei and, therefore, more easily formed => pK_a of H_a is lowered

8. [acetic acid +H⁺ → vs. protonation on OH (charge localized) vs. protonation on C=O with two resonance structures]

charge localized

charge delocalized, => > stability therefore, more favored CA species

9. a. **5**. [five resonance structures of styryl cation with :NMe₂]

b. **5**. [five resonance structures of cyclopentadienyl anion]

c. **4**. [four resonance structures of pyrrolyl cation] not [structure with nitrogen 'sextetted']

nitrogen 'sextetted'

1.3 Resonance

d. **2**.

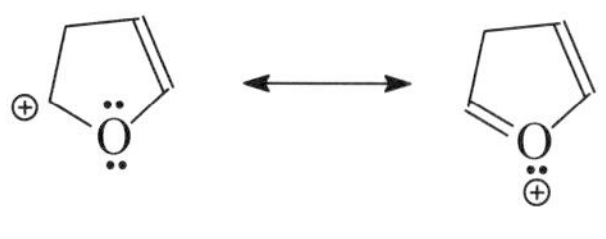

e. **3**.

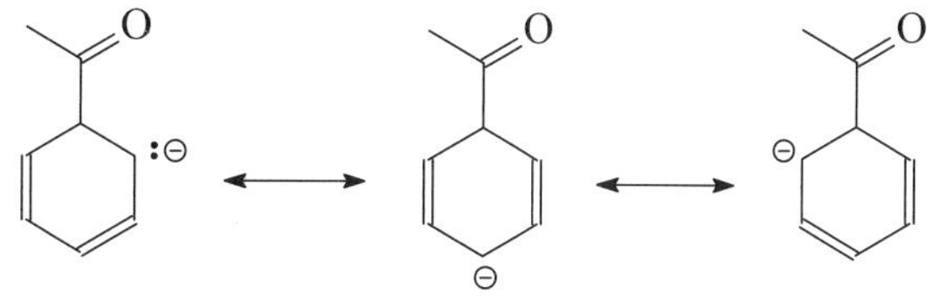

f. **4**.

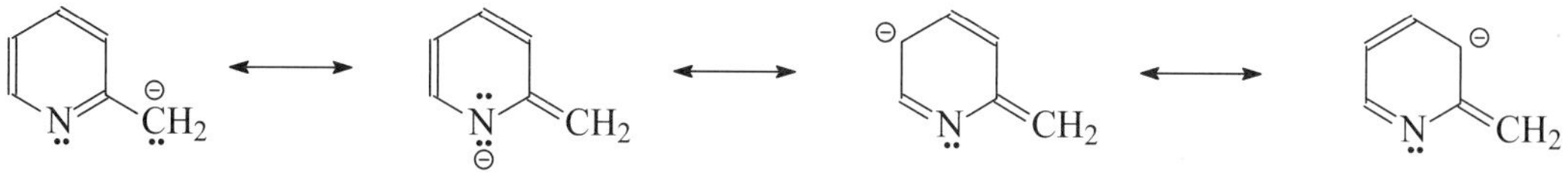

g. **4**.

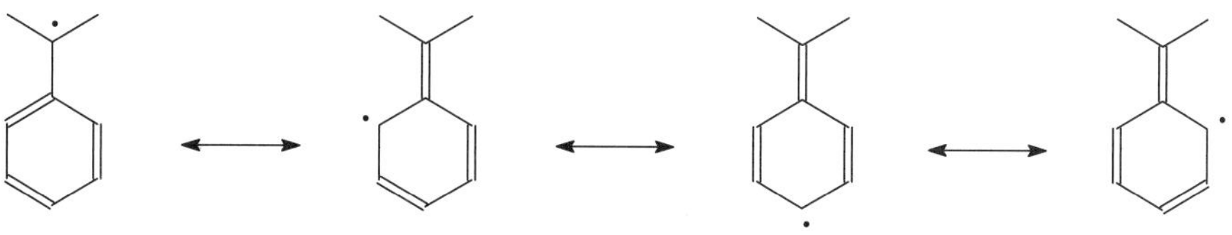

h. **4**.

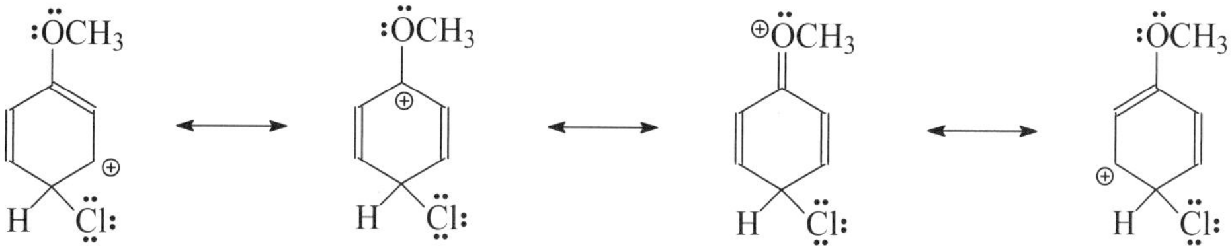

i. **2**.

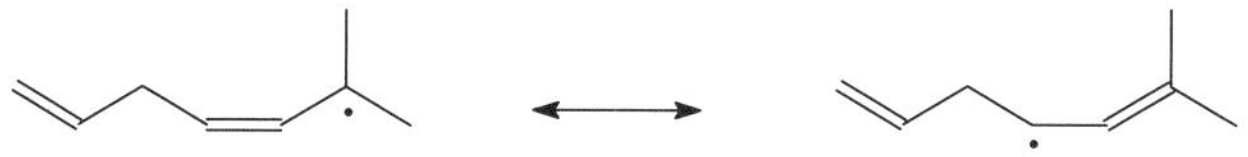

10. 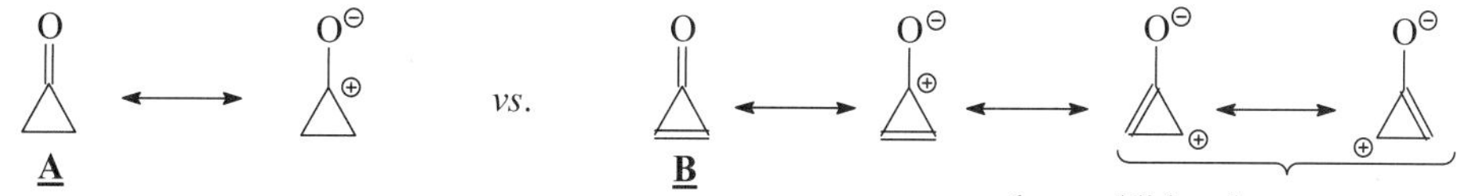 **A** vs. **B**

these additional resonance structures increase both $\underline{\varepsilon}$ and $\underline{d}$ ($\mu = \varepsilon \times d$); therefore, $\mu_B > \mu_A$

11.

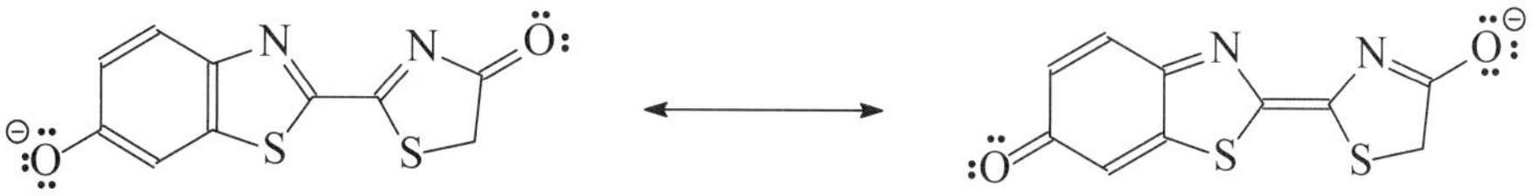

CB of oxyluciferin

1.3 Resonance

CHAPTER 2
ALKANES

2.1 General

1. highest mp: <u>4</u>. bicyclo[2.2.2]octane (most spherical)

2. highest bp: <u>1</u>. <u>n</u>-pentane (least branched)

3.

   eicosane, mp 37^0 dodecahedrane, mp 420^0

   spherical molecules (dodecahedrane) pack more closely in the solid state than linear (eicosane) ones, therefore requiring more energy to separate (melt) them

4. constitutional isomers for

   a. C_6H_{14}: <u>5</u>.

   b. C_7H_{16}: <u>9</u>.

5. different kinds (constitutional) of hydrogens in

   a. 2,3-dimethylpentane: <u>6</u>. 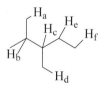   b. 2,4-dimethylpentane: <u>3</u>.

   c. 3-ethylpentane: <u>3</u>.   d. 2,2,4-trimethylpentane: <u>4</u>.

   e. 2,5,5-trimethylheptane: <u>7</u>. 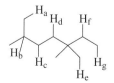   f. 4-ethyl-3,3,5-trimethylheptane: <u>10</u>.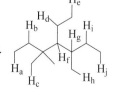

2.2 Nomenclature

1. 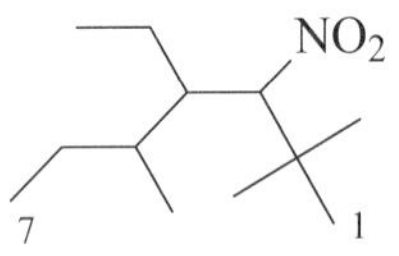 3-nitro-4-ethyl-2,2,5-trimethylheptane

2. 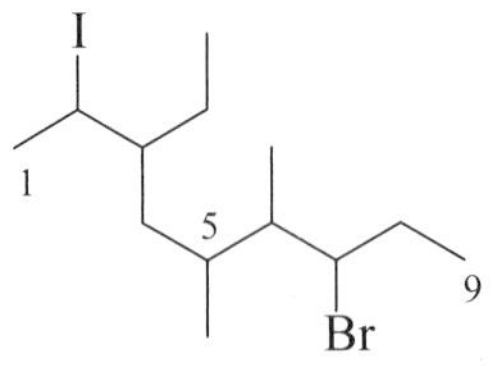 7-bromo-2-iodo-3-ethyl-5,6-dimethylnonane

3. 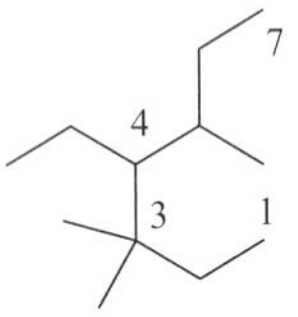 4-ethyl-3,3,5-trimethylheptane

4. 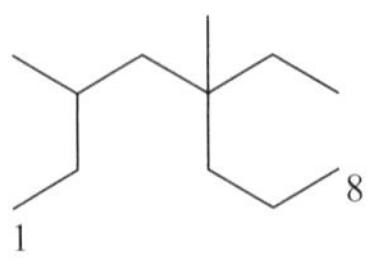 5-ethyl-3,5-dimethyloctane

5. 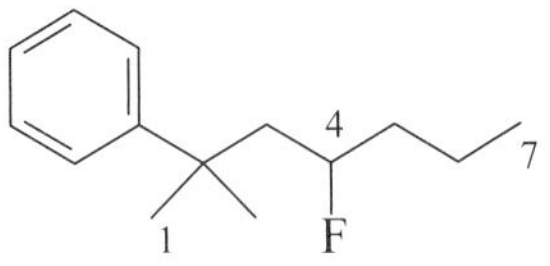 4-fluoro-2-methyl-2-phenylheptane

6. 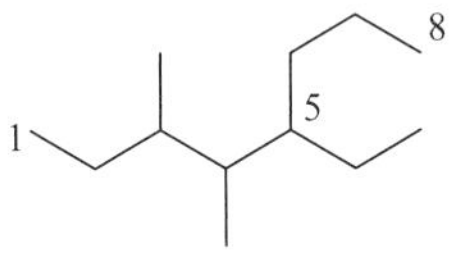 5-ethyl-3,4-dimethyloctane

7. 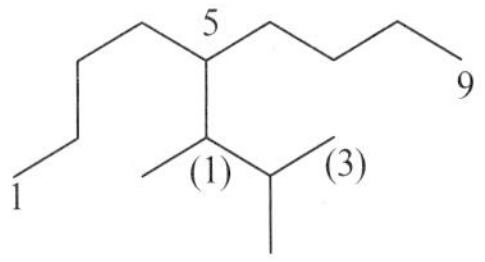 5-(1,2-dimethylpropyl)nonane

8. 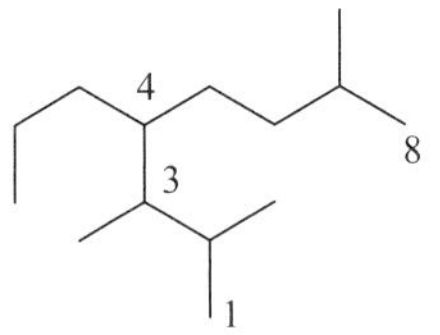 2,3,7-trimethyl-4-*n*-propyloctane
   (choose path with more branching)

9. 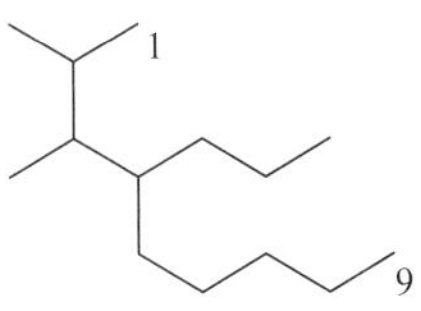 2,3-dimethyl-4-*n*-propylnonane

10. 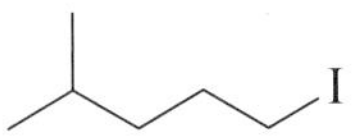 1-iodo-4-methylpentane

11. 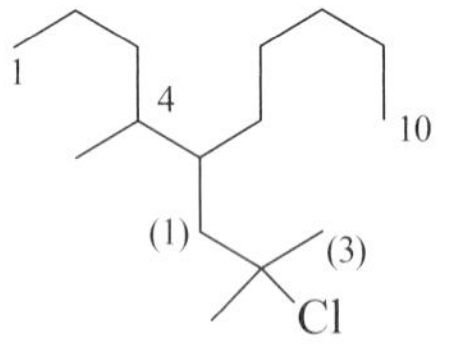 5-(2-chloro-2-methylpropyl)-4-methyldecane

12. 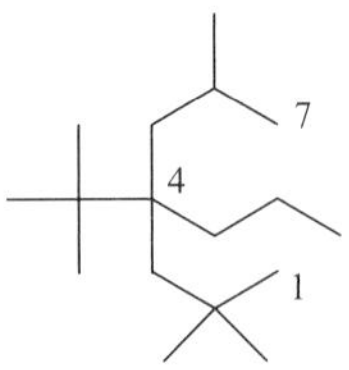 4-*t*-butyl-2,2,6-trimethyl-4-*n*-propylheptane
    not 4-*t*-butyl-4-isobutyl-2,2-dimethylheptane
    (less branching)

Solutions • 253

13. 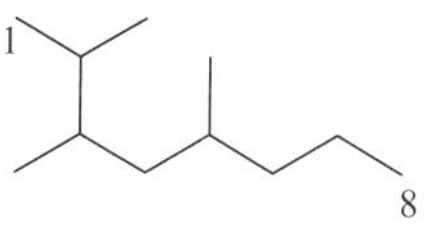 2,3,5-trimethyloctane

14. 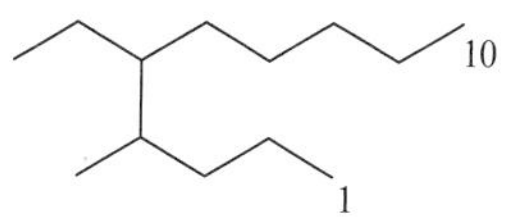 5-ethyl-4-methyldecane

15. 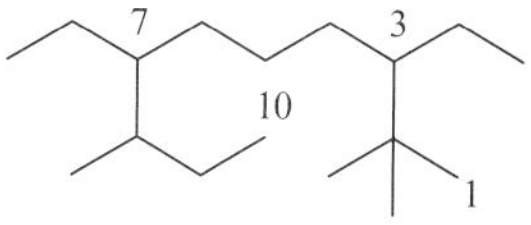 3,7-diethyl-2,2,8-trimethyldecane

16. 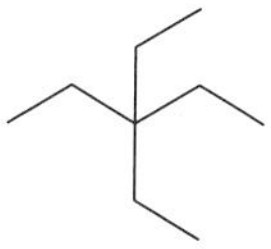 diethylpentane (3,3- not necessary!)

17. a. 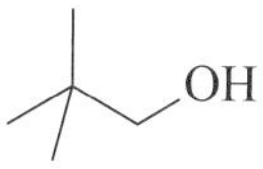   b. 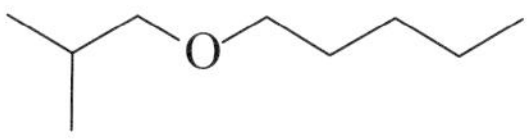   c.

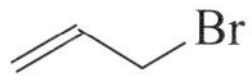

2.3 Conformational analysis, acyclic

1. 3.7 kcal/mol
   −2.0 (2 x 1.0 kcal/mol)
   1.7 kcal/mol

   1.0 kcal/mol

   → 3.7 kcal/mol

   PE — rot'n →

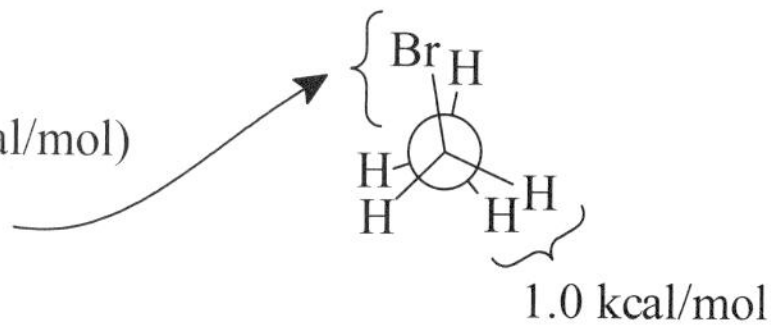

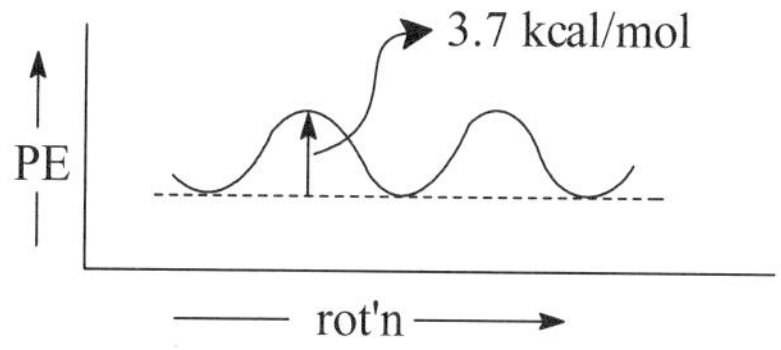

2. a. 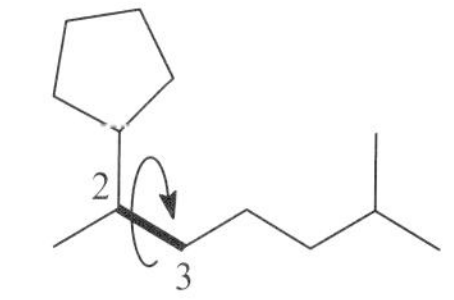 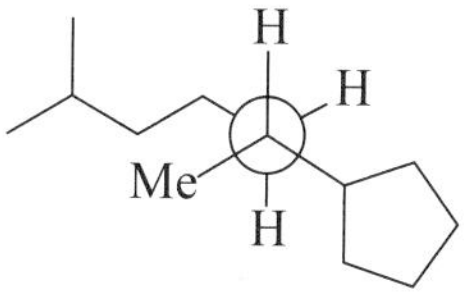 largest R-groups are *anti*

   b. 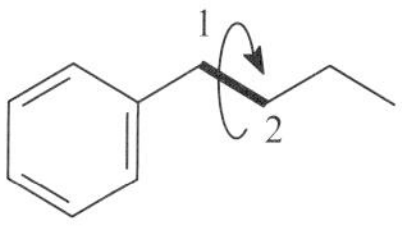 *gauche*: dihedral angle ~ 60°

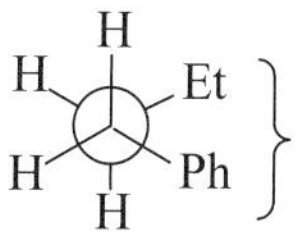

3. a. 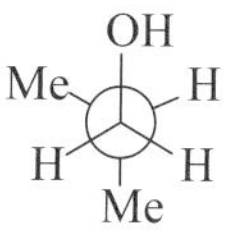 isobutyl alcohol

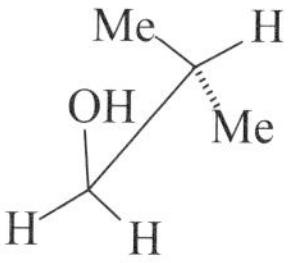

   b. 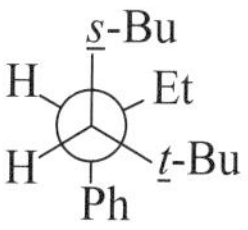 4-*t*-butyl-3-methyl-5-phenylheptane

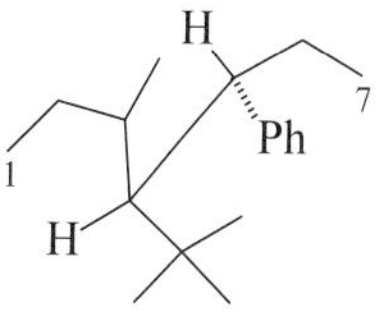

2.3 Conformational analysis, acyclic

4.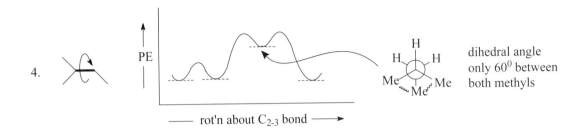

5. intramolecular hydrogen-bond stabilizes a nearly eclipsed conformer for FCH_2CH_2OH

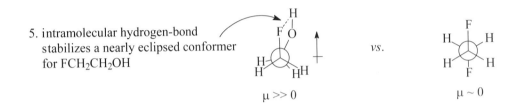

2.3 Conformational analysis, acyclic

CHAPTER 3
CYCLOALKANES

3.1 General

1. highest molecular dipole moment: <u>d</u>.   a.   b.   c. Cl—C≡C—Cl

2. constitutional isomers for
   a. dichlorocyclopentane: <u>3</u>.

      1,1-   1,2-   1,3-

   b. C_6H_{12} that contain a cyclopropyl ring: <u>6</u>.

3. cis/trans stereoisomers for
   a. dichlorocyclopentane: <u>2 pairs</u>.

      1,2-   + trans-   1,3-   + cis-

   b. diphenylcyclohexane: <u>3 pairs</u>.

      1,2- + cis-   1,3- + trans-   1,4- + cis-

   c. 2-chloro-4-ethyl-1-methylcyclohexane: <u>4</u>.

4. different kinds (constitutional and geometric) of hydrogens in

   a. 1-ethyl-1-methylcyclopropane: <u>5</u>.   b. allylcyclobutane: <u>9</u>.

   H_a and H_b are cis- and trans- to methyl,
   and so are 'different'

   c. methylcyclobutane: <u>6</u>.   d. chlorocyclopentane: <u>5</u>.   e. vinylcyclopentane: <u>8</u>.

         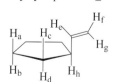

5. least strained: <u>a</u>.

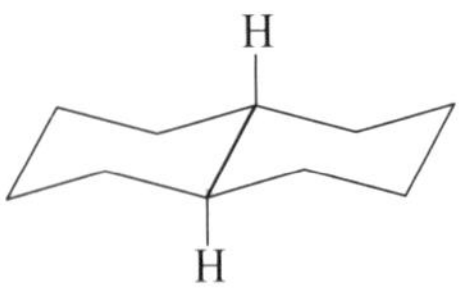

trans-
no strain

b. 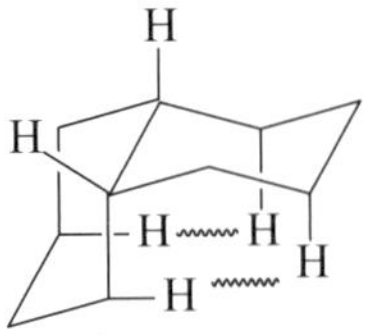
cis-
diaxial strain

c. 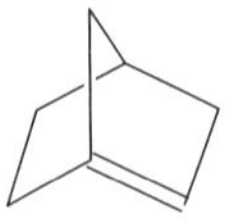
ring strain prevents
alkene from being planar -
violates Bredt's rule

d. 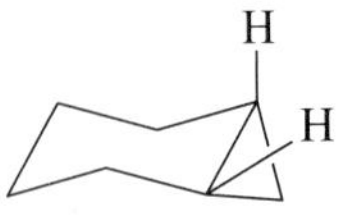
angle strain

6.

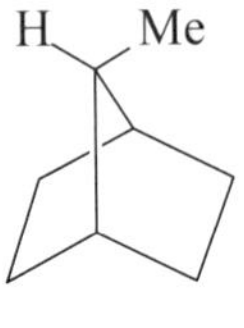

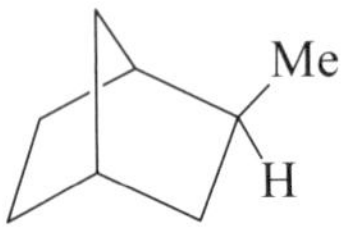

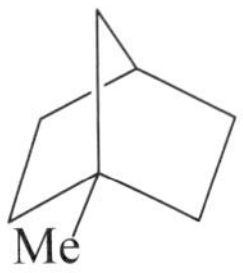

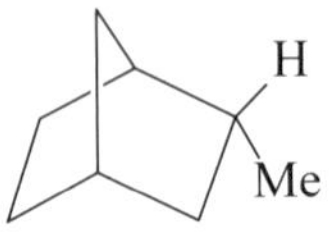

exo-   endo-
stereoisomers

7. only <u>2</u>.

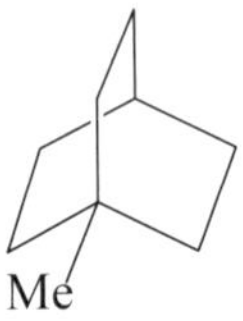

+   =   =

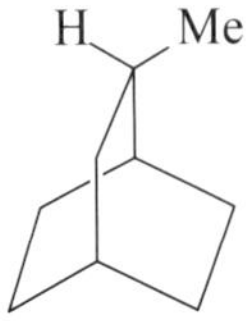

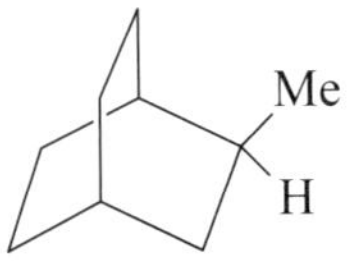

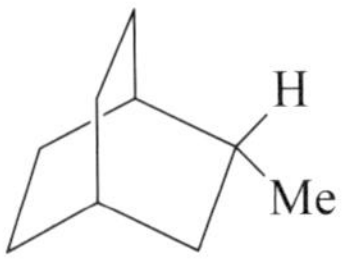

identical structures (*for now!*)

3.2 Nomenclature

1.

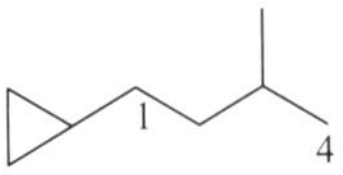

1-cylclopropyl-3-methylbutane

2.

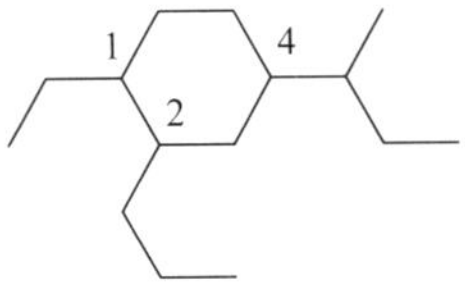

4-<u>s</u>-butyl-1-ethyl-2-<u>n</u>-propylcyclohexane

3.

<u>t</u>-pentylcyclopentane

(1,1-dimethylpropyl)cyclopentane

2-cyclopentyl-2-methylbutane

4.

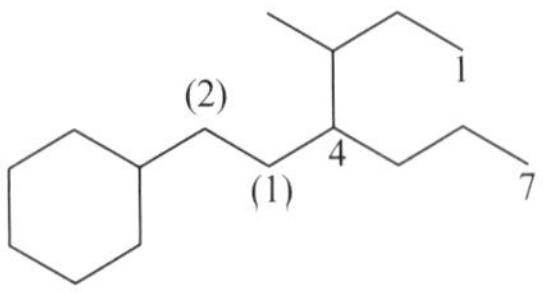

4-(2-cyclohexylethyl)-3-methylheptane

3.2 Nomenclature

5. 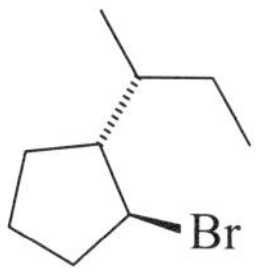 *trans*-1-bromo-2-*s*-butylcyclopentane

6. 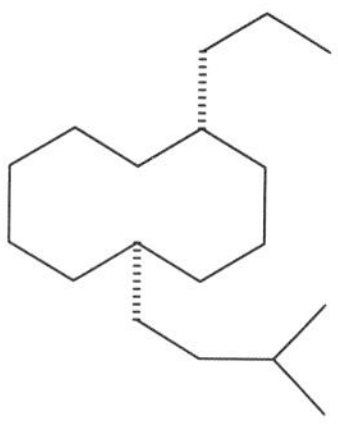 *cis*-1-isopentyl-5-*n*-propylcyclodecane

7. 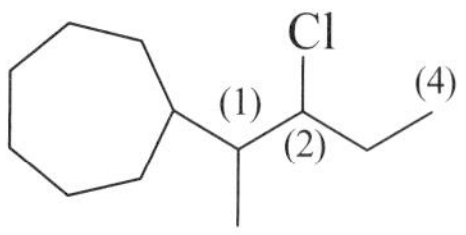 (2-chloro-1-methylbutyl)cycloheptane

8. 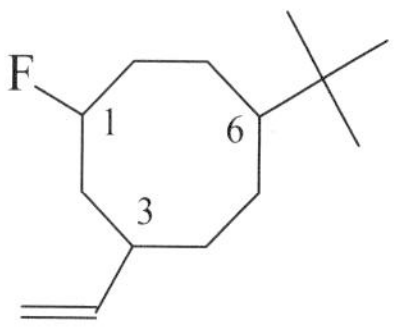 1-fluoro-6-*t*-butyl-3-vinylcyclooctane

9. 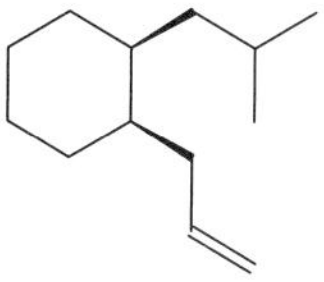 *cis*-1-allyl-2-isobutylcyclohexane

10. 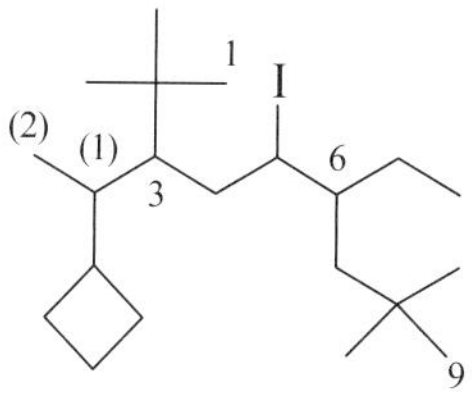 5-iodo-3-(1-cyclobutylethyl)-6-ethyl-2,2,8,8-tetramethylnonane

11. 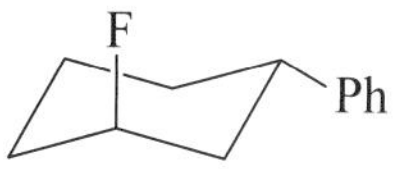 *trans*-1-fluoro-3-phenylcyclohexane

12. 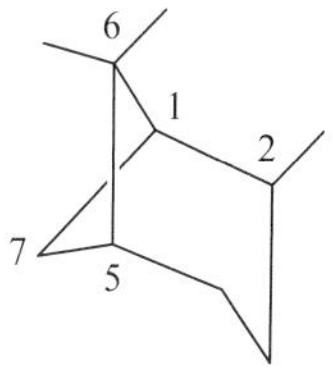 2,6,6-trimethylbicyclo[3.1.1]heptane

13. 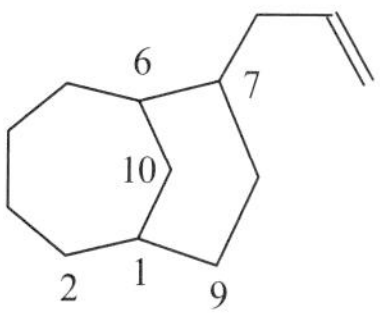 7-allylbicyclo[4.3.1]decane

14. 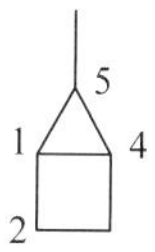 5-methylbicyclo[2.1.0]pentane

15. 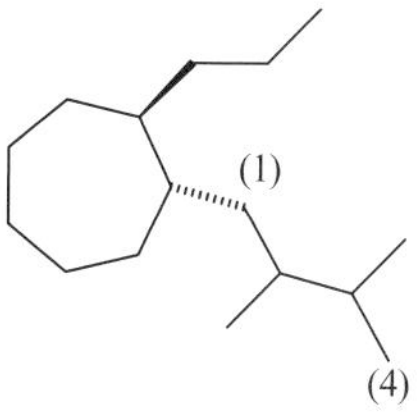 *trans*-1-(2,3-dimethylbutyl)-2-*n*-propylcycloheptane

16. 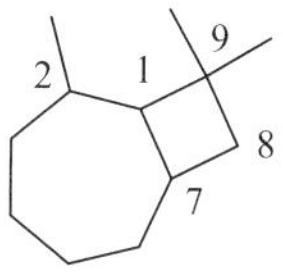 2,9,9-trimethylbicyclo[5.2.0]nonane

3.2 Nomenclature

3.3 Conformational analysis, cyclic

1. most stable conformer:

  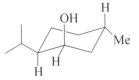

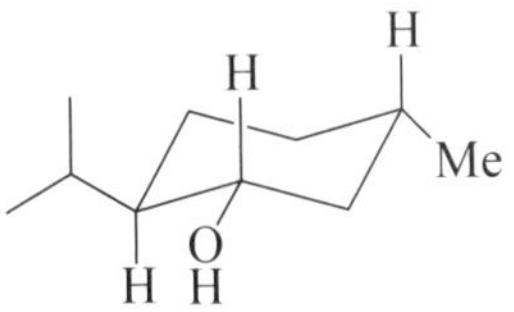

menthol   *neomenthol*

2. a. 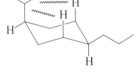 degenerate structures (same energy) therefore, $K_{eq} = 1$

b.

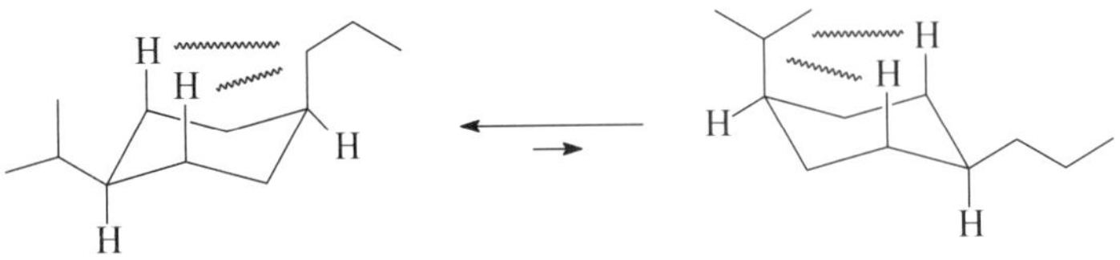

> 1,3-diaxial interactions => less stable conformer therefore, $K_{eq} < 1$

c.

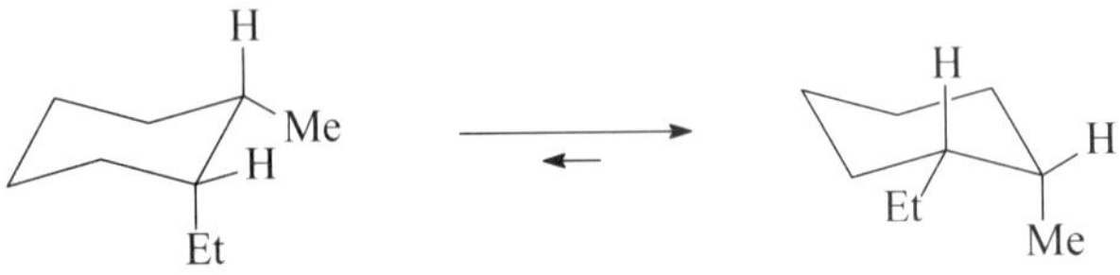

larger ethyl group in more stable equatorial position therefore, $K_{eq} > 1$

3. most negative ΔH_{comb} (=> least stable):

a.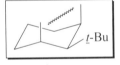

least stable 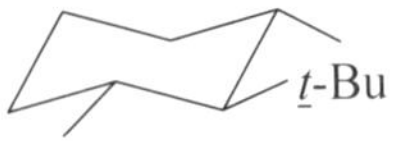   (most stable, therefore, *least* negative ΔH_{comb})

b.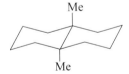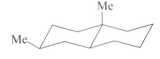

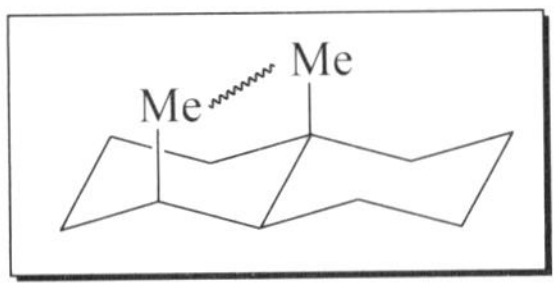

> dimethyl repulsion therefore, least stable

c.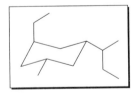

least stable 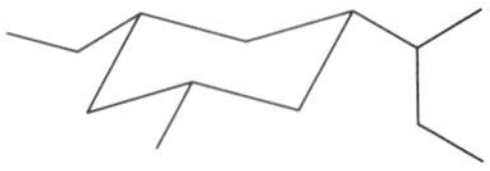   (most stable)

3.3 Conformational analysis, cyclic

4. least negative ΔH_comb (=> most stable):

a.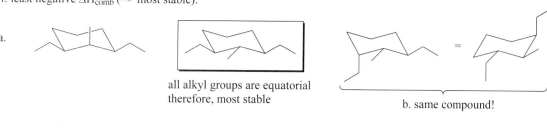

all alkyl groups are equatorial therefore, most stable

b. same compound!

5. most stable conformer for:

[trans- vs. cis- structures with Cl and t-Bu, leading via :B (-HCl) to t-Bu-cyclohexene]

t-butyl group prevents 'flipping,' so chlorine cannot assume the axial position in the *trans*-isomer therefore, the *cis*-isomer would react more rapidly

6. [trans- F/OH cyclohexane] vs. [cis- F/OH cyclohexane] → [twist-boat with OH----F]

twist-boat stabilized by intramolecular hydrogen-bonding; not possible in chair conformer (or in *trans*-isomer)

7. a. [structures 1 and 2 of sugar]

1   **2**

b. configuration **2** is less stable (one substituent must be axial) and would burn with a more negative ΔH_comb

8. [dimethyl structure] → [Me, Me chair]

5.4 kcal/mol less stable than

5.4 kcal/mol
-0.9
-0.9 } (2 Me-H 1,3-diaxial strain interactions)
―――――
3.6 kcal/mol (Me-Me 1,3-diaxial strain interaction)

9. a. number of *cis/trans* stereoisomers: **8**.

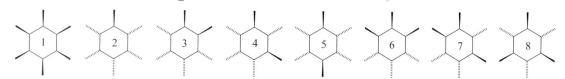

3.3 Conformational analysis, cyclic

b. for conformational chair-chair flipping, $K_{eq} = 1$ for configurations <u>1</u>, <u>3</u>, and <u>5</u>: 3e/3a ⇌ 3a/3e

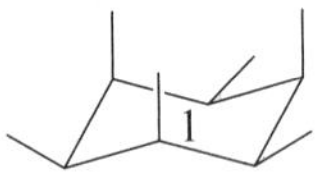

1	2	3	4
3e/3a	4e/2a	3e/3a	5e/1a

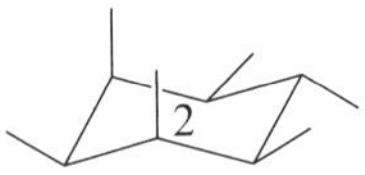

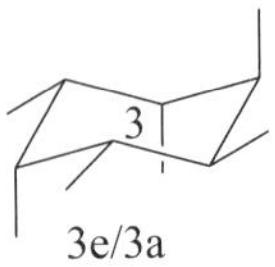

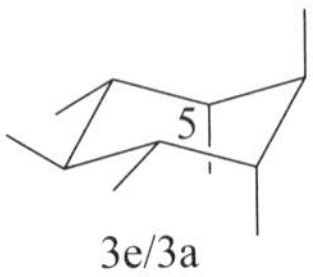 5	6	7	8
3e/3a	4e/2a	4e/2a	6e/0a

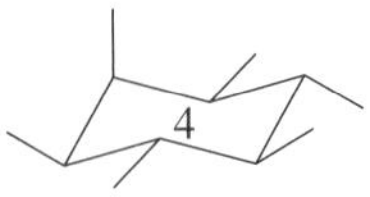

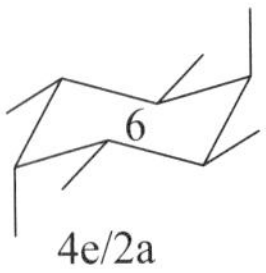

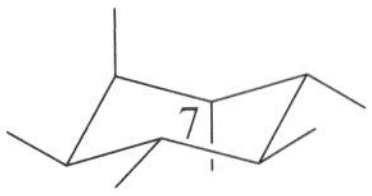

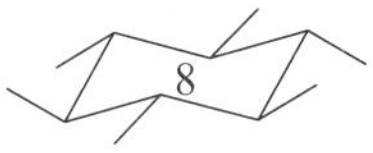

c. least stable: <u>1</u>.   *three* 1,3-diaxial steric interactions exist between two methyl groups (only *one* such interaction exists in configurations <u>3</u> and <u>5</u>)

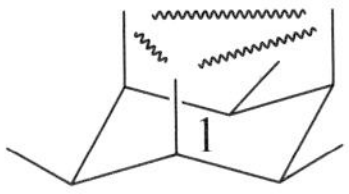

d. least likely to flip: <u>8</u>.

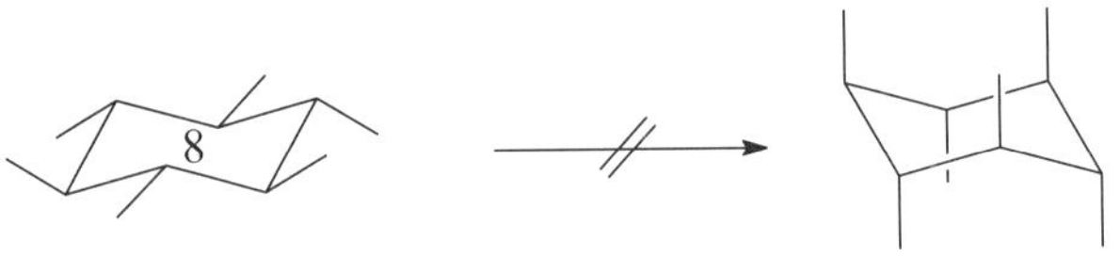

all methyls are equatorial ↛ all axial!

3.3 Conformational analysis, cyclic

CHAPTER 4
REACTION BASICS

1.
   a. addition
   b. oxidation [O]
   c. substitution
   d. substitution
   e. elimination
   f. reduction [H]
   g. oxidation [O]
   h. addition
   i. reduction [H]
   j. rearrangement
   k. oxidation [O]
   l. substitution
   m. elimination
   n. addition
   o. reduction [H]
   p. reduction [H]
   q. rearrangement
   r. elimination
   s. substitution
   t. reduction [H]

2. a.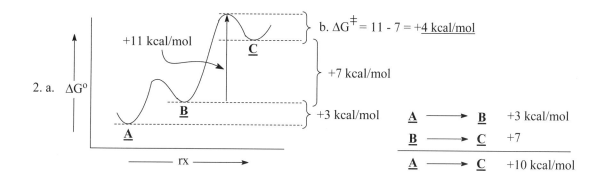

   b. $\Delta G^{\ddagger} = 11 - 7 = $ +4 kcal/mol

A	→ **B**	+3 kcal/mol
B	→ **C**	+7
A	→ **C**	+10 kcal/mol

3. a.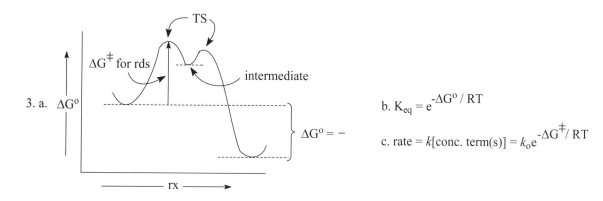

   b. $K_{eq} = e^{-\Delta G^{\circ} / RT}$

   c. rate = k[conc. term(s)] = $k_o e^{-\Delta G^{\ddagger}/RT}$

4. a.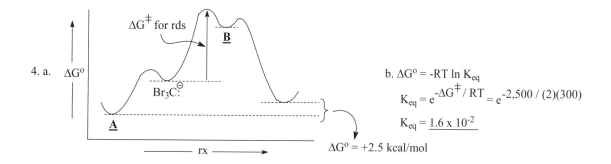

   b. $\Delta G^{\circ} = -RT \ln K_{eq}$

   $K_{eq} = e^{-\Delta G^{\ddagger}/RT} = e^{-2,500/(2)(300)}$

   $K_{eq} = $ 1.6 x 10^{-2}

   $\Delta G^{\circ} = $ +2.5 kcal/mol

5. a. type of reaction: <u>rearrangement</u>;   mechanism: <u>polar / ionic</u>.

b.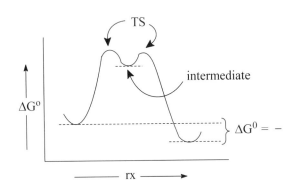

c. $K_{eq} = [\underline{B}] / [\underline{A}] = 75\% / 25\% = \underline{3.0}$; $\Delta G^\circ = -RT \ln K_{eq}$.

d. nucleophiles: $\underline{A}$, H_2O, and $\underline{B}$.

e.

(reaction coordinate diagram showing TS, intermediate, ΔG°, $\Delta G^0 = -$, rx)

6. (cyclohexane with methyl and I → rds, $-I^\ominus$ → carbocation → fast, +MeOH, $-H^\oplus$ → cyclohexane with methyl and OMe)

7.   a. pericyclic   b. free radical   c. pericyclic   d. polar / ionic

     e. pericyclic   f. polar / ionic   g. free radical   h. polar / ionic

     i. pericyclic   j. polar / ionic

CHAPTER 5
ALKENES AND CARBOCATIONS

5.1 General

1. a. (E)-3-chloromethyl-4-_s_-butyl-2-methyl-3-octene

   b. _trans_-5-phenyl-2-pentene

   c. 3-_n_-propyl-1-nonene

   d. 6-methylbicyclo[5.2.0]-3,8-nonadiene

2. a. (Z)-   b. (E)-   c. (Z)-   d. (Z)-   e. (Z)-   f. (Z)-

3. a. number of alkenes: <u>4</u>.

   b. least negative ΔH$_{hydrogenation}$: most stable (trisubstituted)

4. number of _geometric_ isomers: <u>4</u>.

   trans, trans-   _cis, cis-_   _trans, cis-_   _cis, trans-_

5. most stable carbocation: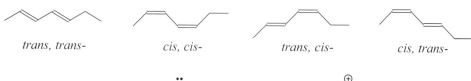

6. a. no. deg. unsat: $C_{17}H_{36} - C_{17}H_{20(18+3-1)} = H_{16}$ => $H_{16}/2 = 8$ deg.
   hydrogenation: $C_{17}H_{30}F_3NO - C_{17}H_{18}F_3NO = H_{12}$ => $H_{12}/2 = 6$ DB
   no. rings = 8 - 6 DB = <u>2</u>.

   fluoxetine

   b. no. deg. unsat: $C_{17}H_{36} - C_{17}H_{16(18+1-3)} = H_{20}$ => $H_{20}/2 = 10$ deg.
   no. DB = 10 - 4 rings = <u>6</u>.

   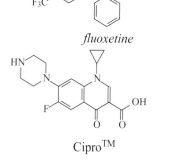
   Cipro^{TM}

264 • Chapter 5 Alkenes and Carbocations

c. no. deg. unsat: $C_{28}H_{58} - C_{28}H_{34(35-1)} = H_{24} => H_{24}/2 = 12$ deg.
no. rings = 12 - 5 DB -1 TB(= 2 DB!) = <u>5</u>.

d. no. deg. unsat: $C_{17}H_{36} - C_{17}H_{10(14-0-4)} = H_{26} => H_{26}/2 = 13$ deg.
no. DB = 13 - 3 rings = <u>10</u>.

e. no. deg. unsat: $C_{19}H_{40} - C_{19}H_{16(17+2-3)} = H_{24} => H_{24}/2 = 12$ deg.
no. rings = 12 - 8 DB = <u>4</u>.

f. no. deg. unsat: $C_{19}H_{40} - C_{19}H_{20(20+1-1)} = H_{20} => H_{20}/2 = 10$ deg.
hydrogenation: $C_{19}H_{32}FNO_3 - C_{19}H_{20}FNO_3 = H_{12} => H_{12}/2 = 6$ DB
no. rings = 10 - 6 DB = <u>4</u>.

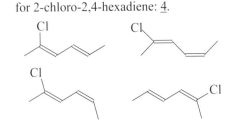

7. number of stereoisomers for 2,4-hexadiene: <u>3</u>;   for 2-chloro-2,4-hexadiene: <u>4</u>.

8. a. (Z)-3-methyl-2-phenyl-2-hexene

  b. propylene dichloride (note: no double bond)

  c. styrene bromohydrin

  d. *trans*-cyclohexene glycol

  e. isobutylene epoxide

5.1 General

9.

[Reaction scheme: methylenecyclobutane + H⁺ → cyclobutylcarbinyl cation **1** → (1,2-R: shift) → cyclopentyl cation **2** with H → (1,2-H: shift) → 1-methylcyclopentyl cation **3** → (+Cl⁻) → 1-chloro-1-methylcyclopentane]

[Reaction coordinate diagram showing three transition states labeled **1**, **2**, **3** with ΔG⁰ = − (negative, exergonic overall)]

10. a.

[CH₂=CH–CH=CH–C(CH₃)=CH₂ type diene + H⁺ (proton removed as shown with H⁺ leaving) → allylic/pentadienyl cation CA with arrows showing resonance ↔ most important contributing resonance structure (tertiary allylic cation at the isopropylidene carbon)]

most important contributing resonance structure

b.

[PhCH₂–CH=CH₂ + H⁺ → PhCH₂–CH⁺–CH₃ (secondary) → 1,2-H: shift → Ph–CH⁺–CH₂CH₃ benzylic cation]

most stable intermediate: benzylic carbocation

11.

[2-adamantyl cation with adjacent H's → ~H:⁻ ✗ → would give 1-adamantyl-like rearranged cation]

rigidity of the carbon skeleton prevents carbocation from being planar

12. neither regiospecificity nor stereospecificity: <u>a</u>.

[CH₃CH=CHCH₂CH₃ + HF → CH₃CHF–CH₂CH₂CH₃ + CH₃CH₂–CHF–CH₂CH₃]

b. [CH₂=CHCH₂CH₂CH₃ + Cl₂ / (XS) NaBr → CH₃CH₂CH₂–CHBr–CH₂Cl]   *regiospecific*

c. [cyclobutane-ene + Cl₂/H₂O, *anti*-add'n → trans-2-chlorocyclobutanol]   *stereospecific*

d. [ethylcyclopropene + D₂/Pt, *syn*-add'n → cis-1,2-dideuterio-ethylcyclopropane]   *stereospecific*

266 • Chapter 5 Alkenes and Carbocations

13. β-pinene $\xrightarrow{+H^\oplus}$ [carbocation intermediate] $\xrightarrow{-H^\oplus}$ α-pinene

more highly substituted double bond => more stable olefin therefore, $K_{eq} \gg 1$

5.2 Reactions

1. Ph–CH(CH₃)–CH=CH₂ $\xrightarrow[+H^\oplus]{HCl}$ Ph–CH(CH₃)–CH⁺–CH₃ $\xrightarrow{1,2\text{-}H{:}^\ominus \text{ shift}}$ Ph–C⁺(CH₃)–CH₂–CH₃ $\xrightarrow{+Cl^\ominus}$ Ph–C(Cl)(CH₃)–CH₂–CH₃

2. 1-nitrocyclohexene $\xrightarrow{HI}$ [cyclohexyl cation with NO₂] → trans-1-iodo-2-nitrocyclohexane

3. (1-isopropenyl)cyclopentane $\xrightarrow{+H^\oplus}$ [tertiary cation] $\xrightarrow{1,2\text{-}R{:}^\ominus \text{ shift}}$ [ring-expanded cation] $\xrightarrow[-H^\oplus]{+H_2O}$ 1,1-dimethylcyclohexan-2-ol

4. cyclopentane $\xrightarrow{1.\ Cl_2,\ \Delta}$ chlorocyclopentane $\xrightarrow[MeOH]{2.\ KOMe}$ cyclopentene $\xrightarrow[CHCl_3]{3.\ Br_2}$ trans-1,2-dibromocyclopentane

5. 3-ethyl-1-(trimethylammonio)cyclohex-1-ene $\xrightarrow{HI}$ [cation] $\xrightarrow{1,2\text{-}H{:}^\ominus \text{ shift}}$ [cation] → iodo product with NMe₃⁺ and Et

6. 1-fluorocyclopentene $\xrightarrow{HF}$ [cation ↔ fluoronium resonance] → 1,1-difluorocyclopentane

7. vinylcyclohexane $\xrightarrow{DCl}$ [cation with D] $\xrightarrow{1,2\text{-}H{:}^\ominus \text{ shift}}$ [tertiary cation with D] → 1-chloro-1-(2-deuterioethyl)cyclohexane

8. ![structure] →HBr→ [cation] →1,2-R:⁻ shift→ [cation] →again!→ [cation] → product with Br

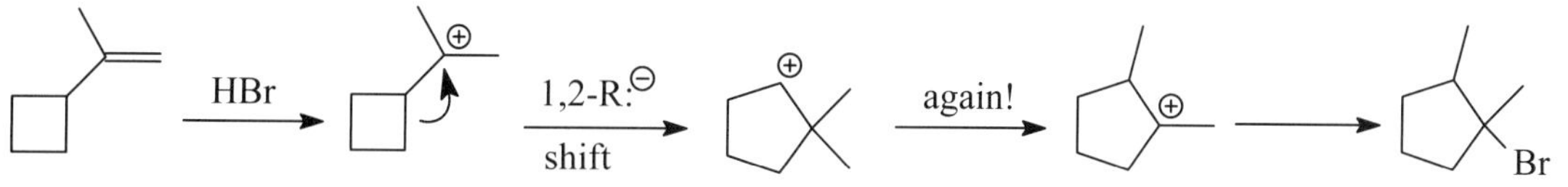

9. [alcohol] →+H⁺→ [oxocarbenium intermediate] → [cyclic oxonium] →−H⁺→ [cyclic ether]

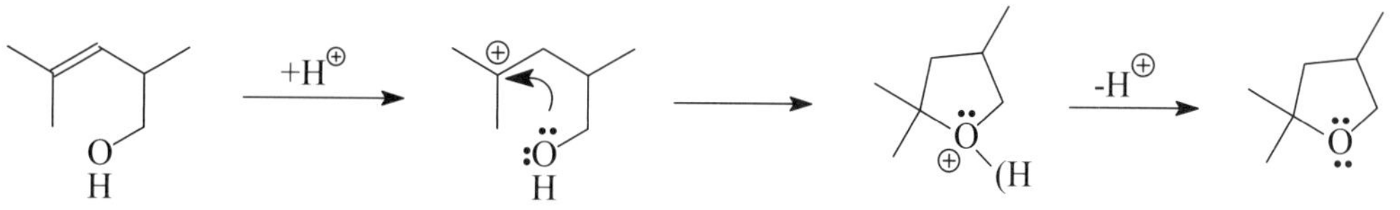

10. [cyclopentenyl-CCl₃] →HBr→ [cation-CCl₃] → [cyclopentyl-CCl₃ with Br]

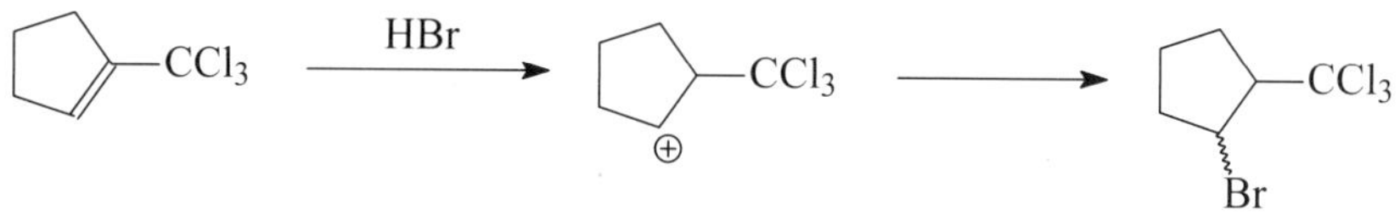

11. [cyclobutene-Et] →DBr→ [cation with Et, D, H] → *anti*-add'n + *syn*-add'n = racemic mixture

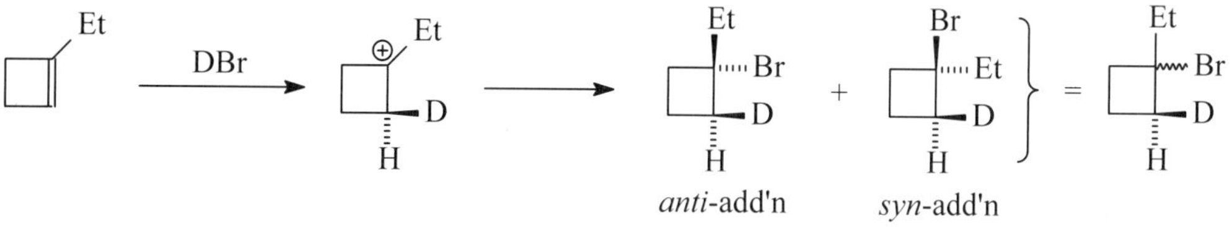

12. [cyclopentene] →1. H₂/Pd→ [cyclopentane] →2. Br₂/hν→ [cyclopentyl-Br]

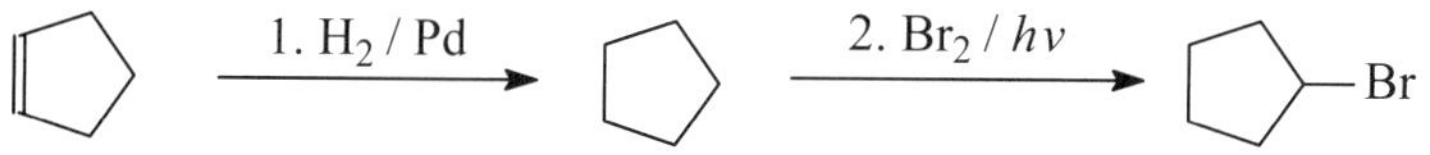

13. [methylenecyclohexane] →+H⁺→ [tertiary cation] →+EtOH, −H⁺→ [cyclohexyl-OEt, Me]

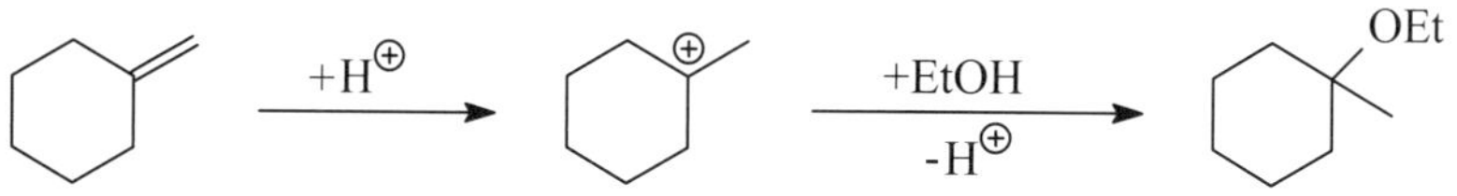

14. [MeO-cyclopentene-D] →HF→ [MeO cation with D] → *syn*- product + *anti*- product

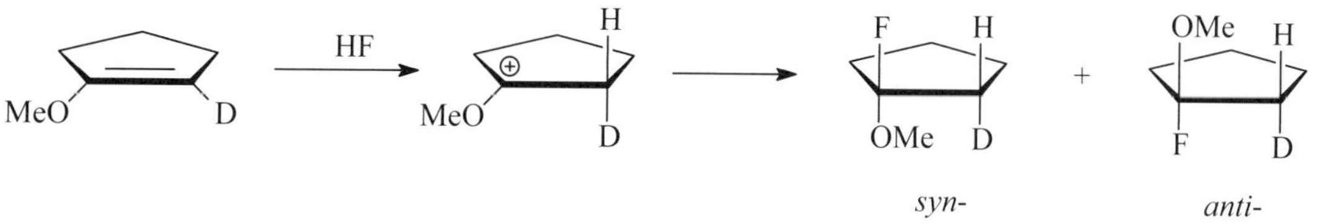

15. [Cl-propene] →HI→ [chloronium-like cation] ↔ [resonance structure] → [Cl, I product]

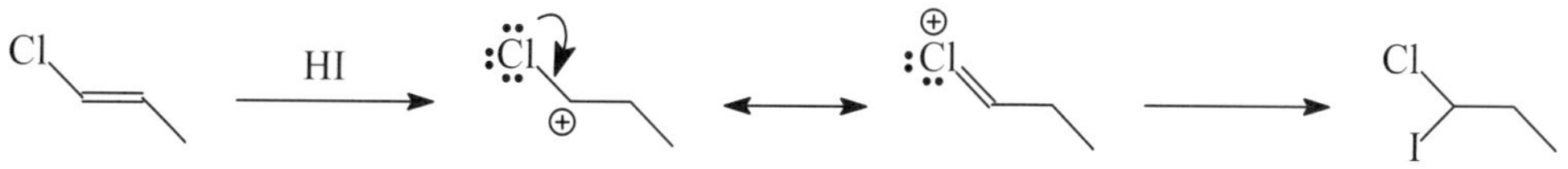

16. [methylcyclopropene] →Cl₂/H₂O→ [Cl, OH cyclopropane product]

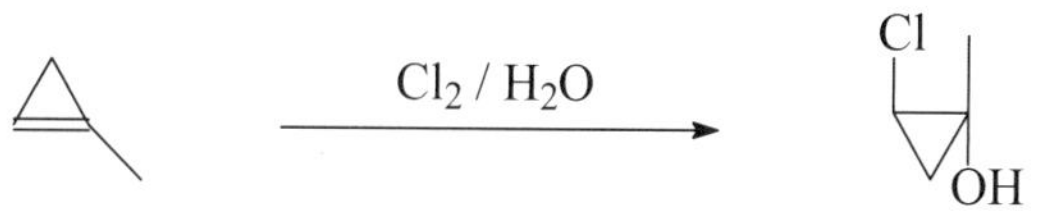

17. [methylenecyclopentane] →1. B₂D₆→ [cyclopentyl-D, BD₂] →2. H₂O₂, ⁻OH→ [cyclopentyl-D, OH]

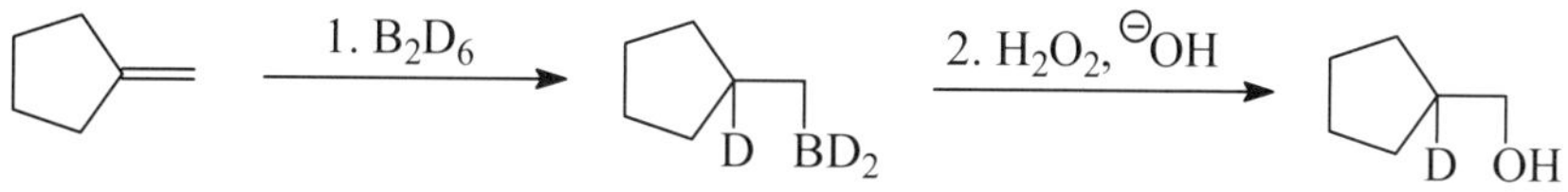

5.2 Reactions

18. $CH_2=CHCH_3 \xrightarrow{Cl_2}$ [cyclic chloronium ion with $\delta+$ on carbons, $\ominus I$ attacking] $\longrightarrow$ $CH_3CH(I)CH_2Cl$

19. 1-ethylcyclohexene $\xrightarrow{\text{1. Hg(OAc)}_2,\text{ PhOH}}$ (1-ethyl-1-OPh-2-HgOAc cyclohexane) $\xrightarrow{\text{2. NaBH}_4}$ 1-ethyl-1-phenoxycyclohexane

20. $H_2C=C=CH_2 \xrightarrow[\text{Zn(Cu)}]{\text{(XS) CH}_2\text{I}_2}$ [methylenecyclopropane] $\longrightarrow$ *spiropentane*

21. AcO-norbornene $\xrightarrow{\text{1. KMnO}_4,\ \ominus OH}$ AcO-norbornane-diol (OH, OH) $\xrightarrow{\text{2. HIO}_4}$ AcO-cyclopentane-(CHO)(CHO)

22. (decalin with methylenated ring) $\xrightarrow[\text{2. H}_3O^{\oplus},\text{ Zn}]{\text{1. O}_3}$ cyclohexane with –CH$_2$C(O)CH$_3$ and –CH$_2$CHO substituents

23. Ph-CH(CH$_3$)-CH=CH$_2$ $\xrightarrow[\text{+O–O+}]{\text{HBr}}$ Ph-CH(CH$_3$)-CH•-CH$_2$Br $\xrightarrow[\text{−Br•}]{\text{+HBr}}$ Ph-CH(CH$_3$)-CH(H)-CH$_2$Br

    note: opposite regioselectivity than HBr *without* peroxide - and no rearrangement

24. trans-hex-3-ene $\xrightarrow[hv]{\text{H}_2\text{C−N}_2}$ *trans*-1,2-diethylcyclopropane + N$_2$

25. cyclopentyl bromide $\xrightarrow[\text{−HBr}]{\text{1. base}}$ cyclopentene $\xrightarrow[\text{3. NaHSO}_3]{\text{2. OsO}_4}$ *cis*-cyclopentane-1,2-diol

26. 3-methyl-1-butene $\xrightarrow{\text{I−N}_3}$ [iodonium ion with $\ominus N_3$ attacking] $\longrightarrow$ (CH$_3$)$_2$CH-CH(N$_3$)-CH$_2$I with azide :N=N$^{\oplus}$=N:$^{\ominus}$

5.2 Reactions

270 • Chapter 5 Alkenes and Carbocations

34.

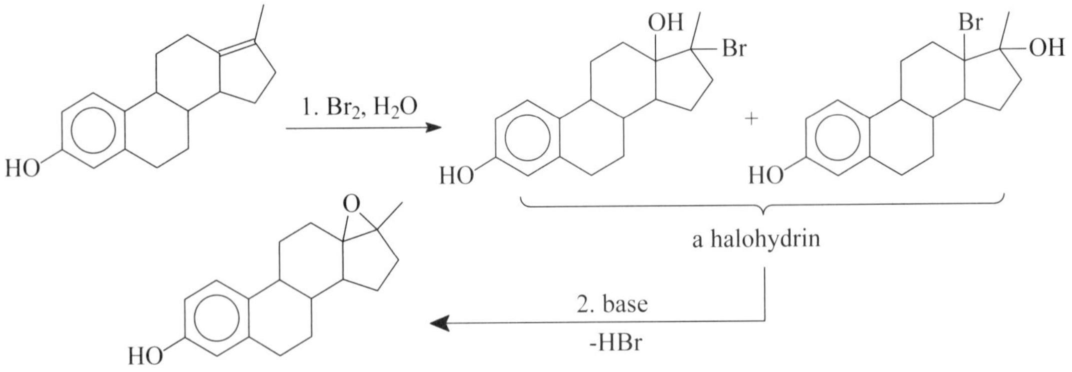

35.

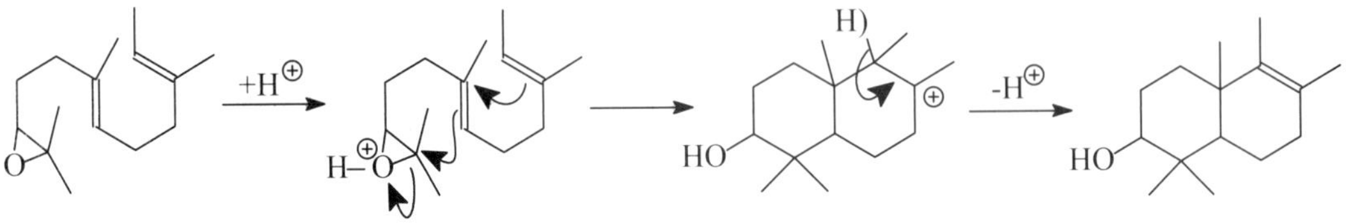

36.

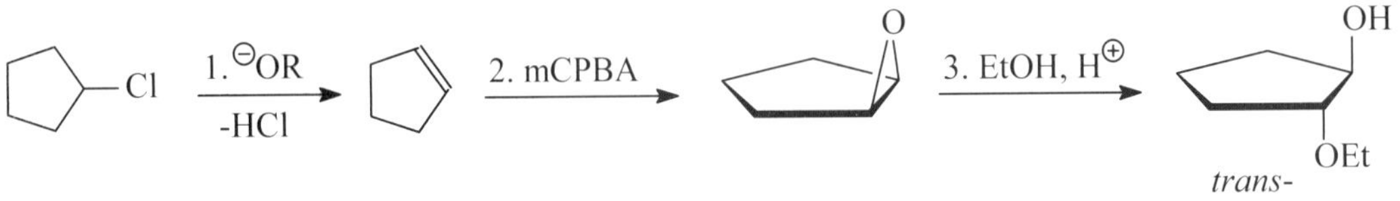

37.

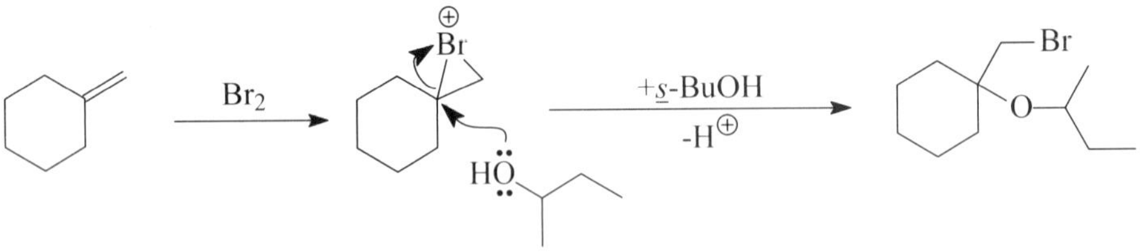

38.

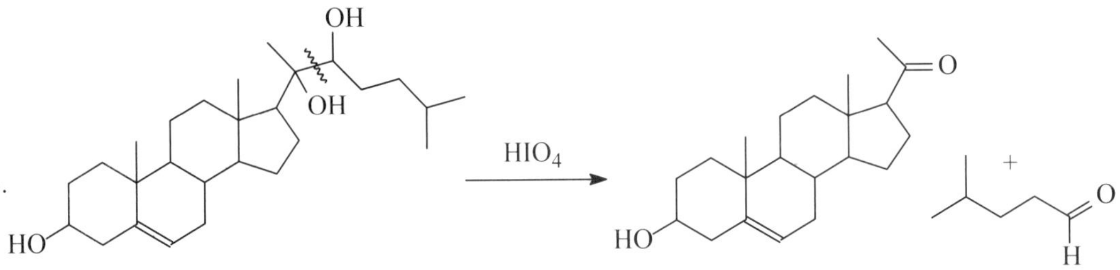

39.

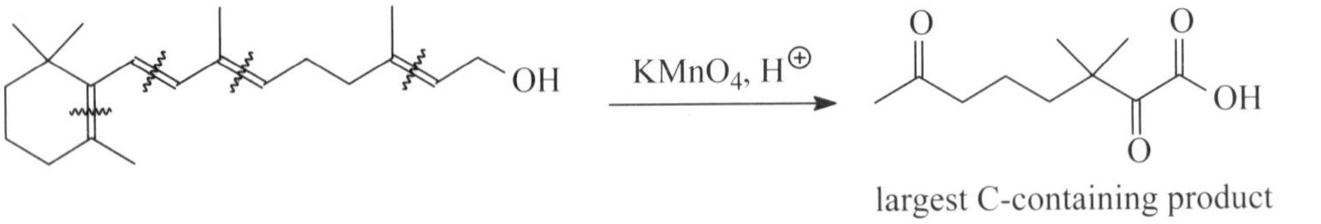

5.2 Reactions

40. [Reaction scheme: substrate with isopropyl group and acetate-bearing bicyclic ether, treated with 1. OsO₄ 2. NaHSO₃ to give tetraol, then 3. HIO₄ to give an aldehyde/ketone product plus 4-oxopentanal.]

41. [Reaction: vinylcyclopropane + HCl/+H⁺ → cyclopropylcarbinyl cation → 1,2-R: shift → cyclobutyl cation with H → 1,2-H: shift → cyclobutyl cation → cyclobutyl chloride + methylenecyclobutane]

42. [Ph–CHCl(H) with :CH₂⁻ Li⁺ (butyllithium), −H⁺ → Ph–C̈H–Cl carbanion, −Cl⁻ → [PhC̈H] carbene → with cyclohexene → norcarane with Ph]

a 1,1- (or α-) elimination to produce a carbene

43. [Reaction of dichlorodienyl ester with 3-phenoxybenzyl group + (CH₃)₂CI₂ (1 equiv), Zn(Cu) → cyclopropane product (permethrin-like structure)]

more electron-rich (nucleophilic) double bond

5.3 Syntheses

1. [Dihydronaphthalene + HBr → benzylic cation → 1,2-H: shift → more stable benzylic cation → +Br⁻ → 1-bromotetralin]

2. [Cyclohexanol, 1. H₂SO₄, −H₂O → cyclohexene, 2. HCl → chlorocyclohexane]

3. [t-BuCl, 1. KOH, EtOH, −HCl → isobutylene, 2. HF → t-BuF]

272 • Chapter 5 Alkenes and Carbocations

4. Cyclobutyl-Cl →(1. ⁻OMe, MeOH, -HCl)→ cyclobutene →(2. H₂/Pt)→ cyclobutane

5. cyclopentane →(1. Cl₂, hv)→ chlorocyclopentane →(2. NaOEt/EtOH, -HCl)→ cyclopentene →(3. D₂/Pd)→ trans-1,2-dideuteriocyclopentane

6. cycloheptane →(1. Cl₂, Δ)→ chlorocycloheptane →(2. KO-t-Bu, t-BuOH)→ cycloheptene →(3. KMnO₄, H⁺)→ HO₂C-(CH₂)₅-CO₂H

7. 2-methyl-1-pentene →(Cl₂, (XS) NaBr)→ [chloronium ion intermediate with Br⁻ attack] → 2-bromo-2-methyl-1-chloropentane

8. PhCH₂-CH=CH₂ →(1. O₃; 2. Zn, H⁺)→ PhCH₂CHO

9. H₂C=CH₂ →(1. CH₂N₂, hv)→ cyclopropane →(2. Br₂, hv)→ bromocyclopropane

10. cyclopentanol →(1. H₂SO₄, -H₂O)→ cyclopentene →(2. CH₂I₂, Zn(Cu))→ bicyclo[3.1.0]hexane

11. 1,2-dihydronaphthalene →(1. Hg(OAc)₂, H₂O; 2. NaBH₄)→ 2-tetralol (no rearrangement)

    →(H₃O⁺)→ secondary carbocation →(1,2-H:⁻ shift)→ benzylic carbocation →(+H₂O, -H⁺)→ 1-tetralol

12. bromo-bicyclic →(1. KOMe, MeOH)→ bicyclic alkene →(2. O₃; 3. Zn, H⁺)→ cyclooctane-1,5-dione

13. cyclohexane →(1. Br₂, hv)→ bromocyclohexane →(2. NaOH)→ cyclohexene →(3. DBr)→ trans-1-bromo-2-deuteriocyclohexane

5.3 Syntheses

14. 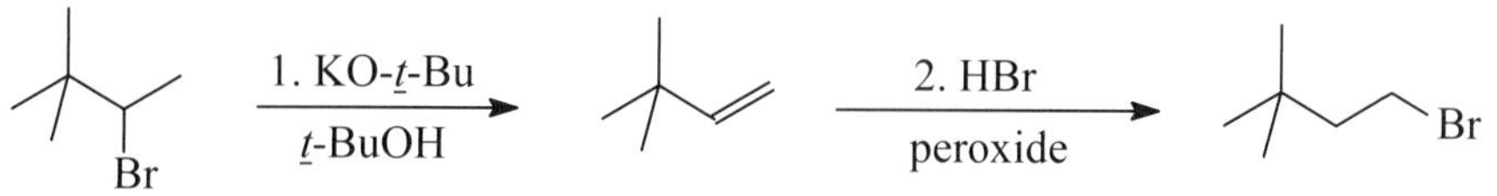 (structure with Br) →[1. KO-t-Bu / t-BuOH] (alkene) →[2. HBr / peroxide] (alkyl bromide)

15. 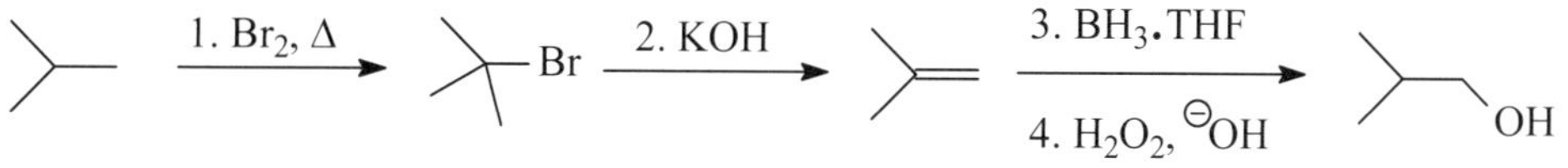 (isobutane) →[1. Br₂, Δ] (t-Bu-Br) →[2. KOH] (isobutylene) →[3. BH₃·THF; 4. H₂O₂, ⁻OH] (isobutanol)

16. 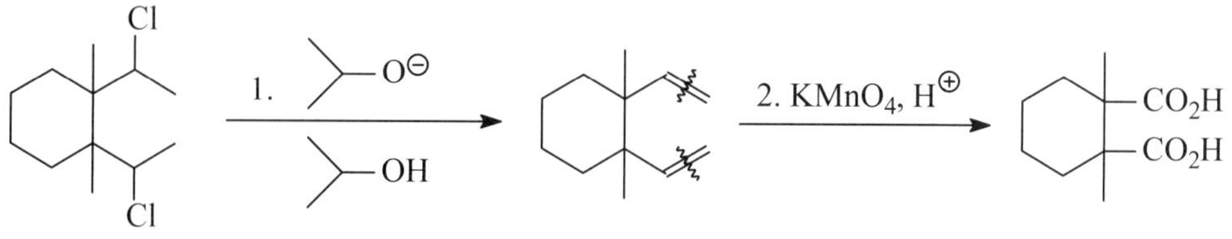 (dichloride on cyclohexane) →[1. iPrO⁻ / iPrOH] (diene) →[2. KMnO₄, H⁺] (dicarboxylic acid)

17. 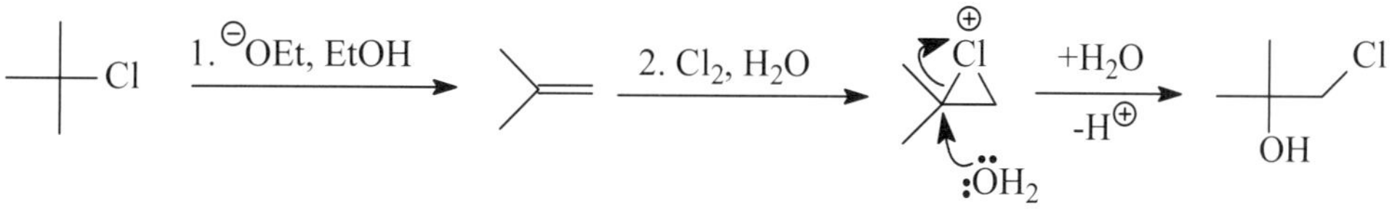 (t-Bu–Cl) →[1. ⁻OEt, EtOH] (isobutylene) →[2. Cl₂, H₂O] (chloronium with :OH₂) →[+H₂O, −H⁺] (chlorohydrin)

18. 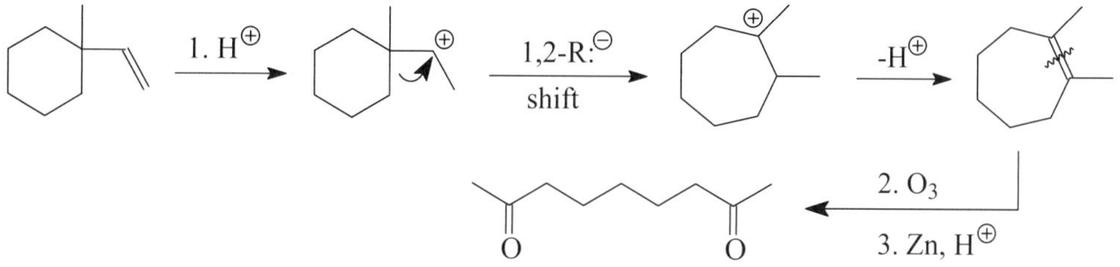 (1-methyl-1-vinylcyclohexane) →[1. H⁺] (cation) →[1,2-R:⁻ shift] (ring-expanded cation) →[−H⁺] (cycloheptene) →[2. O₃; 3. Zn, H⁺] (diketone)

19. 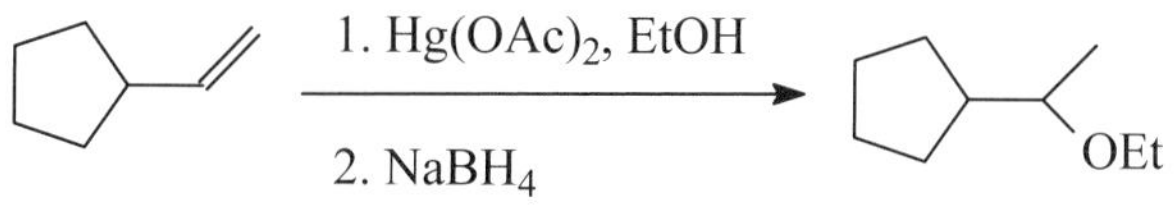 (vinylcyclopentane) →[1. Hg(OAc)₂, EtOH; 2. NaBH₄] (ether with OEt)   (not EtOH, H⁺ ⇒ rearrangement)

20. 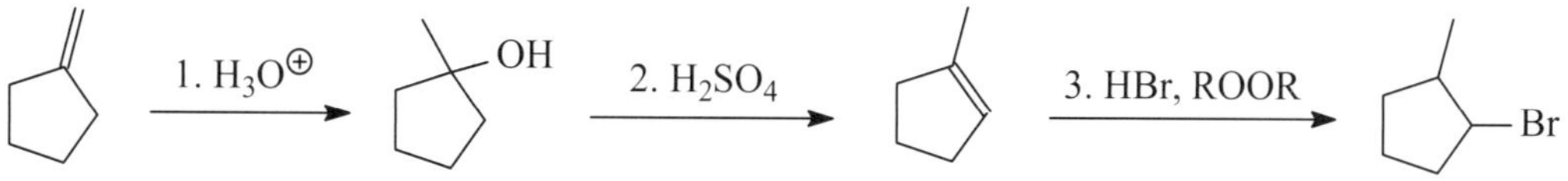 (methylenecyclopentane) →[1. H₃O⁺] (tertiary alcohol) →[2. H₂SO₄] (methylcyclopentene) →[3. HBr, ROOR] (bromide)

21. 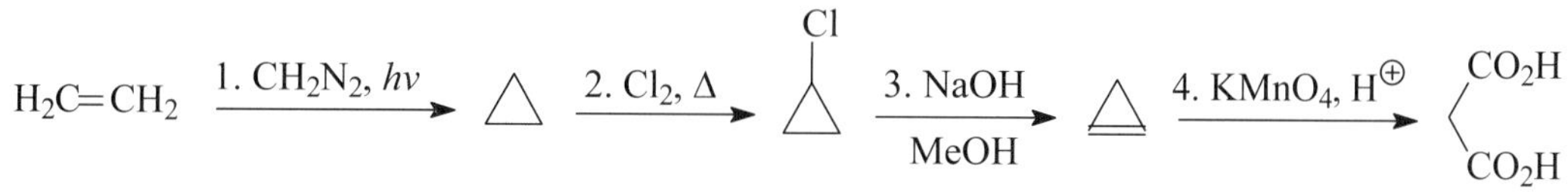 H₂C=CH₂ →[1. CH₂N₂, hv] (cyclopropane) →[2. Cl₂, Δ] (chlorocyclopropane) →[3. NaOH / MeOH] (cyclopropene) →[4. KMnO₄, H⁺] (malonic acid, CO₂H / CO₂H)

5.3 Syntheses

274 • Chapter 5 Alkenes and Carbocations

22. (CH₃)₃C–Br $\xrightarrow[\text{MeOH}]{1.\ \text{KOMe}}$ (CH₃)₂C=CH₂ $\xrightarrow{2.\ H_3O^\oplus}$ (CH₃)₃C–OH

(CH₃)₂C=CH₂ + (CH₃)₃C–OH $\xrightarrow{3.\ H^\oplus}$ ((CH₃)₃C)₂O or (CH₃)₂C=CH₂ $\xrightarrow[\text{4. NaBH}_4]{3.\ \text{Hg(OAc)}_2,\ \underline{t}\text{-BuOH}}$ ((CH₃)₃C)₂O

23. cyclobutanol $\xrightarrow{1.\ H_2SO_4}$ cyclobutene $\xrightarrow{2.\ CH_2I_2,\ Zn(Cu)}$ bicyclo[1.1.0]butane derivative

5.4 Mechanisms

1. [alkene] $\xrightarrow{+H^\oplus}$ [cation] $\xrightarrow[\text{shift}]{1,2\text{-R:}^\ominus}$ [rearranged cation] $\xrightarrow{-H^\oplus}$ [dimethylcyclopentene]

2. [methylcyclohexene] $\xrightarrow{+H^\oplus}$ [cation] $\xrightarrow{-H^\oplus}$ [dimethylcyclohexene]

3. [diene] $\xrightarrow{+H^\oplus}$ [allyl cation] ↔ [allyl cation] → [oxonium with MeOH] $\xrightarrow{-H^\oplus}$ [OMe product]

4. [methylenecycloheptane with methyls] $\xrightarrow{+H^\oplus}$ [cation] $\xrightarrow[\text{shift}]{1,2\text{-R:}^\ominus}$ [cation] $\xrightarrow{-H^\oplus}$ [isopropenyl cyclohexane]

5. [norbornene-CO₂H] $\xrightarrow[-I^\ominus]{+I_2}$ [iodonium] → [cation with OH] $\xrightarrow{-H^\oplus}$ [iodolactone]

5.4 Mechanisms

5.4 Mechanisms

12.

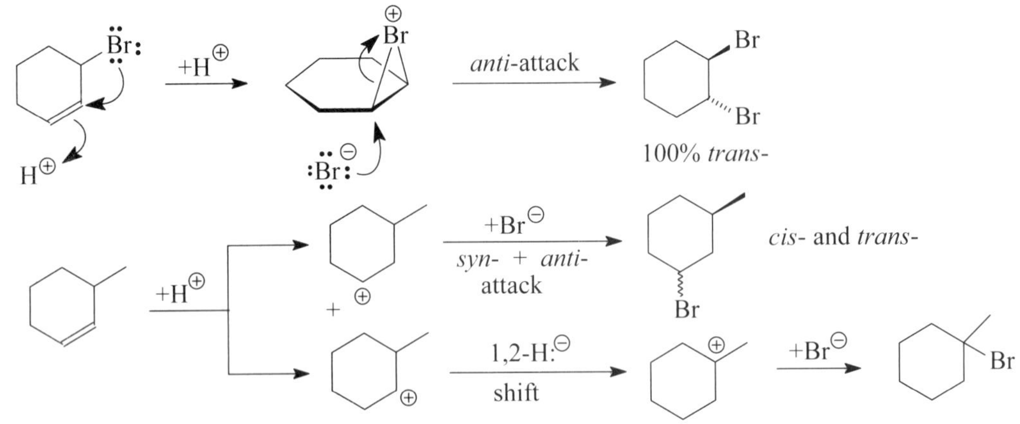

13. **A** is C_{14} => other aldehyde is CH_2O; no. deg. unsat: $C_{15}H_{32} - C_{15}H_{24} = H_8$ => $H_8/2 = 4$ deg. hydrogenation: $C_{15}H_{28} - C_{15}H_{24} = H_4$ => $H_4/2 = 2$ DB; therefore, 2 rings are present

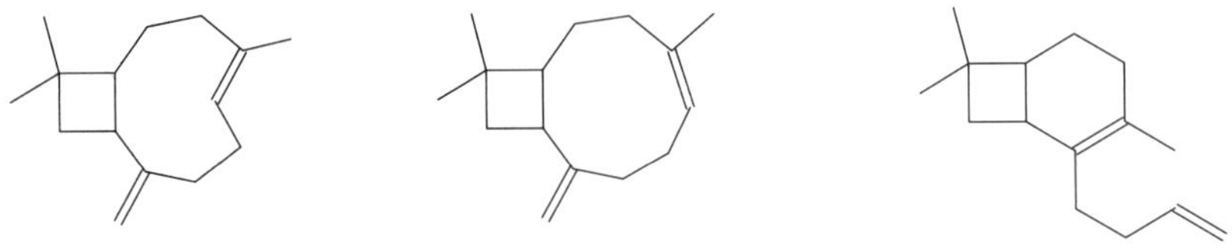

caryophyllene   *isocaryophyllene*   incorrect: cannot exist in *cis/trans* forms

14.

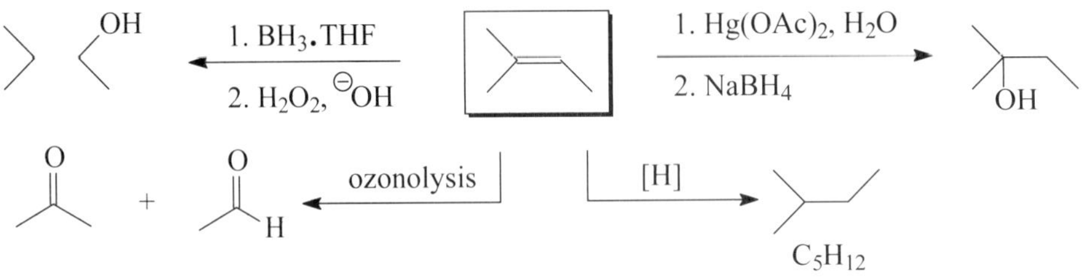

15.

A
isoprene

16. no. deg. unsat: $C_{10}H_{22} - C_{10}H_{16} = H_6$ => $H_6/2 = 3$ deg.

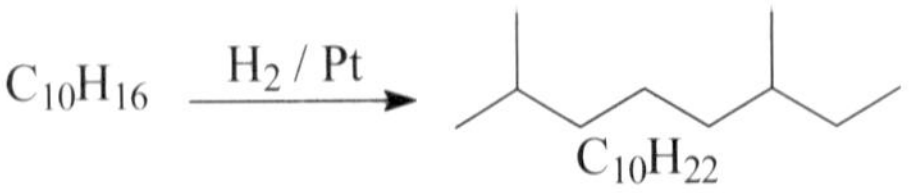

$C_{10}H_{16}$ $\xrightarrow{H_2/Pt}$ $C_{10}H_{22}$ hydrogenation: $C_{10}H_{22} - C_{10}H_{16} = H_6$ => $H_6/2 = 3$ DB

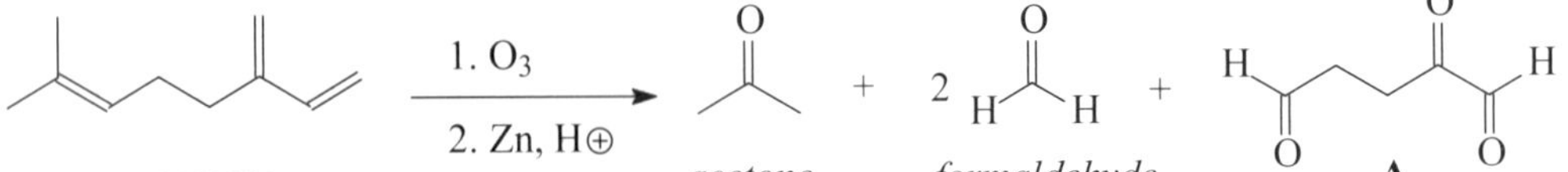

myrcene   *acetone*   *formaldehyde*   **A**

5.4 Mechanisms

17. hydrogenation: $C_{10}H_{20} - C_{10}H_{16} = H_4 \Rightarrow H_4/2 = 2$ DB   **A** = [4-isopropenyl-1-methylcyclohexene structure]

B   **A**

other ozonolysis product: H-CHO (formaldehyde)

18. a. n [methyl methacrylate structure] $\longrightarrow$ *poly(methyl methacrylate)*

b. [cationic polymerization mechanism of styrene with $+H^\oplus$, showing propagation steps] $\Longrightarrow$ *etc.*

19. [cationic polymerization of tert-butyl vinyl ether with $+H^\oplus$, showing propagation steps] *etc.* $\Longleftarrow$

20. $H_2C=\overset{\oplus}{N}=\overset{\ominus}{\underset{..}{N}}: \longleftrightarrow H_2\overset{..}{C}-\overset{\oplus}{N}\equiv N: \longrightarrow [:CH_2] + N_2$

[:CH_2] + [diazomethane] $\longrightarrow$ [cyclic intermediate with N=N] $\longrightarrow H_2C=CH_2 + :N\equiv N:$

21. $CH_2N_2 \xrightarrow[-N_2]{h\nu} H_2C: +$ [benzene] $\longrightarrow$ [norcaradiene intermediate] $\longrightarrow$ [cycloheptatriene]

22. [cyclopentylidenecyclopentane] $\xrightarrow{+H^\oplus}$ [cation] $\xrightarrow{\text{1,2-R:}^\ominus \text{ shift}}$ [spiro cation] $\xrightarrow{\text{1,2-R:}^\ominus \text{ shift}}$ [decalin cation with H] $\xrightarrow{-H^\oplus}$ [octahydronaphthalene]

23. [starting tricyclic structure with exocyclic methylene] $\xrightarrow{+H^\oplus}$ [tertiary cation] $\xrightarrow{\text{1,2-R:}^\ominus \text{ shift}}$ [rearranged cation] (H) $\xrightarrow{-H^\oplus}$ [final alkene product]

5.4 Mechanisms

278 • Chapter 5 Alkenes and Carbocations

24.

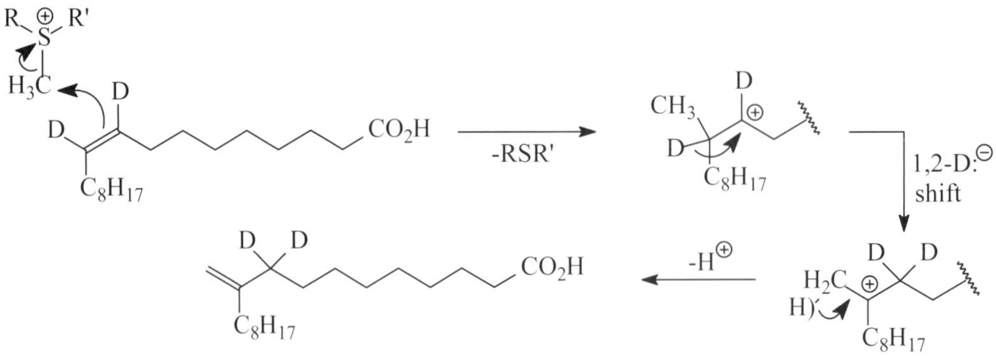

25.

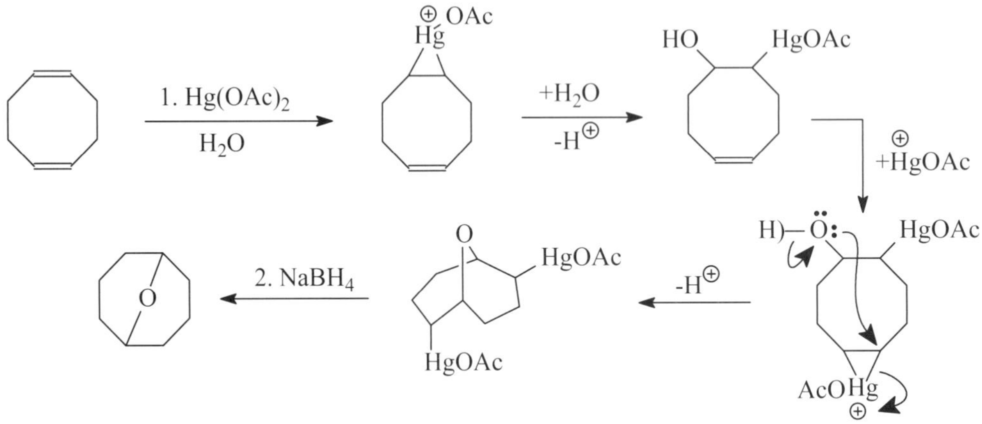

26.

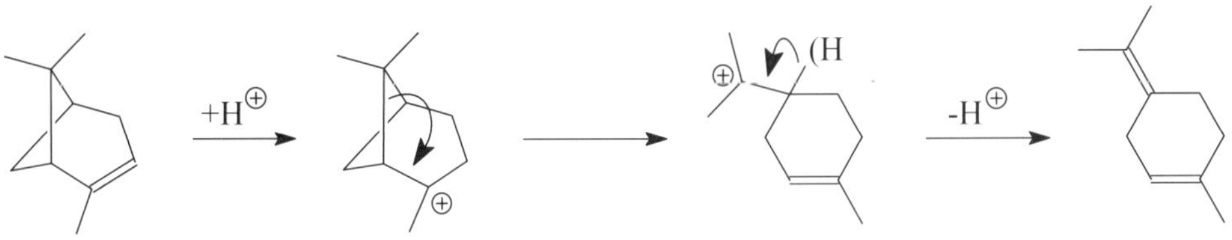

27. 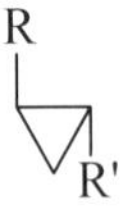 => *trans*- DB 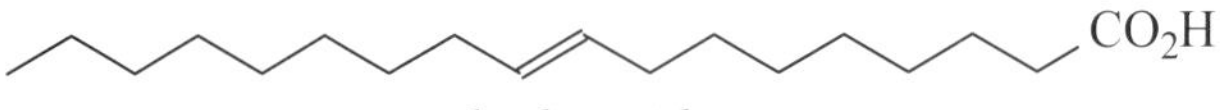   *elaidic acid*

28. a. 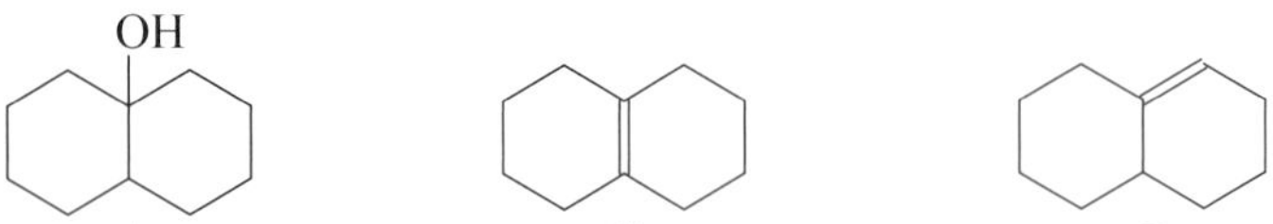 **A**, **D**, **C**

b.

A $\xrightarrow{Br_2}$ NR (deep red color of bromine persists);   **B** or **C** $\xrightarrow{Br_2}$ color discharged

or Baeyer test:

A $\xrightarrow{KMnO_4}$ NR (purple color of $MnO_4^{\ominus}$ persists);   **B** or **C** $\xrightarrow{KMnO_4}$ brown ppt (MnO_2) forms

5.4 Mechanisms

29. 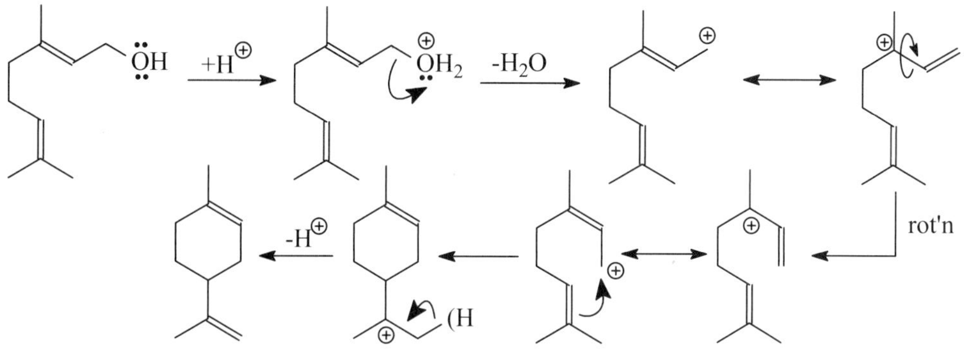 [mechanism scheme for acid-catalyzed cyclization of geraniol to limonene: +H⁺, -H₂O, carbocation resonance, rot'n, cyclization, -H⁺]

30. a., b. 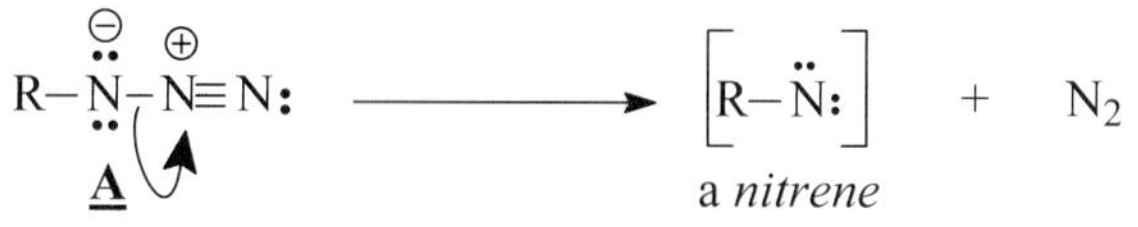 R—N⁻—N⁺≡N: ⟶ [R—N:] + N₂
              A   a *nitrene*

  c. *retention of configuration* suggests a concerted (or pericyclic) mechanism

31. 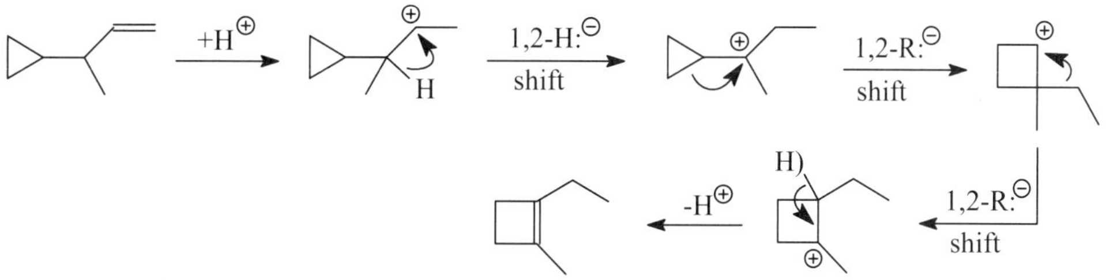 [mechanism: cyclopropyl alkene + H⁺ → carbocation, 1,2-H:⁻ shift, 1,2-R:⁻ shift (ring expansion to cyclobutyl cation), 1,2-R:⁻ shift, -H⁺ → 1-ethyl-2-methylcyclobutene]

32. 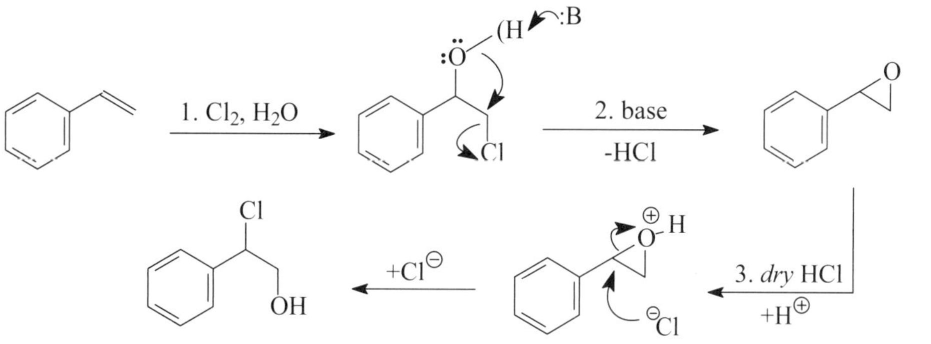 [styrene 1. Cl₂, H₂O → chlorohydrin; 2. base, -HCl → styrene oxide; 3. dry HCl, +H⁺ → protonated epoxide + Cl⁻ → 2-chloro-2-phenylethanol]

33. 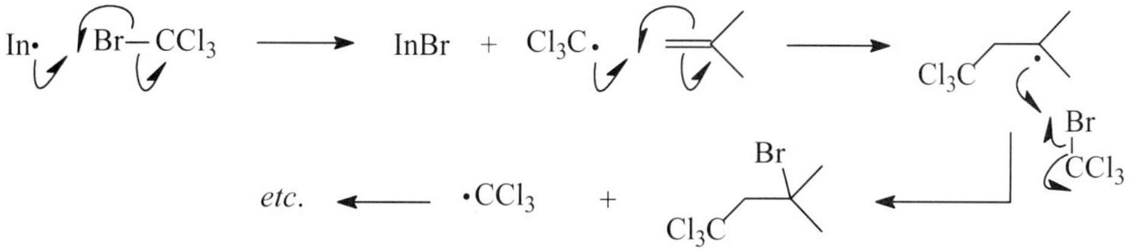 In• Br—CCl₃ ⟶ InBr + Cl₃C• + alkene ⟶ Cl₃C—CH₂—C(CH₃)₂•

    Cl₃C—CH₂—C(CH₃)₂—Br + •CCl₃ ⟵

    etc. ⟵ •CCl₃ + Cl₃C—CH₂—C(CH₃)₂—Br

(this mechanism is similar to the addition of HBr in the presence of peroxides)

5.4 Mechanisms

280 • Chapter 5 Alkenes and Carbocations

34. a. no. deg. unsat: $C_{16}H_{34} - C_{16}H_{30} = H_4 \Rightarrow H_4/2 = 2$ deg.
    hydrogenation: $C_{16}H_{34}O - C_{16}H_{30}O = H_4 \Rightarrow H_4/2 = 2$ DB

   c. $\underline{E} \Rightarrow$ a *cis*- DB at C_{12}; $\underline{F} \Rightarrow$ a *trans*- DB at C_{10}

$$\underline{A} \xrightarrow[\text{2. Zn, H}^{\oplus}]{\text{1. O}_3} \underline{B} + \underline{C} + \underline{D}$$

35. [mechanism: path a gives *not observed* product via $-D^{\oplus}$; path b via 1,2-D:$^{\ominus}$ shift gives *observed* product via $-H^{\oplus}$]

   therefore, path <u>b</u> is favored

36. The rigidity of the bicyclic structure in the conjugate base of $\underline{A}$ prevents delocalization of the negative charge onto oxygen: such a contributing resonance structure would violate Bredt's rule (the olefinic region cannot be planar). Loss of this stabilization prevents the carbanion from forming (pK$_a$ of $\underline{A}$ is raised relative to cyclohexanone), as required by the proposed mechanism, and therefore prevents hydrogen-deuterium exchange.

5.4 Mechanisms

CHAPTER 6
ALKYNES

6.1 Reactions

1. CH₃CH=CH−C≡C−H → 1. NaH → CH₃CH=CH−C≡C:⁻ → 2. D₂O → CH₃CH=CH−C≡C−D

2. CH₃(CH₂)₅C≡CH → H⁺, HgSO₄, PhOH → CH₃(CH₂)₅C(OPh)=CH₂

3. Ph−C≡CH → 1. B₂H₆; 2. H₂O₂, HO⁻ → (Z)-PhCH=CHOH → taut → PhCH₂CHO

4. CH₃CH₂CH₂CH₂Cl → 1. ⁻OMe (E2); 2. Cl₂ → CH₃CH(Cl)CH(Cl)CH₃ → 3. (XS) NaNH₂; 4. BH₃·THF; 5. H₂O₂, ⁻OH → CH₃CH₂CH=CHOH → taut → CH₃CH₂CH₂CHO

5. 1-methyl-1-chlorocyclopentane → RC≡C:⁻ → 1-methylcyclopentene + RC≡CH   3° R-X => elimination, not substitution!

6. cyclohexyl−C≡CH → 1. Li / NH₃ → cyclohexyl−CH=CH₂ → 2. HBr, ROOR (peroxide effect) → cyclohexyl−CH₂CH₂−Br

7. (CH₃)₂CH−C≡CH → 1. H₂ / Pd(Pb) → (CH₃)₂CH−CH=CH₂ → 2. BH₃; 3. H₂O₂, ⁻OH → (CH₃)₂CHCH₂CH₂OH

8. CH₃(CH₂)₇C≡CH → 1. NaH; 2. CH₃(CH₂)₁₂Cl; 3. Lindlar catalyst (cis-[H]) → cis-CH₃(CH₂)₇CH=CH(CH₂)₁₂CH₃

9. CH₃CH₂CHCl₂ → 1. (XS) NaNH₂ → Et−C≡C:⁻ → 2. H₃O⁺, HgSO₄ → CH₃CH₂C(OH)=CH₂ → taut → CH₃CH₂COCH₃

10. (CH₃)₃C−C≡CH → Cl₂ / H₂O → (CH₃)₃C−C(OH)=CHCl → taut → (CH₃)₃C−CO−CH₂Cl

6.1 Reactions

11. Ph—≡≡ $\xrightarrow{\text{1. (XS) HI}}$ Ph-CI₂-CH₃ $\xrightarrow{\text{2. Zn(Cu)}}$ [PhC̈CH₃] + cyclopentene → bicyclic product with Me, Ph

12. 2,6-disubstituted aryl-C(Cl)₂CH₃ with OMe $\xrightarrow{\text{1. (XS) NaNH}_2}$ Ar–C≡C:⁻ $\xrightarrow{\text{2. D}_2\text{O}}$ Ar–C≡C–D

13. HC≡C–CH₂OH $\xrightarrow[\text{(2 equiv)}]{\text{1. LiNH}_2}$:⁻C≡C–CH₂–Ö:⁻ $\xrightarrow[\text{(1 equiv)}]{\text{2. }n\text{-C}_5\text{H}_{11}\text{Br}}$ n-C₅H₁₁–C≡C–CH₂–O⁻ $\xrightarrow{\text{3. H}^+}$ n-C₅H₁₁–C≡C–CH₂OH

(less stable anion therefore, more reactive)

14. HC≡CH $\xrightarrow[\text{2. }n\text{-Pr-I}]{\text{1. NaNH}_2 \text{ (1 equiv)}}$ HC≡C–n-Pr $\xrightarrow{\text{3. NaNH}_2}$ n-Pr–C≡C:⁻ $\xrightarrow[\text{3° RX!}]{\text{4. }t\text{-BuCl}}$ isobutylene + n-Pr–C≡C–H (elimination, not alkylation)

6.2 Syntheses

1. CH₃CH₂–CHBr–CHBr–CH₂CH₃ $\xrightarrow{\text{1. (XS) NaNH}_2}$ Et–C≡C–Et $\xrightarrow{\text{2. KMnO}_4, \text{H}^+}$ EtCOOH

2. HC≡CH $\xrightarrow[\text{2. }n\text{-PrBr}]{\text{1. NaH (1 equiv)}}$ n-Pr–C≡CH $\xrightarrow{\text{3. H}_2/\text{Pd(Pb)}}$ cis-alkene $\xrightarrow{\text{4. HBr, R}_2\text{O}_2}$ n-alkyl–Br

3. CH₂=CHCl $\xrightarrow{\text{1. (XS) NaNH}_2}$:⁻C≡C:⁻ $\xrightarrow{\text{2. H}^+, \text{HgSO}_4, \text{MeOH}}$ CH₂=CH–OMe

4. HC≡CH $\xrightarrow[\text{3. NaH}]{\text{1. NaH (1 equiv); 2. Et-I}}$ Et–C≡C:⁻ $\xrightarrow{\text{4. }n\text{-Bu-I}}$ Et–C≡C–n-Bu $\xrightarrow[\text{trans-[H]}]{\text{5. Li/NH}_3}$ trans-3-octene

5. (CH₃)₃C−C≡CH →[1. H₂ / Pd(Pb)] (CH₃)₃C−CH=CH₂ →[2. HCl] carbocation →[1,2-R:⁻ shift] rearranged cation → tertiary chloride product

6. cyclohexyl-CH=CH-CH₃ →[1. Br₂ / 2. (XS) NaNH₂] cyclohexyl-C≡CH →[3. H₂ / Lindlar, cis-[H]] cyclohexyl-CH=CH-CH₃ (cis)

7. HC≡CH →[1. NaNH₂ / 2. Me-Br] HC≡C-CH₃ →[3. H₂ / Pd(Pb)] cis-CH₂=CH-CH₃ (cis-2-butene) →[4. CH₂I₂, Zn(Cu)] cyclopropane (dimethyl)

8. HC≡C-CH₃ →[1. Li / NH₃ or H₂ / Lindlar] CH₂=CH-CH₃ →[2. HBr, (t-Bu)₂O₂] CH₃CH₂CH₂Br

9. (CH₃)₃C−C≡CH
   →[H⁺, HgSO₄, H₂O] (CH₃)₃C−C(OH)=CH₂ →[taut] (CH₃)₃C−C(=O)−CH₃
   →[1. BH₃·THF / 2. H₂O₂, ⁻OH] (CH₃)₃C−CH=CH−OH →[taut] (CH₃)₃C−CH₂−CHO
   →[1. Li / NH₃] trans-alkene →[2. O₃ / 3. Zn, H⁺] (CH₃)₃C−CH₂−CHO

10. Ph−CH=CH₂ →[1. Cl₂ / 2. (XS) NaNH₂] Ph−C≡C:⁻ →[3. Et-I] Ph−C≡C−Et →[4. Li / NH₃, trans-[H]] Ph−CH=CH−Et (trans)

11. Ph−C≡C−Ph
    →[1. H₂ / Pd(Pb)] cis PhCH=CHPh →[2. CH₂N₂, hν] 1,2-diphenylcyclopropane
    →[1. Na / NH₃] trans PhCH=CHPh →[2. O₃ / 3. Zn, H⁺] PhCHO

12. Et−C≡C−Et →[1. Na / NH₃] trans Et−CH=CH−Et →[2. CHCl₃, KO-t-Bu ⇒ [:CCl₂]] dichlorocyclopropane (Et, Cl / Cl, Et)

6.2 Syntheses

284 • Chapter 6 Alkynes

13. (hexyl chloride) $\xrightarrow[\text{3. (XS) NaNH}_2]{\text{1. KOH} \atop \text{2. Cl}_2}$ (pent-1-ynide) $\xrightarrow[\text{5. Li / NH}_3]{\text{4. EtCl}}$ (trans n-Bu, Et alkene) $\xrightarrow[\text{base}]{\text{CHCl}_3}$ (dichlorocyclopropane with n-Bu and Et)

14. $HC\equiv CH \xrightarrow[\text{2. }n\text{-pentyl chloride}]{\text{1. NaH (1 equiv)}} HC\equiv C-n\text{-pentyl} \xrightarrow{\text{3. H}_3O^{\oplus}\ \text{HgSO}_4}$ (enol with OH) $\xrightarrow{\text{taut}}$ (methyl ketone)

15. $HC\equiv CH \xrightarrow[\text{3. NaNH}_2]{\text{1. NaNH}_2\ (\text{1 equiv}) \atop \text{2. }n\text{-decyl bromide}} n\text{-decyl}-C\equiv C:^{\ominus} \xrightarrow[\text{5. H}_2\text{/ Lindlar} \atop \text{6. mCPBA}]{\text{4. 1-bromo-5-methyl-hexane}}$ (epoxide product)

16. (pent-1-yne) $\xrightarrow[\text{2. }n\text{-Pr-I}]{\text{1. NaNH}_2}$ (oct-4-yne) $\xrightarrow[\text{4. Zn, H}^{\oplus} \atop \text{(or 3. KMnO}_4, \text{H}^{\oplus})]{\text{3. O}_3}$ 2 (butanoic acid)

6.3 Mechanisms

1. (mechanism showing iodonium formation on alkyne of sugar, intramolecular OH attack, -H⁺, giving bromo-iodo vinyl ether product)

2. (mechanism showing protonation of OH, -H₂O to carbocation, alkyne attack forming ring, +H₂O, -H⁺, tautomerization ~H⁺ to give acetyl-bicyclic ketone)

6.3 Mechanisms

6.3 Mechanisms

CHAPTER 7
STEREOCHEMISTRY

7.1 General

1. chiral molecules: <u>a</u>, <u>b</u>, <u>f</u>, <u>h</u>, and <u>l</u>.

2. a. <u>3</u>.   b. <u>6</u>. 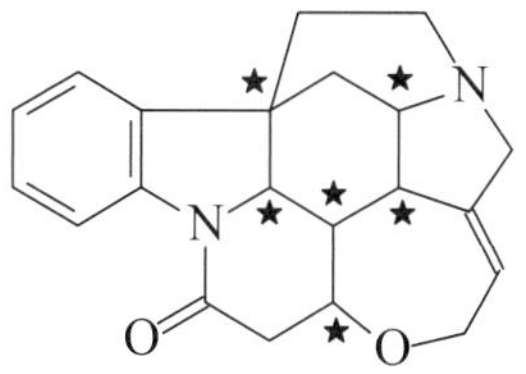   c. <u>4</u>. 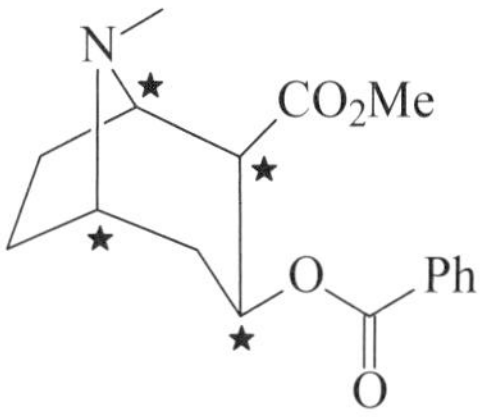   d. <u>2</u>.

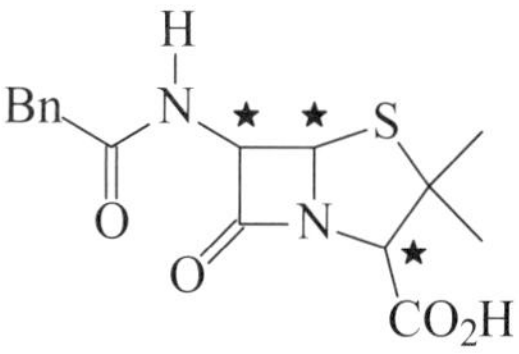

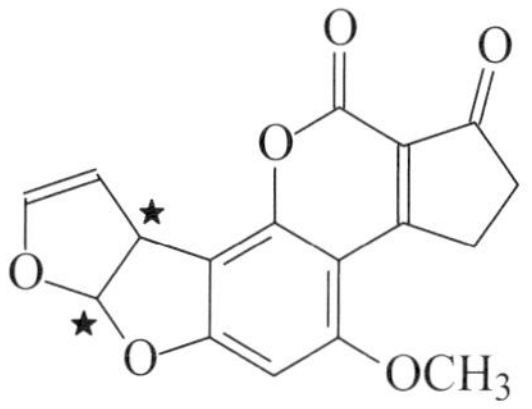

3. (structures shown with stereochemistry assignments)

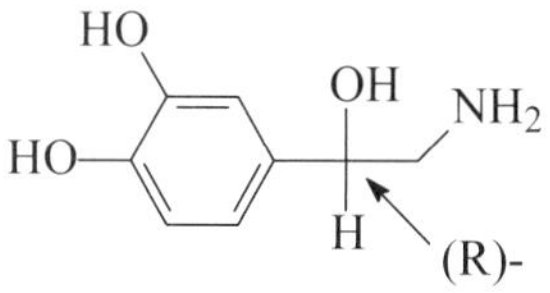

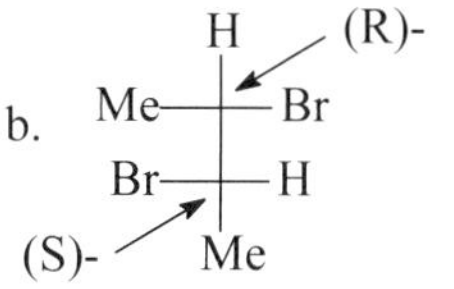

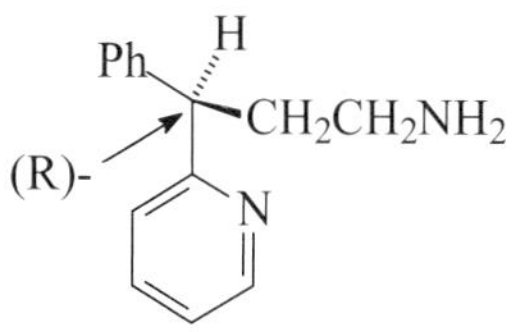

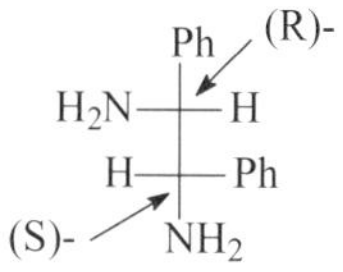

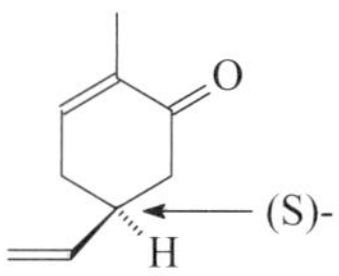

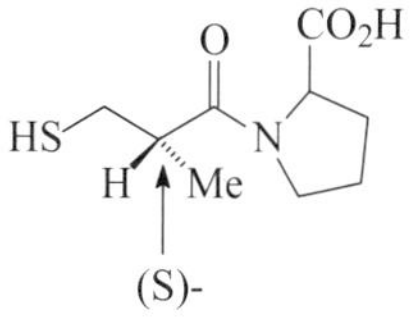

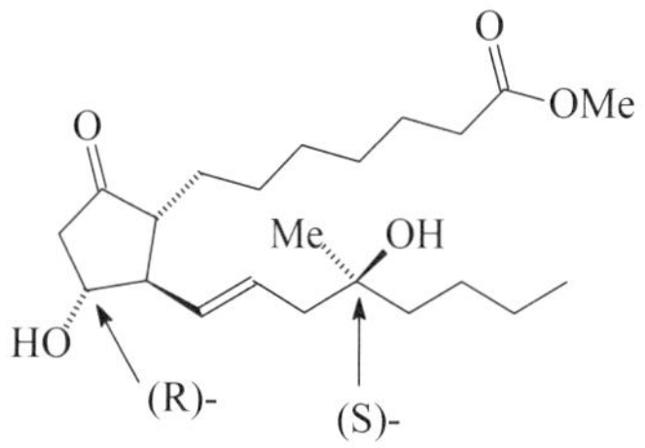

4.
a. enantiomers	b. enantiomers	c. diastereomers	d. enantiomers
e. identical	f. diastereomers	g. enantiomers	h. diastereomers
i. enantiomers	j. enantiomers		

7.1 General

5. a. **8**.

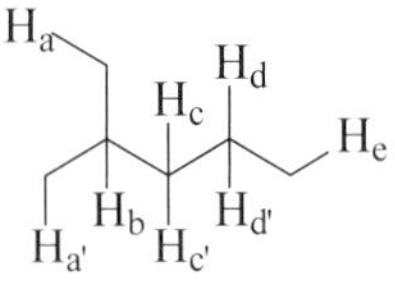

a, a'; c, c'; d, d'
are enantiomeric

b. **7**.

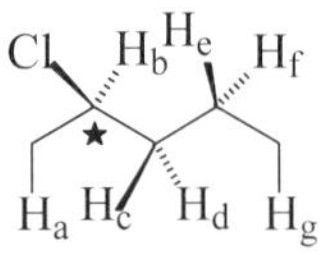

c, d; e, f are
diastereomeric

c. **7**.

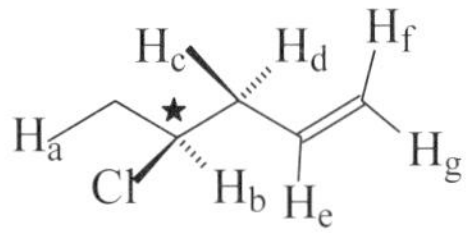

c, d; f, g are
diastereomeric

6. a. (R)-3-chloro-4-phenyl-1-butene

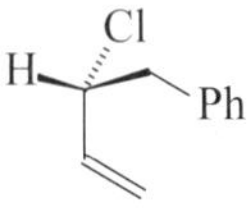

b. (S)-1-chloro-2-propanol

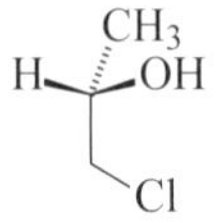

c. (R)-4-bromo-5-methyl-4-*n*-propyl-1-heptene

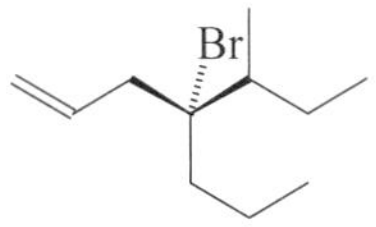

d. (R)-1-iodoethyl methyl ether

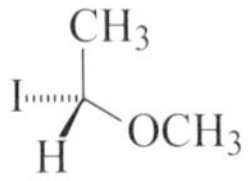

e. (3S, 4R)-3-*s*-butyl-4-isopropyl-1,6-heptadiene

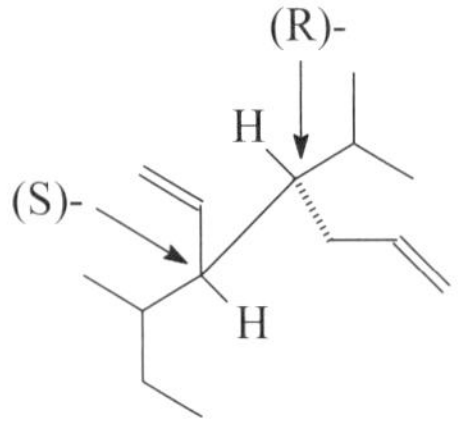

f. *meso*-3,4-dimethylhexane

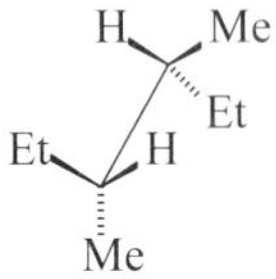

7. a. **4** pairs.

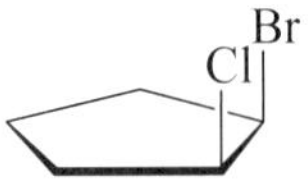

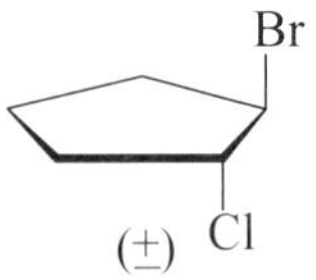

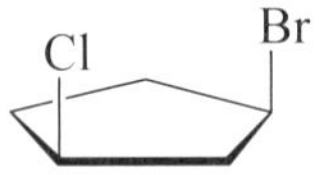

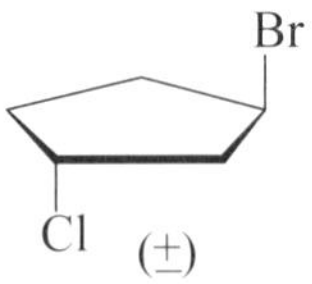

b. **5**.

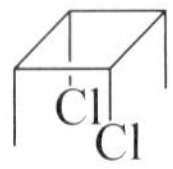

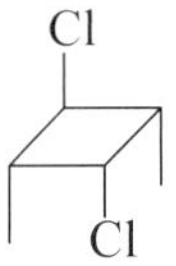

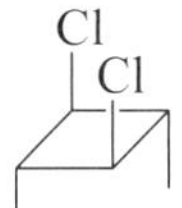

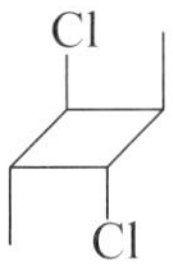

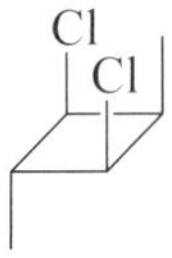

c. **2** pairs.

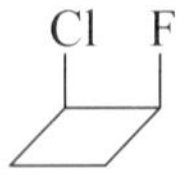

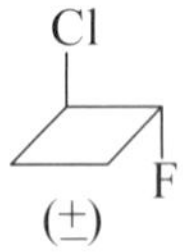

7.1 General

d. 2 *meso*-isomers, 1 pair of enantiomers.

```
    Et              Et              Et
H───OH         H───OH          H───OH
H───Cl         Cl───H          H───Cl
H───OH         H───OH          HO───H
    Et              Et              Et
  meso-          meso'-            (±)
```

e. 2 *meso*-isomers.

```
     Me              Me
H────Cl         H────Cl
H────Cl         Cl───H
H────Cl         Cl───H
H────Cl         H────Cl
     Me              Me
```

8. a.

```
    CHO                              CHO
H────OH                          H────OH
HO───H        epimerize at C₄    HO───H
HO───H       ─────────────►      H────OH
   CH₂OH                            CH₂OH
 enantiomer of A                  D-xylose
```

b. i.

```
         H                    (S)-
MeHN─────Me              MeHN   Me
H────OH        =              H
    Ph                   H   OH
                            Ph   (R)-
```

ii. ee = +10° / +40° = 25% (+), 75% (±)
therefore, % (+) = 25% + (75% / 2) = <u>63%</u>

9. ee = +68° / +170° = 40% (+), 60% (±)
therefore, % (−) = 60% / 2 = <u>30%</u>

no. chiral carbons: <u>4</u>.

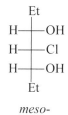

10.

7.1 General

11. a. ED b. D c. ED d. ED e. D f. D

    g. E h. ED

12. 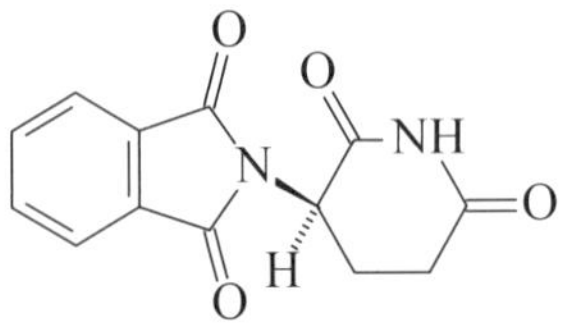 (R)-thalidomide

13. 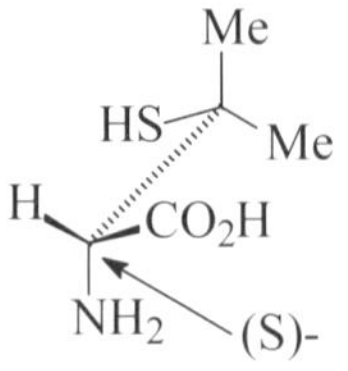 (S)-

14. a. 11.

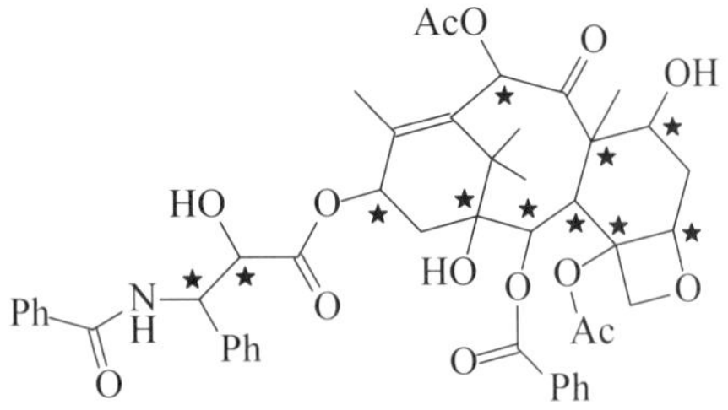

    b. ee = +24° / +120° = 20% (+), 80% (±)
    therefore, % (+) = 20% + (80% / 2) = 60%

15. 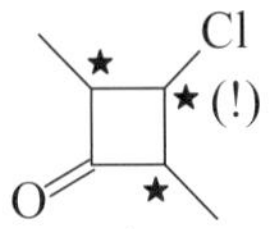 **A**

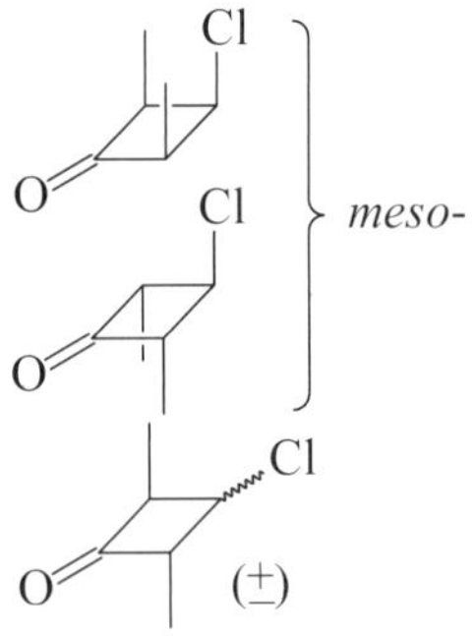

    meso-

    (±)

    3 chiral carbons, 2 meso stereoisomers, 1 pair of enantiomers

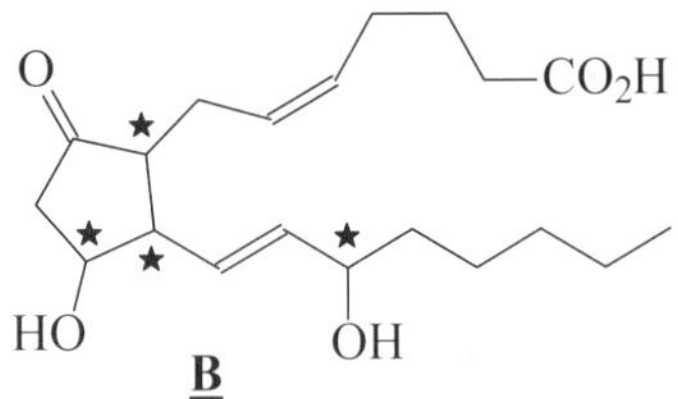

    B

    no. stereoisomers: $2^n = 2^4 = 16$.

16. a. no. stereoisomers: $2^3 = 8$.

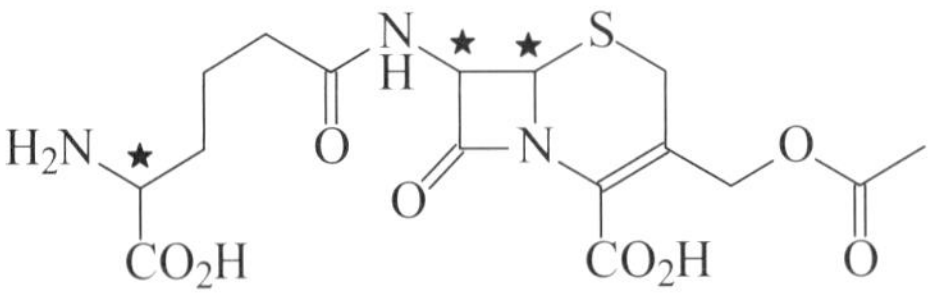

    b. ee = +82° / +103° = 80% (+), 20% (±); therefore, % (−) = 20% / 2 = 10%

7.1 General

7.2 Reactions and stereochemistry

1. [Reaction scheme: (Me)(Ph)C=C(H)(Me) with 1. OsO$_4$, 2. NaHSO$_3$ gives syn-add'n diol; with Br$_2$, H$_2$O gives anti-add'n bromohydrin. Each shown with its equivalent Fischer-like projection + enantiomer.]

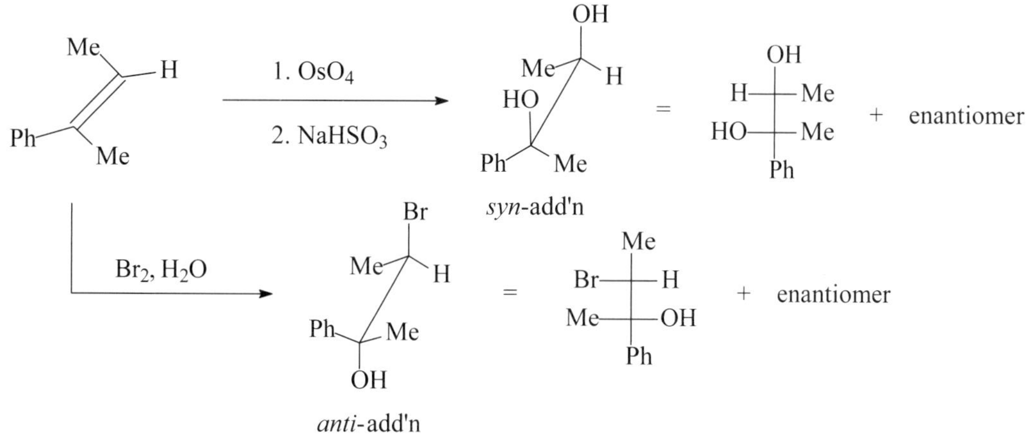

2. a. *anti*-add'n. [Fischer projection with Me, A, H on top C and B, H, Me on bottom C → sawhorse → rot'n → – A–B → *cis*- alkene with Me and H groups]

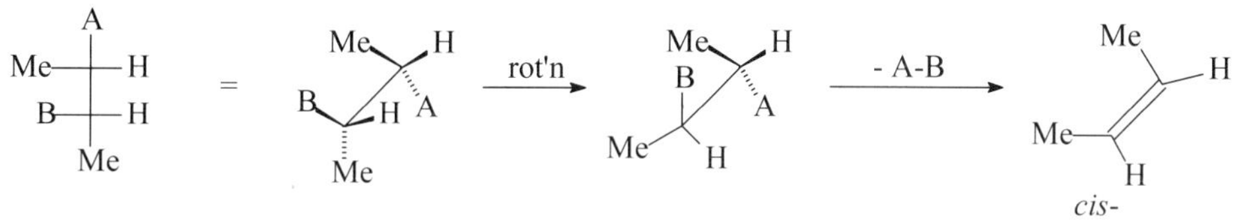

   b. *syn*-add'n. [Fischer projection with Me, C, D on top C and Me, C, D on bottom C → sawhorse → rot'n → – C–C → *trans*- alkene]

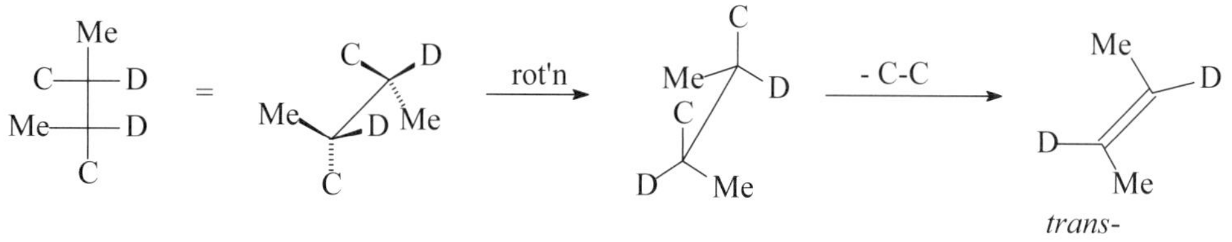

   c. *anti*-add'n. [Fischer projection with CO$_2$H, DO, H on top C and D, H, CO$_2$H on bottom C → sawhorse → rot'n → – D$_2$O → (E)- alkene HO$_2$C–CH=CH–CO$_2$H]

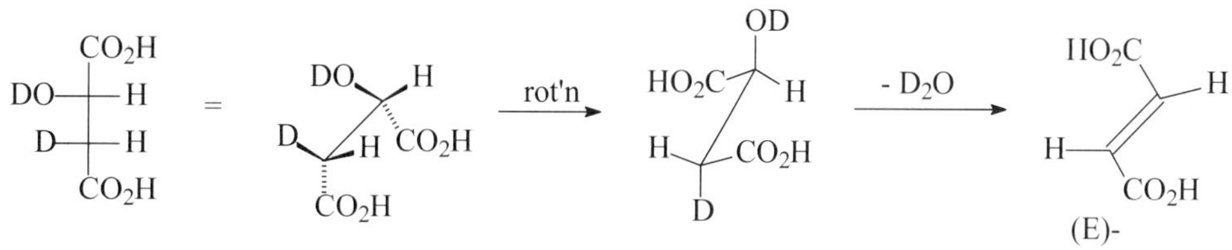

3. a. [Ph,Cl,Et-substituted alkene + H⁺ → secondary cation → 1,2-H:⁻ shift → new cation; then + Br⁻ gives two diastereomers (2 fractions: each E); + Cl⁻ gives meso + chiral (2 fractions: 1 M + 1 E)]

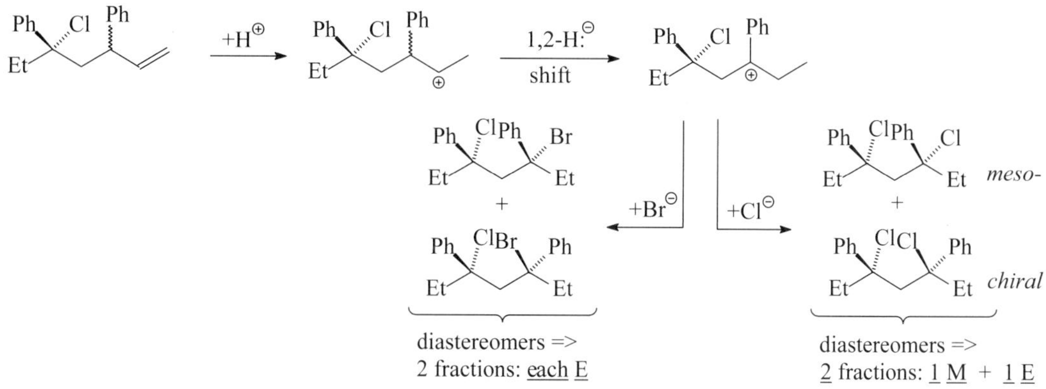

b. CH=CH(Et)(Et) —Br2, anti-add'n→ (Br,H,Et)(H,Et,Br) + (H,Et,Br)(Br,H,Et) } enantiomers => 1 fraction: R

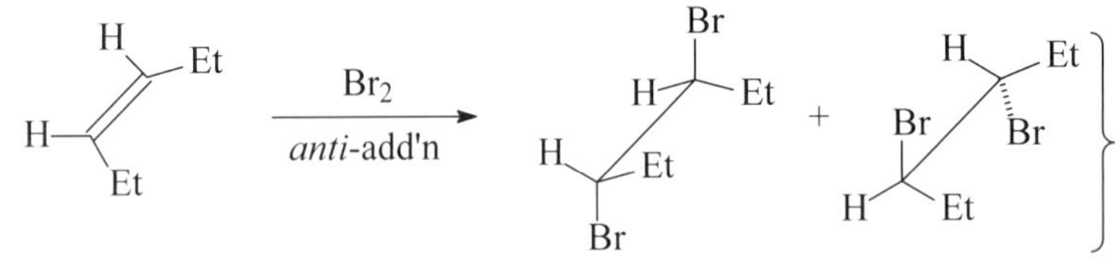

c. [bicyclic alkene] —KMnO4, H⊕→ [cyclopentane with CO2H, two Me, CO2H] => 1 fraction: M

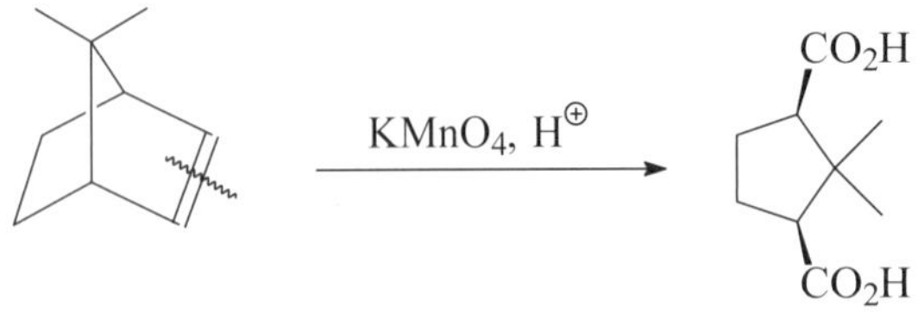

d. CH2=CH–CH(*)(Ph)–CH3 —+H⊕→ CH3–CH(*)(Ph)–⊕CH2? ... —1,2-H:⊖ shift→ ⊕ Ph ... —+I⊖→ (I,Ph)(Ph,I) } enantiomers => 1 fraction: R

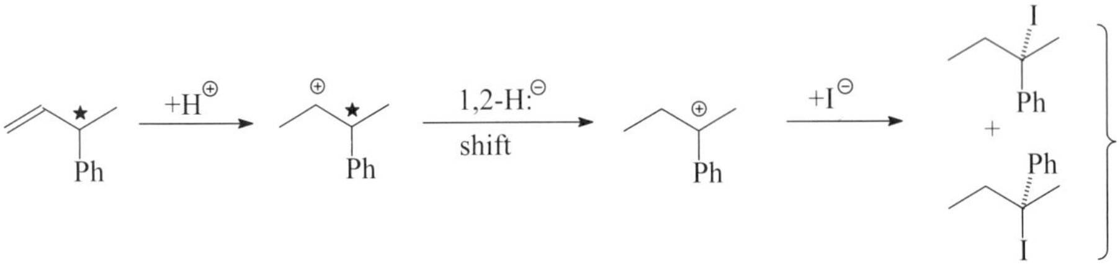

e. CH2=CH–CH2–CHCl(Me) —HF→ (F,H / H,Cl,Me) + (H,F / H,Cl,Me) } diastereomers => 2 fractions: each E

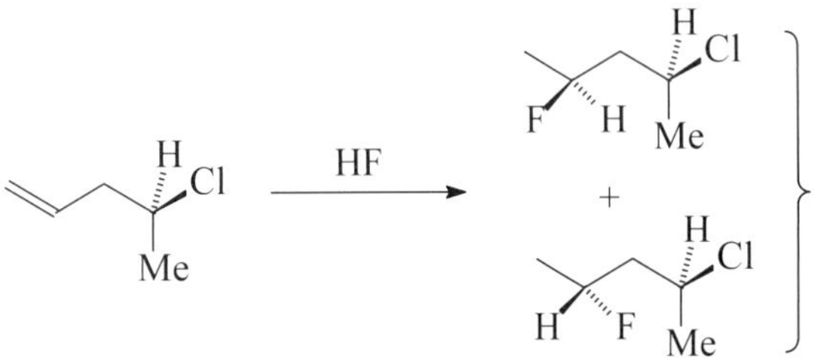

f. [dimethylcyclohexene] —D2/Ni, syn-add'n→ [cyclohexane with Me, Me, D, D] + [cyclohexane with Me, Me, D, D] } diastereomers => 2 fractions: each M

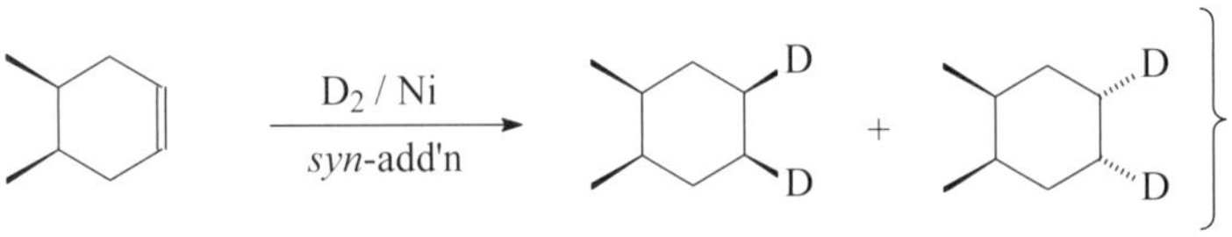

g. [Me-cyclopentane-vinyl, cis] —+H⊕→ [cyclopentyl cation with Et] —1,2-H:⊖ shift→ [cyclopentyl cation] —+MeOH, −H⊕→ (Me,H / OMe,Et) + (Me,H / Et,OMe) } diastereomers => 2 fractions: each E

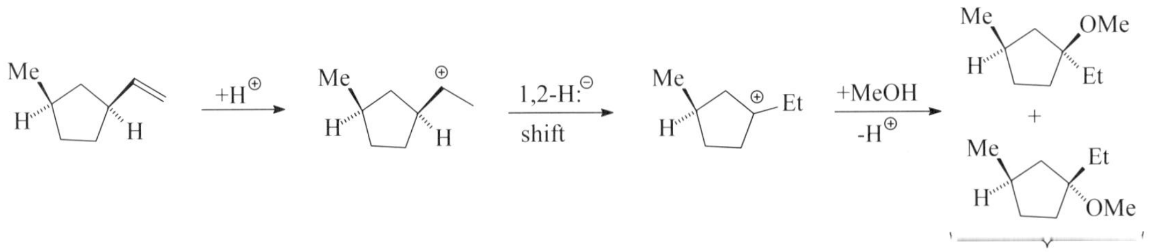

h. [1,2-dimethylcyclohexene] —1. BH3·THF 2. H2O2, ⊖OH, syn-add'n→ [cyclohexane H,OH] + [cyclohexane H,OH] } enantiomers => 1 fraction: R

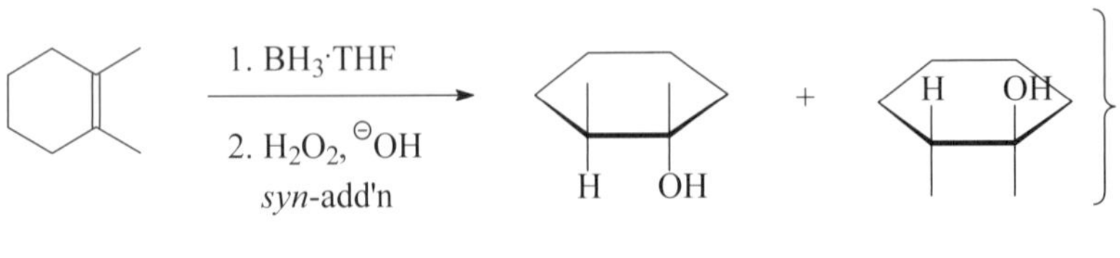

7.2 Reactions and stereochemistry

i. [reaction scheme showing cyclohexane with Et, OH, and vinyl substituents undergoing H₃O⁺/+H⁺, 1,2-H:⁻ shift, then +H₂O/-H⁺ to give meso- and chiral diastereomers]

diastereomers => 2 fractions: 1 M + 1 E

H₂ / Pt, syn-add'n → 1 fraction: E

1. Hg(OAc)₂, H₂O
2. NaBH₄

diastereomers => 2 fractions: each E

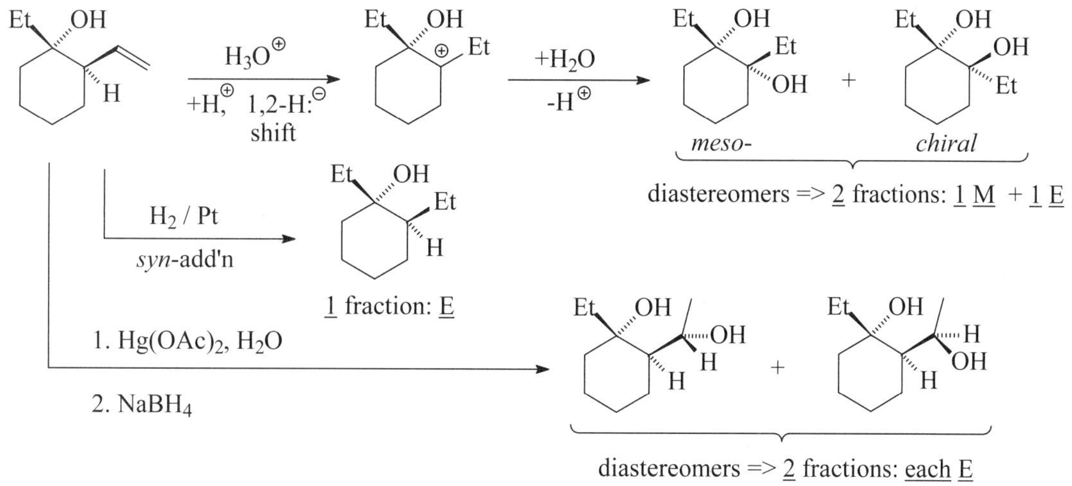

j. [norbornene]
1. OsO₄
2. NaHSO₃
→ diastereomers => 2 fractions: each M

1. mCPBA
2. H₃O⁺
→ enantiomers => 1 fraction: R

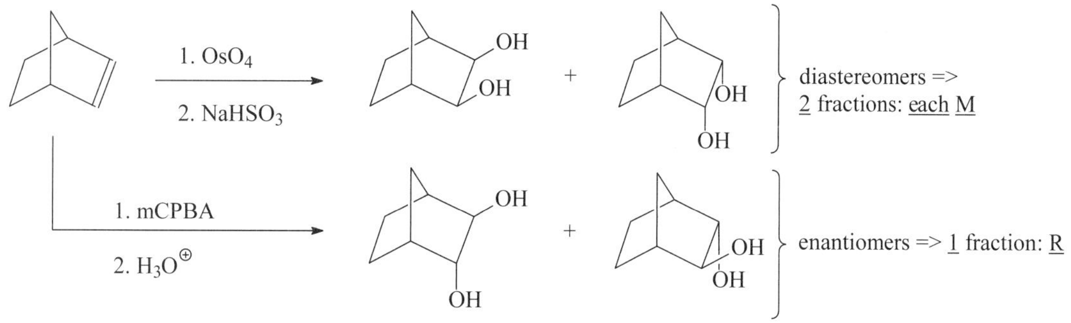

k. [dihydrofuran] +H⁺ → [oxocarbenium] +MeOH/-H⁺ → enantiomers => 1 fraction: R

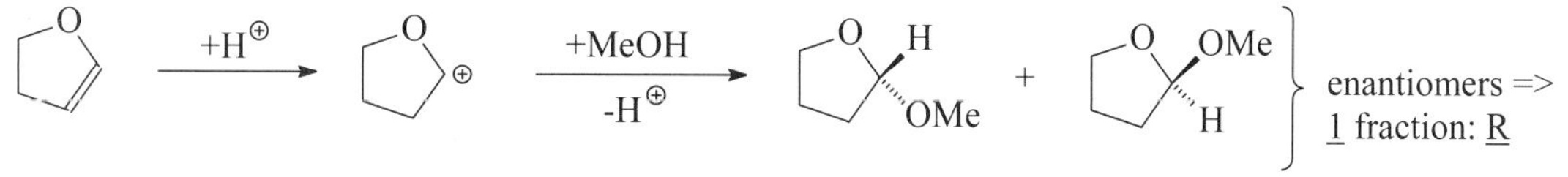

4. a. Ph—≡—Ph
1. H₂ / Lindlar catalyst, syn-add'n → cis-stilbene
2. Br₂, anti-add'n → racemate

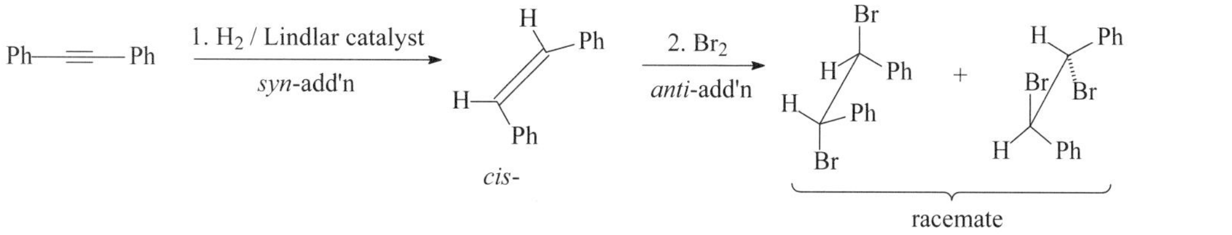

note: reduction of the alkyne to a *trans*-olefin, followed by Br₂ addition, would yield the *meso*-dibromide (*see next problem*)

7.2 Reactions and stereochemistry

b. 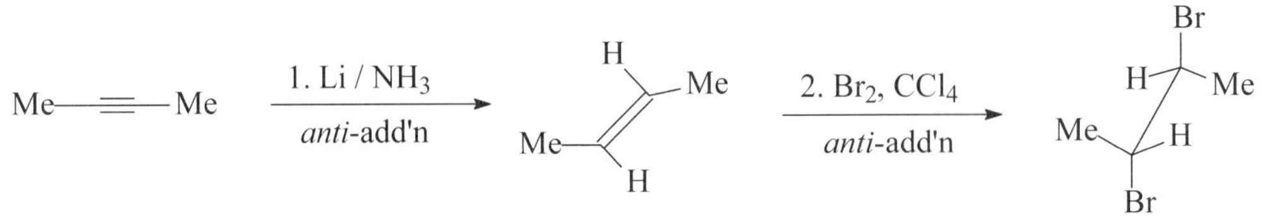 Me—≡—Me $\xrightarrow[\text{anti-add'n}]{\text{1. Li / NH}_3}$ (E-2-butene: Me/H on one C, H/Me on other) $\xrightarrow[\text{anti-add'n}]{\text{2. Br}_2\text{, CCl}_4}$ (meso-2,3-dibromobutane)

meso-

c. 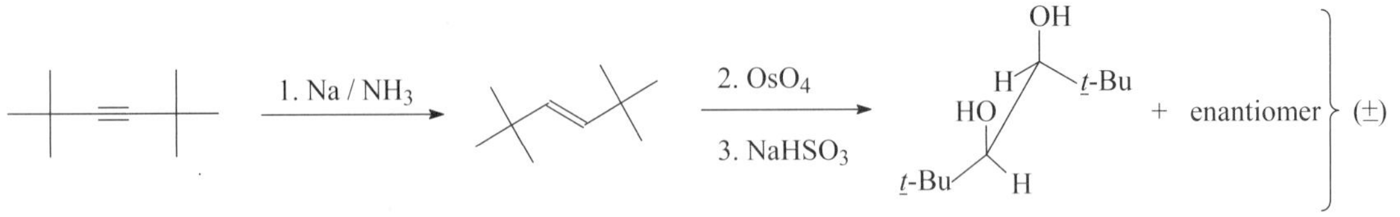 (di-*t*-Bu acetylene) $\xrightarrow{\text{1. Na / NH}_3}$ (E-alkene) $\xrightarrow[\text{3. NaHSO}_3]{\text{2. OsO}_4}$ (diol with HO, H on one C; *t*-Bu; HO, H on other; *t*-Bu) + enantiomer } (±)

d. 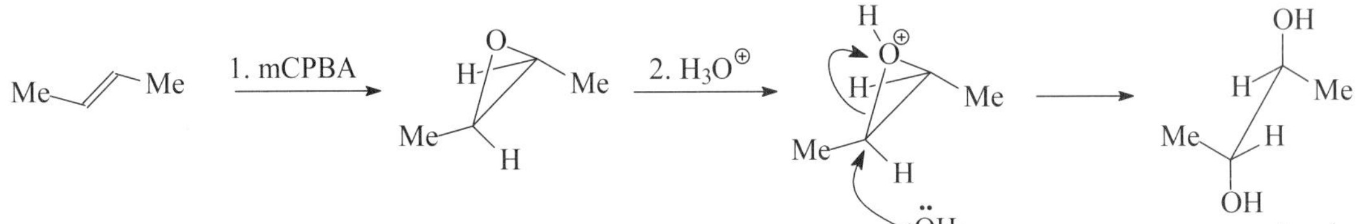 Me—CH=CH—Me $\xrightarrow{\text{1. mCPBA}}$ (epoxide: H, Me / Me, H) $\xrightarrow{\text{2. H}_3\text{O}^\oplus}$ (protonated epoxide with :OH$_2$ attacking) → (diol: H/OH, Me; Me, H/OH)

meso-glycol

5. a. $2^2 = \underline{4}$.   b. diastereomers   c. cannot predict (actual $[\alpha]_D = +62°$)

d. i. <u>2</u>. 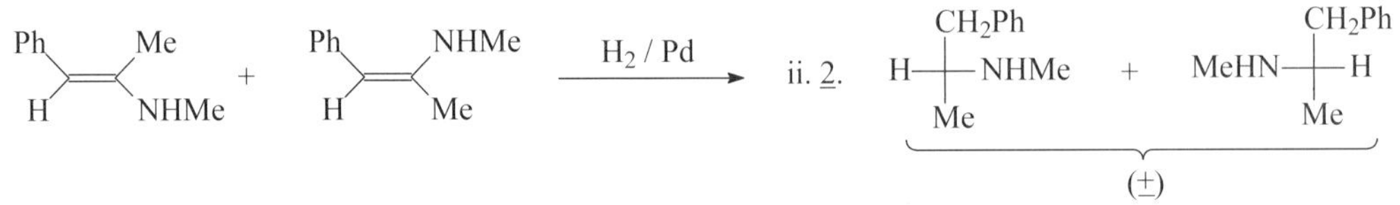 Ph\Me, H/NHMe + Ph\NHMe, H/Me $\xrightarrow{\text{H}_2\text{ / Pd}}$ ii. <u>2</u>. CH$_2$Ph / H—NHMe / Me + CH$_2$Ph / MeHN—H / Me   (±)

7.2 Reactions and stereochemistry

CHAPTER 8
ALKYL HALIDES AND RADICALS

8.1 Reactions

1. [Structure of 3-chloro compound with H and Cl, labeled with 2*, 1*, 1*+1 meso, 1*, 2*] $\xrightarrow{Cl_2, h\nu}$ $\underline{9}$ fractions } 7* are optically active; 2 are optically inactive

2. [pentane] $\xrightarrow{Br_2, h\nu}$ [1-bromopentane] + [2-bromopentane (±)] + [3-bromopentane]

   (no. H) × (reactivity) =   6 × 1.0 = 6.0   4 × 82 = 328   2 × 82 = 164

   % *racemic* 2-bromopentane = 328/(6 + 328 + 164)*100 = 66%; therefore, % (R)- = $\underline{33\%}$

3. [isopentane] $\xrightarrow{1.\ Br_2, \Delta}$ [t-Br] $\xrightarrow{2.\ Mg}$ [t-MgBr] $\xrightarrow{3.\ D_2O}$ [t-D]

4. [1-methylcyclohexanol] $\xrightarrow{1.\ conc\ HCl}$ [1-methylcyclohexyl chloride] $\xrightarrow[3.\ CuI]{2.\ Li}$ (R)$_2$CuLi (*Gilman reagent*) $\xrightarrow{4.\ allyl\ iodide}$ [1-methyl-1-allylcyclohexane]

5. [cyclobutane] $\xrightarrow[\substack{2.\ Li\\3.\ CuI}]{1.\ Cl_2, h\nu}$ (cyclobutyl)$_2$CuLi $\xrightarrow{4.\ allyl\ iodide}$ [vinylcyclobutane] $\xrightarrow{5.\ HI}$ [1-methyl-1-iodocyclopentane]

   [cyclobutylethyl cation H] $\xrightarrow[\text{shift}]{1,2\text{-R:}^\ominus}$ [methylcyclopentyl cation] $\xrightarrow[\text{shift}]{1,2\text{-H:}^\ominus}$ [methylcyclopentyl cation]

6. [propane] $\xrightarrow{1.\ Br_2, \Delta}$ [i-PrBr] $\xrightarrow{2.\ Mg}$ [i-PrMgBr] $\xrightarrow{3.\ Ph-\equiv-H}$ Ph–C≡C:$^\ominus$ + [propane]

7. [PhBr] $\xrightarrow[2.\ CuI]{1.\ Li}$ Ph$_2$CuLi $\xrightarrow{3.\ \underline{n}\text{-PrBr}}$ Ph-Pr $\xrightarrow{4.\ NBS, R_2O_2}$ [PhCHBrEt] $\xrightarrow{5.\ KOH}$ [Ph-CH=CH-CH$_3$] $\xrightarrow[H_2O]{6.\ Br_2}$ [Ph-CH(OH)-CHBr-CH$_3$]

8.1 Reactions

8.2 Syntheses

8.3 Mechanisms

1.

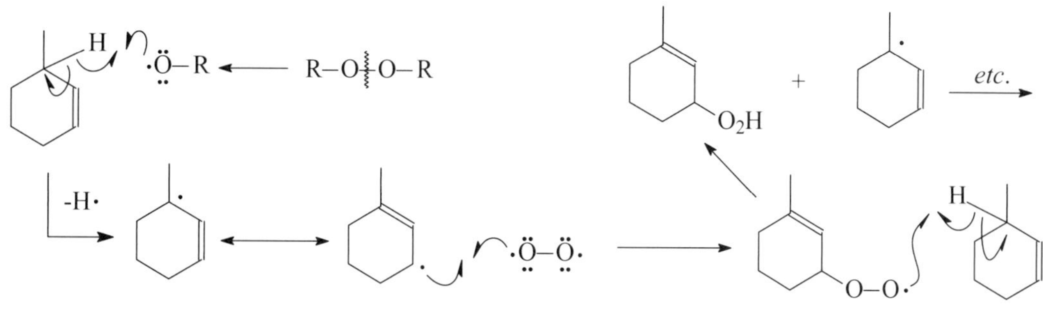

2.

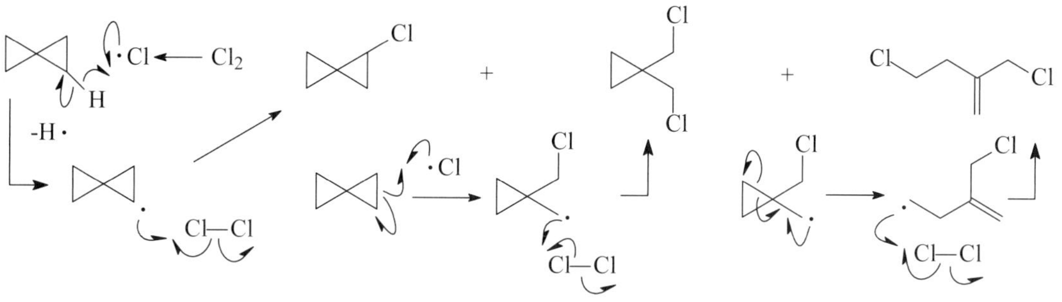

3.

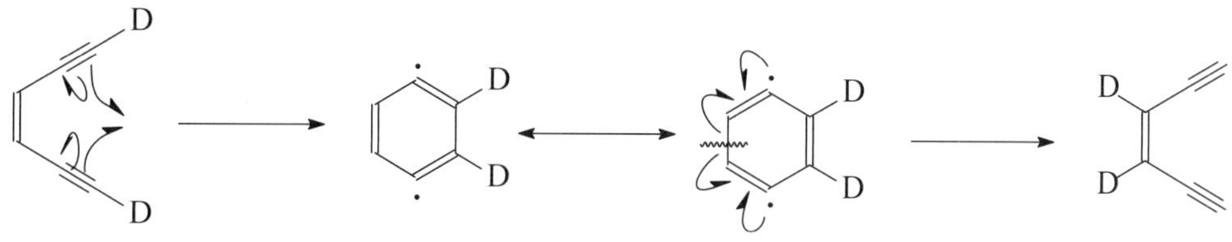

4.

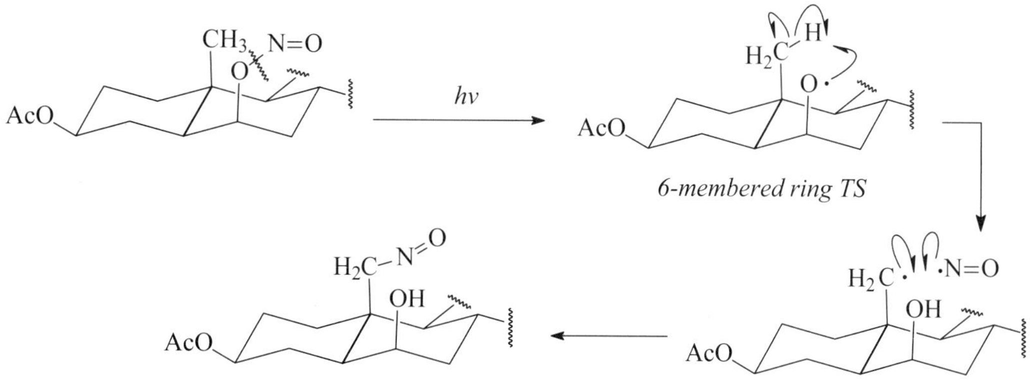

8.3 Mechanisms

298 • Chapter 8 Alkyl Halides and Radicals

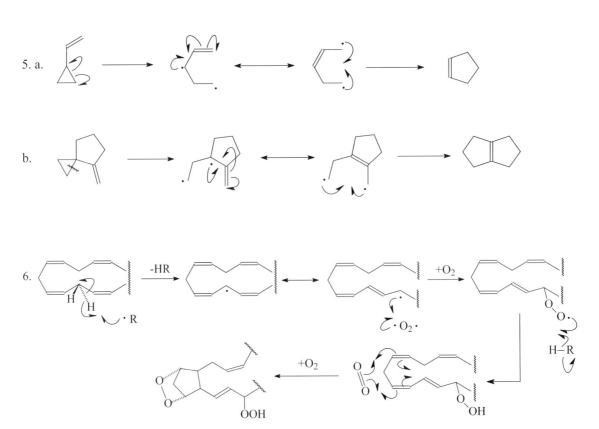

8.3 Mechanisms

CHAPTER 9
S$_N$1, S$_N$2, E1, AND E2 REACTIONS

9.1 General

1. faster reaction: <u>b</u>. AcO$^\ominus$ + (allyl)–Cl $\xrightarrow{S_N2}$ (allyl)–OAc

   solvent effect: acetate in HMPA (polar aprotic) is more nucleophilic than in ethanol (polar protic); the later H-bonds to acetate, thereby dampening its nucleophilicity

2. poorest leaving group: <u>b</u>. 'leavability' parallels the acidity of the CA of the leaving group; :NH$_3$ is the weakest CA (of the choices, <u>c</u> has the best leaving group)

3. stronger nucleophile: <u>a</u>. Et$_3$N: ⇌ :NEt$_3$ (inversion between b forms) rapid pyramidal inversion lowers nucleophilicity; such "flipping" is impossible with <u>a</u>

4. most reactive by an S$_N$2 pathway: <u>c</u>. least sterically crowded target carbon; note that even though <u>a</u> is primary, it is neopentyl-like, which generally never undergoes an S$_N$2 reaction

5. solvent that will maximize the rate of reaction: <u>a</u>. Et$_3$N: + (propyl)–Br $\xrightarrow{S_N2}$ (propyl)–$\overset{\oplus}{N}$Et$_3$ Br$^\ominus$

   polar solvents (<u>a</u> or <u>b</u>) stabilize the developing charge in the TS; the amine is more nucleophilic in DMSO (polar aprotic) than methanol (polar protic):

   Et$_3$N:----H–O–Me   } H-bonding stabilizes the amine, thereby increasing ΔG‡
             <u>b</u>

6. more reactive by an E2 pathway: <u>b</u>.

   <u>a</u>: no *trans*-diaxial hydrogen available
   <u>b</u>: *anti*-periplanar TS possible

7. approximate k_H / k_D: <u>c</u>. Ph–CH(Br)–CH$_2$H(D) + $^\ominus$O-t-Bu $\xrightarrow{E2}$ Ph–CH=CH$_2$

   a carbon-hydrogen (deuterium) bond is broken in the rate-determining-step; therefore, a primary hydrogen kinetic isotope effect (~7) is observed

8. reaction to yield the more stereochemically pure product: **b**.

a. [cis/trans 1-Et-1-Br-4-Et-cyclohexane] or [diastereomer] $\xrightarrow[S_N1]{MeOH}$ [carbocation intermediate] $\xrightarrow[-H^{\oplus}]{+MeOH}$ [Et, OMe product]

either diastereomer would give the same ratio of diastereomeric ethers because of a common intermediate

b. (R)- $\xrightarrow[S_N2]{^{\ominus}OMe}$ (S)- or (S)- $\xrightarrow[S_N2]{^{\ominus}OMe}$ (R)-

stereospecific: either enantiomer gives an optically pure, but different, ether

9. change in rate of reaction: **b**.

Ph$_2$CHBr $\xrightarrow[S_N1 \; rds]{EtOH}$ [Ph$_2$CH$^{\oplus}$] $\xrightarrow[-H^{\oplus}]{+EtOH}$ Ph$_2$CHOEt

rate = k[RX] ; changing the concentration of EtOH has no effect on the rate

10. a. [isopropyl bromide] $\xrightarrow[\sim100\% \; E2!]{^{\ominus}O\text{-}t\text{-Bu}}$ [propene]   vs.   [n-butyl bromide] $\xrightarrow{^{\ominus}OMe}$ S$_N$2 > E2

b. [1-methyl-1-bromocyclohexane] $\xrightarrow{^{\ominus}OH}$ [methylenecyclohexane]   vs.   [methylcyclohexyl bromide, 2°] $\xrightarrow{^{\ominus}OH}$

3° RX => ~100% E2   2° RX => S$_N$2 + E2

c. [sec-butyl bromide] $\xrightarrow{^{\ominus}OR}$ [ether] + [alkene]   vs.   [sec-butyl bromide] $\xrightarrow{^{\ominus}SR}$ [thioether] >> [alkene]

RS$^{\ominus}$ is a better nucleophile, and weaker base, than RO$^{\ominus}$; therefore, S$_N$2 / E2 ratio is larger for RS$^{\ominus}$

9.1 General

11. expected primary hydrogen kinetic isotope effect: <u>b</u>.

a. [structure with H, Cl, H(D)] → KO-*t*-Bu, E2 - Hofmann → [isobutylene] no carbon-hydrogen (deuterium) bond is broken in rds; therefore, $k_H / k_D \sim 1$

b. [structure with Cl, H(D), H(D)] → KOH, E2 - Zaitsev → [alkene with H(D)] a carbon-hydrogen (deuterium) bond is broken in rds; therefore, $k_H / k_D \sim 7$

c. [structure with H, Cl, (D)H, H(D)] → KOMe → [product]—OMe + [alkene] S_N2 + E2

no carbon-hydrogen (deuterium) bond is broken in rds

12. [structure with SO₂, OCH₃, TsO, C:⁻, H] → if *intra*molecular → [TS structure with δ-]

unfavorable TS: Nu----R----L bond angle not linear

however, if bimolecular → [two-molecule structure] → [linear TS structure]

a linear TS is possible => < $\Delta G^{\ddagger}$; therefore, an *inter*molecular reaction is kinetically favored

9.2 Reactions

1. [octyl bromide] —KOMe, MeOH, S_N2 > E2→ [octyl methyl ether]

2. [tertiary iodide] —⁻OEt, E2 - Zaitsev→ [trisubstituted alkene]

3. [sec-butyl chloride] + ⁻O-t-Bu —E2, Hofmann→ [1-butene]

4. [sec-butyl bromide] + :N⁻(R)₂ —E2, Hofmann→ [3-methyl-1-butene]

302 • Chapter 9 S_N1, S_N2, E1, and E2 Reactions

5. CH₃CH₂CH₂CH₂CH₂CH₂I $\xrightarrow[S_N2]{KCN / DMF}$ CH₃CH₂CH₂CH₂CH₂CH₂—C≡N:

6. (1-methylcyclopentyl chloride) $\xrightarrow[S_N1 > E1]{MeOH, RT}$ (methylcyclopentyl cation)⊕ $\xrightarrow[-H^⊕]{+MeOH}$ (1-methyl-1-methoxycyclopentane)

7. (2-bromo-2-methylbutane) $\xrightarrow[E1 > S_N1]{refluxing\ EtOH}$ (tert-carbocation)⊕ $\xrightarrow[Zaitsev]{-H^⊕}$ (2-methyl-2-butene)

8. (2-chlorotetrahydropyran) $\xrightarrow[S_N1]{-Cl^⊖}$ (oxocarbenium resonance) ↔ (oxocarbenium resonance) $\xrightarrow[-H^⊕]{+HOAc}$ (2-acetoxytetrahydropyran) *racemate*

9. CH₃CH₂CH₂Br + Me₂N̈H $\xrightarrow[S_N2 > E2]{}$ CH₃CH₂CH₂—N⊕(Me)(Me)(H) $\xrightarrow{-H^⊕}$ CH₃CH₂CH₂—N̈(Me)(Me)

10. (CH₃)₂CH—Br + ⊖O-*t*-Bu $\xrightarrow{E2}$ CH₂=CHCH₃

11. (2-iodopentane) + ⊖O—C(=O)CH₃ $\xrightarrow[S_N2 > E2]{}$ (3-pentyl acetate, OAc)

12. Cl—CH₂CH₂CH₂CH(Cl)CH₃ $\xrightarrow[S_N2]{⊖SH\ (1\ equiv)}$ HS—CH₂CH₂CH₂CH(Cl)CH₃

    reactivity: 1⁰ > 2⁰

13. (3-chloro-1-methylcyclohexene) $\xrightarrow[S_N1]{EtOH}$ (allylic cation) ↔ (allylic cation) $\xrightarrow[-H^⊕]{+EtOH}$ (product with OEt) + (product with OEt)

14. (1-bromo-1-phenyl-2-methylpropane... Ph on C) $\xrightarrow[S_N1]{-AgBr}$ (secondary cation, Ph) $\xrightarrow[shift]{1,2-H:^⊖}$ (tertiary cation, Ph) $\xrightarrow{⊖OAc}$ (Ph, OAc product)

9.2 Reactions

9.2 Reactions

15. Me—Et, Me—H, Cl—H, Me (wedge structure) = Newman projection with Cl, Me, H, Et, H, Me →(rot'n)→ anti-periplanar TS with H, Me, Et / H, Me, Cl →($^{\ominus}$OEt, -HCl, E2)→ Me/Et and H/Me alkene

anti-periplanar TS

16. Chlorocyclohexane with *t*-Bu and Me = chair with Me, H axial and Cl axial →($^{\ominus}$OMe, -HCl, E2)→ cyclohexene with *t*-Bu and Me

trans-diaxial TS → Hofmann olefin

17. Deuterated *t*-Bu cyclohexyl chloride = chair conformation with H, D, D positions and Cl axial →(-HCl, E2)→ cyclohexene product with *t*-Bu and two D

(C-H bond *slightly* weaker than C-D bond)

18. CH$_3$, H—D, Br—H, Ph = Newman with H, D, CH$_3$, Br, Ph →(rot'n)→ H, Me, D / H, Ph, Br →(-HBr, E2)→ Me/D and H/Ph alkene

19. Cyclohexyl compound with Me, Br, and (CH$_2$)$_3$OH →(-Br$^{\ominus}$, S$_N$1)→ carbocation with OH →→ protonated cyclic ether →(-H$^{\oplus}$)→ bicyclic ether (methyl-substituted)

20. Allyl iodide (CH$_2$=CHCH$_2$CH(I)CH$_3$) →(:PPh$_3$, S$_N$2)→ $^{\oplus}$PPh$_3$ I$^{\ominus}$ salt

↓ (*t*-BuOH, S$_N$1)

allylic cation →(1,2-H:$^{\ominus}$ shift)→ rearranged cation ↔ resonance cation → two ether products with O-*t*-Bu

21. EtS—CHMe—Cl →(MeOH, S$_N$1)→ sulfonium cation ↔ resonance → +MeOH, -H$^{\oplus}$ → EtS—CHMe—OMe

22. Iodide with NMe$_2$ group →(-I$^{\ominus}$, intra-S$_N$2)→ pyrrolidinium salt I$^{\ominus}$

304 • Chapter 9 S_N1, S_N2, E1, and E2 Reactions

23. $HSe:^{\ominus}$ + [cyclobutyl-Cl] $\xrightarrow{S_N2 \gg E2}$ [cyclobutyl-SeH]

24. Ph-CH(OH)-CH$_2$-N(H)(Me) $\xrightarrow[S_N2]{PhCH_2Cl\ (1\ equiv)}$ Ph-CH(OH)-CH$_2$-N(Me)(CH$_2$Ph)

25. $Me_3\overset{\oplus}{N}$-CH$_2$-CH$_2$-OH, Me-$\overset{\oplus}{O}$(Me)-Me $\xrightarrow[S_N2]{-Me_2O,\ -H^{\oplus}}$ $Me_3\overset{\oplus}{N}$-CH$_2$-CH$_2$-O-Me $\ ^{\ominus}BF_4$

26. (R)-sec-butanol $\xrightarrow[ret]{1.\ TsCl}$ (R)-sec-butyl-OTs $\xrightarrow[S_N2,\ inv]{2.\ ^{\ominus}OH}$ (S)-sec-butanol

27. ^{i}Pr-$\overset{\oplus}{S}$(Et)(Me), :NH$_2$-Ph $\xrightarrow{S_N2}$ Et-S-^{i}Pr + PhNH(Me)

28. [3,4-dihydroxyphenyl-CH(OH)-CH$_2$-NHMe] + Br-CH$_2$-CH$_2$-F (better leaving group) (1 equiv) $\xrightarrow{S_N2}$ [3,4-dihydroxyphenyl-CH(OH)-CH$_2$-N(Me)-CH$_2$-CH$_2$-F]

29. [1-chloronorbornane] $\xrightarrow{refluxing\ MeOH}$ NR! {rigidity of bicyclic structure prevents formation of a planar carbocation; backside attack impossible; *Bredt's rule* precludes double bond at bridgehead}

30. (E)-2-iodo-3-fluoro-2-butene $\xrightarrow[S_N2]{^{\ominus}SePh\ (XS)}$ (E)-2-iodo-3-(phenylseleno)-2-butene (vinyl halides unreactive under S_N2 conditions)

31. H$_2$N-C$_6$H$_4$-C(=O)-O-CH$_2$-CH$_2$-NEt$_2$ (more nucleophilic nitrogen) $\xrightarrow[S_N2]{EtBr\ (1\ equiv)}$ H$_2$N-C$_6$H$_4$-C(=O)-O-CH$_2$-CH$_2$-$\overset{\oplus}{N}$Et$_3$ Br$^{\ominus}$

9.2 Reactions

32. $Ph-C\equiv CH$ $\xrightarrow{\text{1. NaNH}_2}$ $Ph-C\equiv C:^{\ominus}$ $\xrightarrow[\text{E2 > S}_N\text{2}]{\text{2. cyclohexyl-Br}}$ $Ph-C\equiv CH$ + cyclohexene

33. (iodo-spiro[4.4]) $\xrightarrow[\text{E1}]{\text{MeOH}}$ (spiro cation) $\xrightarrow[\text{shift}]{\text{1,2-R:}^{\ominus}}$ (hydrindane cation) $\xrightarrow{-H^{\oplus}}$ (hydrindene)

34. PhOTs + $^{\ominus}$S—C≡N: ⟶ NR! (aryl tosylates unreactive under S$_N$ conditions)

35. (bicyclic bromide) $\xrightarrow[\text{S}_N\text{1}]{\text{HOAc, RT}}$ (cation) $\xrightarrow[\text{shift}]{\text{1,2-R:}^{\ominus}}$ (ring-expanded cation) $\xrightarrow[-H^{\oplus}]{+\text{HOAc}}$ (hydrindanyl OAc)

36. (cyclohexyl-SMe$_2$) $\xrightarrow[\text{S}_N\text{2}]{\text{1. MeI}}$ (sulfonium) $\xrightarrow[-\text{Me}_2\text{S, E1}]{\text{2. refluxing EtOH}}$ (cyclohexyl cation) $\xrightarrow{-H^{\oplus}}$ cyclohexene

37. (steroid with α-Br, α-Me at C6, dienone) $\xrightarrow[\text{E2}]{^{\ominus}\text{O-}t\text{-Bu}}$ (exocyclic methylene product)   Hofmann olefin

38. 4,4'-bipyridine :N⟨⟩—⟨⟩N: $\xrightarrow[\text{S}_N\text{2}]{\text{(XS) MeI}}$ Me—N$^{\oplus}$⟨⟩—⟨⟩N$^{\oplus}$—Me   2 I$^{\ominus}$

39. a. *anti*- H, Br, Me on bicyclic lactone $\xrightarrow[\text{E2 - Zaitsev}]{^{\ominus}\text{OEt}}$ (endocyclic trisubstituted alkene lactone)

b. *syn*- H, Br, Me on bicyclic lactone $\xrightarrow[\text{E2 - Hofmann}]{^{\ominus}\text{OEt}}$ (exocyclic methylene lactone)

9.3 Syntheses

1. Ph-CH(Ph)-CH(Br)-Me →[H₂O, Ag⁺ / S_N1] Ph-CH(Ph)-CH⁺-Me →[1,2-H:⁻ shift] Ph-C⁺(Ph)-CH₂-Me →[+H₂O, -H⁺] Ph-C(Ph)(OH)-CH₂-Me

   →[NaOMe, MeOH / S_N2] Ph-CH(Ph)-CH(OMe)-Me

   →[LDA or KO-*t*-Bu / E2 - Hofmann] Ph-CH(Ph)-CH=CH₂

2. [bromocyclohexane with t-Bu and Me substituents] →[MeOH, Δ / E1] [carbocation] →[-H⁺] [trisubstituted cyclohexene with t-Bu]

   (not ⁻OMe! => E2 - Hofmann olefin)

3. [trans-bromo-methyl-D cyclohexane] = [chair with D, H, Br, Me] →[⁻OEt, -DBr, E2] [3-methylcyclohexene] Hofmann olefin

   (not HOEt, Δ! => E1 - Zaitsev olefin)

4. [Me,H / Me,I / Ph Newman or Fischer] = [staggered] →[rot'n] [eclipsed-like] →[KOEt, EtOH / E2] (E)-Me(Me)C=C(Ph)(H)

5. [1-methyl-1-(iodomethyl)cyclohexane] →[EtOH, Δ / E1] [primary? cation on CH₂] →[1,2-R:⁻ shift] [ethylcyclohexyl cation] →[-H⁺] [1-ethylcyclohexene]

6. cyclopropyl-CH₂-Br →[PhOH, AgNO₃ / S_N1 + E1] cyclopropyl-CH₂⁺ →[1,2-R:⁻ shift] [cyclobutyl cation] →[+PhOH, -H⁺ or, -H⁺] cyclobutyl-OPh + cyclobutene

7. [1-methylcyclohexyl chloride] →[PhCO₂H / S_N1] [1-methylcyclohexyl cation] →[+PhCO₂H, -H⁺] [1-methylcyclohexyl benzoate]

9.3 Syntheses

8. (CH₃)₂CHCHBrCH₃ →[1. KO-*t*-Bu, *t*-BuOH; E2 - Hofmann]→ (CH₃)₂CHCH=CH₂ →[2. HBr, ROOR]→ (CH₃)₂CHCH₂CH₂Br

9. (iPr)₂CH₂ →[1. Br₂, *hv*]→ (CH₃)₃C-CH(iPr) with Br on quaternary C →[2. EtOH, RT; S_N1]→ (CH₃)₂C(OEt)-CH(iPr)...

   (not ⁻OEt! => 100% E2)

10. decalinyl bromide →[1. MeOH, Δ; E1]→ octalin (mixture) →[2. O₃; 3. Zn, HCl]→ cyclodecane-1,6-dione

    decalinyl bromide →[1. KO-*t*-Bu, *t*-BuOH; E2 - Hofmann]→ octalin (Hofmann isomer) →[2. KMnO₄, H⁺]→ 2-(2-carboxyethyl)cyclohexanecarboxylic acid

11. Ph₂CHCH₃ →[1. NBS, peroxide; 2. NaOMe (E2)]→ Ph₂C=CHPh (Ph₂C=CHCH...) →[3. HBr, peroxide; 4. KO-*t*-Bu (E2 - Hofmann)]→ Ph₂C(H)CH=CH₂... → Ph/Ph CH-CH=CH₂

12. PhCH(OTs)CH₃ →[1. KOH, EtOH; E2]→ PhCH=CHCH₃ →[2. H₃O⁺]→ PhCH(OH)CH₂CH₃ →[3. TsCl]→ PhCH(OTs)CH₂CH₃

13. methylcyclopentane →[1. Br₂, Δ]→ 1-bromo-1-methylcyclopentane →[2. KOMe; E2]→ 1-methylcyclopentene →[3. BH₃·THF; 4. H₂O₂, ⁻OH]→ trans-2-methylcyclopentanol

14. norbornylmethanol →[SOCl₂, -HCl]→ [chlorosulfite intermediate] →[-SO₂]→ norbornylmethyl chloride

    [S_N2 reactivity slow (neopentyl-like); avoid S_N1 (rearrangement)]

15. $H_2C=CH_2$ →[1. Br₂, CCl₄]→ BrCH₂CH₂Br →[2. HSCH₂CH₂SH, base, S_N2]→ BrCH₂CH₂SCH₂CH₂S⁻ →[-Br⁻, S_N2]→ 1,4-dithiane

9.4 Mechanisms

4. Mechanism showing Et$_2$N: attacking C-Cl carbon, displacing Cl$^-$ via S$_N$2 to form aziridinium intermediate with Et, Et on N$^+$ and CH$_2$CH$_2$Ph substituent; then HO$^-$ attacks via S$_N$2 to give product with :NEt$_2$ and CH$_2$OH, CH$_2$CH$_2$Ph.

5.
$$\text{α-Cl carboxylate} \xrightarrow[\text{inv}]{\text{intra-S}_N2} [\text{α-lactone intermediate}] \xrightarrow[\text{inv}]{\text{inter-S}_N2} \text{α-OH carboxylate} \quad \text{double inv = net ret}$$

(in conc $^\ominus$OH, product is [inverted α-OH carboxylate], formed by initial inter-S$_N$2)

6. Thiol/allylic iodide substrate → DMF, –I$^\ominus$, S$_N$1 → sulfonium/allyl cation resonance structures → –H$^\oplus$ → 6-membered S-containing ring (thiopyran) with vinyl group.

7. Pyridinium N-oxide with butyl bromide: –Br$^\ominus$, S$_N$2 → N-O-butyl pyridinium intermediate → pyridine deprotonates α-CH → butyraldehyde + pyridinium.

8.
trans-**A** (HO and Cl trans on cyclohexane) ⇌ chair with axial OH attacking axial C–Cl, –Cl$^\ominus$, intra-S$_N$2 with inv possible → protonated bicyclic oxonium → –H$^\oplus$ → **B** (bicyclic ether)

cis-**A** ⇌ conformer → ∦→ **B**, intra-Walden inv not possible

therefore, inter-S$_N$2 with inv (by $^\ominus$OH) → trans-diol

9.
$$\text{ClCH}_2\text{CH}_2\text{-S-CH}_2\text{CH}_2\text{Cl} \xrightarrow[-\text{Cl}^\ominus]{\text{rds}} \text{episulfonium}^+ \xrightarrow[-\text{H}^\oplus]{+\text{H}_2\text{O}} \text{ClCH}_2\text{CH}_2\text{-S-CH}_2\text{CH}_2\text{OH} \xrightarrow{\text{repeat}} \text{HOCH}_2\text{CH}_2\text{-S-CH}_2\text{CH}_2\text{OH}$$

9.4 Mechanisms

310 • Chapter 9 S_N1, S_N2, E1, and E2 Reactions

9. (*cont.*) similarly,

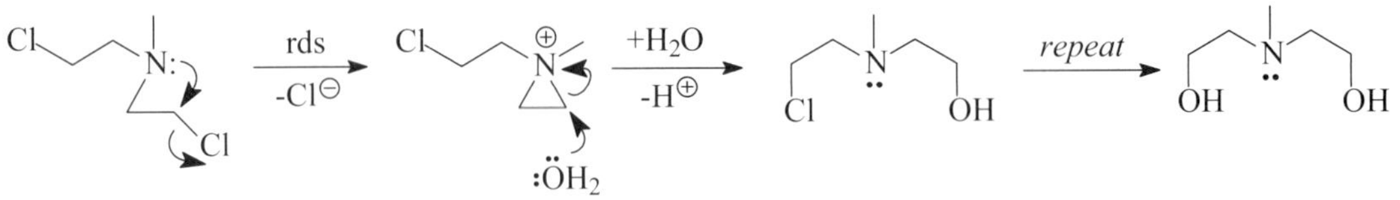

10. **I**: intra-S_N2 (NGP) not possible | therefore, inter-S_N2: → *trans*-enantiomer only

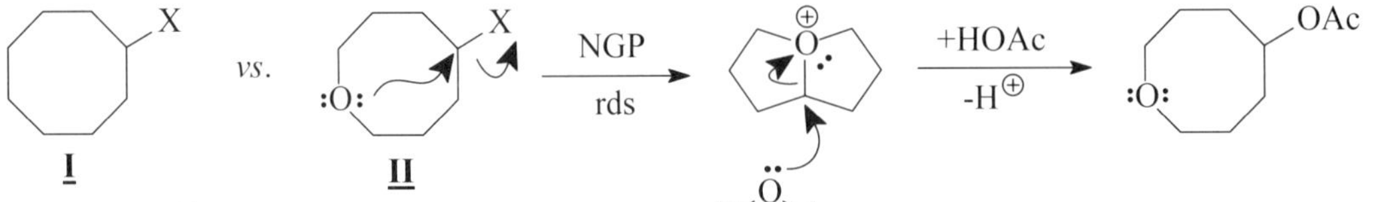

    vs.

    II: NGP, $-TsO^{\ominus}$ → +HOAc, $-H^{\oplus}$ → *racemate*

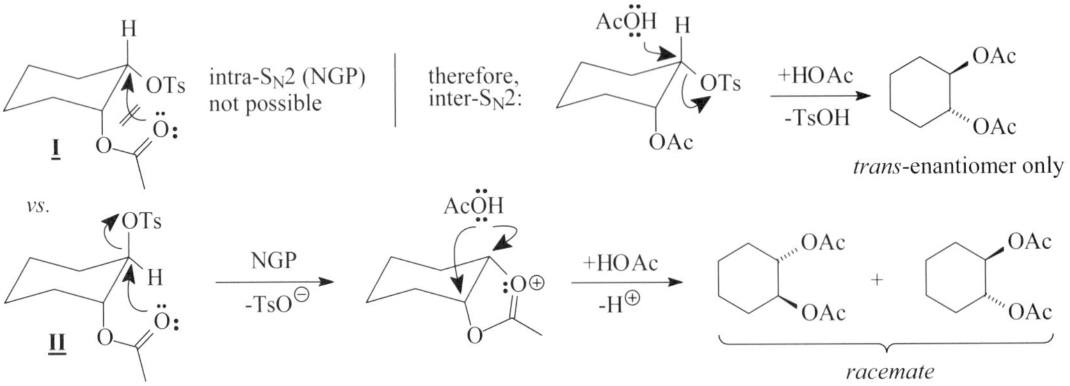

11. NGP, $-^{\ominus}OTs$ → [intermediate] → $-H^{\oplus}$ → 40% / 60%

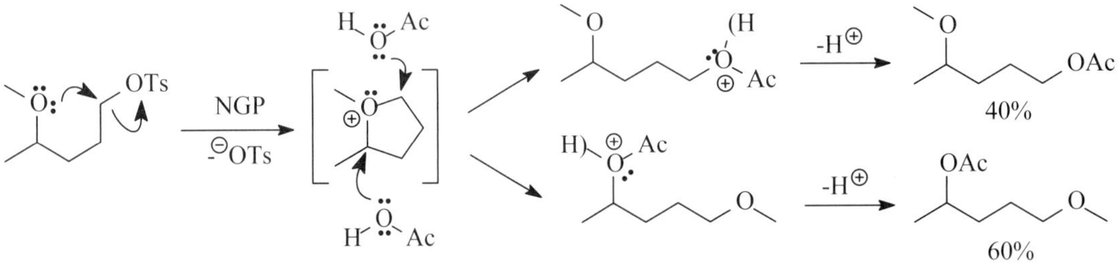

12. **I**: NGP not possible – therefore, no kinetic enhancement

    vs. **II**: NGP, rds → +HOAc, $-H^{\oplus}$ →

13. NGP, rds, k → [less strained intermediate $\rightarrow$ < $\Delta G^{\ddagger}$ therefore, $k \gg k'$] → +EtOH, $-H^{\oplus}$ → EtO~~~OH

    vs. NGP, rds, k' → +EtOH, $-H^{\oplus}$ → EtO~~~OH

9.4 Mechanisms

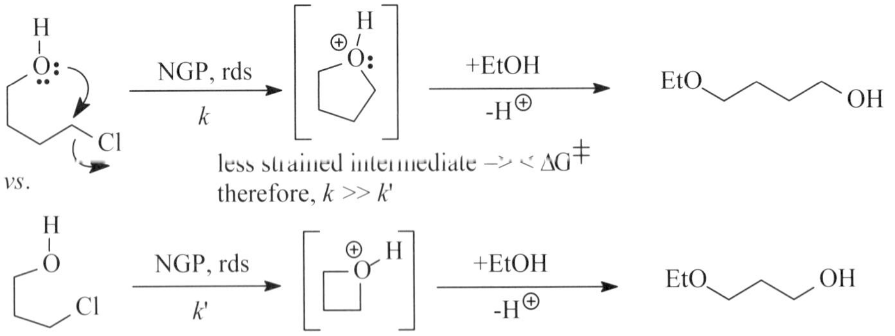

14. 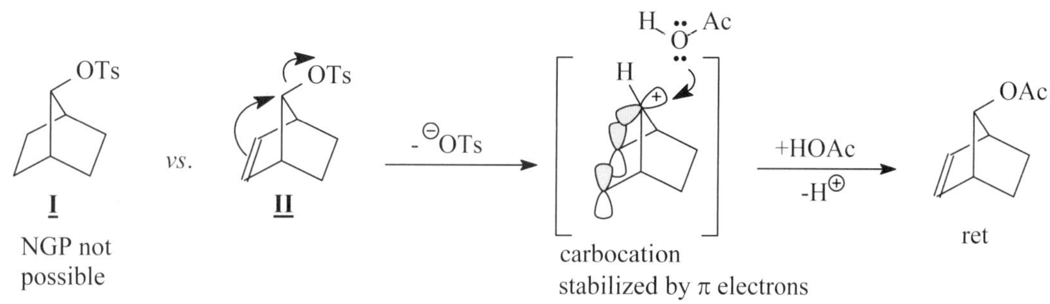 **I** NGP not possible vs. **II** $\xrightarrow{-^{\ominus}OTs}$ [carbocation stabilized by π electrons] $\xrightarrow[-H^{\oplus}]{+HOAc}$ ret (OAc product)

15. 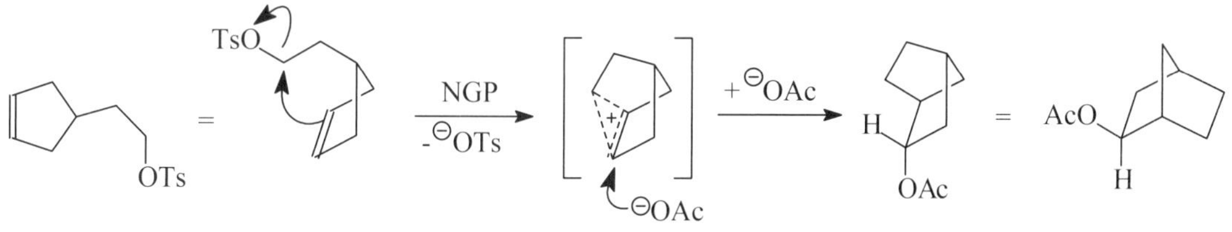 (cyclopentenyl-ethyl OTs) $\xrightarrow[-^{\ominus}OTs]{NGP}$ [bridged cation] $\xrightarrow{+^{\ominus}OAc}$ norbornyl-OAc

16. RNA: $\xrightarrow[-H^{\oplus}\ NGP]{dil\ ^{\ominus}OH}$ → $\xrightarrow{-R'O^{\ominus}}$ → cyclic phosphate intermediate $\xrightarrow{+^{\ominus}OH}$ → *2'-phosphate* or *3'-phosphate*

DNA: NGP not possible; therefore, more stable in *dil* base

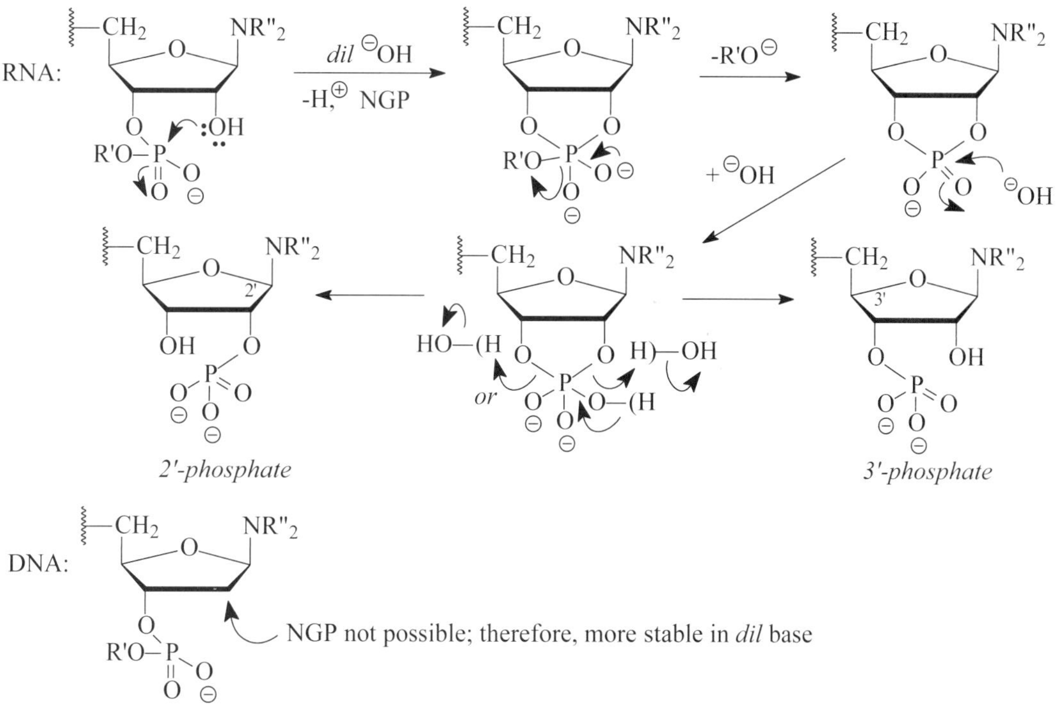

17. 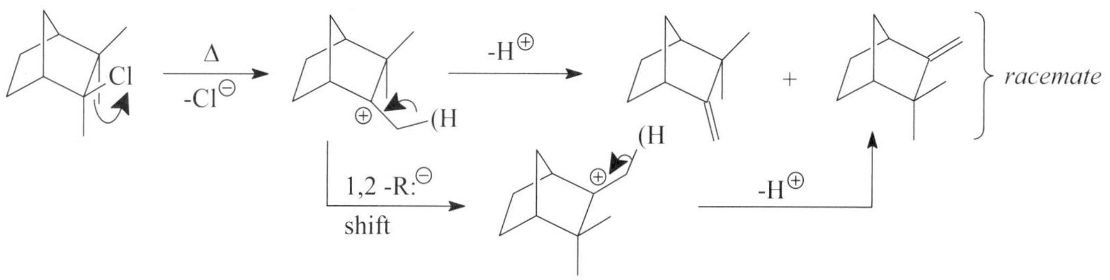 $\xrightarrow[-Cl^{\ominus}]{\Delta}$ → $\xrightarrow{-H^{\oplus}}$ products + $\}$ *racemate*

  $\downarrow$ 1,2 -R:$^{\ominus}$ shift → $\xrightarrow{-H^{\oplus}}$

9.4 Mechanisms

312 • Chapter 9 S_N1, S_N2, E1, and E2 Reactions

18. [mechanism showing conversion of ATP-like ribose with adenine, via cyclic phosphate intermediate, losing PP_i = $HO-P(O)(O^-)-O-P(O)(O^-)-O^-$]

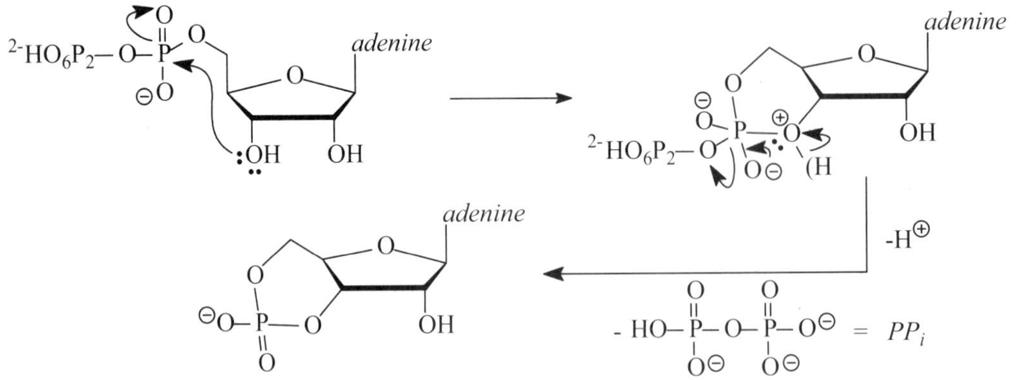

19. a. [mechanism with OPP groups, -BH⁺, -OPP, forming diene-OPP product]

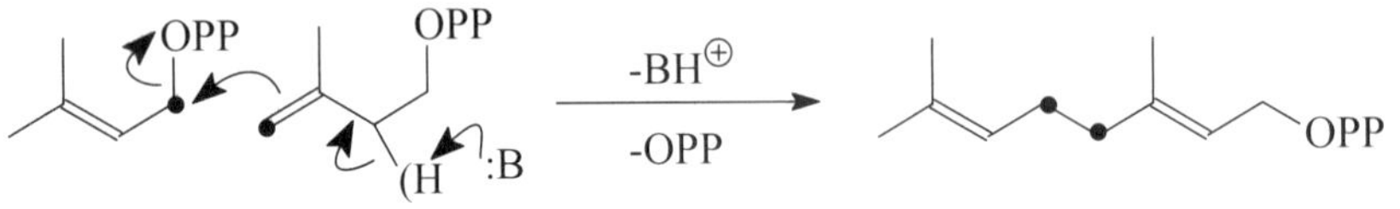

b. [mechanism starting from geraniol + H⁺, -H₂O, forming carbocation, resonance, rotation, cyclization, +H₂O, -H⁺ to give α-terpineol]

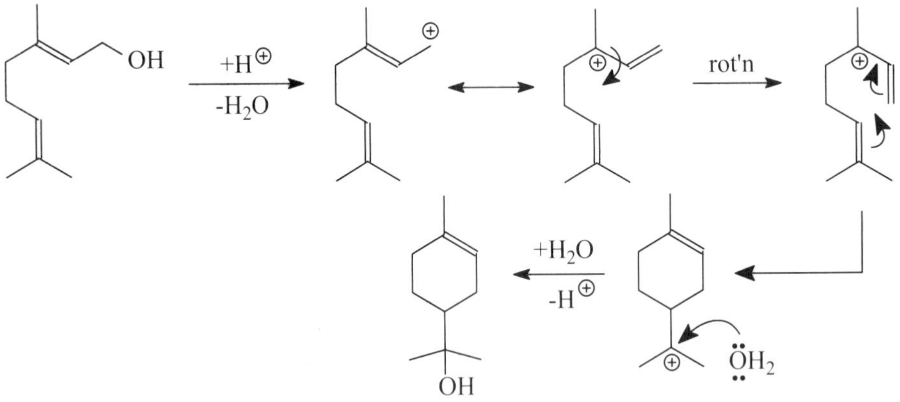

c. coupling mechanism:

[farnesyl-OPP + I-PP, -H⁺, -OPP, showing bond formed to give extended polyprenyl-OPP]

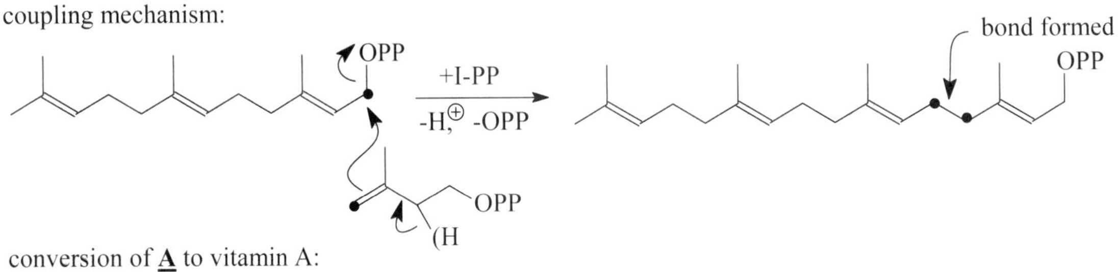

conversion of **A** to vitamin A:

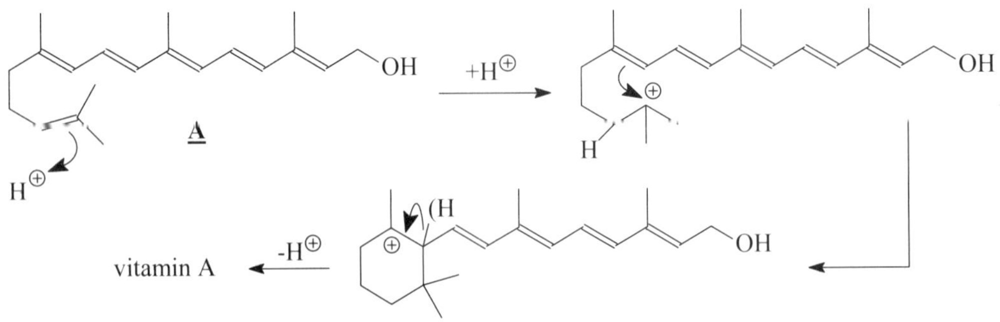

[mechanism showing **A**, +H⁺, cyclization, -H⁺ to give vitamin A]

9.4 Mechanisms

9.4 Mechanisms

CHAPTER 10
NMR

1. [structure: 2-methyl-2-butene]
2. [structure: dichloroacetaldehyde diethyl acetal]
3. [structure: 2-methyl-2,4-pentanediol]
4. [structure: 2-methyl-1-phenyl-2-propanol]

5. [structure: methyl dimethoxyacetate]
6. [structure: 4-bromophenyl ethyl ether]
7. [structure: 1-chloro-2,2-difluoropropane]
8. [structure: indane]

9. [structure: (1-bromoethyl)benzene]
10. [structure: 1,3-dichloro-2-methyl-2-butene]
11. [structure: (3-bromopropyl)benzene]
12. [structure: 3-phenoxypropanoic acid]

13. [structure: trans-1-bromo-3,3-dimethyl-1-butene] $J \Rightarrow$ *trans-*
14. [structure: 2,5-dimethyl-2,4-hexadiene]
15. [structure: 3-fluoro-3-hydroxy-3-methyl-2-butanone...]
16. [structure: 3-hydroxy-4,4-dimethylpentanal]

17. [structure: 4-(2-methyl-1-propenyloxy)benzaldehyde]

18. H_a and H_b are diastereomeric protons; ^{19}F ($I = 1/2$)
therefore, max multiplicity for $H_{a\,or\,b}$ = doublet x doublet x doublet = <u>8 lines</u>

19. 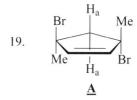   vs.   [structure B]

A

methylene protons are identical
=> singlet

B

H_a => doublet; H_b => doublet
(appears as a multiplet)

20.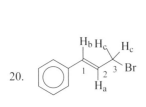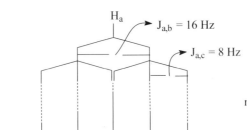

multiplicity: <u>5 lines</u> (pentuplet)

21. a. [structure with Cl, F, Hc, Ha, Hb, Cl]   highest field proton is H_a
$J_{a,F}$ => <u>doublet</u>

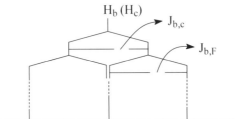

b. H_b and H_c are diastereomeric: $J_{b,c} \approx J_{b,F}$ => triplet

22. [Ph-CH(Hb,Hc)-CH(Ha)(NH2)-CO2H structure] H_b and H_c are diastereomeric and independently couple with H_a; if $J_{a,b} \neq J_{a,c}$, H_a would appear as a doublet x doublet = <u>4 lines</u> (assuming no coupling through nitrogen)

23. [t-butyl acetate structure]

24. a. H_a (lowest field proton): doublet x doublet => <u>4 lines</u>

  b. C_7 and C_8 are diastereomeric carbons; therefore, <u>8</u> chemical shifts

[cholesterol structure with numbered side chain and starred carbons, labeled Ha, Hb, Hc, DO]

25. [mechanism: Br-C(Me)2-C(Me)2-Br + SbF5 → -SbF5Br⁻ → cation with :Br: → bromonium ion with Me groups]

all methyls are equivalent therefore, appear as a singlet

26. ^{31}P (I = 1/2), n_P = 2; therefore, $2nI + 1$ = <u>3 (triplet)</u>

[(i-Pr-O)2P(=O)-CH(Ha,Ha)-P(=O)(O-i-Pr)2 structure]

27. a. *amplitude*: signal at δ -16.1 is highest amplitude because most molecules (66%) contain Pt with I = 0 (no further spin-spin coupling with H_a is observed, *i.e.*, $J_{H,Pt}$ = 0)

multiplicity: ^{31}P (I = 1/2), so $J_{H,?}$ = $J_{H,P}$ > 0, n_P = 2
therefore, 2nI + 1 = <u>triplet</u>

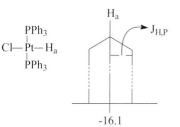

b. *amplitude*: signals at δ -13.6 and -19.6 arise from fewer (34%) molecules containing ^{195}Pt (I = 1/2)

multiplicity: H_a now undergoes spin-spin coupling with both P and Pt to give a <u>doublet of triplets</u> ($J_{H,Pt}$ >> $J_{H,P}$)

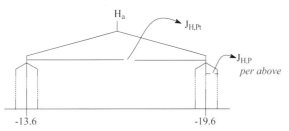

c. H_a is very *highly shielded*, essentially a hydride, because of the polarization of the Pt-H bond (much higher electron density around H_a than typically encountered in C-H bonds)

28.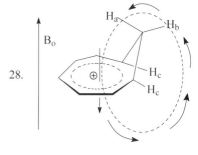

Because of magnetic anisotropy of the aromatic ring current, H_a experiences diamagnetic (and H_b paramagnetic) lines of force relative to applied field B_o. Therefore, H_a is more shielded (and H_b deshielded) than normally observed in hydrocarbon protons on sp^3 carbons.

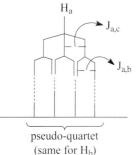

pseudo-quartet
(same for H_b)

29. *multiplicities*:

H–^{11}B–H (with H above and below, charge ⊖)   vs.   H–^{10}B–H (with H above and below, charge ⊖)

^{11}B (I = 3/2)
therefore, 2nI + 1 = <u>quartet</u>
(higher amplitude)

^{10}B (I = 3)
therefore, 2nI + 1 = <u>septet</u>
(lower amplitude)

relative *amplitudes* of quartet/septet reflect the natural abundance of ^{11}B/^{10}B = 80%/20%

CHAPTER 11
CONJUGATED SYSTEMS

11.1 Reactions

1. 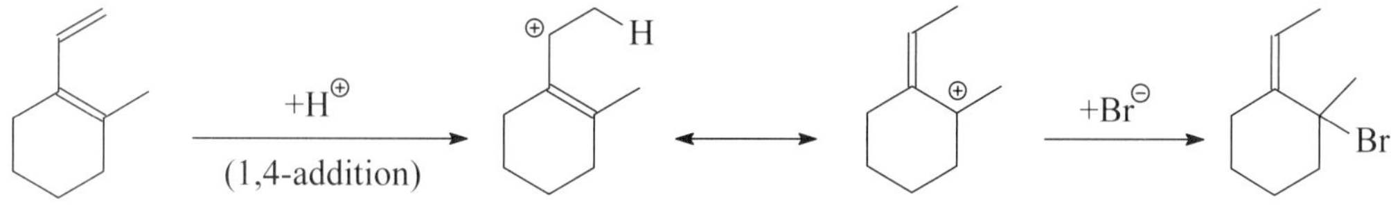 [vinyl methylcyclohexene] →(+H⁺, 1,4-addition)→ [carbocation intermediate with H] ↔ [resonance structure] →(+Br⁻)→ [brominated product]

2. 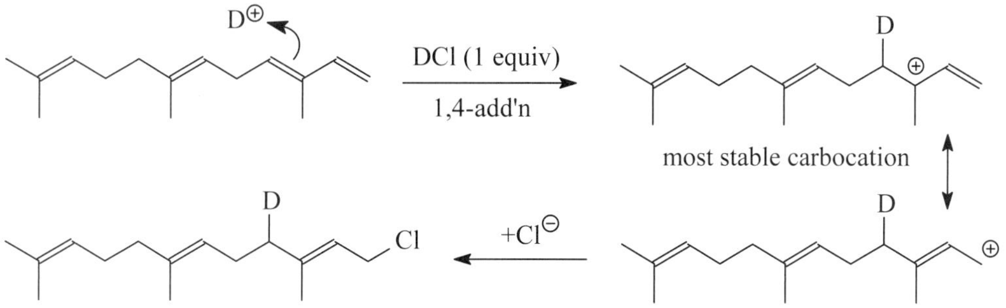 [polyene starting material with D⁺] →(DCl (1 equiv), 1,4-add'n)→ [carbocation with D, most stable carbocation] ↔ [resonance structure with D] →(+Cl⁻)→ [final product with D and Cl]

3. 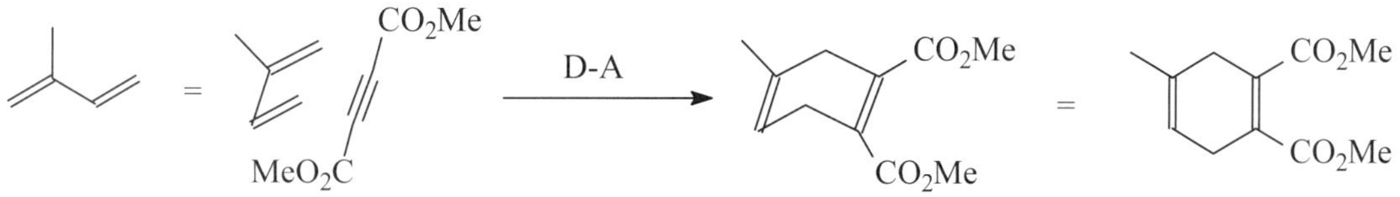 [isoprene] = [s-cis isoprene] + [dimethyl acetylenedicarboxylate] →(D-A)→ [cyclohexadiene diester] = [rearranged cyclohexadiene diester]

4. [bicyclic ketone] →(retro-D-A)→ [dienone intermediate] = [open-chain dienone]

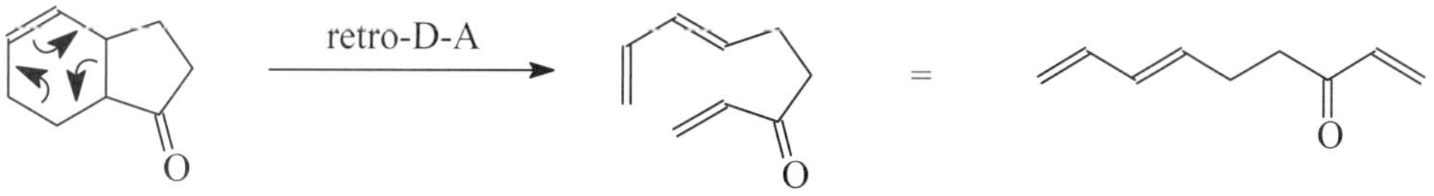

5. 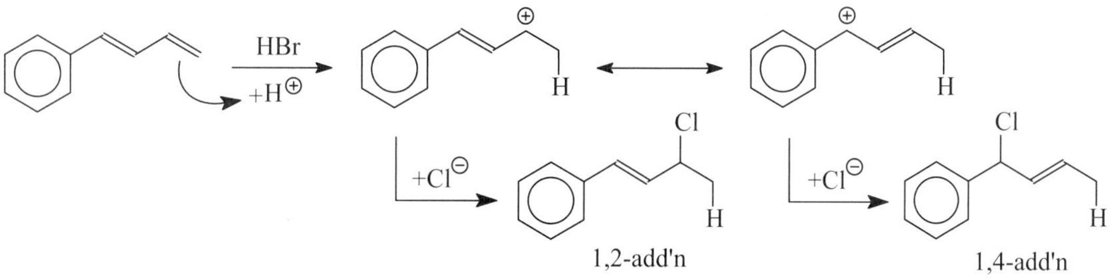 [1-phenyl-1,3-butadiene] →(HBr, +H⊕)→ [allyl cation intermediate] ↔ [benzylic cation resonance structure]
   →(+Cl⁻)→ [1,2-add'n product]   →(+Cl⁻)→ [1,4-add'n product]

   product of thermodynamic control - more conjugated system than 1,4-adduct, therefore, more stable

320 • **Chapter 11** Conjugated Systems

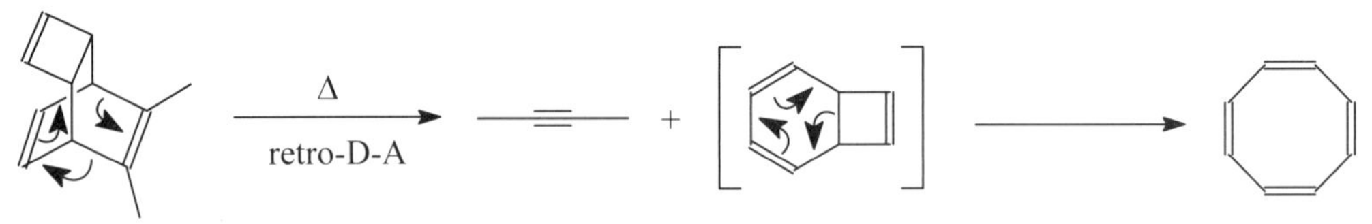

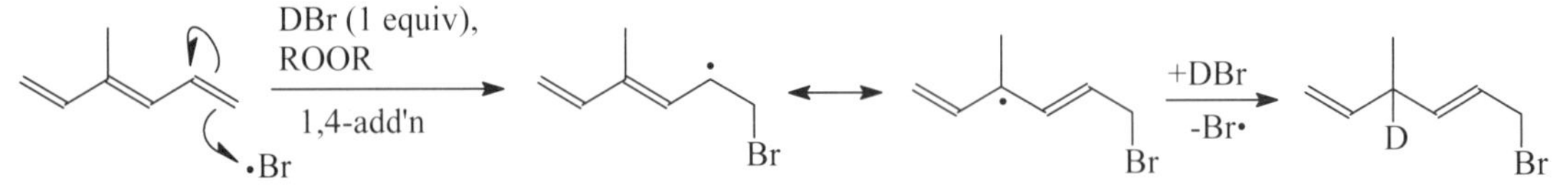

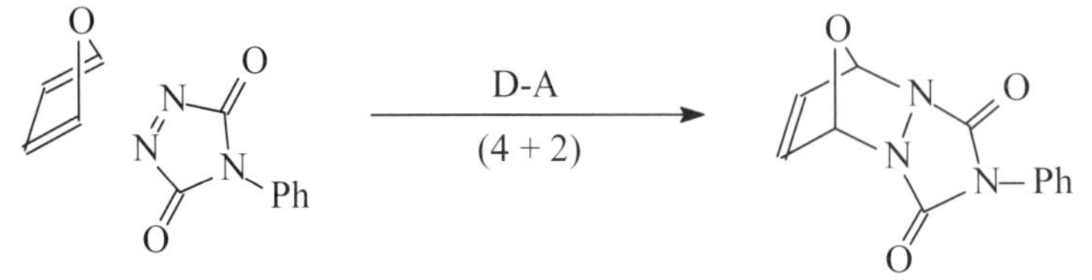

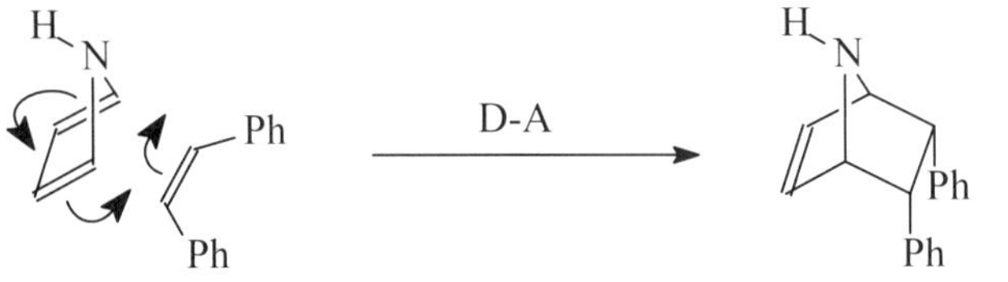

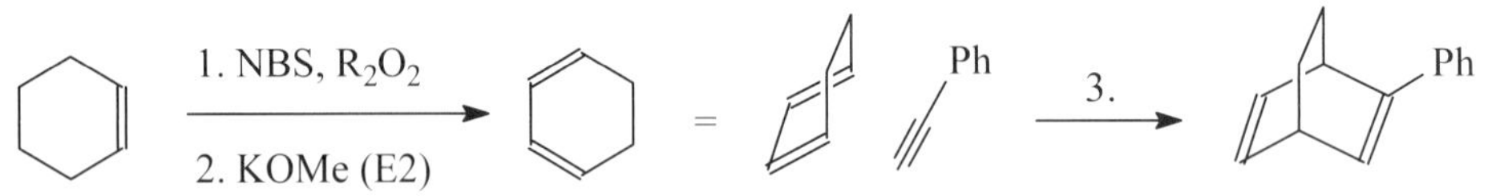

11.1 Reactions

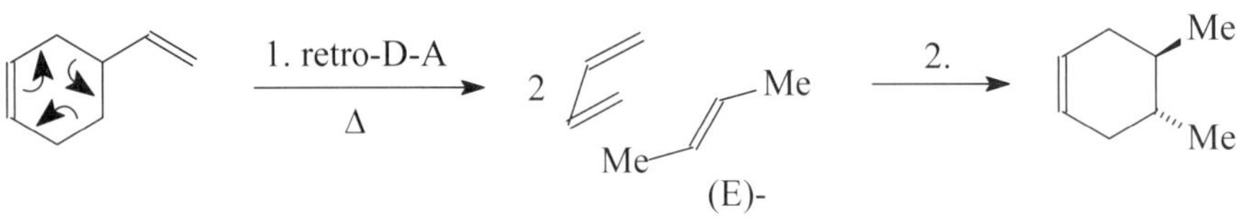

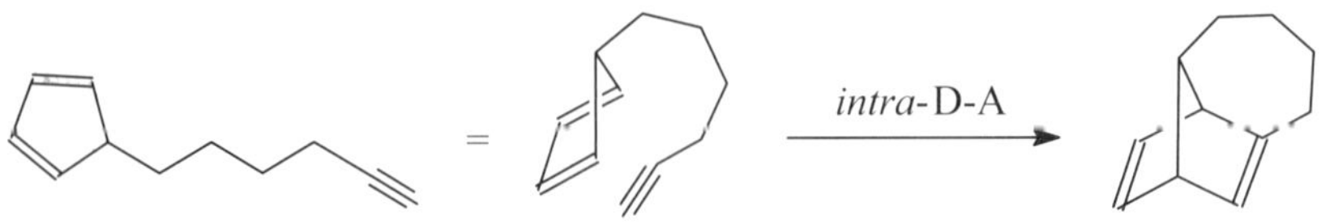

11.1 Reactions

322 • Chapter 11 Conjugated Systems

19.

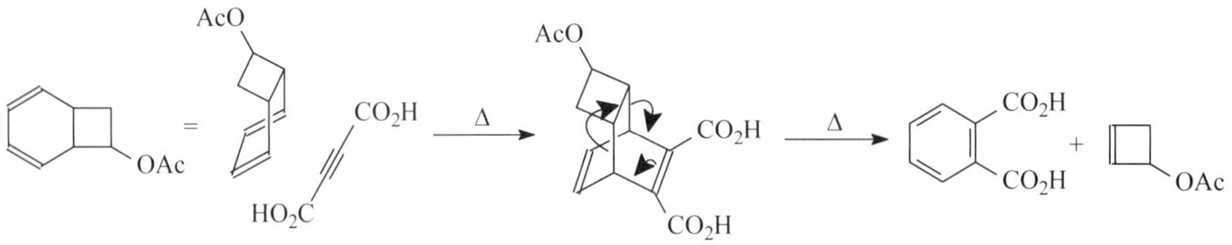

20.

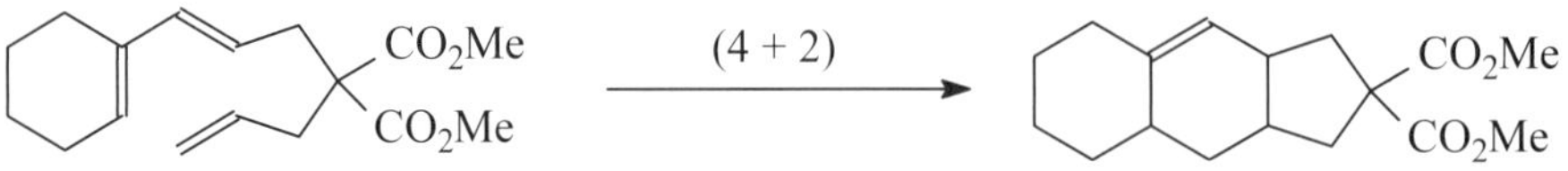

21.

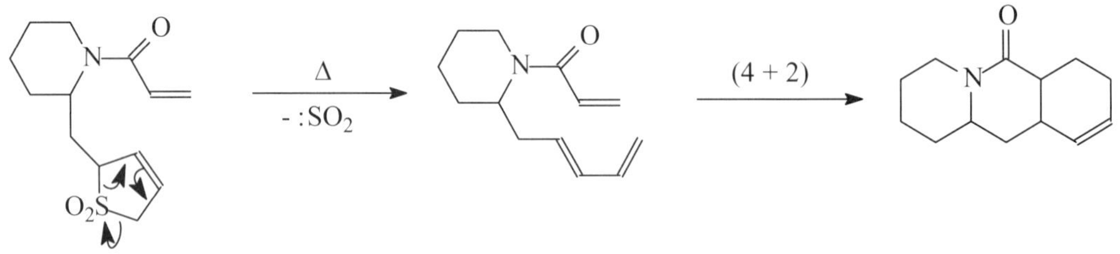

22.

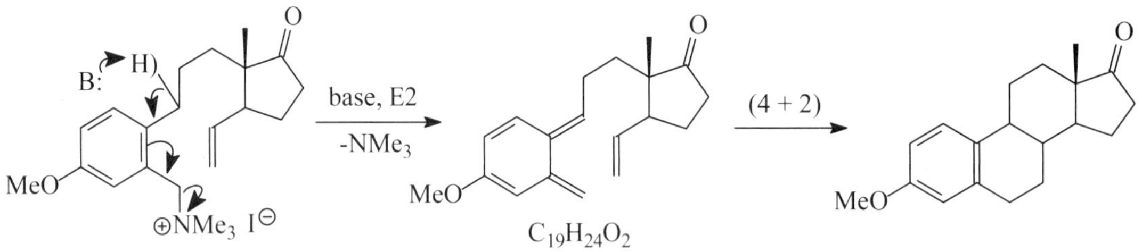

23.

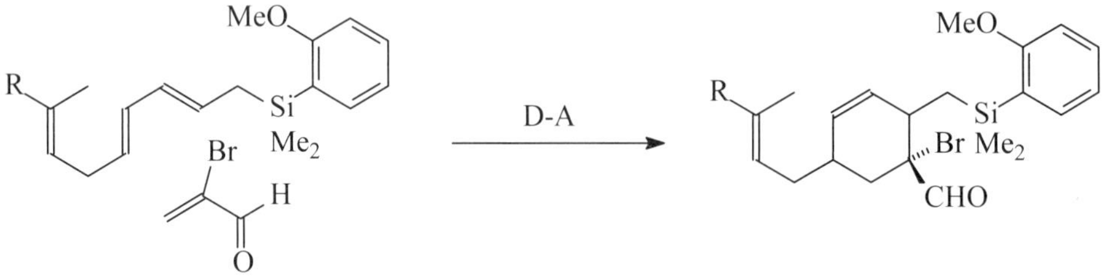

24.

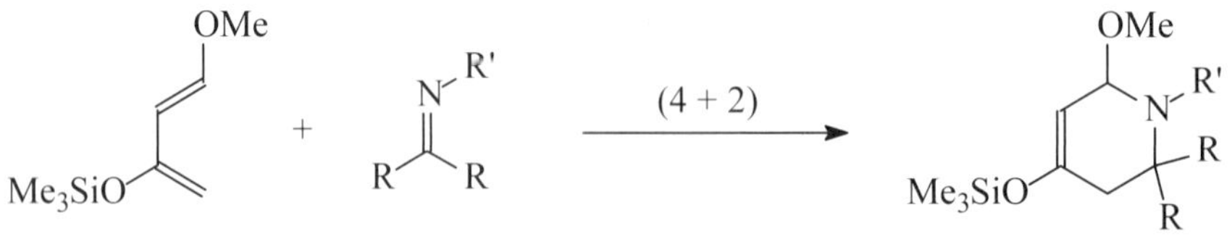

11.1 Reactions

11.2 Syntheses

1. 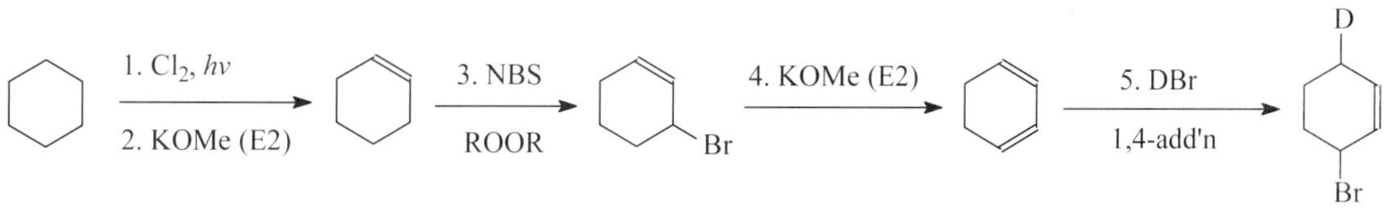
1. Cl_2, *hv*
2. KOMe (E2)
3. NBS
ROOR
Br
4. KOMe (E2)
5. DBr
1,4-add'n
D
Br

2. 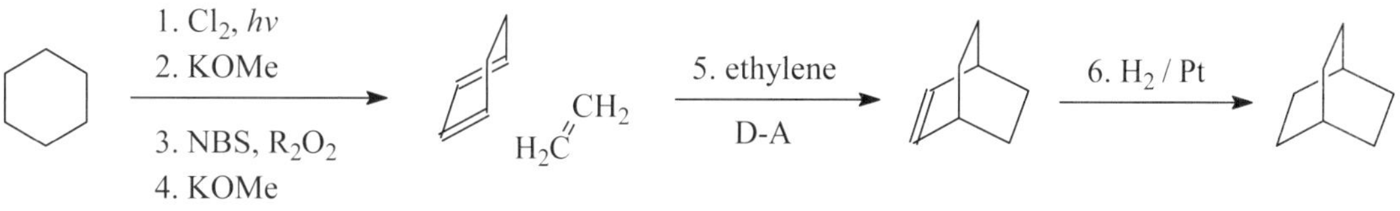
1. Cl_2, *hv*
2. KOMe
3. NBS, R_2O_2
4. KOMe
CH_2
H_2C
5. ethylene
D-A
6. H_2 / Pt

3. 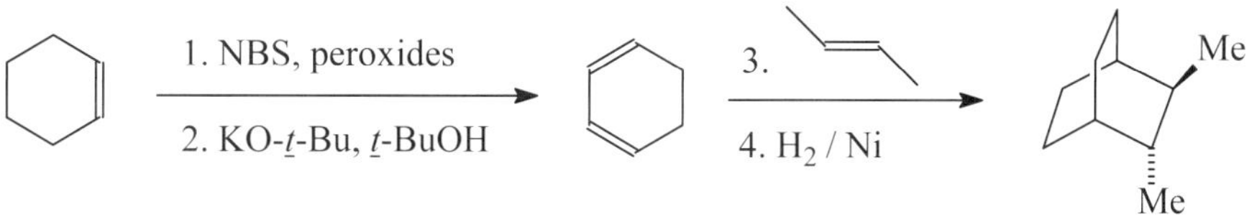
1. NBS, peroxides
2. KO-*t*-Bu, *t*-BuOH
3.
4. H_2 / Ni
Me
Me

4. 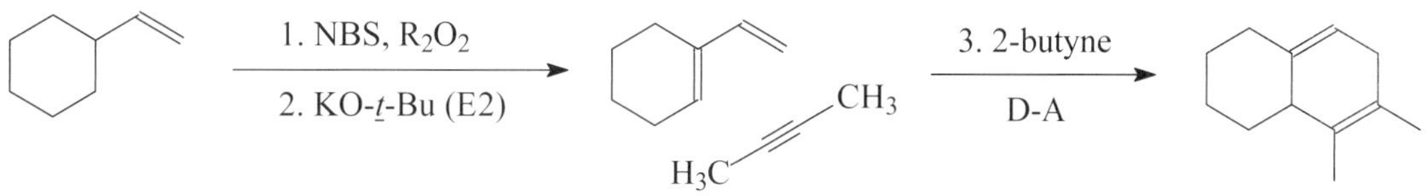
1. NBS, R_2O_2
2. KO-*t*-Bu (E2)
CH_3
H_3C
3. 2-butyne
D-A

5. 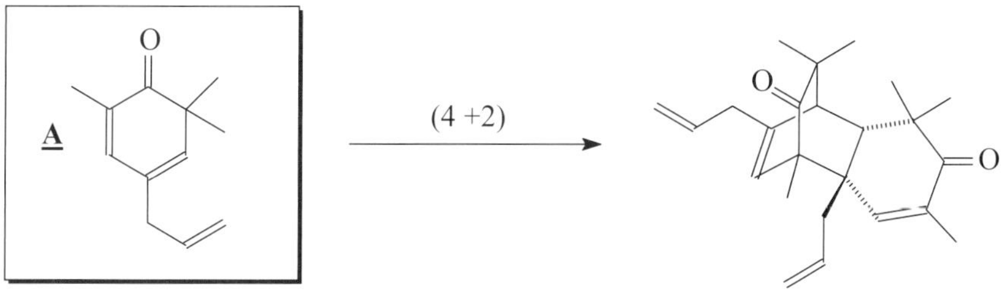
A
O
(4 +2)
O
O

6. 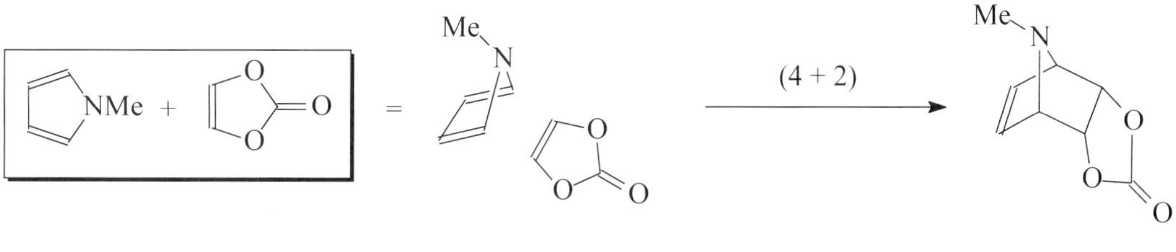
NMe +
O
O
O
=
Me
N
O
O
O
(4 + 2)
Me
N
O
O
O

7. 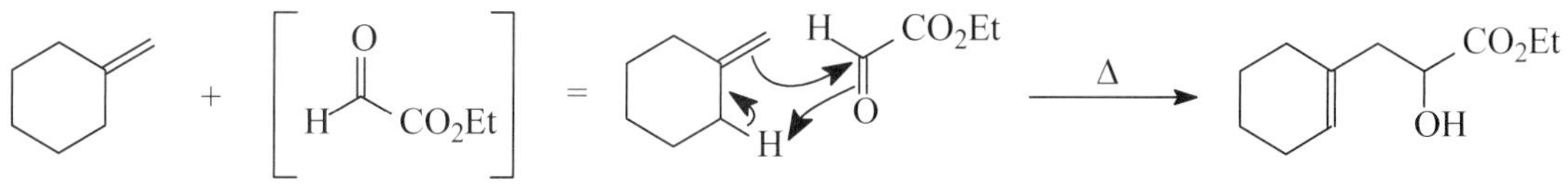
+
O
H
CO_2Et
=
H
CO_2Et
O
H
Δ
CO_2Et
OH

11.3 Mechanisms

1. [Phenolphthalein structure] → (> pH 8.5, −H⁺) → [ring-opened quinoid structure]

   - sp³ carbon prevents conjugation from one ring to the other two; therefore, absorption occurs at shorter wavelengths (UV in this case)

   - → conjugation results in a 'red shift;' absorption occurs at a longer wavelength, moving into the VIS, and the molecule, therefore, is "colored"

2. [cyclohexene with vinyl group] →(Δ, retro-D-A)→ [isoprene] + [butadiene] →(Δ, D-A)→ [limonene-type product]

3. a. [benzocyclobutene] →(Δ)→ [o-quinodimethane] + maleic anhydride →(4+2)→ [tetrahydronaphthalene anhydride]

3. b. [benzocyclobutene-tethered cyclopentanone with vinyl] →(Δ)→ [o-quinodimethane intermediate] →(Δ)→ [steroid-like tetracyclic ketone]

4. a. [pyridine]   longest λ_{max}: n ⟶ π*

   b. *low* probability   A = ε · c · d   ε (molar absorptivity) only ~ 10 − 100 for n ⟶ π* transitions (vs. > 10,000 for π ⟶ π* transitions)

   c. [pyridinium]   no non-bonding electrons in the CA of pyridine; therefore, <u>no</u> n ⟶ π* transition

5. 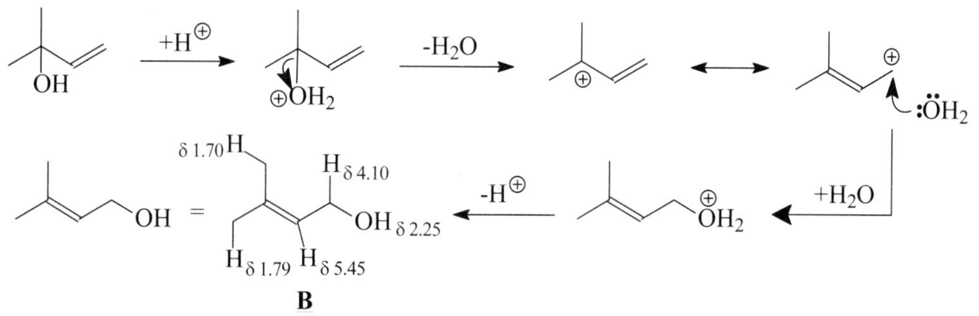 [Mechanism showing acid-catalyzed reaction: isoprenol + H⁺ → protonated form → −H₂O → allyl cation resonance structures → +H₂O → protonated intermediate → −H⁺ → product **B** with NMR assignments: H δ 1.70, H δ 4.10, H δ 1.79, H δ 5.45, OH δ 2.25]

6. 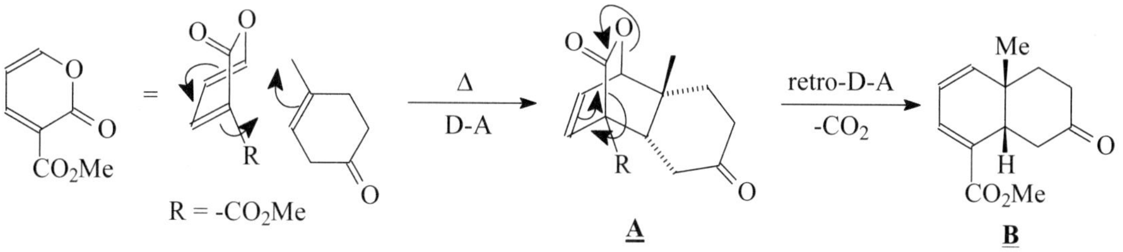 [Diels-Alder reaction of 2H-pyran-2-one with CO₂Me group and methylcyclohexenone (R = −CO₂Me), Δ D-A → intermediate **A** → retro-D-A, −CO₂ → product **B** with Me, CO₂Me groups]

7. 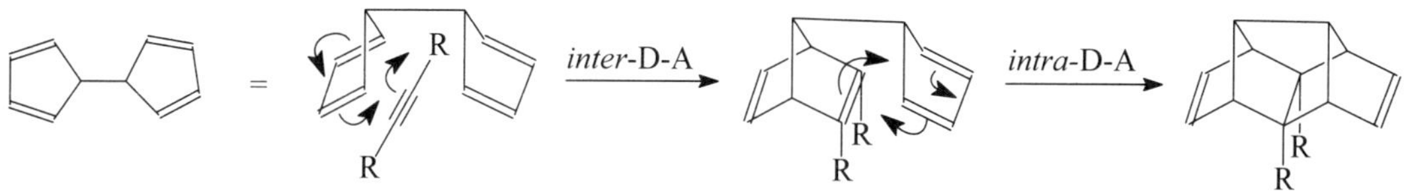 [Dicyclopentadiene-type with alkyne R groups, inter-D-A → intermediate → intra-D-A → cage product with R groups]

8. 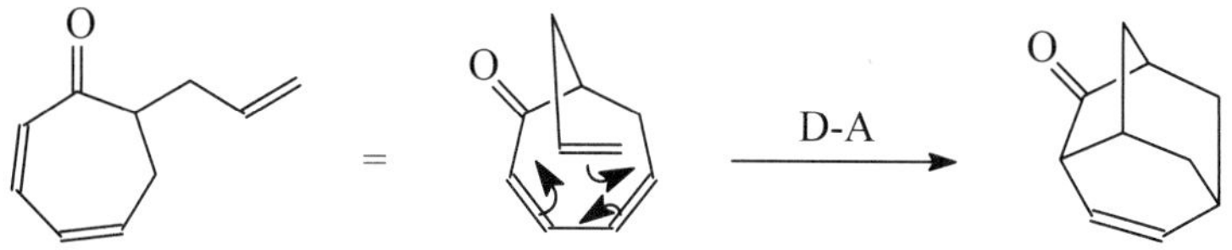 [Cycloheptenone with allyl group → D-A → bicyclic ketone product]

9. 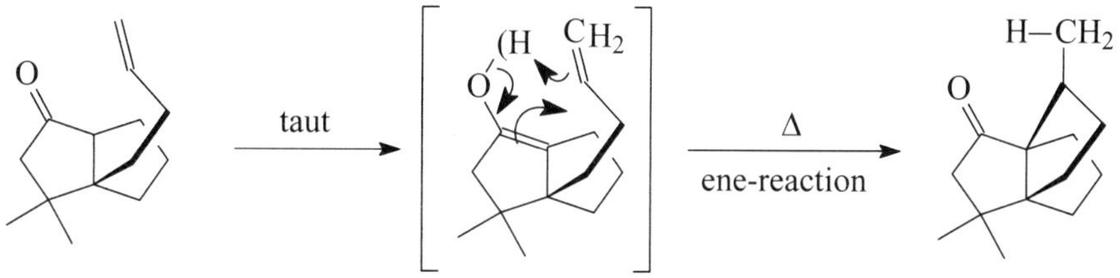 [Ketone with pendant vinyl group → taut → enol intermediate with (H), CH₂ → Δ, ene-reaction → product with H−CH₂]

10. 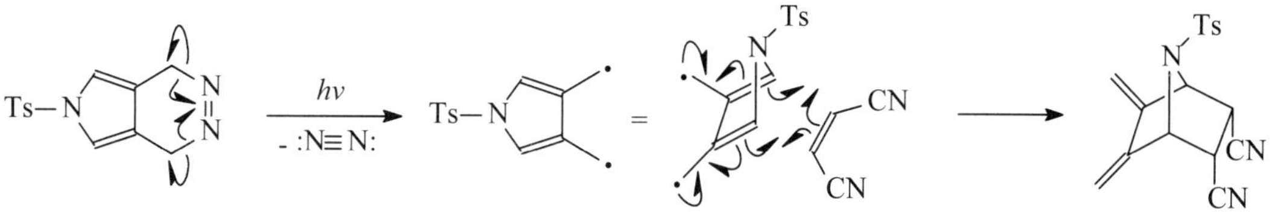 Ts−N pyrroline-fused diazo compound, hv, $-:N\equiv N:$ → Ts−N diradical = resonance structure + maleonitrile (CN, CN) → Ts−N bicyclic product with two CN groups and exocyclic methylenes

11.3 Mechanisms

326 • Chapter 11 Conjugated Systems

11.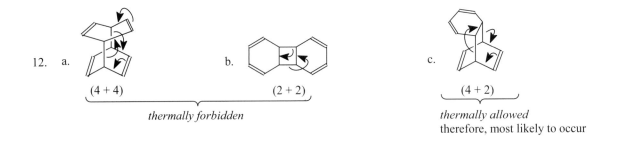

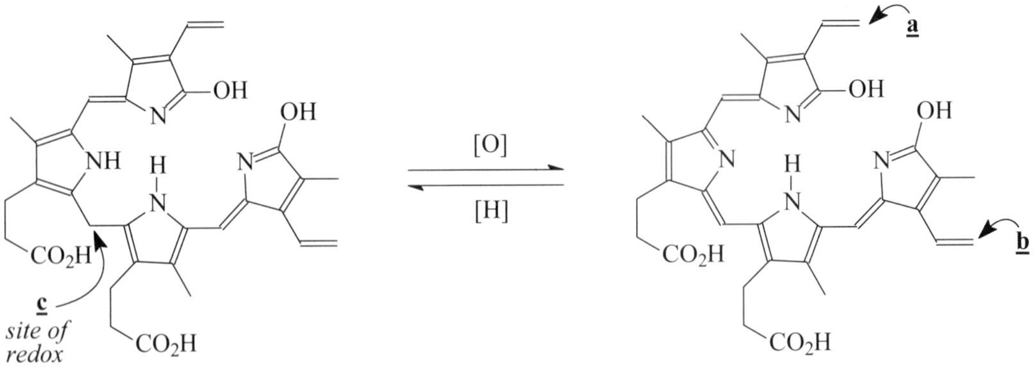

bilirubin (red) ⇌ [O]/[H] ⇌ biliverdin (green)

Increased conjugation promotes a 'red shift' in λ_{max}, causing the color of pigments to move toward the green-blue end of the VIS spectrum:
- biliverdin is conjugated from C_a to C_b, whereas bilirubin's conjugation is disrupted (as a consequence of reduction) at C_c
- biliverdin, therefore, absorbs at longer wavelengths (red) than bilirubin; alternatively, biliverdin is transparent to shorter wavelengths (green).

12. a. (4 + 4)   b. (2 + 2)   c. (4 + 2)

a, b: *thermally forbidden*

c: *thermally allowed* therefore, most likely to occur

11.3 Mechanisms

CHAPTER 12
AROMATICS

12.1 General

1. The following compounds obey the Hückel (4n +2) rule and would be expected to have aromatic character:

★ ↷ = lone pairs of electrons are in a p orbital (other lone pairs are in sp² orbitals)

b., 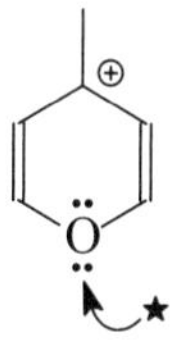 d.,  f., 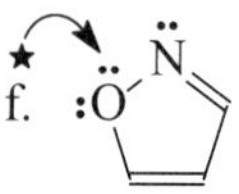 i., 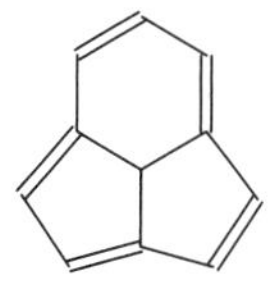 j. 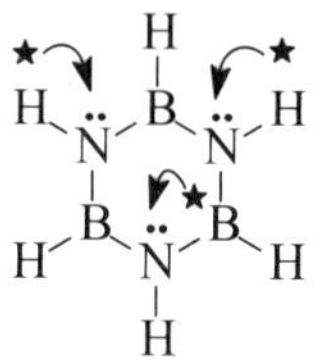 [structures shown]

k. 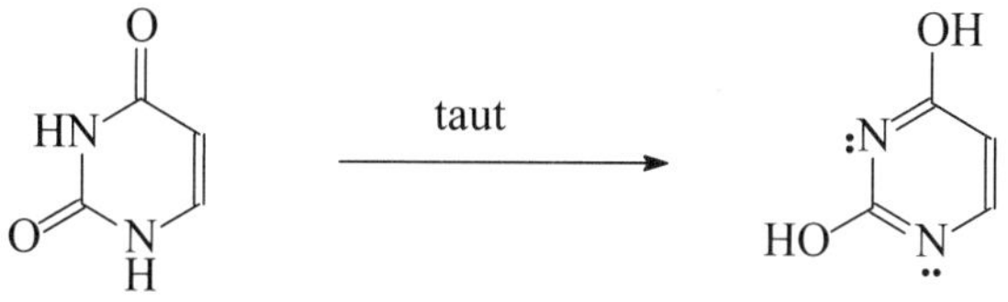 [uracil structure] —taut→ [dihydroxypyrimidine structure] (6 π electrons)

m. product: [cyclobutadiene] —2 Li→ 2 Li⁺ [cyclobutadiene dianion]²⁻ (6 π electrons)

n. product: [pentalene precursor with H₃C:⁻ and :CH₃⁻] —2 MeLi / −2 CH₄→ 2 Li⁺ [pentalene dianion] (10 π electrons)

o. product: [dibromocyclopentane-cyclopropylidene] —Zn→ [intermediate with ZnBr] —−ZnBr₂→ [calicene] ↔ [cyclopentadienide-cyclopropenylium resonance]

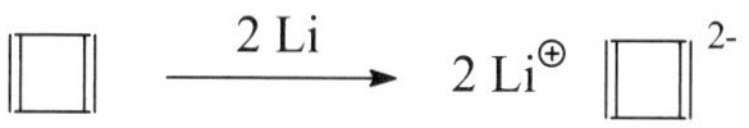

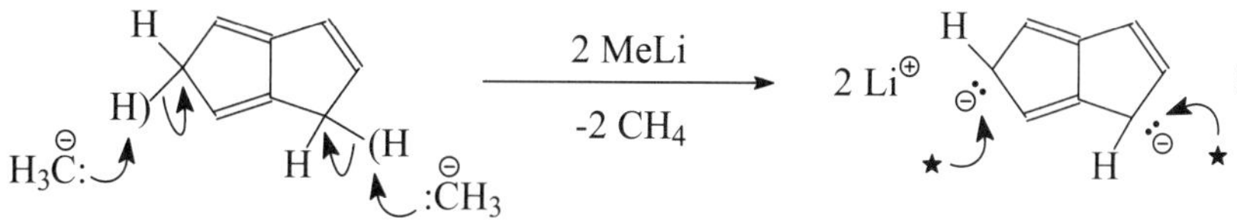

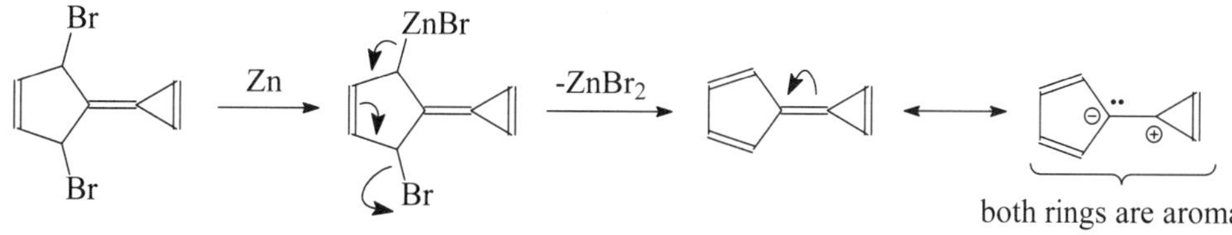

both rings are aromatic
μ >> 0!

Note: 1. carbocation from the reaction of [chloroindene] —SbF₅ / −SbF₅Cl⁻→ [indenyl cation] (8 π electrons ≠ 4n+ 2)

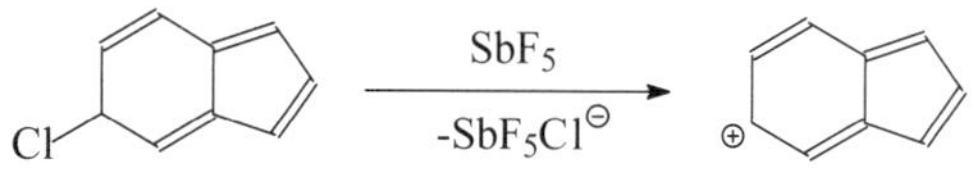

12.1 General

328 • Chapter 12 Aromatics

2. largest μ: <u>a</u>.

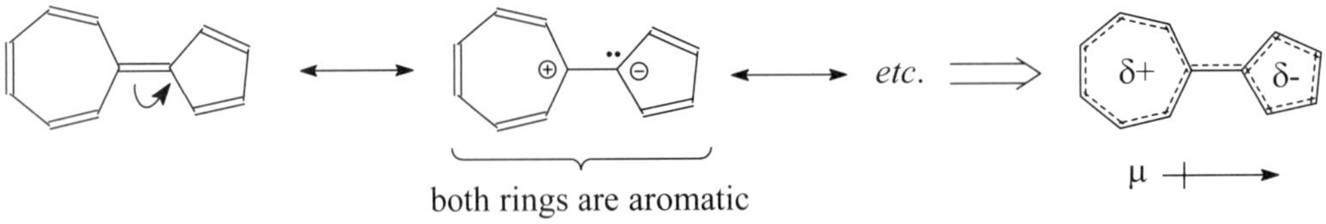

both rings are aromatic

note: flow of electrons in either direction in <u>b</u> or <u>c</u> would result in one ring being aromatic and the other anti-aromatic, thereby lessening the benefit of charge separation and lowering μ.

3. 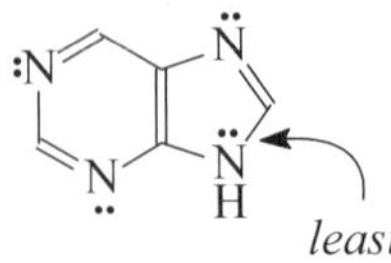 *least* basic (delocalized, part of aromatic ring current) 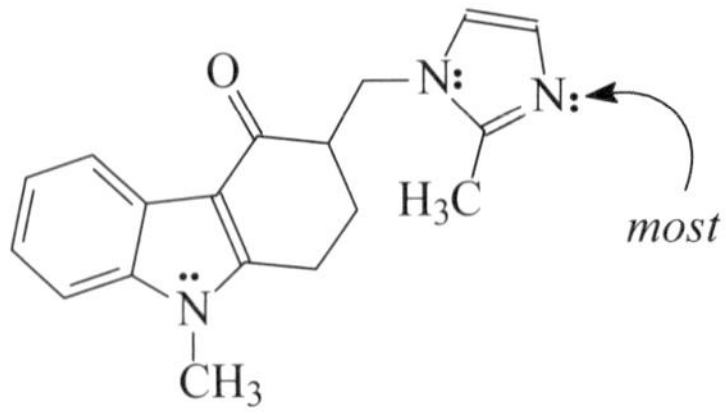   *most* basic (localized)

4. least stable: <u>b</u>.

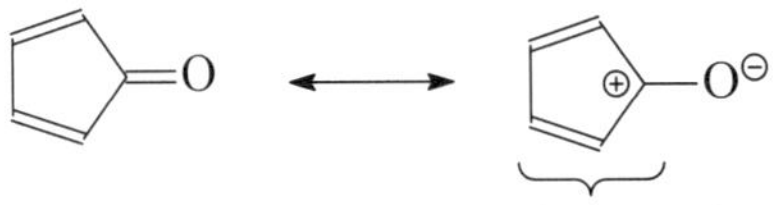

anti-aromatic

<u>a</u> and <u>c</u> have corresponding aromatic, and therefore stabilizing, contributing resonance structures

5. most acidic: <u>d</u>.

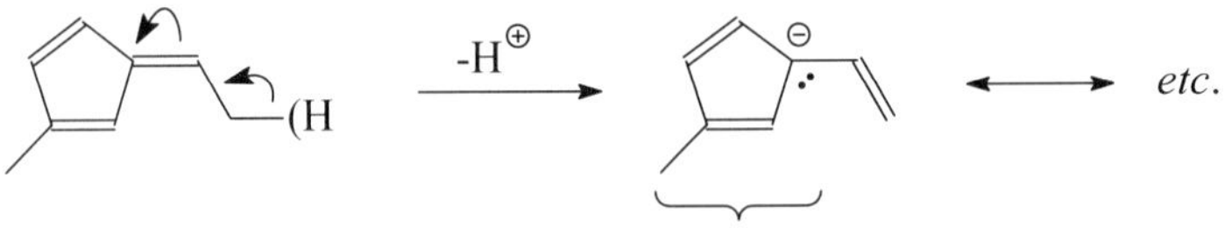

aromatic CB

loss of a proton from <u>a</u>, <u>b</u>, or <u>c</u> would produce a resonance-stabilized, *but not aromatic*, CB

6. most likely to undergo an S_N1 reaction: <u>a</u>.

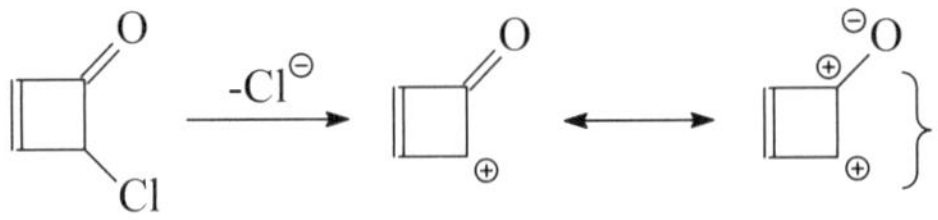

aromatic (4n + 2) contributing structure stabilizes carbocation

7. a. 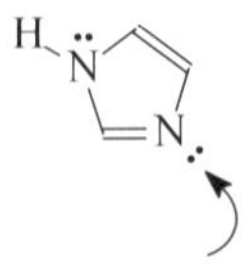 *more* basic (localized)   b. *most* basic ... aromatic   c. *more* basic (localized)

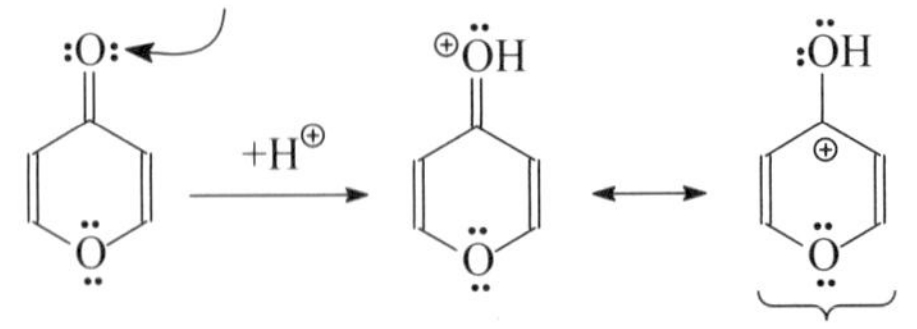

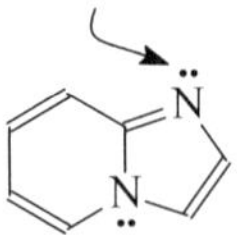

12.1 General

8.

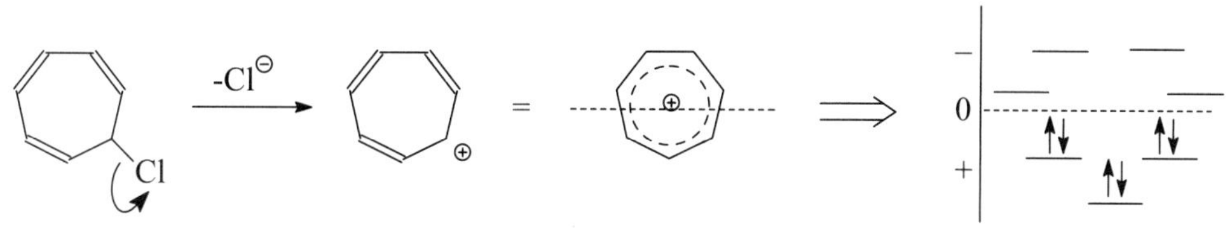

aromatic - all protons are equivalent ⇐ all bonding MOs are filled

9. largest molecular dipole moment: <u>b</u>.

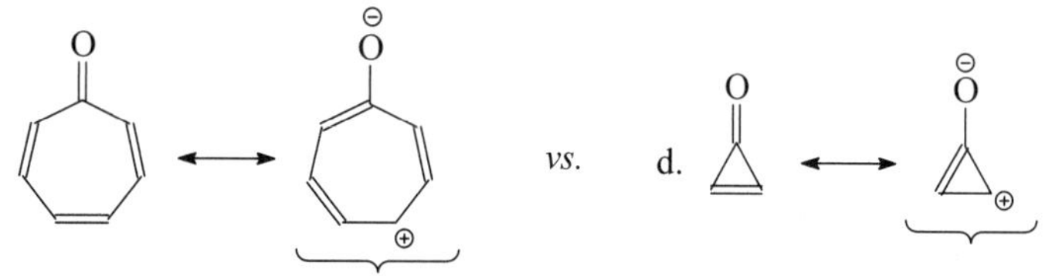

b. aromatic contributor with longer charge separation (>d)

vs. d. aromatic contributor with shorter charge separation

$\mu = \varepsilon \cdot d$

- aromaticity promotes charge separation in <u>b</u> and <u>d</u> (but not <u>a</u> or <u>c</u>), thereby increasing ε
- charge separation distance (d) is greater in <u>b</u> than <u>d</u>

10.

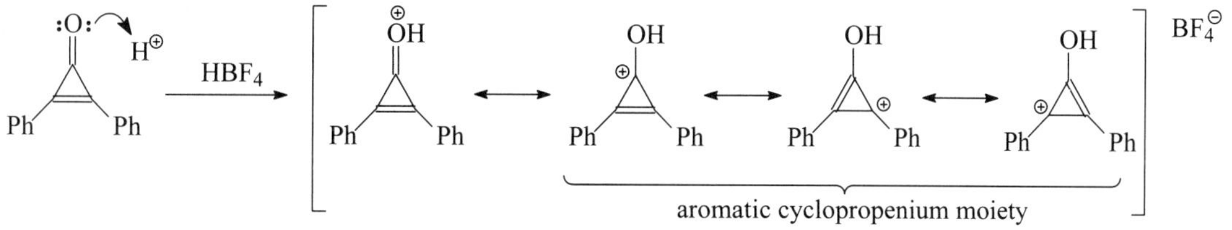

aromatic cyclopropenium moiety

11.

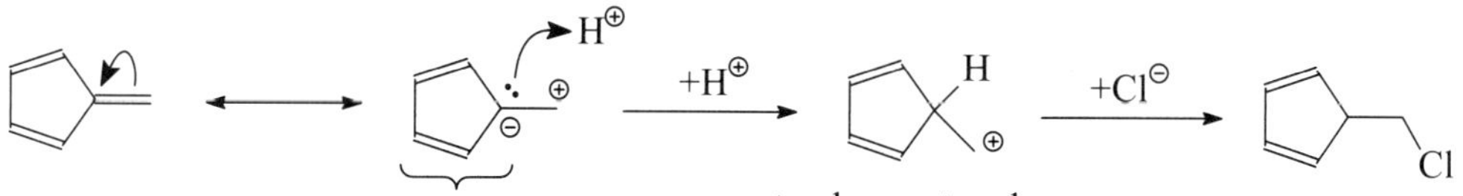

aromatic

note: does not undergo a 1,2-H:⁻ shift!

12.

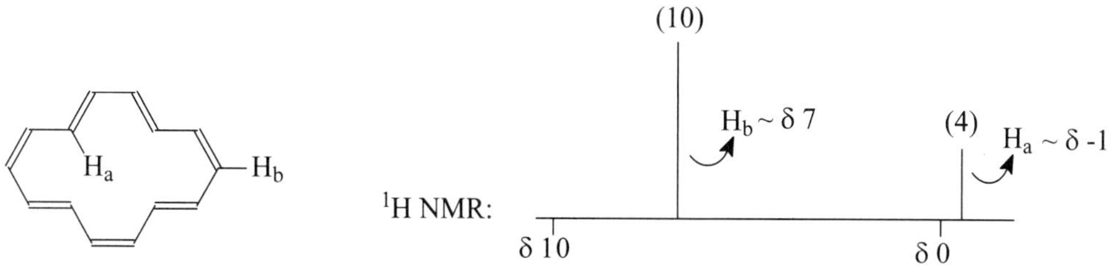

¹H NMR:

$H_b \sim \delta\ 7$ (10)

$H_a \sim \delta\ -1$ (4)

magnetic anisotropy causes the four H_a protons to be highly shielded (above TMS) and the ten H_b protons to be deshielded (into the aromatic region)

12.1 General

12.2 Reactions

1. PhNHAc —fuming sulfuric acid→ 4-HO₃S-C₆H₄-NHAc + *o*-isomer

2. *o*-cresol —HONO$_2$ / H$_2$SO$_4$→ 2-methyl-4-nitrophenol (no! avoid 1,2,3-subst'n at the position ortho to both OH and Me)

3. PhPH$_2$ (*o*-, *p*-director) —H$_2$SO$_4$, SO$_3$ / +H$^⊕$→ PhPH$_2$H$^+$ (*m*-director) → *m*-H$_3$P-C$_6$H$_4$-SO$_3$H

4. Pyrrole —Cl$_2$ / Fe→
   - 2-substitution: three resonance structures of the intermediate (more important set of contributing resonance structures) → −H$^⊕$ → 2-chloropyrrole
   - vs. 3-substitution: two resonance structures (less favorable) ✗

5. PhCH$_2$CH$_2$Br —1. AlCl$_3$→ PhCH$_2$CH$_2^⊕$ —1,2-H:$^⊖$ shift→ PhCH$^⊕$CH$_3$ —PhH, F-C alkylation→ PhCH(CH$_3$)Ph —2. NBS, R$_2$O$_2$→ PhCBr(CH$_3$)Ph —3. KOMe, E2→ Ph(CH$_2$)=CPh (1,1-diphenylethylene... actually PhC(=CH$_2$)Ph)

6. PhCH=CHCH$_2$CH$_3$
   - —Br$_2$ / CCl$_4$→ PhCHBr-CHBr-CH$_2$CH$_3$
   - —Br$_2$, Fe→ 3-Br-C$_6$H$_4$-CHBr-CHBr-CH$_2$CH$_3$
   - —NBS, R$_2$O$_2$→ PhCH=CH-CHBr-CH$_3$ + PhCHBr-CH=CH-CH$_3$

7. $\overset{\delta+}{I}\text{---}\overset{\delta-}{Cl}$ $\underset{\text{FeX}_3}{\longrightarrow}$ $\xrightarrow{-\text{FeX}_3\text{Cl}^{\ominus}}$ [I$^{\oplus}$] + Ph–Se(Me) $\xrightarrow{\text{EAS}}$ o-I-C$_6$H$_4$-SeMe + p-isomer

8. PhN=O $\xrightarrow{\text{Br}_2, \text{FeBr}_3}$ p-Br-C$_6$H$_4$-N=O + o-Br-C$_6$H$_4$-N=O

9. PhSH $\xrightarrow{\text{Cl}_2, \text{FeCl}_3}$ o-Cl-C$_6$H$_4$-SH + p-isomer

10. Pyridine $\xrightarrow{\text{Cl}_2, \text{BF}_3}$

[resonance structures for attack at C-4, labeled *least important*]

[more important set of contributing resonance structures for attack at C-3] $\xrightarrow{-\text{H}^{\oplus}}$ 3-chloropyridine

11. HCHO $\xrightarrow{+\text{H}^{\oplus}}$ H$_2$C=OH$^{\oplus}$ ↔ $^{\oplus}$CH$_2$–OH + 2,4,5-trichlorophenol $\xrightarrow{\text{EAS}}$ (hydroxymethyl trichlorophenol) $\xrightarrow{+\text{H}^{\oplus}}$ (protonated CH$_2$OH$_2^{\oplus}$) $\xrightarrow{-\text{H}_2\text{O}}$ (benzyl cation) $\xrightarrow{\text{EAS}}$ bis(trichlorohydroxyphenyl)methane

12. Benzenediazonium-2-carboxylate $\xrightarrow{1. -\text{CO}_2, -\text{N}_2}$ [benzyne] + cyclohexadiene $\xrightarrow{2. \text{D-A}}$ benzonorbornadiene adduct

12.2 Reactions

332 • Chapter 12 Aromatics

13. 1,3,5-trimethylbenzene + picric acid (2,4,6-trinitrophenol) → a π-complex (face-to-face stacked aromatic rings)

14.

2-fluoro-5-chlorobenzonitrile + :NH$_3$ $\xrightarrow{\text{nucleophilic aromatic subst'n}}$ [Meisenheimer intermediate with F, $^+$NH$_3$, CN, Cl] $\xrightarrow{\text{-HF, add'n - elim mechanism}}$ 2-amino-5-chlorobenzonitrile

15.

3-bromo-N,N-dimethylaniline $\xrightarrow[\text{NAS, benzyne mechanism}]{\text{1. MeLi (-HBr)}}$ benzyne with NMe$_2$ $\xrightarrow{+\text{MeLi}}$ aryl anion with NMe$_2$ and Me $\xrightarrow{\text{2. H}^\oplus}$ 3-methyl-N,N-dimethylaniline

16.

4-chloro(trifluoromethyl)benzene $\xrightarrow[\text{EAS}]{\text{1. HNO}_3\text{, H}_2\text{SO}_4}$ 2-nitro-4-(trifluoromethyl)chlorobenzene $\xrightarrow[\text{NAS, add'n - elim mechanism}]{\text{2. NaOMe, MeOH}}$ 2-nitro-4-(trifluoromethyl)anisole

17.

2,5-dimethyl-2,5-dihydrofuran $\xrightarrow{+\text{H}^\oplus}$ protonated oxocarbenium $\rightarrow$ ring-opened allyl cation with OH $\xrightarrow[\text{EAS}]{\text{toluene}}$ arylated alcohol $\xrightarrow[\text{-H}_2\text{O}]{+\text{H}^\oplus}$ secondary cation $\xrightarrow{\text{EAS}}$ 1,4,5-trimethyldihydronaphthalene

18.

isobutylene $\xrightarrow{+\text{H}^\oplus}$ tert-butyl cation + p-cresol $\xrightarrow{\text{EAS}}$ 2-tert-butyl-4-methylphenol $\xrightarrow{\text{again}}$ 2,6-di-tert-butyl-4-methylphenol

12.2 Reactions

19. 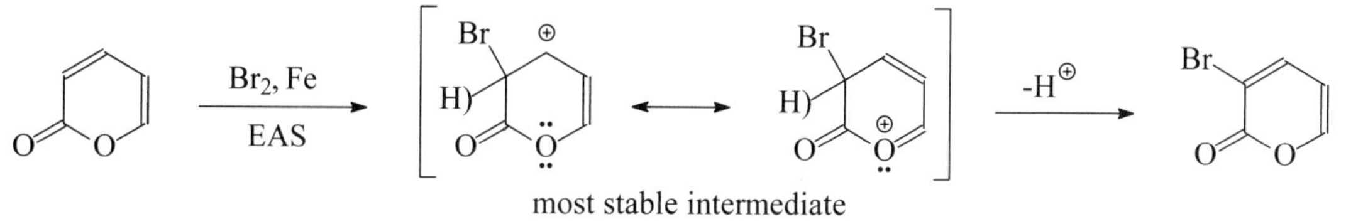 [structure: 2H-pyran-2-one] → Br₂, Fe / EAS → [Br, H on C adjacent to C=O, cation resonance structures — most stable intermediate] → -H⁺ → 3-bromo-2H-pyran-2-one

20. 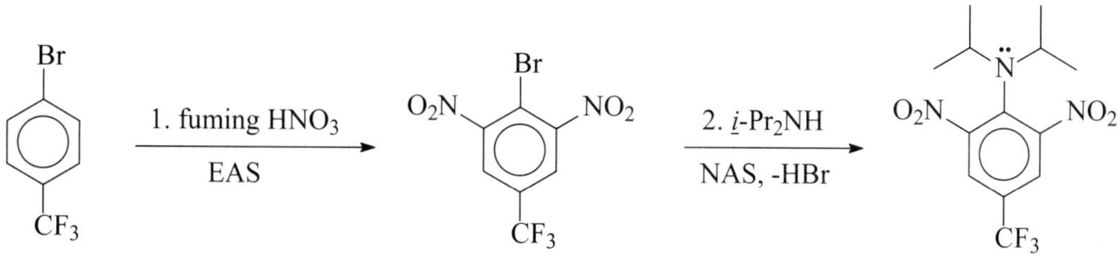 4-bromo-(trifluoromethyl)benzene → 1. fuming HNO₃ / EAS → 2-bromo-1,3-dinitro-5-(trifluoromethyl)benzene → 2. i-Pr₂NH / NAS, -HBr → 2,6-dinitro-4-(trifluoromethyl)-N,N-diisopropylaniline

21. 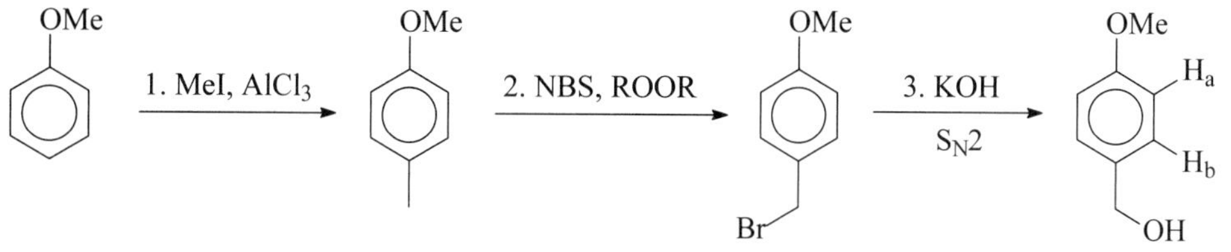 anisole → 1. MeI, AlCl₃ → 4-methylanisole → 2. NBS, ROOR → 4-methoxybenzyl bromide → 3. KOH / S_N2 → 4-methoxybenzyl alcohol (H$_a$, H$_b$ labeled)

¹H NMR: aromatic a-b quartet suggests p-subst'n

22. 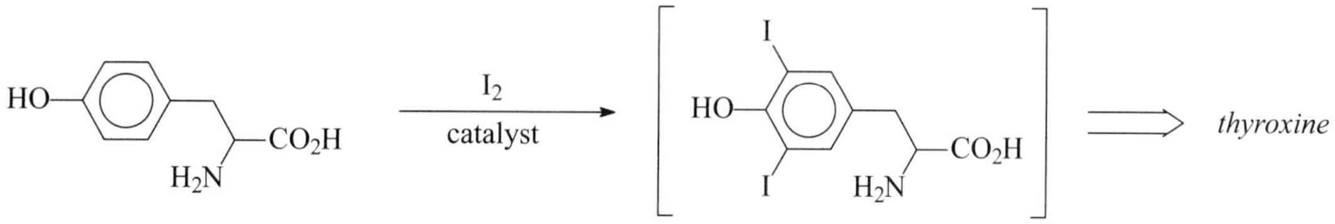 tyrosine → I₂ / catalyst → [3,5-diiodotyrosine] ⟹ *thyroxine*

23. 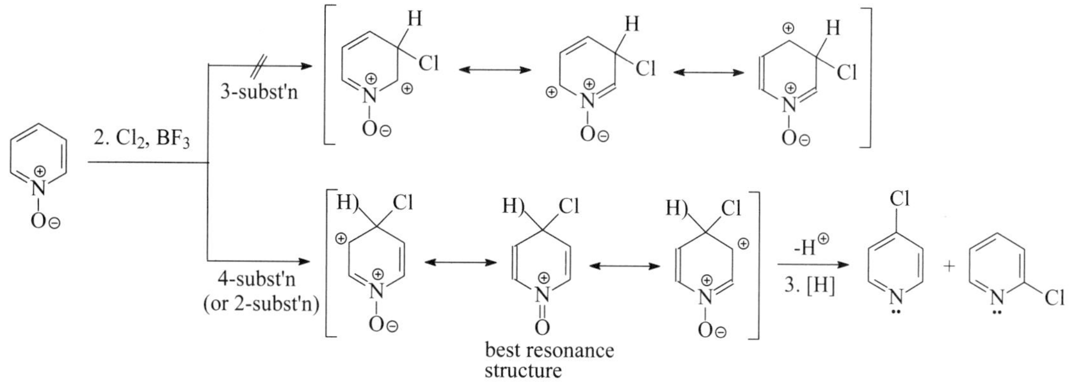 pyridine N-oxide → 2. Cl₂, BF₃

3-subst'n (blocked): resonance structures shown

4-subst'n (or 2-subst'n): resonance structures shown — best resonance structure → -H⁺, 3. [H] → 4-chloropyridine + 2-chloropyridine

12.3 Syntheses

1. Benzene →(1. CH₃I, AlCl₃ / F-C)→ toluene →(2. Cl₂, Fe)→ o-chlorotoluene →(3. H₂/Pt, high T, P)→ 2-chloro-1-methylcyclohexane

2. Benzene →(1. Br₂, Fe; 2. dil D₂SO₄ (EAS))→ 1-bromo-2,3,5-trideuterobenzene →(3. Li)→ 2,3,5-trideutero-phenyllithium →(4. H₂O)→ 1,3,5-trideuterobenzene

3. Benzene →(1. Cl₂, AlCl₃ (×2))→ 1,4-dichlorobenzene →(2. Mg, Et₂O)→ 1,4-bis(MgCl)benzene →(3. D₂O)→ 1,4-dideuterobenzene

4. Benzene →(1. PhCH₂COCl, AlCl₃)→ PhCOCH₂Ph →(2. H₂/Ra-Ni, hydrogenation)→ [PhCH(OH)CH₂Ph] →(followed by hydrogenolysis)→ PhCH₂CH₂Ph

5. Butylbenzene →(1. Cl₂, BF₃; 2. NBS, ROOR)→ 1-(4-chlorophenyl)-1-bromobutane →(3. KOMe, MeOH, E2)→ 1-(4-chlorophenyl)-1-butene

6. Benzene →(1. cyclohexyl chloride, AlCl₃)→ phenylcyclohexane →(2. NBS, ROOR)→ 1-bromo-1-phenylcyclohexane →(3. NaOH, E2)→ 1-phenylcyclohexene

7. 5-methyl-5-hexenyl(?)benzene →(1. O₃; 2. Zn, HCl; 3. O₂ [O])→ 3-phenylpropanoic acid →(4. SOCl₂; 5. AlCl₃ (F-C))→ α-tetralone

12.3 Syntheses

8. 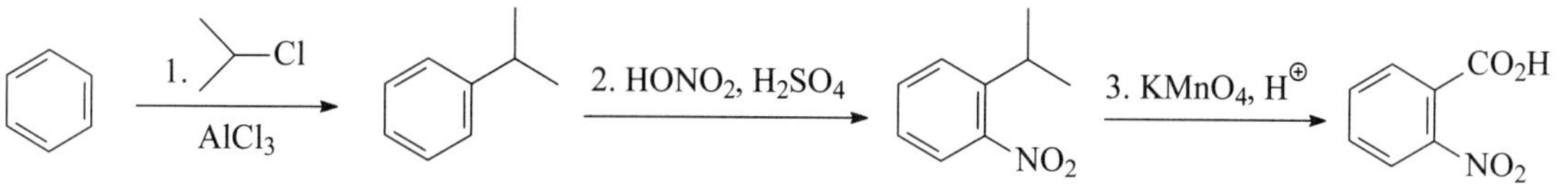
1.
Cl
AlCl3
2. HONO2, H2SO4
NO2
3. KMnO4, H⊕
CO2H
NO2

9. 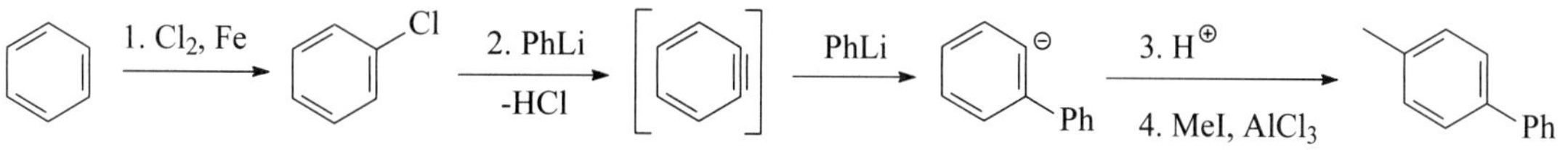
1. Cl2, Fe
Cl
2. PhLi
-HCl
PhLi
Ph
3. H⊕
4. MeI, AlCl3
Ph

10. 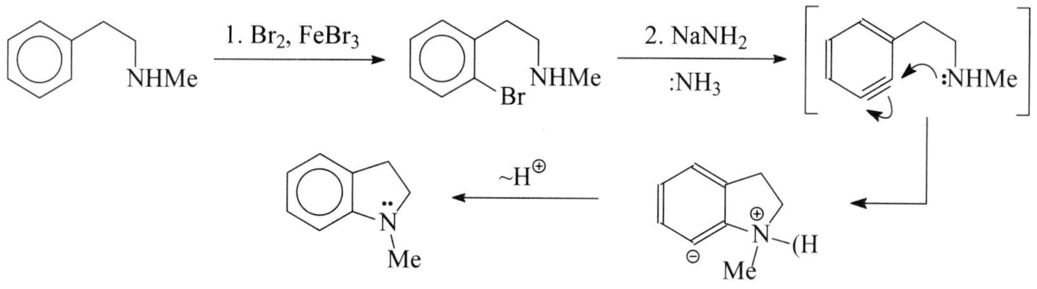
NHMe
1. Br2, FeBr3
Br
NHMe
2. NaNH2
:NH3
:NHMe
~H⊕
N
Me
N
(H
Me

11. 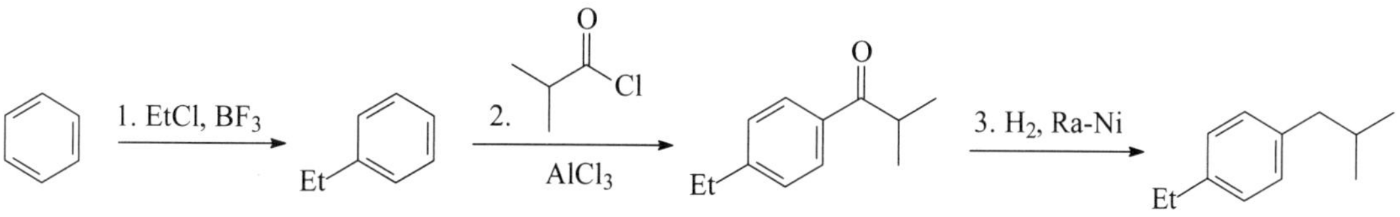
1. EtCl, BF3
Et
2.
O
Cl
AlCl3
O
Et
3. H2, Ra-Ni
Et

12. 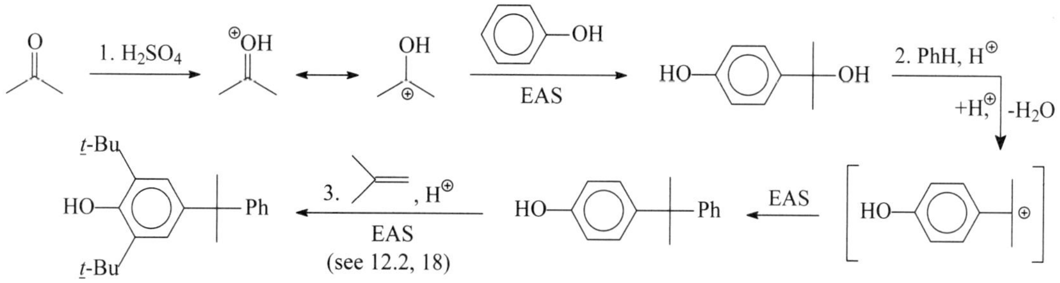
O
1. H2SO4
⊕OH
OH
OH
EAS
HO
OH
2. PhH, H⊕
+H⊕
-H2O
HO
EAS
HO
Ph
3.
, H⊕
EAS
(see 12.2, 18)
t-Bu
HO
Ph
t-Bu

13. 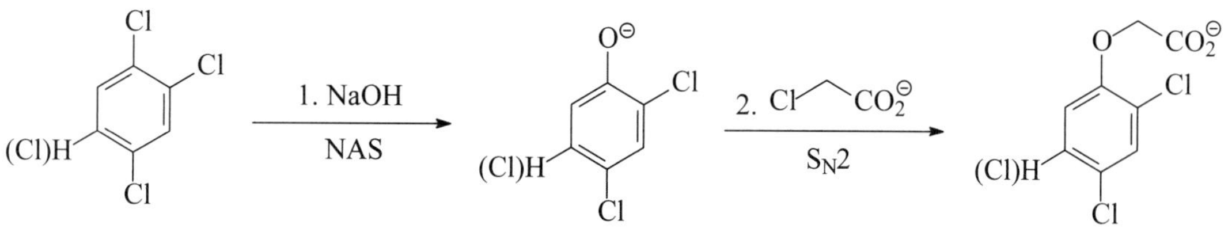
Cl
Cl
(Cl)H
Cl
1. NaOH
NAS
O⊖
Cl
(Cl)H
Cl
2. Cl CO2⊖
SN2
O
CO2⊖
Cl
(Cl)H
Cl

12.4 Mechanisms

12.4 Mechanisms

12.4 Mechanisms

15.–18. *12.4 Mechanisms*

12.4 Mechanisms

CHAPTER 13
ALCOHOLS

13.1 Reactions

1. PhCH₂C(O)CH₃ →[1. NaBH₄ / 2. NH₄Cl]→ PhCH₂CH(OH)CH₃ →[3. PCl₃]→ PhCH₂CHClCH₃ →[4. KO-*t*-Bu / Hofmann, E2]→ PhCH₂CH=CH₂

   PhCH₂CH(OH)CH₃ →[1. H₂/Pd]→ PhCH₂CH(OH)CH₃ →[2. H₂SO₄, E1]→ PhCH=CHCH₃

2. Et₃COH →[1. *i*-PrMgBr]→ Et₃CO⁻ →[2. (2-methylcyclohexyl)Br, Hofmann]→ methylenecyclohexane + Et₃COH

3. (CH₃)₂CHCH(OH)CH₂CH₃ →[1. H₂SO₄, E1]→ 2-methyl-2-pentene →[2. H₃O⁺]→ 2-methyl-2-pentanol

   ↓[HCl, S_N1, +H⁺, −H₂O]

   2° carbocation →[1,2-H: shift]→ 3° carbocation →[+Cl⁻]→ 2-chloro-2-methylpentane

4. 5-hydroxy-2-heptanone →[1. TsCl, −HCl]→ 5-tosyloxy-2-heptanone →[2. NaOAc, S_N2]→ 5-acetoxy-2-heptanone

5. 1-hexen-3-ol →[1. NaH]→ alkoxide →[2. Me−OSO₃Me, −OSO₃Me, S_N2]→ 3-methoxy-1-hexene

6. PhC(O)O-*n*-Bu →[1. PhMgCl, −*n*-BuO⁻]→ [Ph₂C=O] (not isolable) →[PhMgCl]→ Ph₃CO⁻ →[2. H⁺]→ Ph₃COH

   →[1. LiAlH₄, −*n*-BuO⁻]→ [PhCHO] (not isolable) →[LiAlH₄]→ PhCH₂O⁻ →[2. H⁺]→ PhCH₂OH

7. CH₃I →[1. Li, −LiI]→ CH₃Li →[2. (*i*-Pr)₂C=O]→ (*i*-Pr)₂C(CH₃)O⁻ →[3. H⁺]→ (*i*-Pr)₂C(CH₃)OH

13.1 Reactions

13.1 Reactions

13.1 Reactions

13.2 Syntheses

7.

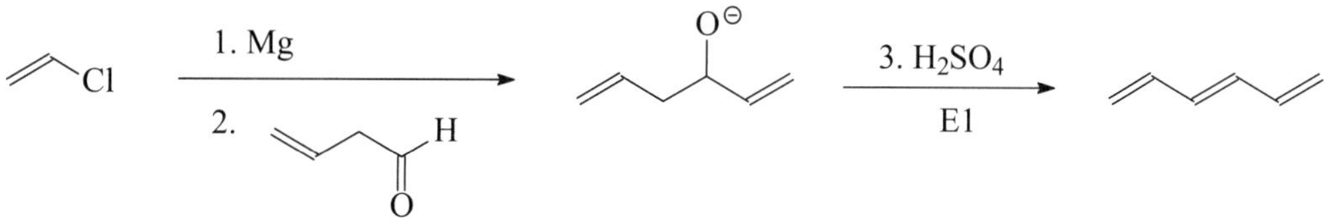

8.

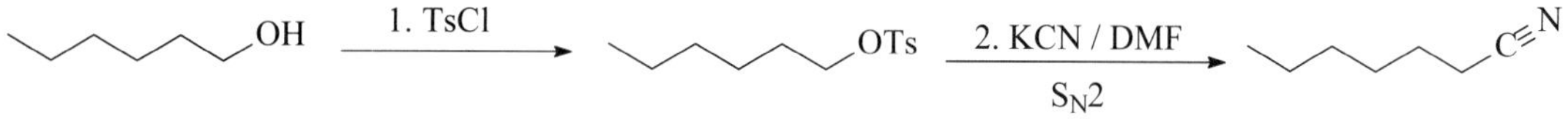

9.

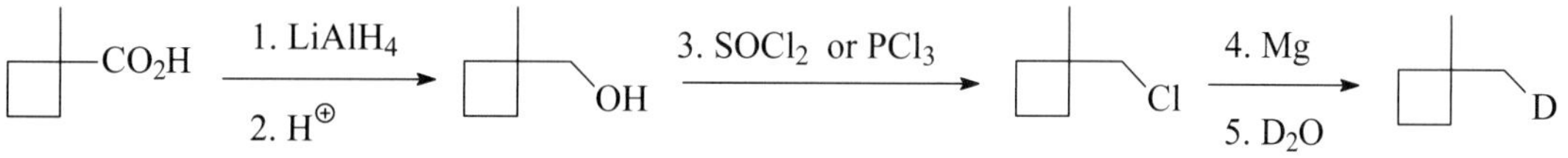

10.

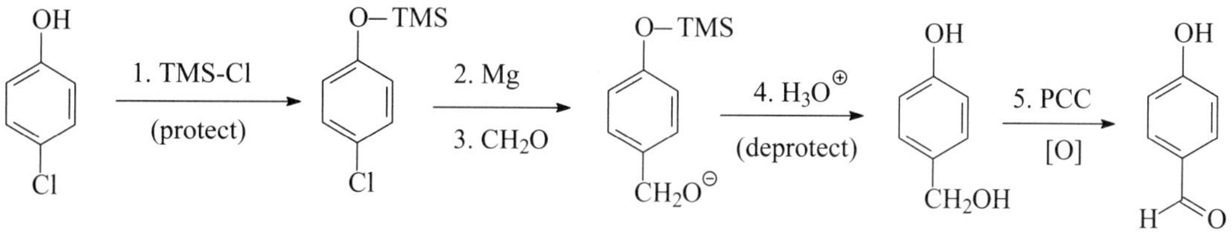

11.

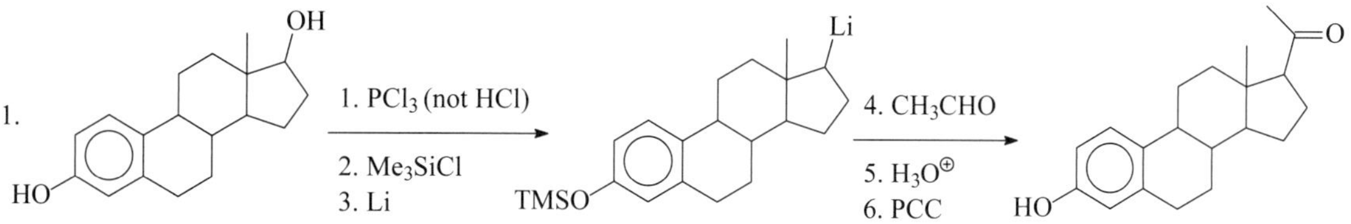

12.

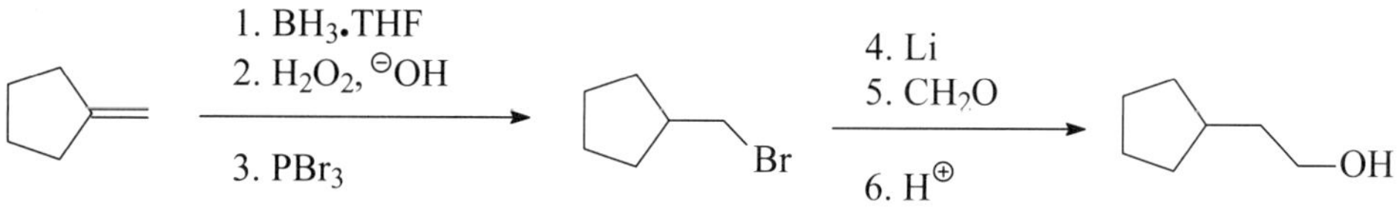

13.

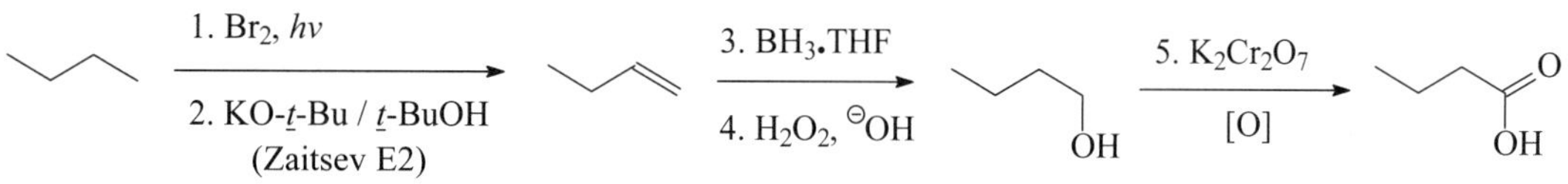

14.

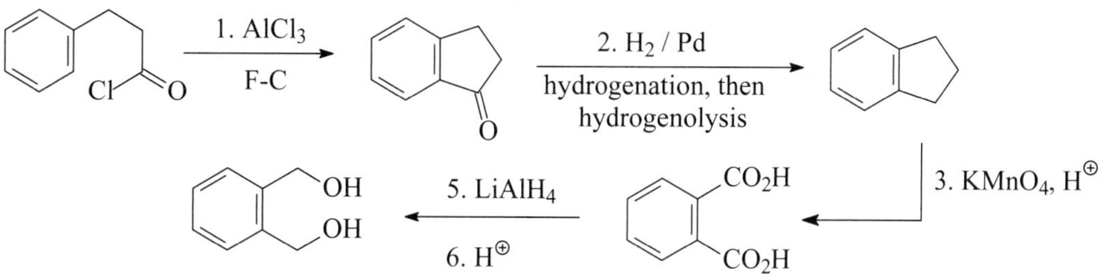

13.2 Syntheses

13.3 Mechanisms

348 • Chapter 13 Alcohols

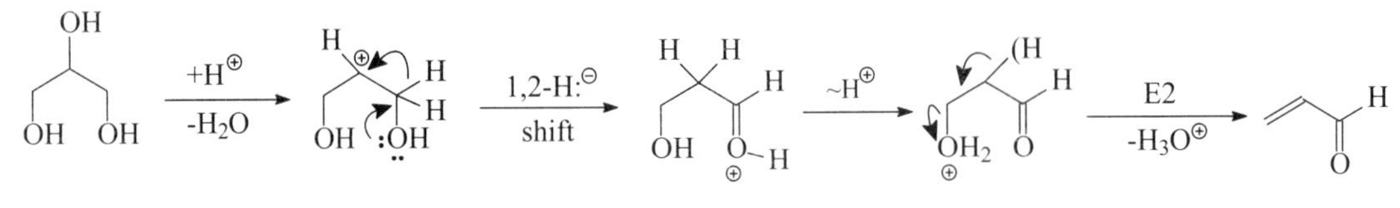

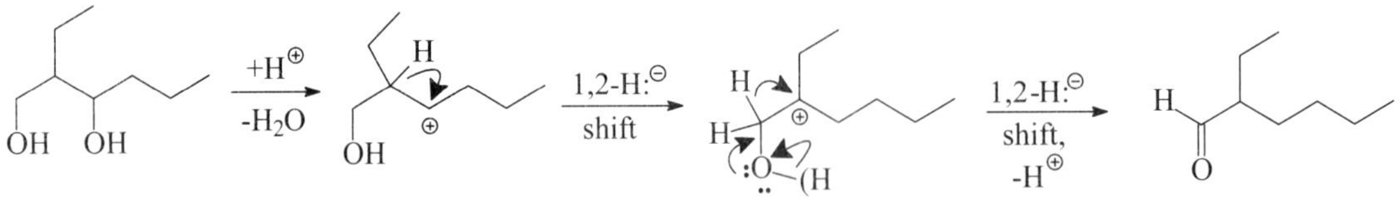

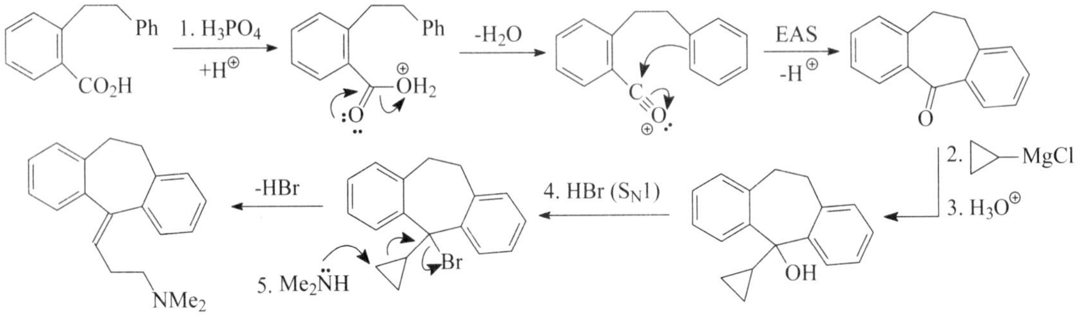

13.3 Mechanisms

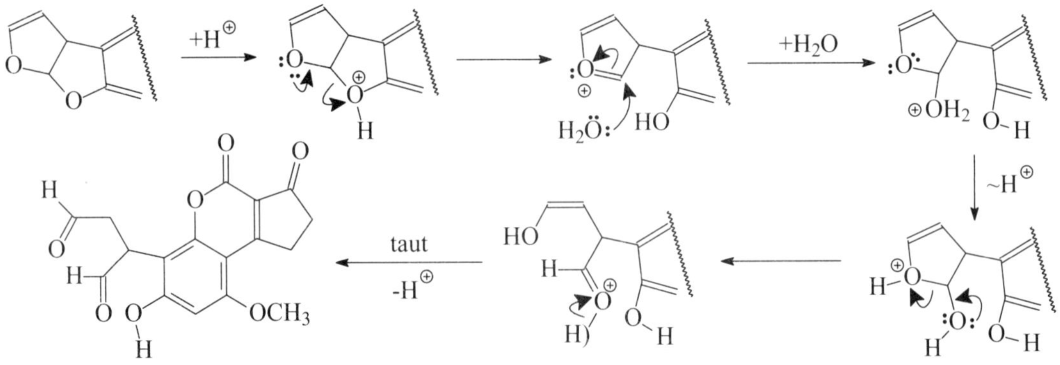

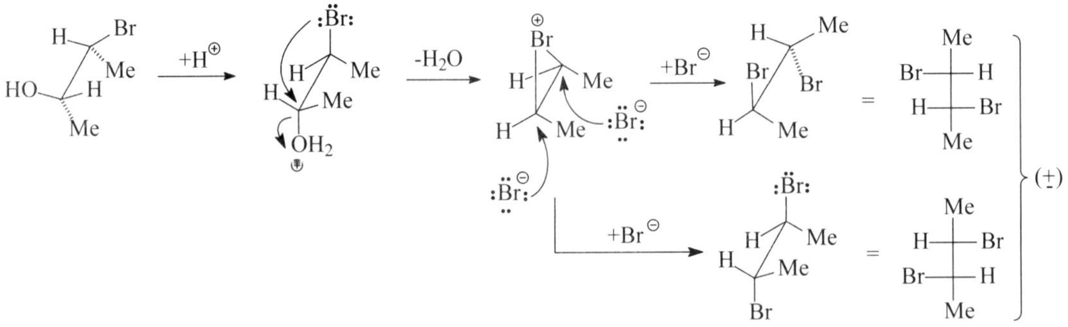

Solutions • 349

11. Top row (trans-): trans-2-bromocyclohexanol $+H^+ \rightarrow$ anti- protonated form (with :Br: and $^+OH_2$) $\xrightarrow{\text{NGP}, -H_2O}$ bromonium ion $\xrightarrow{+Br^-}$ trans-1,2-dibromocyclohexane.

    Bottom row (cis-): cis-2-bromocyclohexanol $+H^+ \rightarrow$ gauche- protonated form $\xrightarrow{S_N2, +Br^-, -H_2O}$ trans-dibromide. NGP not possible.

12. Starting material (α-hydroxy-α-methyl-β-ketoacid type) $\xrightarrow{+H^+}$ protonated ketone $\leftrightarrow$ resonance form $\xrightarrow{\text{1,2-R:}^{\ominus}\text{shift}, -H^+}$ rearranged α-hydroxy acid $\xrightarrow{\text{1. [H]}}$ diol-acid $\xrightarrow{\text{2. }-H_2O, E1}$ enol $\xrightarrow{\text{taut}}$ isobutyryl / 2-methylpropanoic acid derivative (CO_2H ketone).

13. 4-methyl-2-pentanol $\xrightarrow{+H^+, -H_2O}$ 2° carbocation $\rightleftharpoons$ (−H^+/+H^+) 4-methyl-2-pentene $\xrightarrow{+H^+/-H^+}$ 2° cation $\xrightarrow{\text{1,2-H:}^{\ominus}\text{ shift}}$ 3° cation $\xrightarrow{-H^+}$ 2-methyl-2-pentene.

    no energy benefit to 1,2-H:$^{\ominus}$ shift (crossed-out alternative pathway).

14. H_2C-CD_2 with OH, OH (pinacol substrate)

    pathway (1) - E2, $-HOH$ or DOH: gives enols $\rightarrow$ $\xrightarrow{H^+, \text{taut}}$ $HC(=O)CD_2H$ + $H_3C-C(=O)D$ *NOT formed*

    pathway (2) - pinacol, $+H^+, -H_2O$: $\xrightarrow{\text{1,2-H:}^{\cdot}\text{ or D:}^{\cdot}\text{ shift}, -H^+}$ $HC(=O)CD_2H$ + $DH_2C-C(=O)D$ *formed*

    -- therefore, pathway (2) is favored

15. a. dehydration-tautomerization: **A** $\xrightarrow{-DOH}$ enol (D on ring, OH) $\xrightarrow{H^+ \text{ taut}}$ ketone with H (NOT **B**)

    vs.

    pinacol-like: **A** $\xrightarrow{+H^+, -H_2O}$ cation with D's $\xrightarrow{\text{1,2-D:}^{\ominus}\text{ shift}}$ protonated ketone $\xrightarrow{-H^+}$ **B**

    -- therefore, the pinacol-like pathway is favored

13.3 Mechanisms

13.3 Mechanisms

CHAPTER 14
ETHERS

14.1 Reactions

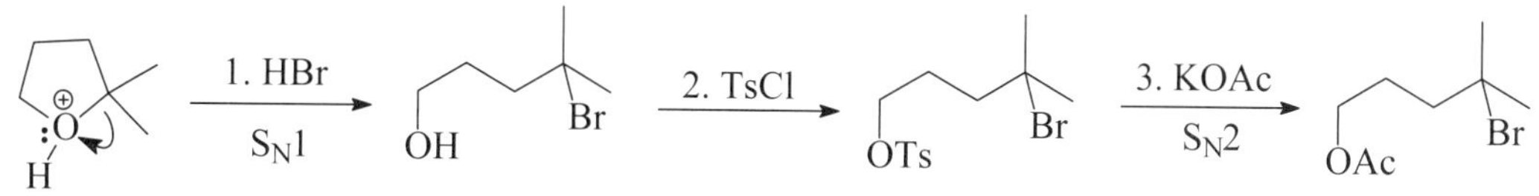

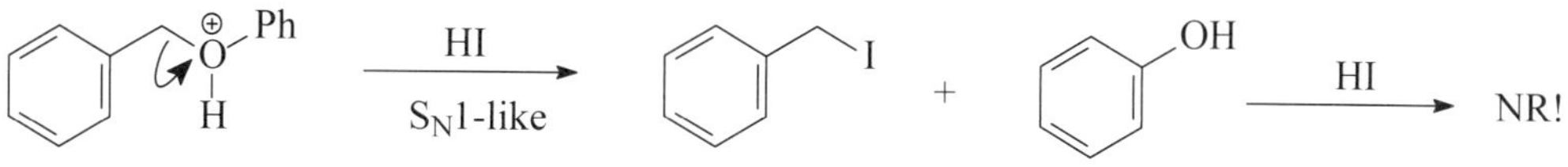

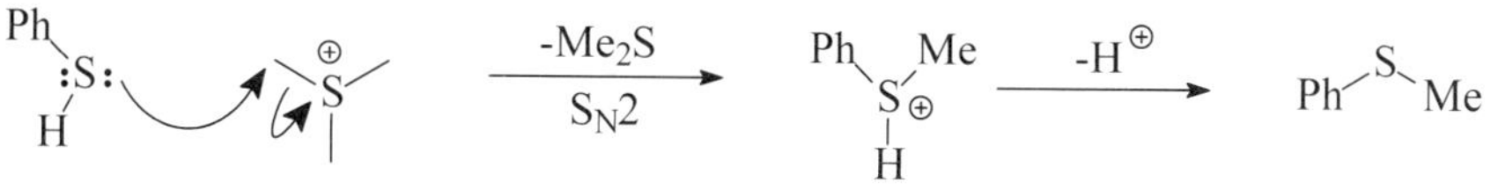

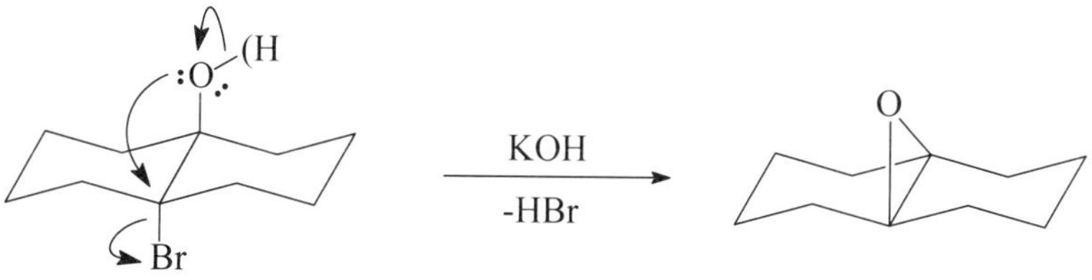

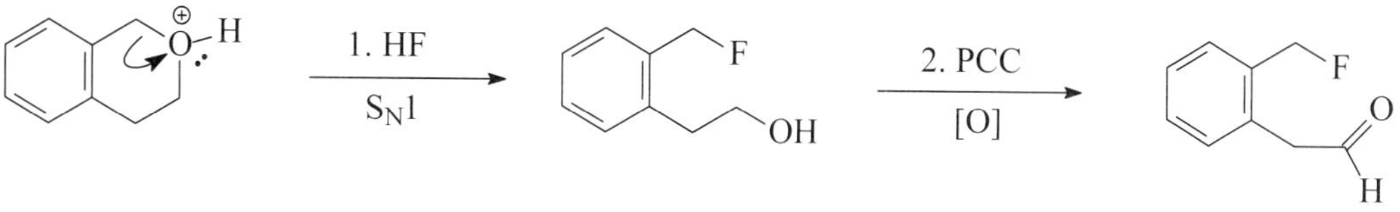

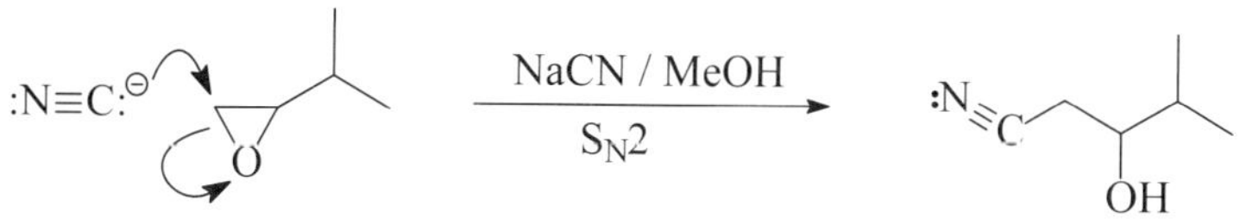

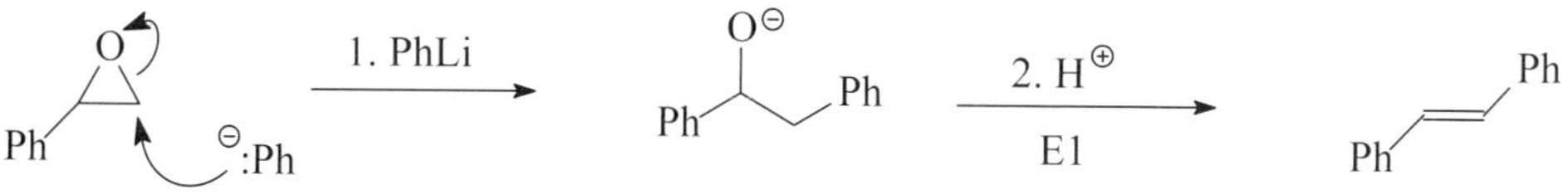

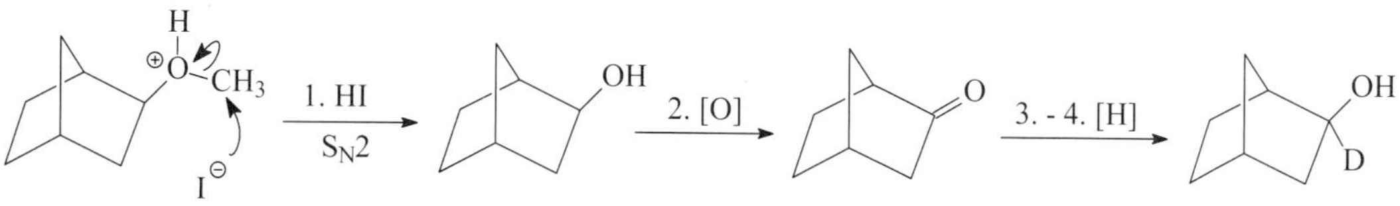

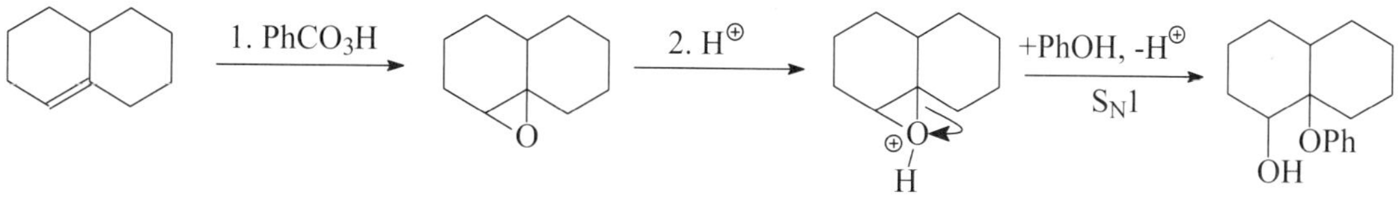

352 • Chapter 14 Ethers

10. 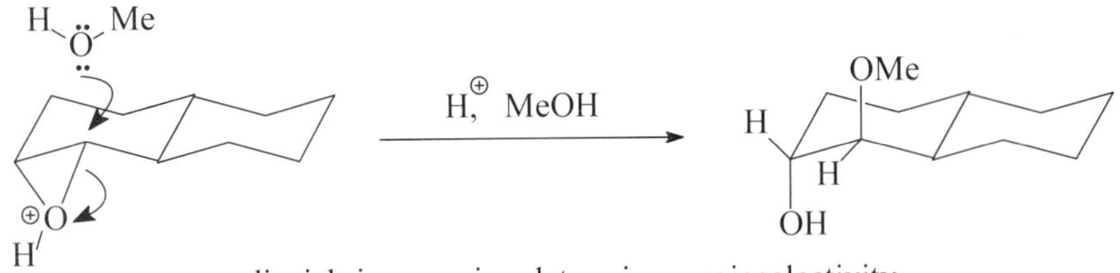 *trans*-diaxial ring opening determines regioselectivity

11.

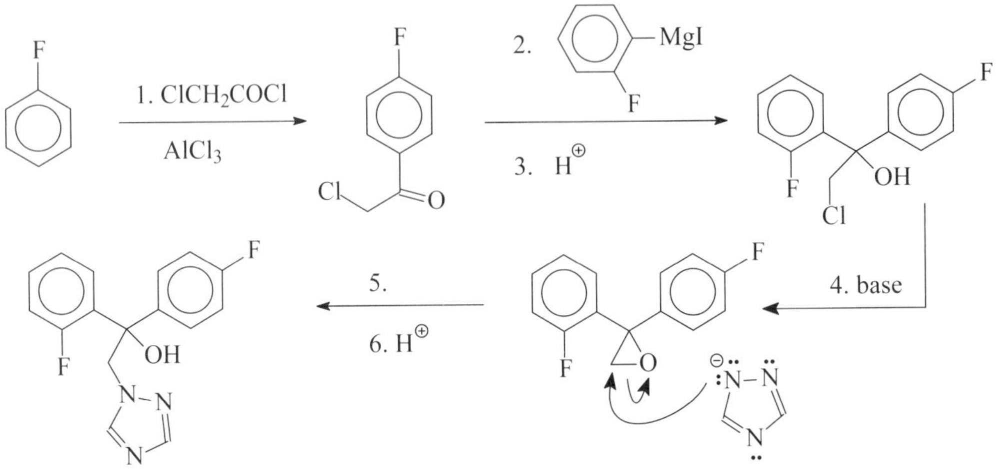

12.

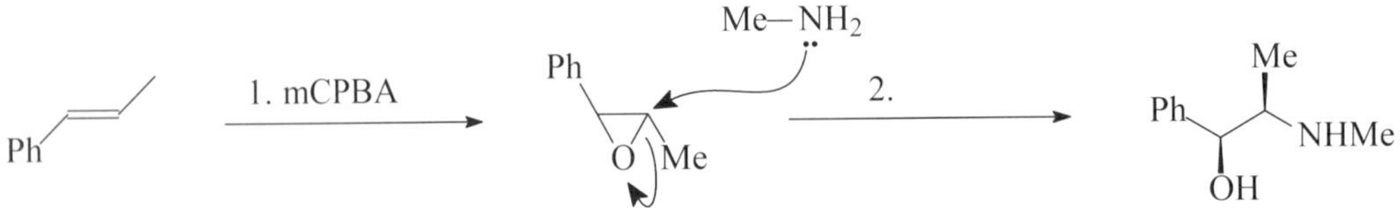

13.

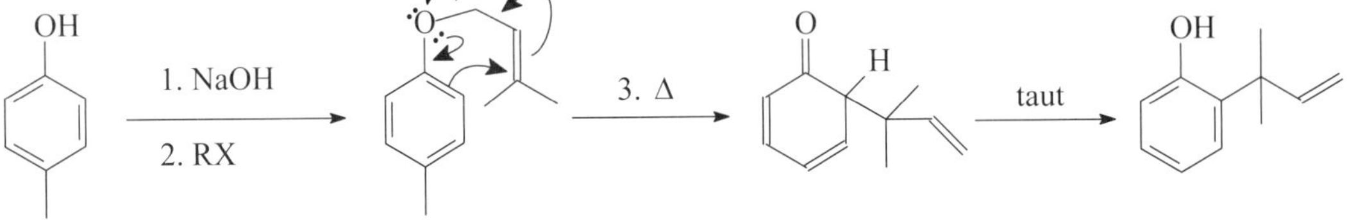

14. a.

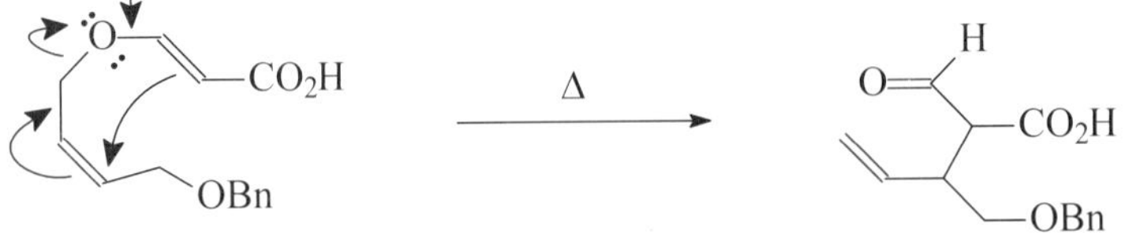

14. b.

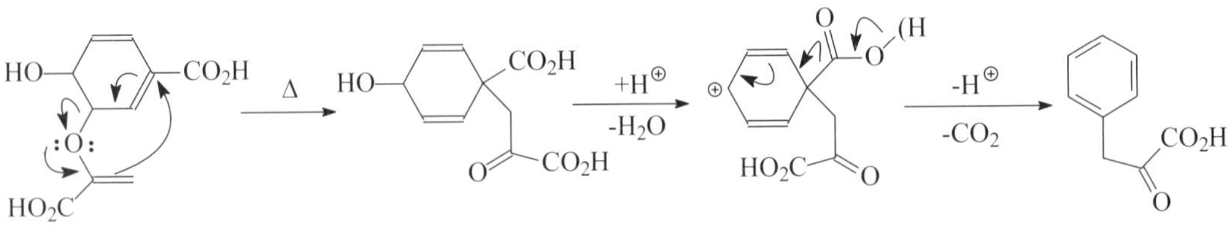

14.1 Reactions

14.2 Syntheses

14.3 Mechanisms

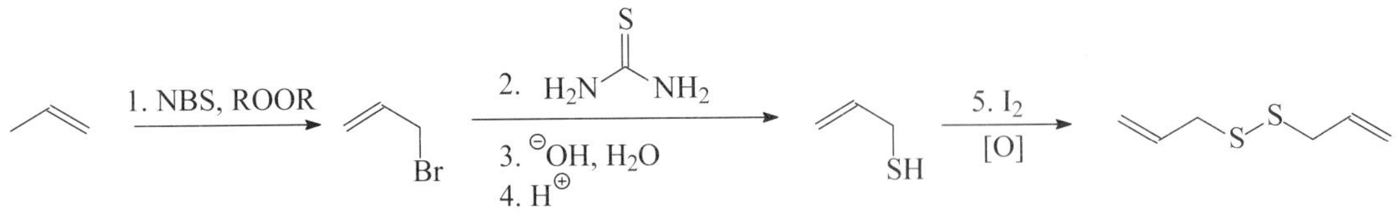

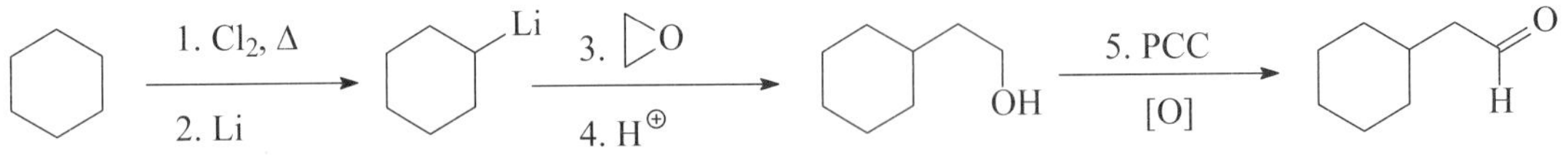

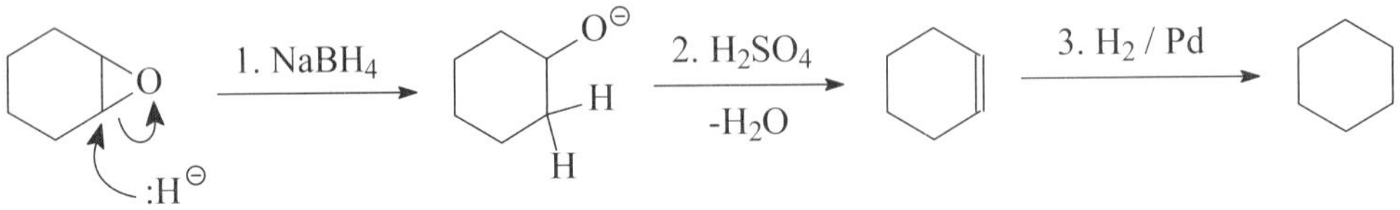

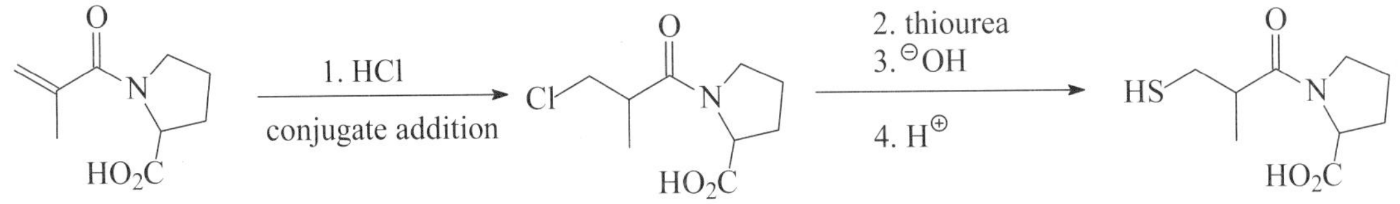

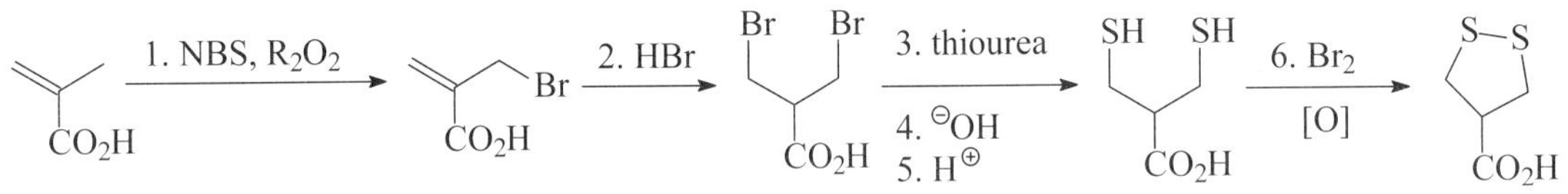

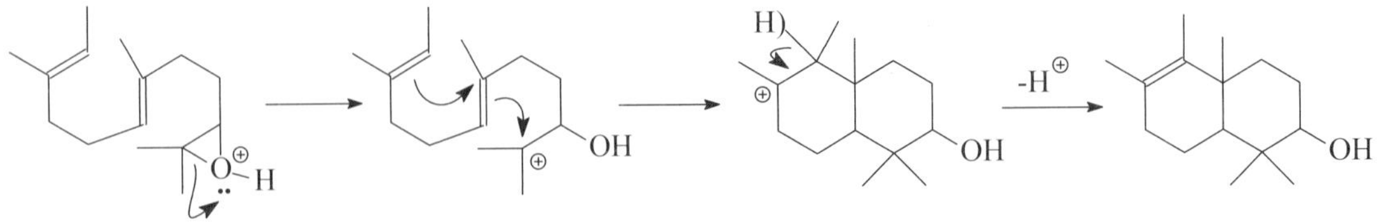

-- see 14.3, 6 for an even more impressive polycyclization!

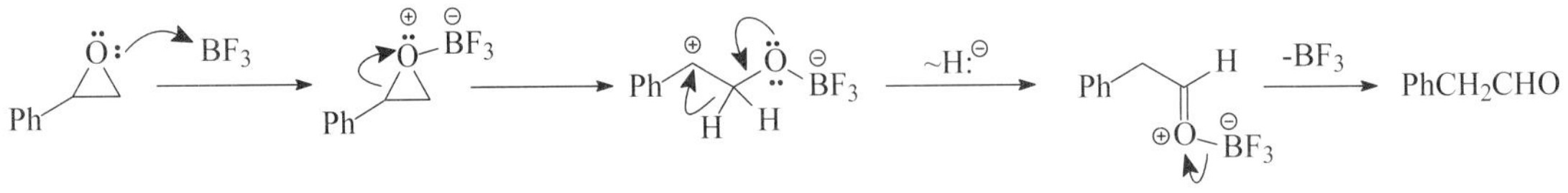

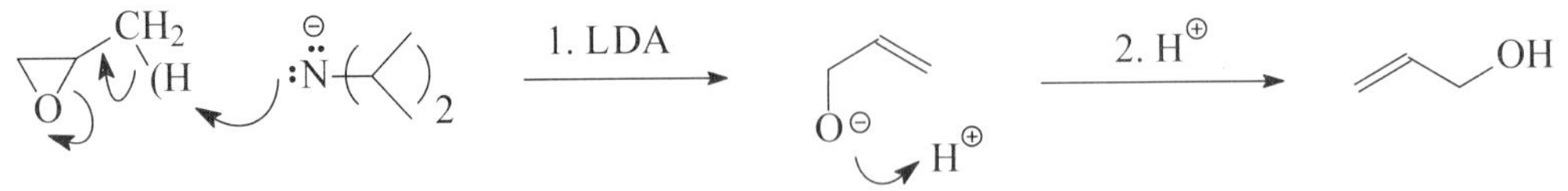

4. 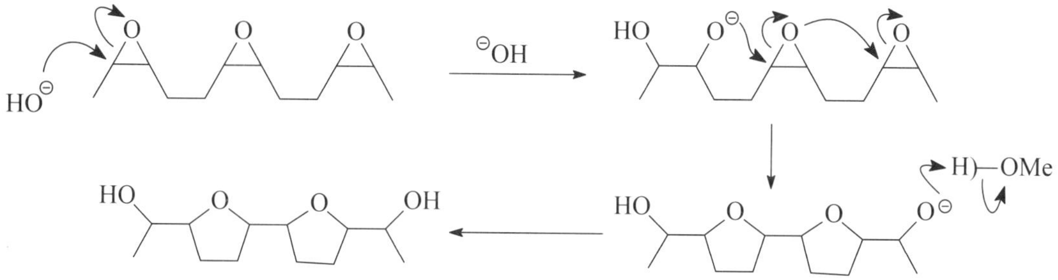

HO⊖
⊖OH
HO
O⊖
H)—OMe
O⊖
OH

5. 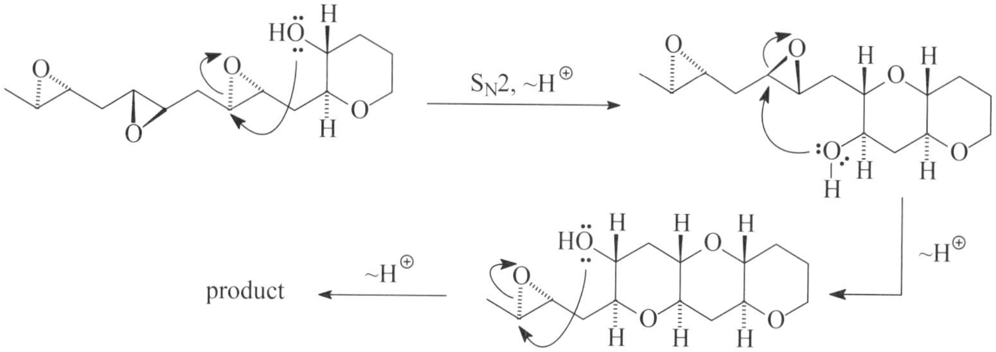

HÖ
H
O
S_N2, ~H⊕
:O
~H⊕
product

6. 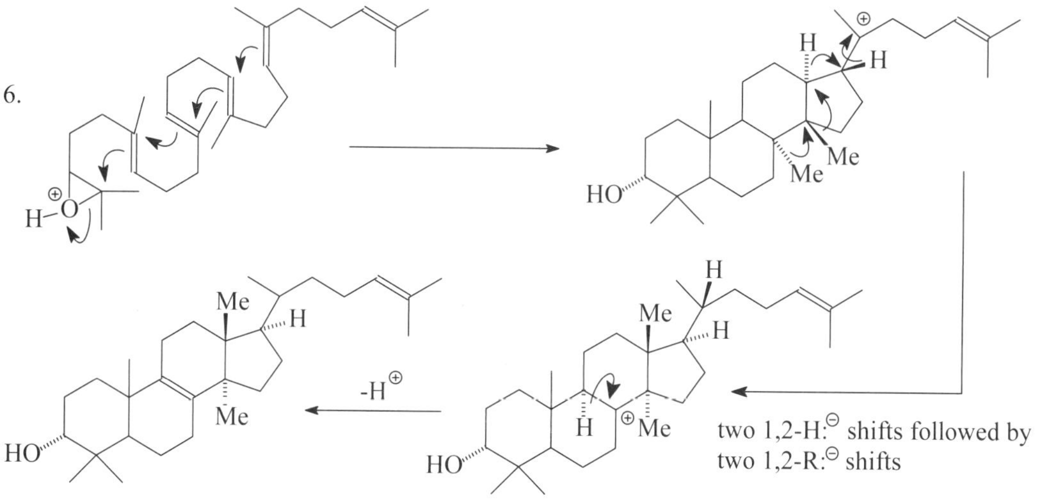

H—O⊕
HO
Me
H
two 1,2-H:⊖ shifts followed by
two 1,2-R:⊖ shifts
-H⊕

7. 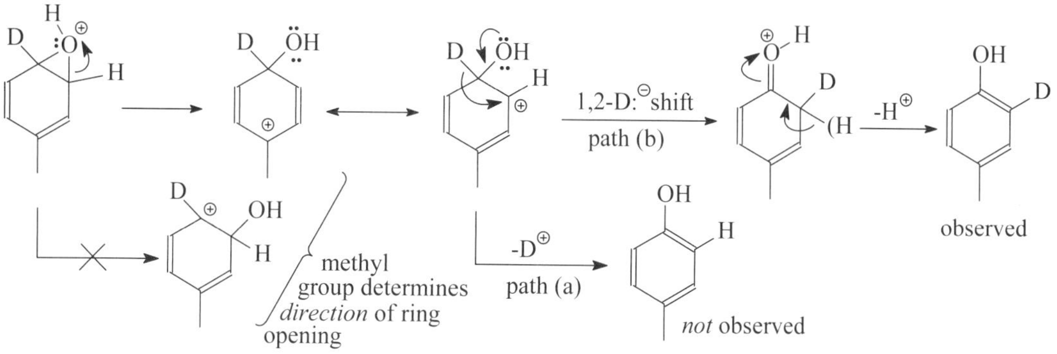

D
H
ÖH
1,2-D:⊖shift
path (b)
-H⊕
OH
observed
methyl
group determines
direction of ring
opening
-D⊕
path (a)
not observed

14.3 Mechanisms

CHAPTER 15
ALDEHYDES AND KETONES

15.1 Reactions

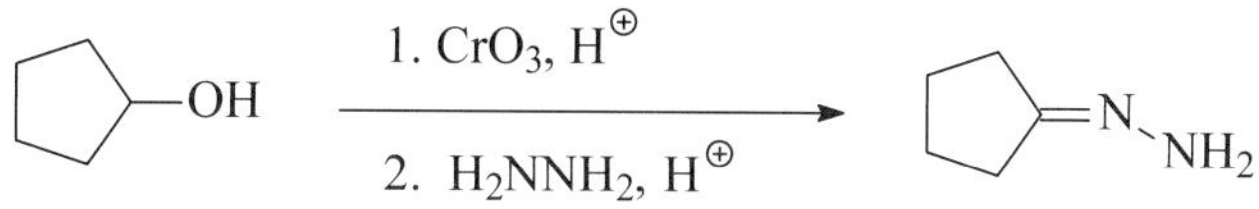

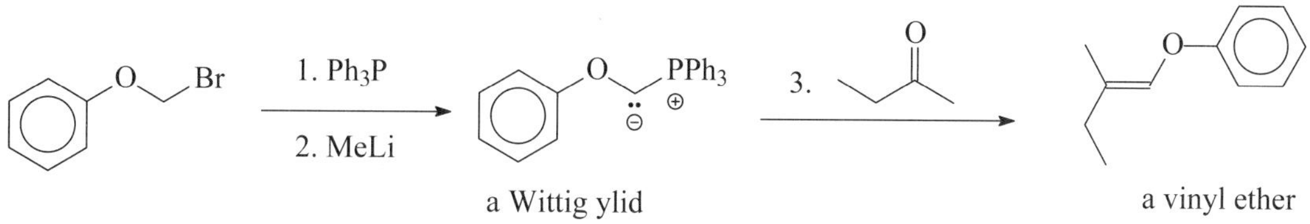

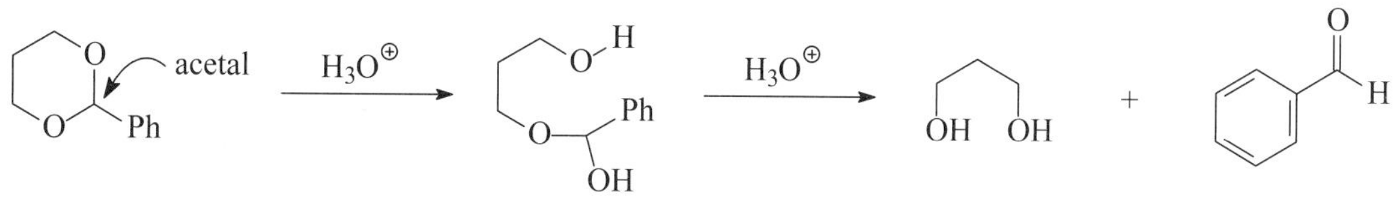

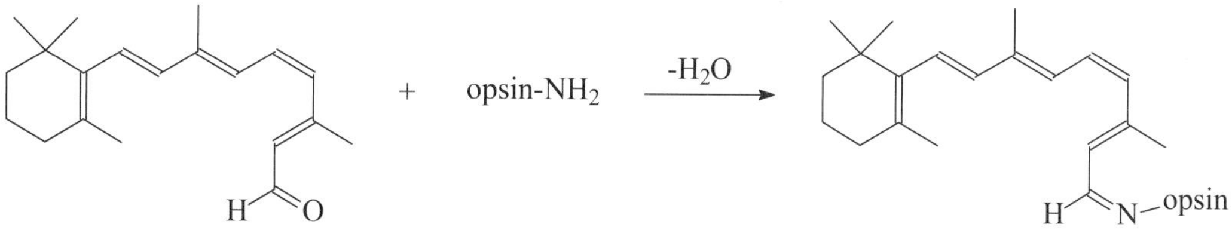

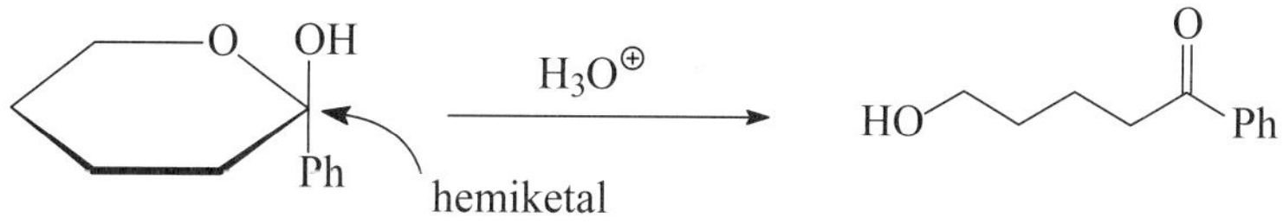

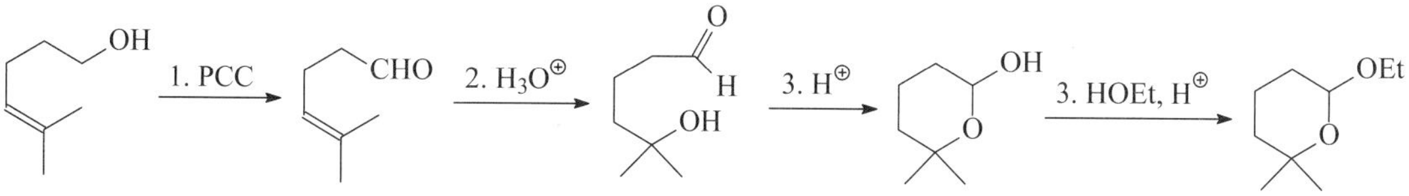

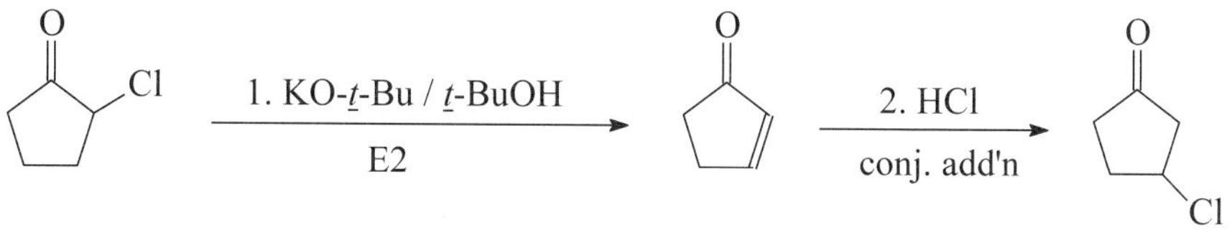

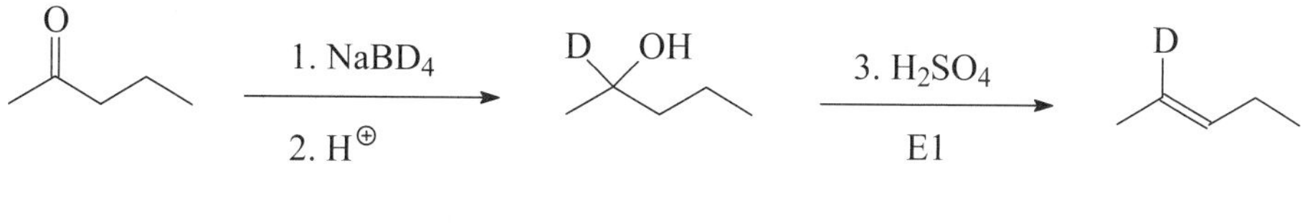

358 • Chapter 15 Aldehydes and Ketones

9. [reaction scheme: cyclohexanol-containing terpene → 1. KMnO₄ [O]; 2. H₂NNH-C(=O)-NH₂ (-H₂O) → semicarbazone intermediate → 3. H₂/Pt [H] → steroid-like product with HN-NH-C(=O)-NH₂]

10. [reaction scheme: methyl 5-oxo-4-methylpentanoate → 1. HOCH₂CH₂OH, H⁺; 2. DIBAH, -78° → aldehyde with dioxolane → 3. Ph₃P=CMe₂; 4. H₃O⁺ → 3-methyl-6-methylhept-6-enal type product]

11. O₂N-C₆H₄-CHO → 1. Ph₃P⁺–CH⁻–CH=CH–(dioxolane); 2. H₃O⁺ → O₂N-C₆H₄-CH=CH-CH=CH-CHO

12. cyclohexanone → 1. Ph₃P=CHOCH₃ → cyclohexylidene-OCH₃ → 2. H₃O⁺ (+H₂O) → cyclohexyl-CH(OH)(OCH₃) → H₃O⁺ (-MeOH) → cyclohexanecarbaldehyde

13. RO-CH₂-X → 1. Ph₃P:; 2. MeLi → Ph₃P⁺-CH⁻OR → 3. butanal → CH₃CH₂CH₂CH=CH-OR → 4. H₃O⁺ (-HOR) → pentanal

14. 2,5-dimethoxy-2,5-dihydrofuran → 1. CH₂I₂, Zn(Cu) → bicyclic dimethoxy cyclopropane-fused THF → 2. H₃O⁺ → cyclopropane-1,2-dicarbaldehyde → 3. Ph₃P=CHC=CH₂ → divinyl-cyclopropane diene product

15. cyclopropyl-CH=N-NH₂ → 1. H₃O⁺ → cyclopropanecarbaldehyde → 2. EtMgI; 3. H₃O⁺ → 1-cyclopropylpropan-1-ol

16. PhO-C(Ph)(Me)-OPh → 1. H₃O⁺ (-2 PhOH) → PhC(=O)Me → 2. H₂NOH (-H₂O) → Ph-C(Me)=N-OH

15.1 Reactions

17.

18.

19.

20.

21.

22.

23.

24.

15.1 Reactions

360 • Chapter 15 Aldehydes and Ketones

25. [structure: 4-hydroxy-3-methoxybenzaldehyde] →(1. D$_2$NND$_2$, $^{\ominus}$OD, D$_2$O / W-K)→ [CHD$_2$ methyl, on ring with OMe, OH] →(2. HI, S$_N$2-like)→ [CHD$_2$ arene with two OH] + MeI

26. [glucuronic acid with O-CH(CN)Ph acetal] →(H$_3$O$^{\oplus}$)→ [glucuronic acid hemiacetal OH] + [HO-C(CN)(Ph)H] → HCN + PhCHO

27. [bicyclic acetal] →(H$_3$O$^{\oplus}$)→ [partial hydrolysis, HO, OH] →(H$_3$O$^{\oplus}$)→ [HOCH$_2$–C(OH)(Me)–CH$_2$CH$_2$–C(O)Me]

28. [CH$_3$COCH$_2$CHO] →(MeOH, H$^{\oplus}$)→ [1715 cm^{-1} C=O, CH$_3$COCH$_2$CH(OMe)(OEt)]
    δ 2.2 (H), 2.8 (H), 4.9 (H), 3.4 (H)

29. [bicyclic methyl acetal structure with OMe] →(H$_3$O$^{\oplus}$)→ [HO–CH$_2$CH$_2$CH$_2$–C(OH)(Me)–O–C(OMe)(Me)–] → [open chain triol with OMe] →(−H$_2$O, −HOMe)→ [HO-CH$_2$CH$_2$CH$_2$–CH(COMe)–CH$_2$–COMe diketone]

30. [safrole: methylenedioxybenzene with allyl] →(H$_3$O$^{\oplus}$)→ [catechol with allyl, two OH] + HCHO

31. [H(O=)C–CH$_2$CH$_2$–CH(NH$_2$)–CO$_2$H] →(1. H$^{\oplus}$, −H$_2$O)→ [1-pyrroline-2-carboxylic acid] →(2. H$_2$/Pt)→ [proline]

32. [21-acetoxy-17-hydroxy-3,3-dimethoxy-pregn-4-en-20-one steroid] →(1. LiAlH$_4$; 2. H$_3$O$^{\oplus}$)→ [17,20,21-triol-4-en-3-one steroid] + EtOH + MeOH

15.1 Reactions

33. [Mechanism: glyoxal + ⁻OH → tetrahedral intermediate → 1,2-H:⁻ shift / intra-Cannizzaro → glycolate intermediate → ~H⁺ → glycolate anion]

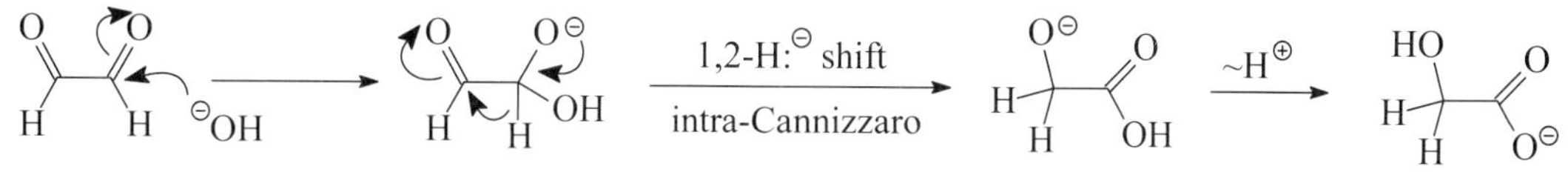

34. [1,1-diacetoxycyclohexane] $\xrightarrow{\text{1. LiAlH}_4 \quad \text{2. H}_3\text{O}^\oplus}$ EtOH + [cyclohexane-1,1-diol] $\xrightarrow{-H_2O}$ cyclohexanone

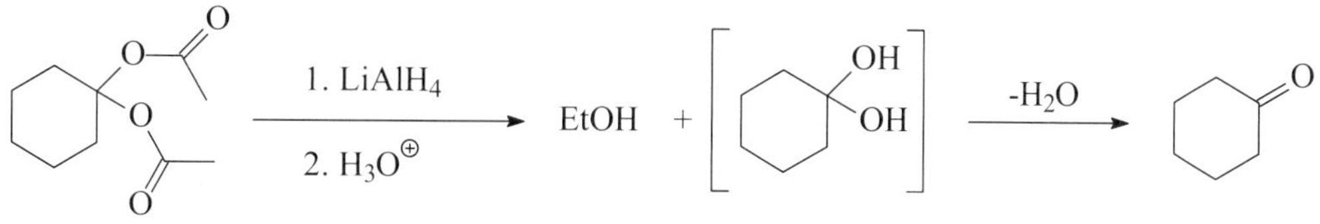

35. [steroid acetonide] $\xrightarrow{H_3O^\oplus}$ [steroid diol] + acetone

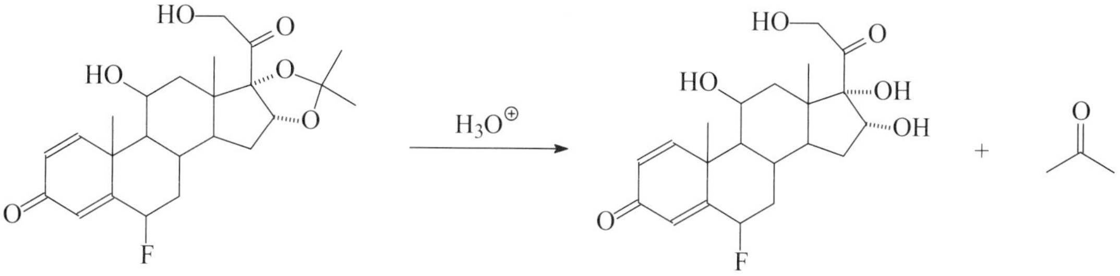

36. cycloheptane-1,2-dione $\xrightarrow{H_2NOH}$ bis-oxime

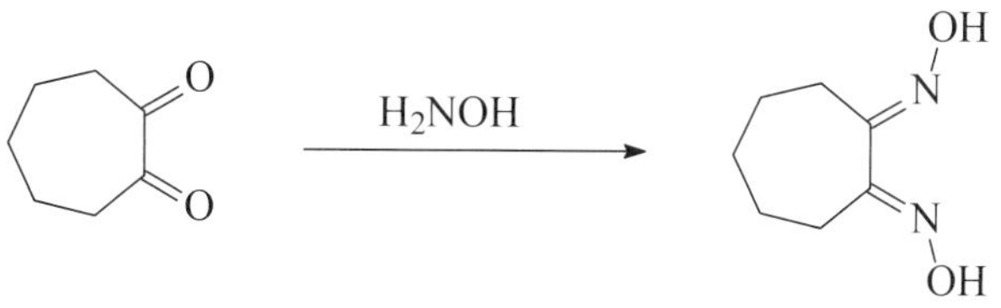

37. [polyoxymethylene acetal chain] $\xrightarrow{H_3O^\oplus}$ n HCHO

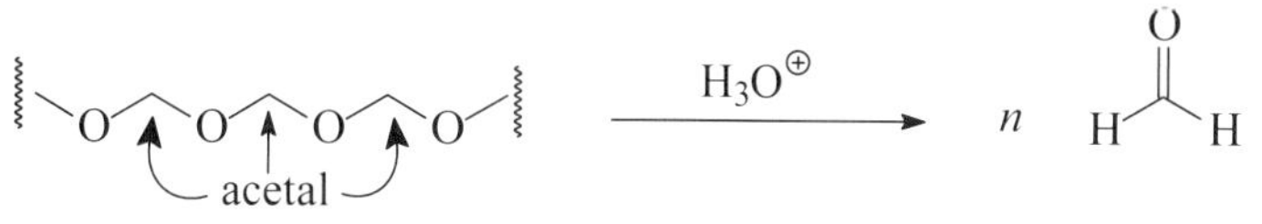

38. a hexose $\xrightarrow{\text{1. NH}_2\text{OH, H}^\oplus}$ oxime $\xrightarrow{\text{2. Ac}_2\text{O}, -H_2O}$ an unstable cyanohydrin $\xrightarrow{-HCN}$ a pentose

15.1 Reactions

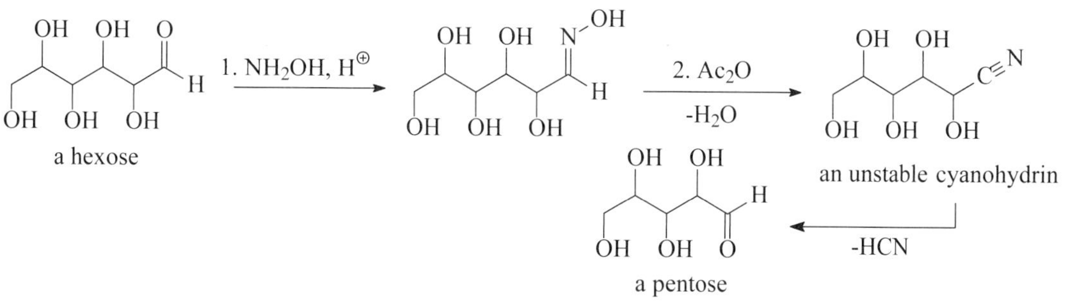

362 • **Chapter 15** Aldehydes and Ketones

39. 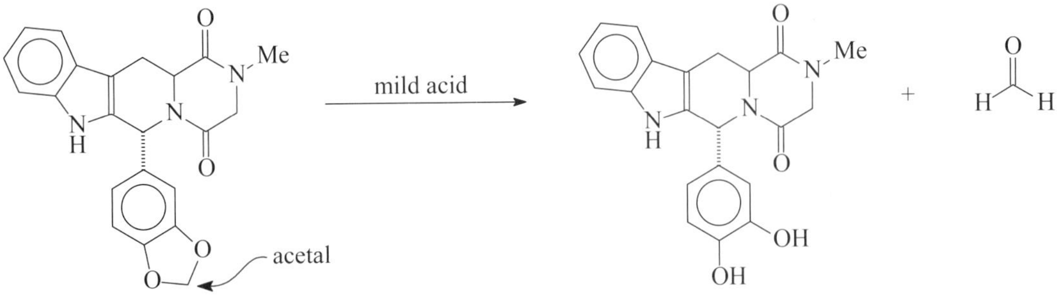 [Structure with acetal group] → mild acid → [diol product] + HCHO

40. 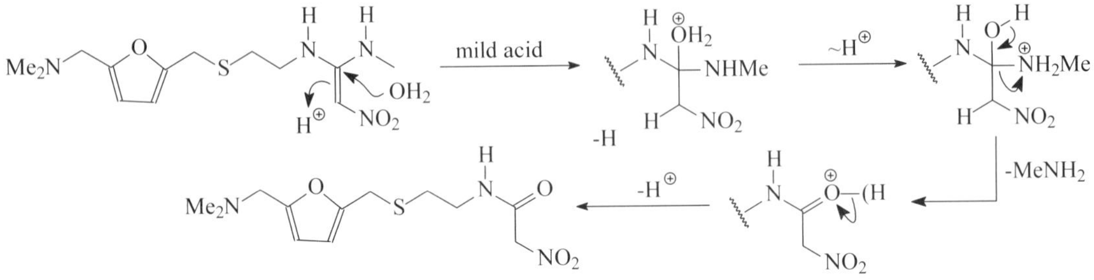 [Furan-containing structure with NO₂ group] → mild acid → [tetrahedral intermediate with OH₂⁺/NHMe] → ~H⁺ → [intermediate with OH/NH₂Me⁺] → -MeNH₂ → [protonated amide] → -H⁺ → Me₂N-furan-CH₂-S-CH₂CH₂-NH-C(=O)-CH₂-NO₂

41. 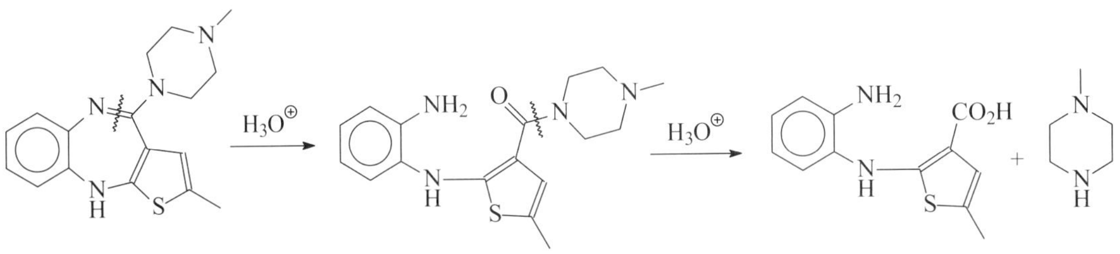 [Benzodiazepine with N-methylpiperazine] → H₃O⁺ → [amide intermediate with NH₂ and piperazine] → H₃O⁺ → [carboxylic acid product] + N-methylpiperazine

42. 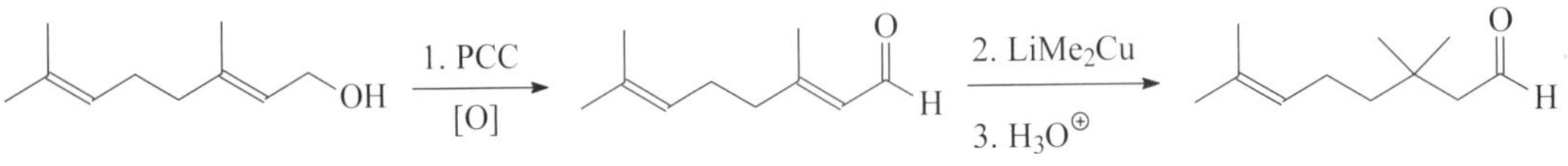 [Geraniol] →(1. PCC [O])→ [geranial, α,β-unsaturated aldehyde] →(2. LiMe₂Cu, 3. H₃O⁺)→ [saturated aldehyde with gem-dimethyl]

43. 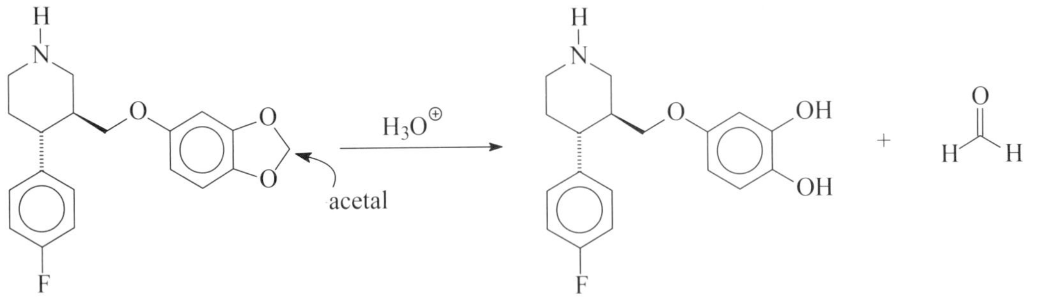 [Piperidine-methylene-dioxyphenyl structure with 4-fluorophenyl group, acetal] → H₃O⁺ → [catechol product] + HCHO

44. 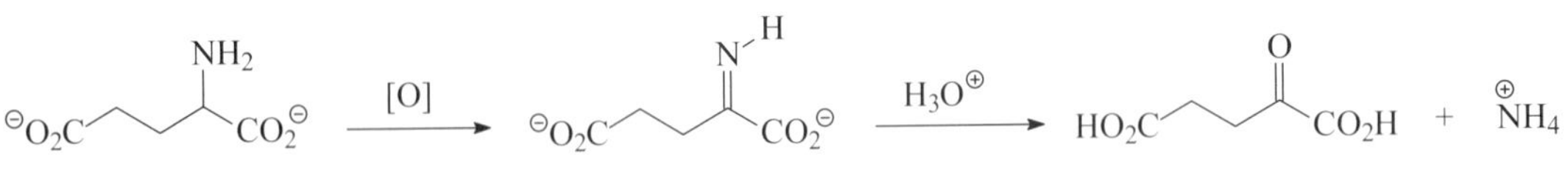 ⁻O₂C-CH(NH₂)-CH₂-CO₂⁻ →[O]→ ⁻O₂C-C(=NH)-CH₂-CO₂⁻ →H₃O⁺→ HO₂C-C(=O)-CH₂-CO₂H + NH₄⁺

15.1 Reactions

45.

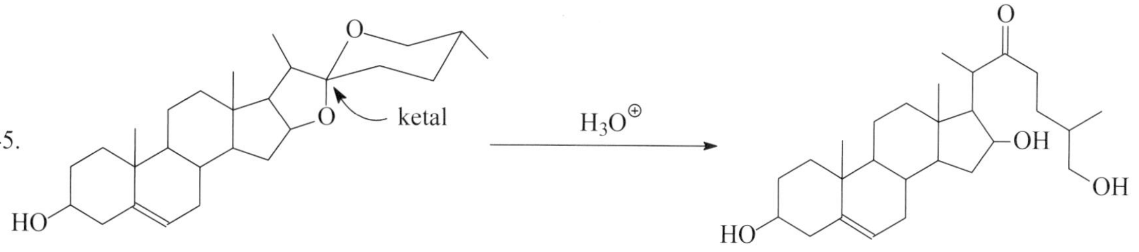

46.

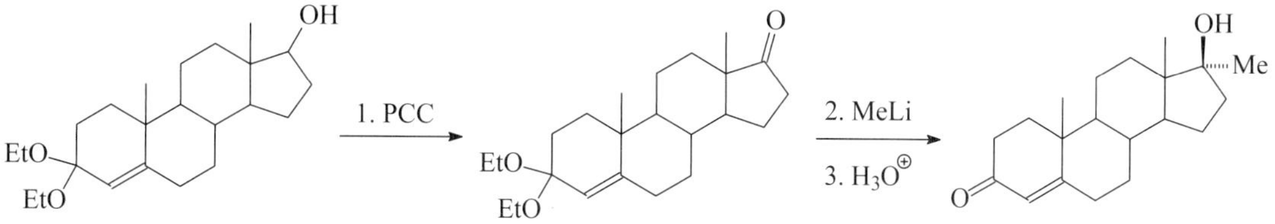

47.

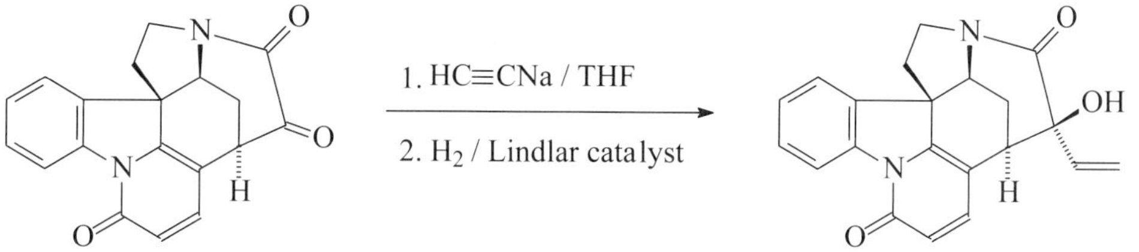

note: addition to ketone, not amide carbonyls

48. a.

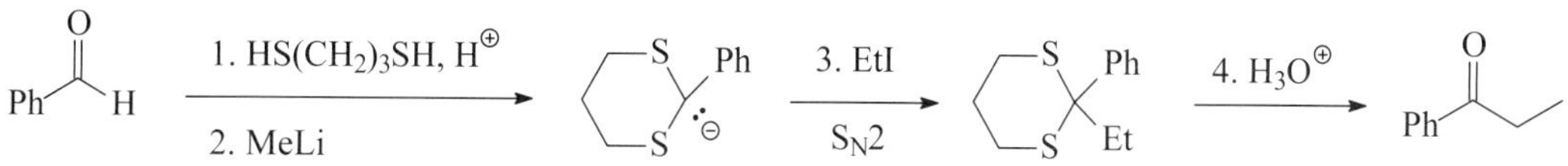

b.

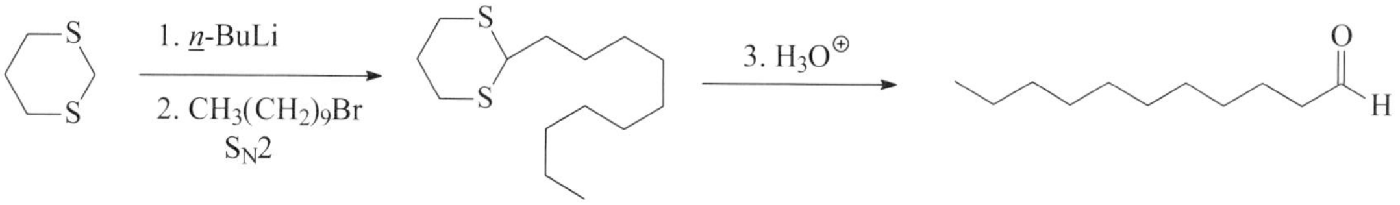

c.

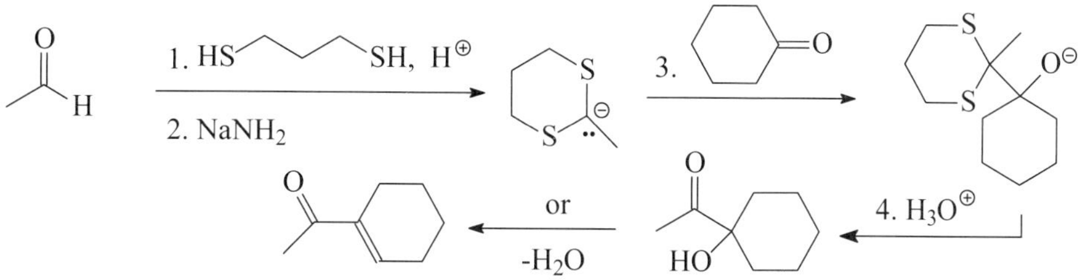

15.1 Reactions

364 • Chapter 15 Aldehydes and Ketones

49. 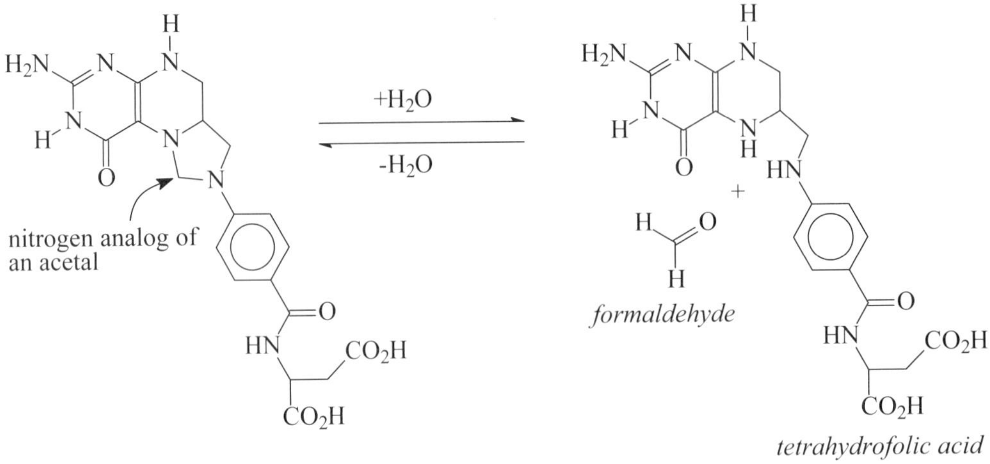 [structure: nitrogen analog of an acetal] ⇌ (+H₂O / −H₂O) tetrahydrofolic acid + formaldehyde

50. 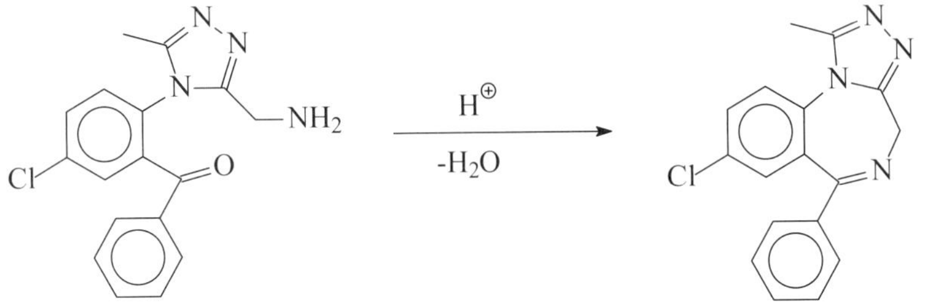 [structure with NH₂, Cl, ketone, triazole] →(H⁺, −H₂O) Xanax™ - (anxiolytic)

15.2 Syntheses

1. 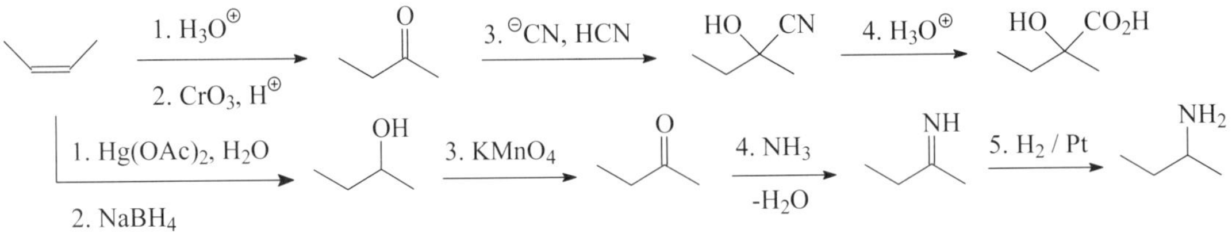 [propene] →(1. H₃O⁺; 2. 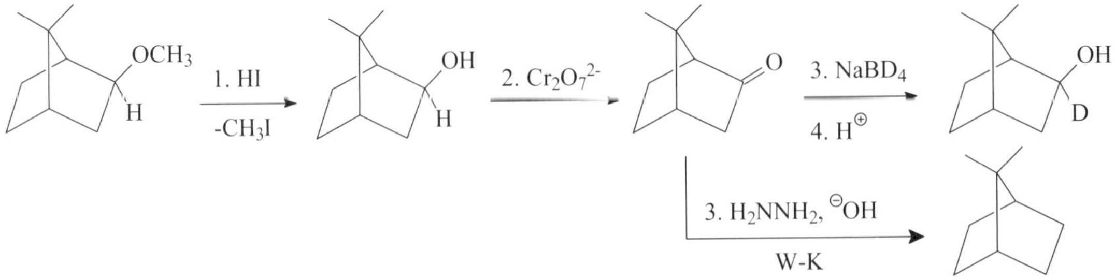 CrO₃, H⁺) [ketone] →(3. ⁻CN, HCN) [cyanohydrin] →(4. H₃O⁺) [α-hydroxy acid]

→(1. Hg(OAc)₂, H₂O; 2. NaBH₄) [2-butanol] →(3. KMnO₄) [2-butanone] →(4. NH₃, −H₂O) [imine] →(5. H₂/Pt) [sec-butylamine]

2. [bornyl methyl ether] →(1. HI, −CH₃I) [bornanol] →(2. Cr₂O₇²⁻) [camphor] →(3. NaBD₄; 4. H⁺) [bornanol-D]

→(3. H₂NNH₂, ⁻OH, W-K) [bornane]

15.2 Syntheses

15.2 Syntheses

366 • Chapter 15 Aldehydes and Ketones

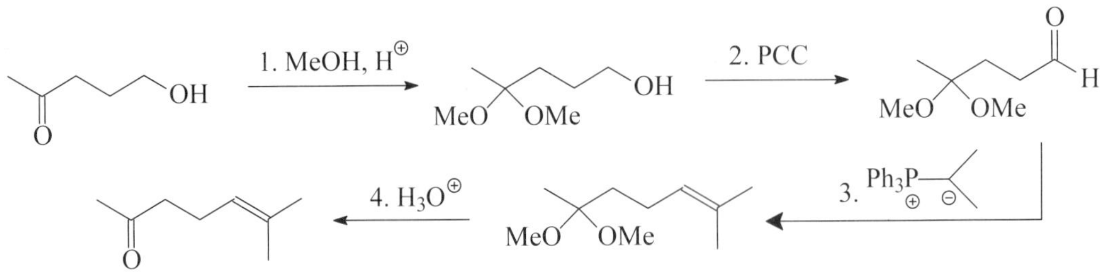

10. (reaction scheme)

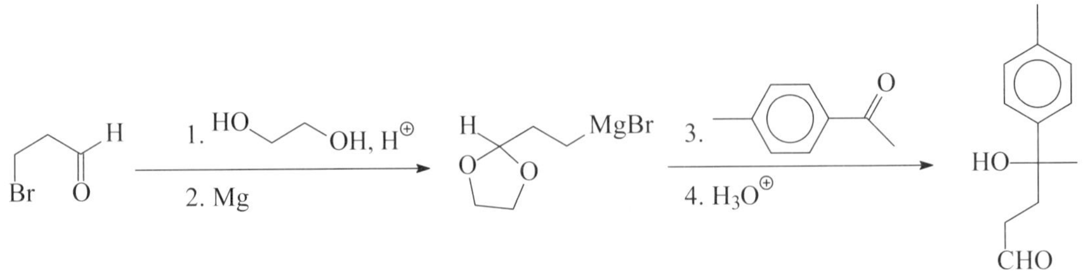

11. (reaction scheme)

12. (reaction scheme)

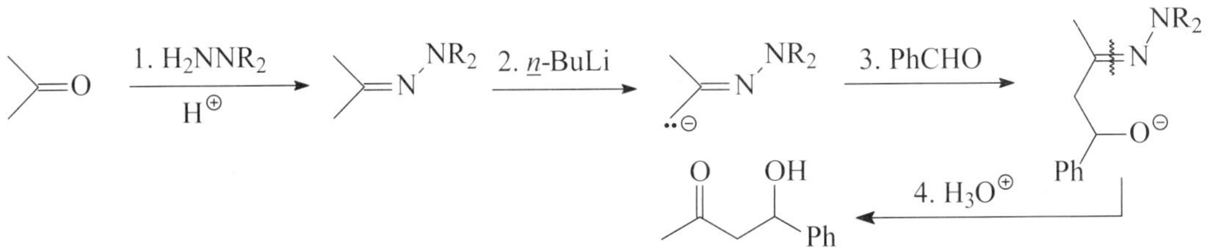

13. (reaction scheme)

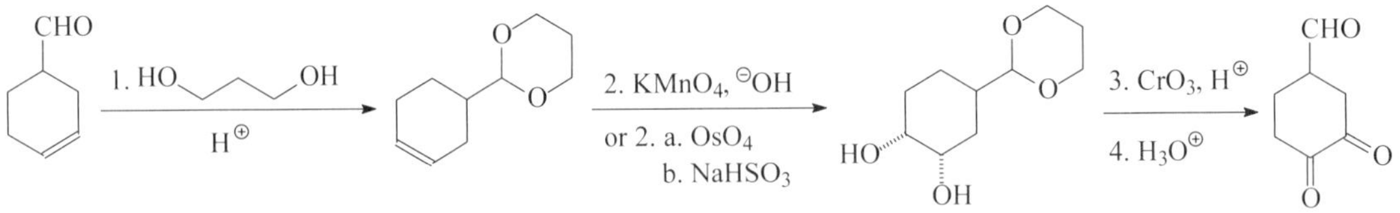

14. (reaction scheme)

15. (reaction scheme)

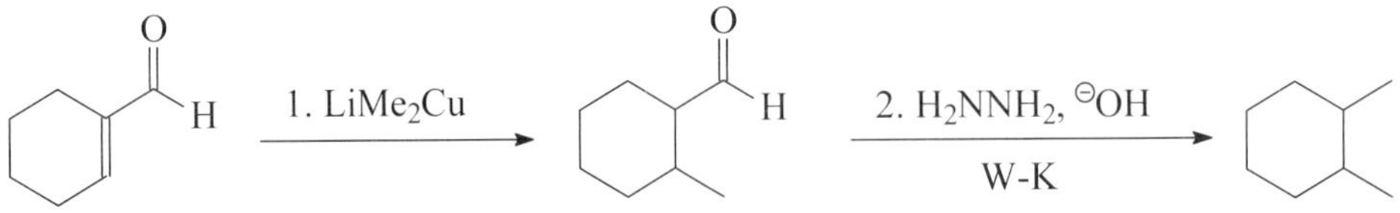

15.2 Syntheses

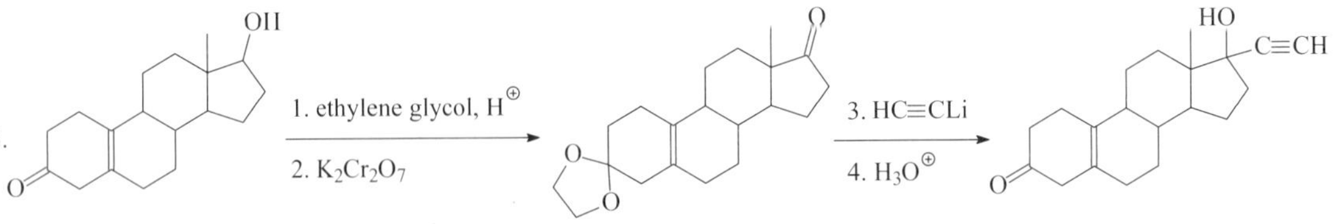

15.2 Syntheses

15.3 Mechanisms

15.3 Mechanisms

15.3 Mechanisms

17.

18.

19.

20. a.

b.

15.3 Mechanisms

15.3 Mechanisms

15.3 Mechanisms

15.3 Mechanisms

34. – 38. *15.3 Mechanisms*

15.3 Mechanisms

43. *(cont.)* / 44. / 45. / 46.

15.3 Mechanisms

15.3 Mechanisms

CHAPTER 16
CARBOXYLIC ACIDS

16.1 Reactions

1. 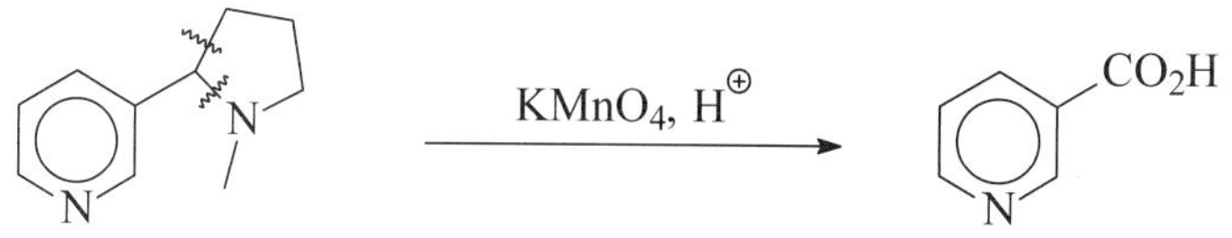 Nicotine + KMnO₄, H⁺ → nicotinic acid (pyridine-3-carboxylic acid)

2. PhCH₂CO₂H →(1. ⁻OH) PhCH₂CO₂⁻ →(Me₃O⁺ BF₄⁻, 2. −Me₂O, S_N2) PhCH₂CO₂Me

3. Acetoacetic acid (CH₃COCH₂CO₂H) →(1. NaBH₄, 2. H⁺) 3-hydroxybutanoic acid (CH₃CH(OH)CH₂CO₂H)

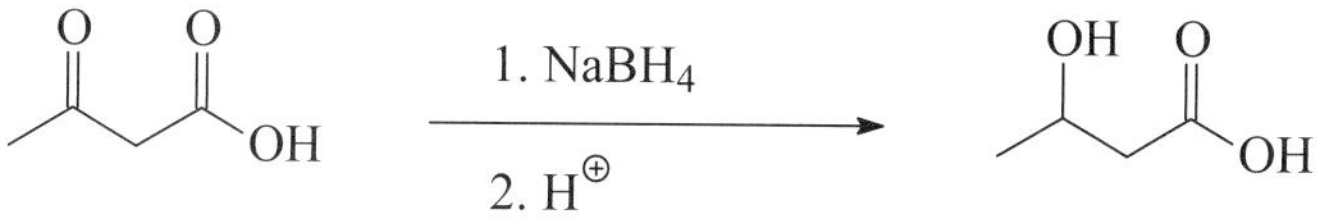

4. 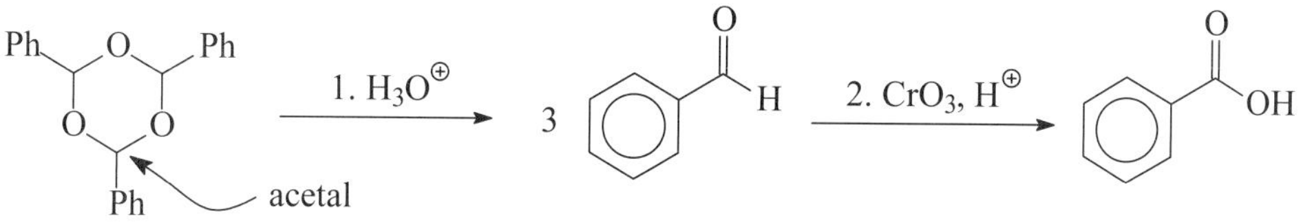 2,4,6-triphenyl-1,3,5-trioxane (acetal) →(1. H₃O⁺) 3 PhCHO →(2. CrO₃, H⁺) PhCO₂H

5. γ-bromo-β-carbon butanoic acid (4-bromobutanoic acid) →(1. ⁻OH) carboxylate →(2. Δ, S_N2) γ-butyrolactone →(3. LiAlH₄, 4. H₃O⁺) 1,4-butanediol

6. 2-naphthol →(1. ⁻OH; 2. CH₂=CHCO₂H, conj. add'n) 3-(2-naphthyloxy)propanoic acid →(3. BH₃; 4. H₃O⁺) 3-(2-naphthyloxy)propan-1-ol

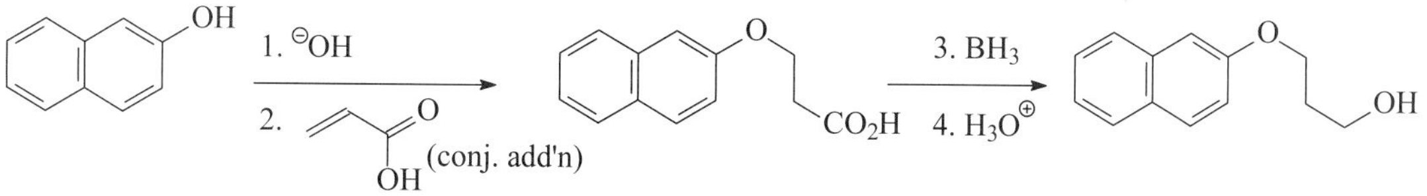

7. 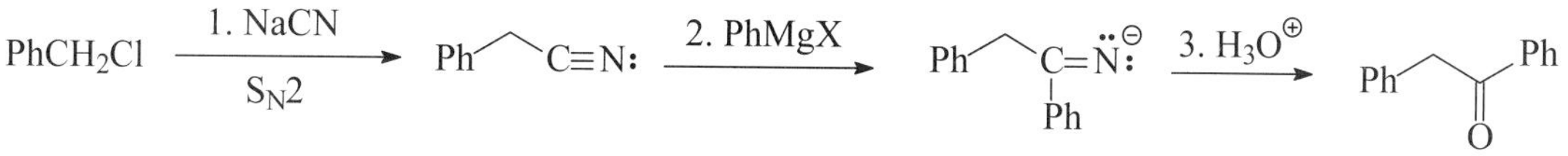 PhCH₂Cl →(1. NaCN, S_N2) PhCH₂C≡N: →(2. PhMgX) PhCH₂C(=N⁻)Ph →(3. H₃O⁺) PhCH₂C(=O)Ph

8. 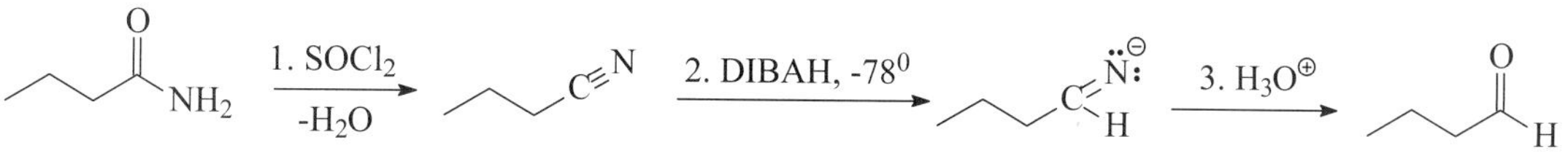 Butyramide (CH₃CH₂CH₂CONH₂) →(1. SOCl₂, −H₂O) CH₃CH₂CH₂C≡N →(2. DIBAH, −78°) CH₃CH₂CH₂CH=N⁻ →(3. H₃O⁺) CH₃CH₂CH₂CHO

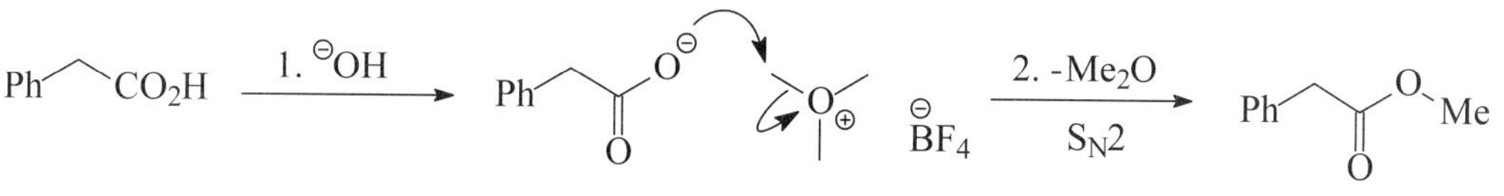

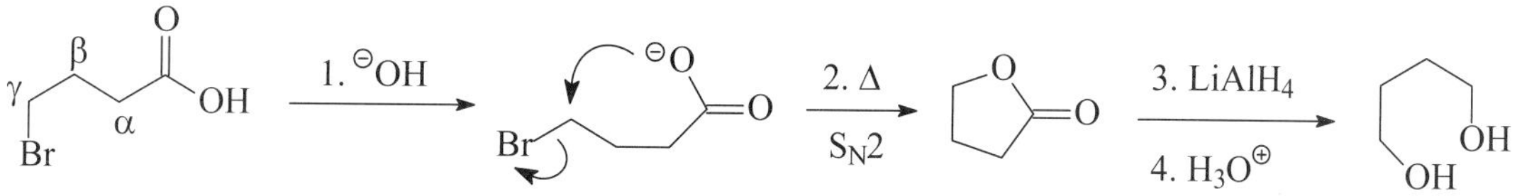

16.1 Reactions

16. [Salicylic acid] → 1. ⁻OH (2 equiv) → [phenolate carboxylate intermediate] (more stable anion, therefore, less reactive) → 2. MeI (1 equiv), S_N2 → [o-methoxybenzoate] → 3. H⁺ → [o-methoxybenzoic acid]

17. CH₃NHSO₂CH₂—[indole]—CH₂CH₂N⁺(CH₃)₂H   ⁻OOC–CH₂–CH₂–COOH   a salt!

16.2 Syntheses

1. PhCO₂H → 1. LiAlH₄; 2. H⁺ → PhCH₂OH → 3. TsCl; 4. KCN (S_N2) → PhCH₂CN → 5. H₃O⁺ → PhCH₂CO₂H

2. CH₂=CHCH₃ → 1. NBS, ROOR; 2. HBr, ROOR → BrCH₂CH₂CH₂Br → 3. NaCN, S_N2 → NCCH₂CH₂CH₂CN → 4. H₃O⁺ → HOOC–CH₂CH₂CH₂–COOH

3. CH₃CH=CHCO₂H → 1. H₃O⁺ (conj. add'n) → CH₃CH(OH)CH₂CO₂H → 2. CrO₃, H⁺ (Jones reagent) → CH₃COCH₂CO₂H → 3. Δ, –CO₂ → CH₃COCH₃

4. o-chloroacetophenone → 1. HOCH₂CH₂OH, H⁺; 2. Mg → [aryl-MgCl with dioxolane] → 3. CO₂; 4. H₃O⁺ → o-acetylbenzoic acid

5. styrene → 1. (CH₃)₂CHCOCl, AlCl₃ → 4-vinyl isobutyrophenone → 2. H₂NNH₂, ⁻OH, W-K → 4-isobutylstyrene → 3. HBr; 4. KCN → 4-isobutyl-α-methylbenzyl cyanide → 5. H₃O⁺ → ibuprofen (2-(4-isobutylphenyl)propanoic acid)

6. CH₃CH₂CH₂CH₂OH $\xrightarrow[\text{2. }n\text{-PrLi}]{\text{1. PCC}}$ CH₃CH₂CH₂-CH(O⁻)-n-Pr $\xrightarrow{\text{3. HCl}}$ CH₃CH₂CH₂-CH(Cl)-n-Pr $\xrightarrow{\text{4. NaCN}}$ CH₃CH₂CH₂-CH(CN)-n-Pr $\xrightarrow{\text{5. NaOH, H}_2\text{O}}$ CH₃CH₂CH₂-CH(CO₂Na)-n-Pr

7. (CH₃)CH(CO₂H)-CH=CH-CO₂H $\xrightarrow[\text{2. CrO}_3, \text{H}^\oplus]{\text{1. H}_3\text{O}^\oplus}$ HO₂C-CH(CH₃)-C(O)-CH₂-CO₂H $\xrightarrow{\text{3. }\Delta\,(-\text{CO}_2)}$ CH₃-CH(CO₂H)-CH₂-C(O)-CH₃ $\xrightarrow[\text{5. H}^\oplus]{\text{4. BH}_3}$ HOCH₂-CH(CH₃)-CH₂-C(O)-CH₃

8. PhCH₂Br $\xrightarrow{\text{1. KCN (S}_N2)}$ PhCH₂CN $\xrightarrow{\text{2. DIBAH, -78°}}$ PhCH₂-CH=N:⁻ $\xrightarrow[\text{hydrolysis}]{\text{3. H}_3\text{O}^\oplus}$ PhCH₂CHO

9. (furan-3,4-dicarbonitrile) $\xrightarrow[\text{2. H}_3\text{O}^\oplus]{\text{1. DIBAH, -78°}}$ (furan-3,4-dicarbaldehyde) $\xrightarrow[-\text{H}_2\text{O}]{\text{3. NH}_2\text{NH}_2, \text{H}^\oplus}$ (hydrazone-aldehyde intermediate) $\xrightarrow{-\text{H}_2\text{O}}$ (fused furan-pyridazine)

10. (norbornadiene) $\xrightarrow{\text{1. KMnO}_4, \text{H}^\oplus}$ HO₂C-CH(CO₂H)-CH(CO₂H)-CO₂H $\xrightarrow[-2\,\text{CO}_2]{\text{2. }\Delta}$ HO₂C-CH₂-CH₂-CH₂-CO₂H

11. RCO₂H $\xrightarrow[\text{2. H}_3\text{O}^\oplus]{\text{1. LiAlH}_4}$ RCH₂OH $\xrightarrow[\text{4. NaCN}]{\text{3. SOCl}_2 \text{ or HCl}}$ RCH₂-CN $\xrightarrow[\text{6. H}_3\text{O}^\oplus]{\text{5. R'MgX}}$ RCH₂COR'

or 5. H₃O⁺, 6. R'Li, 7. H₃O⁺

12. CH₃-C(O)-CH₂-CO₂H $\xrightarrow[\text{2. H}_3\text{O}^\oplus]{\text{1. BH}_3}$ CH₃-C(O)-CH₂-CH₂-OH $\xrightarrow[\text{4. PCC}]{\text{3. HOCH}_2\text{CH}_2\text{OH, H}^\oplus}$ (dioxolane)-CH₂-CHO $\xrightarrow[\text{6. H}_3\text{O}^\oplus]{\text{5. PhMgBr}}$ CH₃-C(O)-CH₂-CH(OH)-Ph

13. CH₃CH₂OH $\xrightarrow[\text{2. Br}_2, \text{CCl}_4]{\text{1. H}_2\text{SO}_4 \text{ (E2)}}$ BrCH₂CH₂Br $\xrightarrow{\text{3. (XS) KCN}}$ NC-CH₂-CH₂-CN $\xrightarrow{\text{4. H}_3\text{O}^\oplus}$ HO₂C-CH₂-CH₂-CO₂H

16.2 Syntheses

16.3 Mechanisms

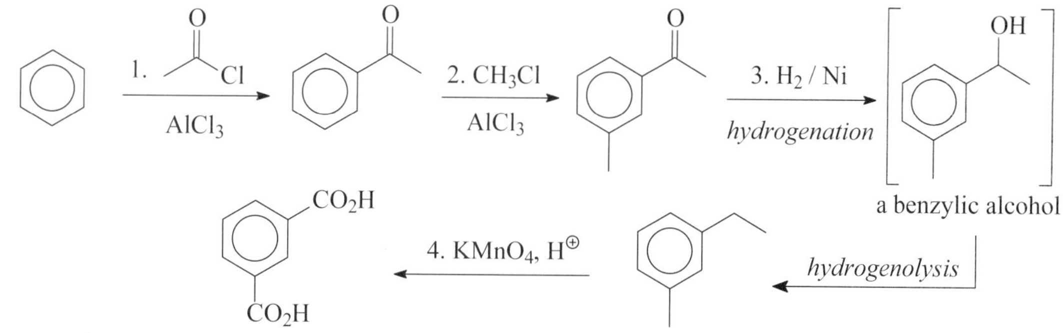

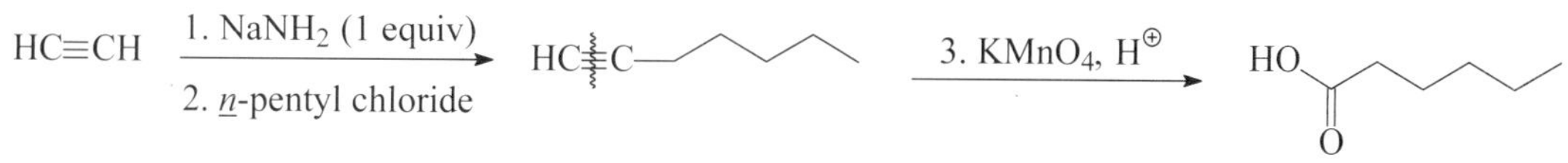

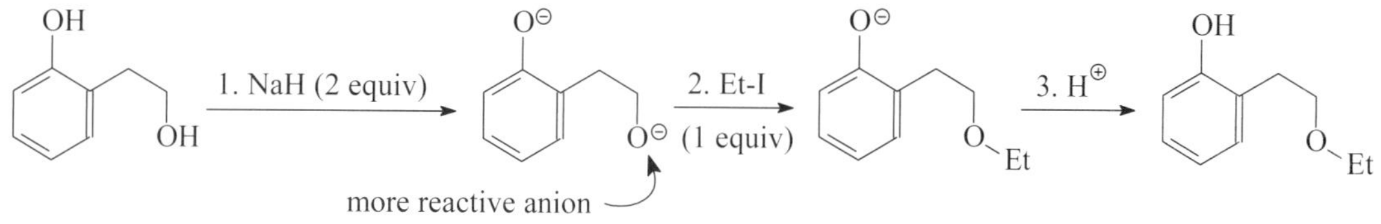

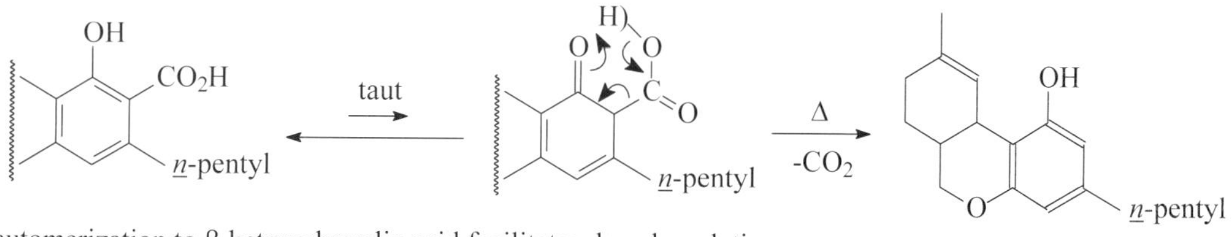

tautomerization to β-ketocarboxylic acid facilitates decarboxylation

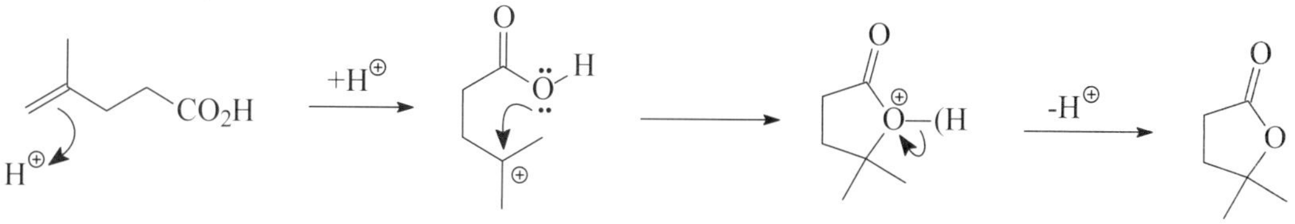

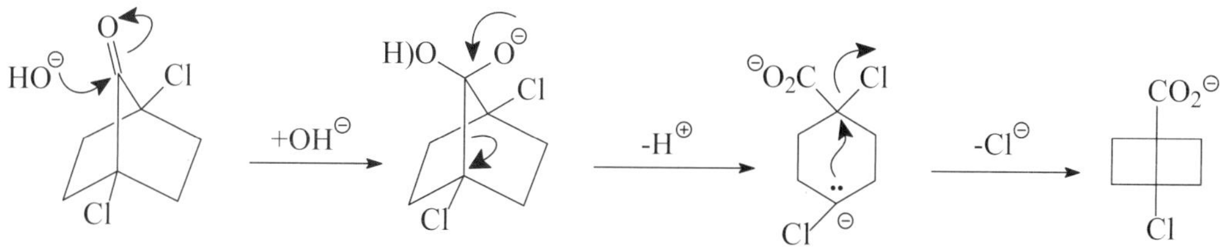

384 • Chapter 16 Carboxylic Acids

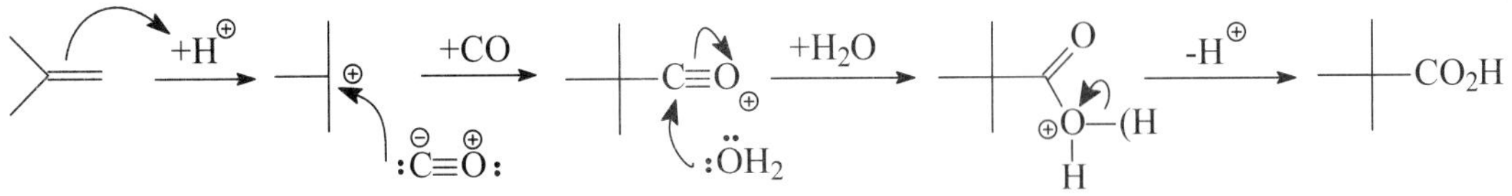

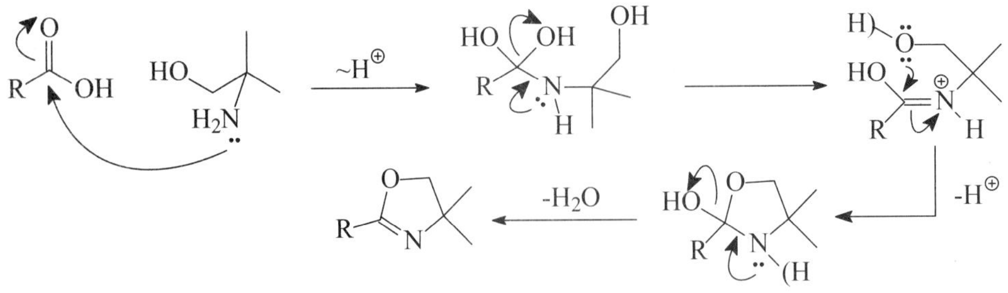

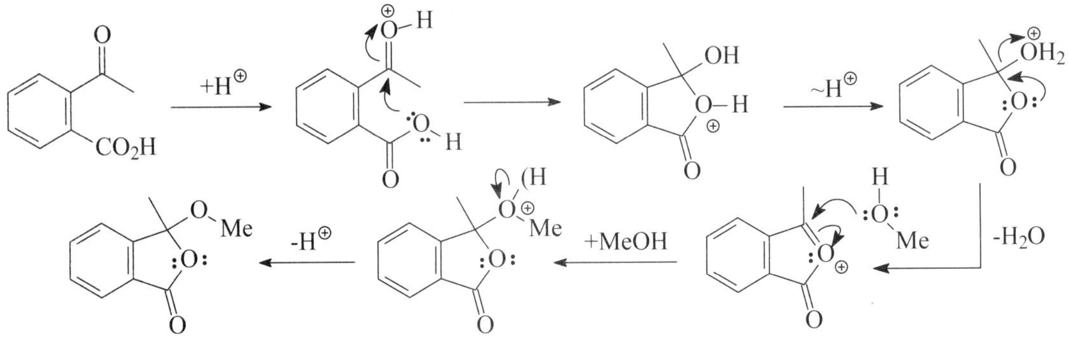

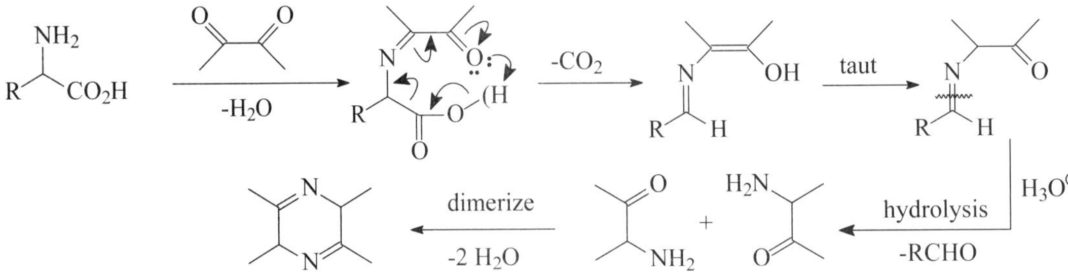

16.3 Mechanisms

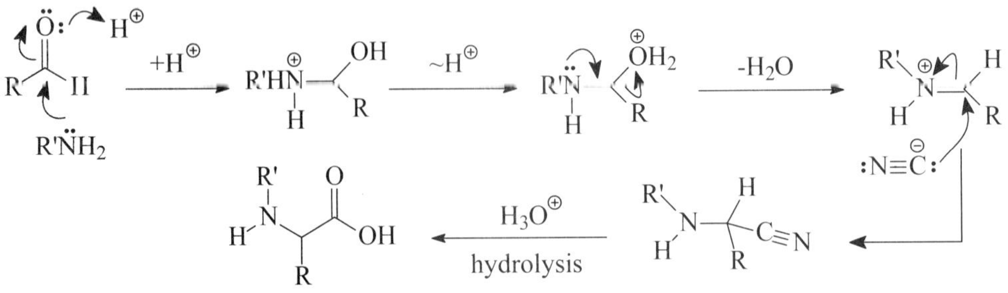

9. [Mechanism scheme showing conversion of 4-hydroxy-1-(carboxymethyl)cyclohexa-2,5-diene-1-carboxylic acid derivative via +H⁺, then -H⁺/-CO₂ to phenylacetic acid derivative.]

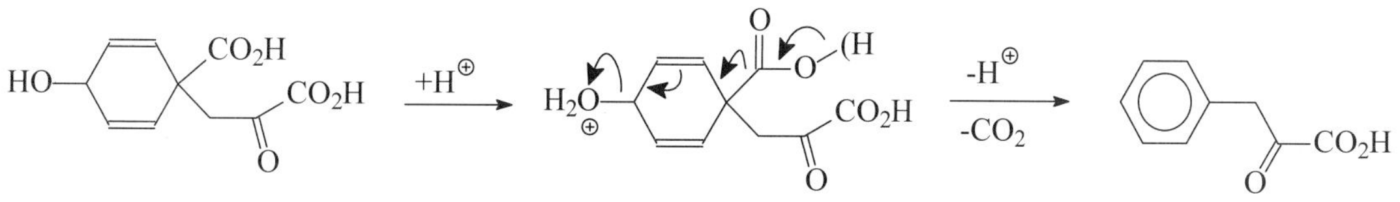

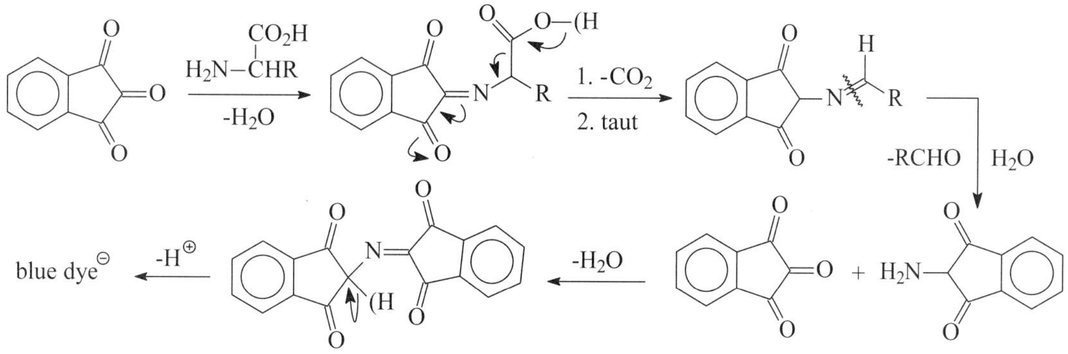

10. [Mechanism scheme showing ninhydrin-type reaction with H₂N–CHR(CO₂H), -H₂O, then 1. -CO₂, 2. taut, then -RCHO/H₂O to give indane-1,2,3-trione + H₂N-indanedione intermediate, then -H₂O, then -H⁺ to give blue dye⁻.]

11. [Mechanism scheme showing NADH reducing pyruvate (⁻O₂C-C(=O)-CH₃) with H–OH to give lactate (⁻O₂C-CH(OH)-CH₃) + NAD⁺.]

12. [Mechanism scheme showing enolate of oxaloacetate, -CO₂, attack on GTP phosphate, -GDP to give phosphoenolpyruvate.]

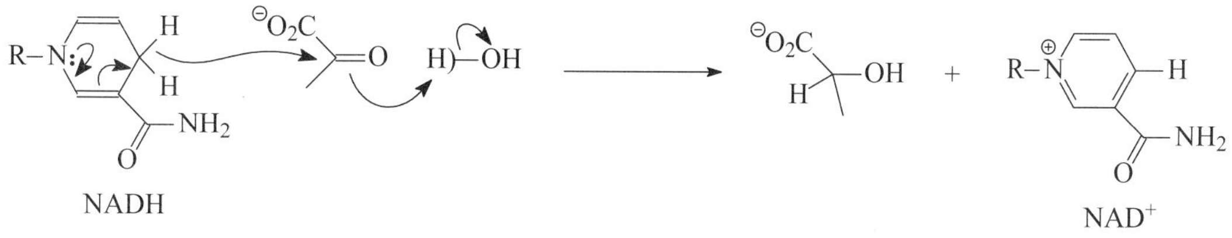

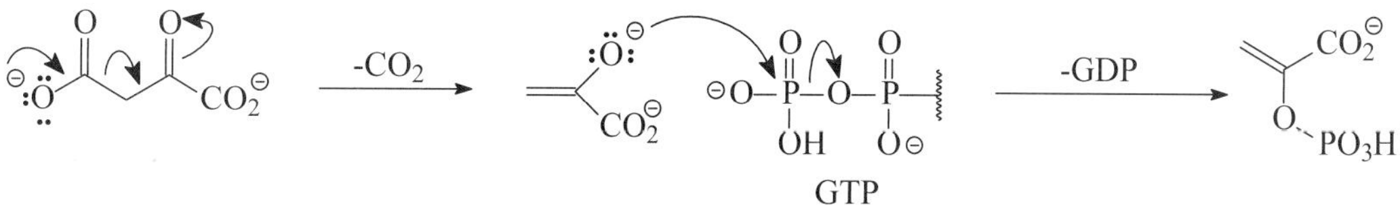

13. [Thiamine/thiazolium mechanism: -H⁺ gives ylide, attack on pyruvate, -CO₂, then tautomerization/protonation, -H⁺ releases acetaldehyde + thiazolium ylide.]

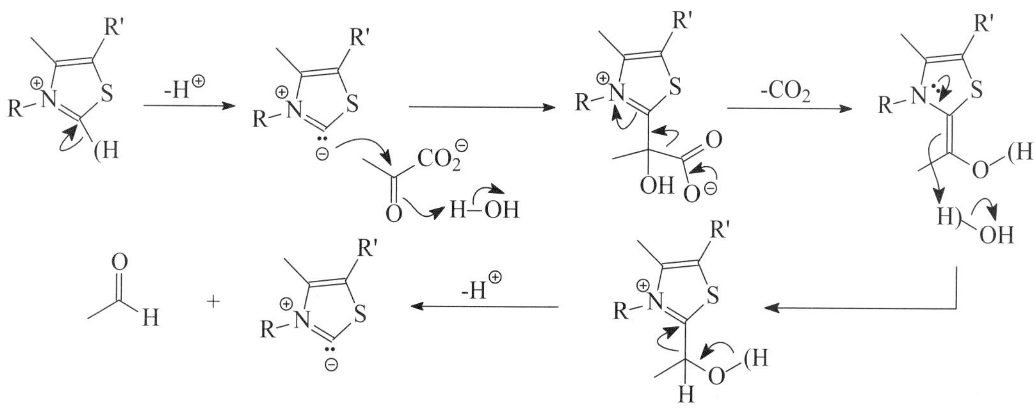

16.3 Mechanisms

14. 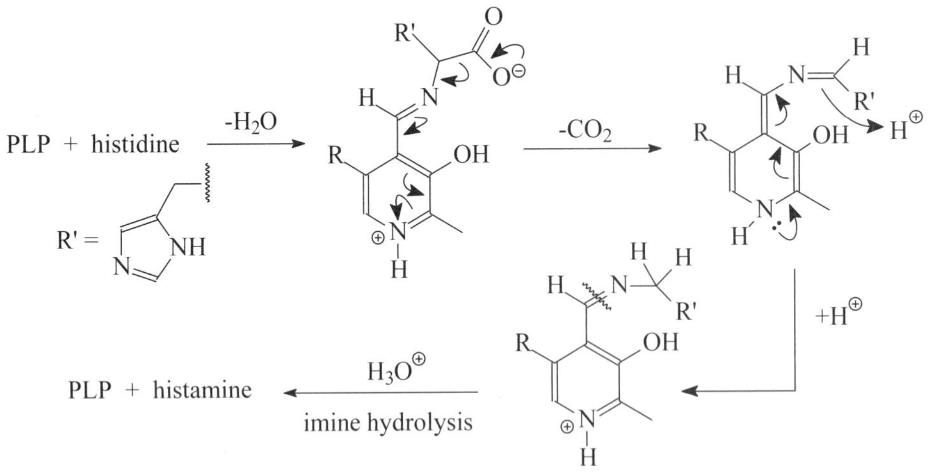 PLP + histidine

R' = (imidazolyl-CH₂–)

$$\text{PLP + histidine} \xrightarrow{-H_2O} \text{[Schiff base intermediate]} \xrightarrow{-CO_2} \text{[quinonoid intermediate]} \xrightarrow{+H^+} \text{[protonated intermediate]} \xrightarrow[\text{imine hydrolysis}]{H_3O^+} \text{PLP + histamine}$$

15. 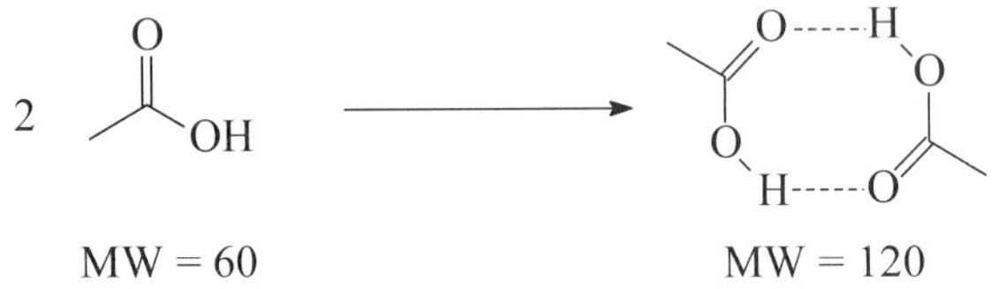 $2\ CH_3COOH \longrightarrow (CH_3COOH)_2$ dimer

MW = 60   MW = 120

intermolecular hydrogen bonding forms a tight dimeric complex in nonpolar solvents

16.3 Mechanisms

CHAPTER 17
CARBOXYLIC ACID DERIVATIVES

17.1 Reactions

1. Cl-C(=O)-C(=O)-OEt (more reactive than ester) + Et$_2$NH (1 equiv) → Et$_2$N-C(=O)-C(=O)-OEt, -HCl

2. (butyric anhydride) + MeNH$_2$ → butyryl-NHMe + butyric acid (CO$_2$H)

3. cyclohexyl carboxylate of cyclopropanol + (XS) EtOH, H$^⊕$ (transesterification) → ethyl cyclohexanecarboxylate + cyclopropanol

4. HO-(CH$_2$)$_3$-OH + Cl-C(=O)-Cl → HO-(CH$_2$)$_3$-O-C(=O)-Cl (-HCl) → cyclic carbonate (1,3-dioxan-2-one), -HCl

5. HO$_2$C-CO$_2$H (oxalic acid) →[1. PCl$_3$ (or SOCl$_2$)] Cl-C(=O)-C(=O)-Cl →[2. LiAlH(O-t-Bu)$_3$; 3. H$^⊕$] OHC-CHO (glyoxal)

6. propionyl diisopropylamide →[H$_3$O$^⊕$, hydrolysis] propionic acid (CO$_2$H) + iPr$_2$NH$_2^⊕$

7. phenyl pentanoate →[1. iPr-MgBr; -$^⊖$OPh] [isopropyl pentyl ketone] (not isolable) →[1. iPr-MgBr; 2. H$_3$O$^⊕$] tertiary alcohol (diisopropyl carbinol) →[or -H$_2$O] alkene

8. glutaric anhydride →[MeOH, H$^⊕$, NAS] MeO$_2$C-(CH$_2$)$_3$-CO$_2$H →[MeOH, H$^⊕$] MeO$_2$C-(CH$_2$)$_3$-CO$_2$Me (dimethyl glutarate)

9. bicyclic-CO$_2$H →[1. LiAlH$_4$; 2. H$^⊕$] bicyclic-CH$_2$OH →[3. (Ac)$_2$O, NAS] bicyclic-CH$_2$-O-C(=O)-Me

17.1 Reactions

17.1 Reactions

10. 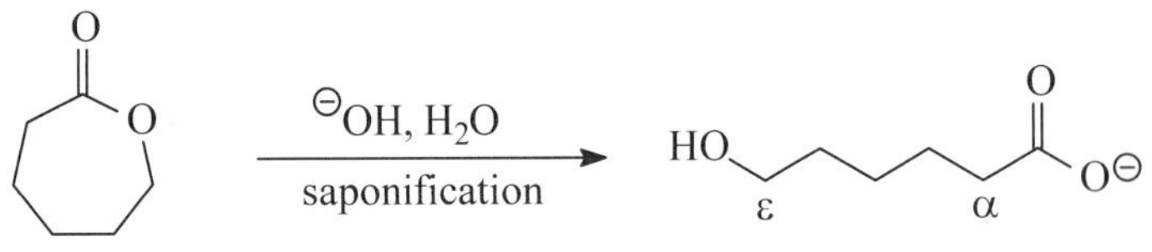 [lactone] + ⁻OH, H₂O (saponification) → HO-(ε)-...-(α)-CO₂⁻

11. 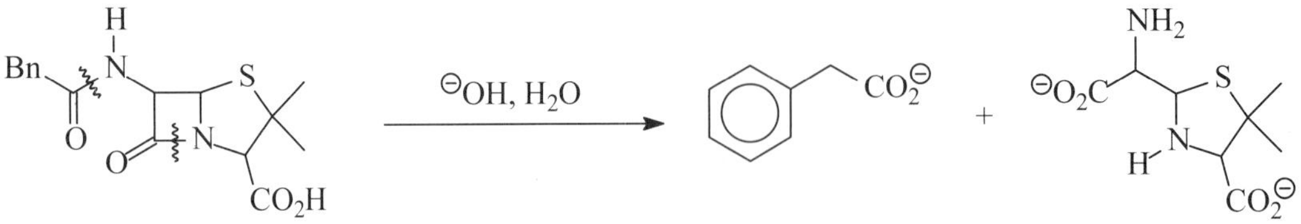 [penicillin-type structure with Bn-C(=O)-NH, β-lactam, S, CO₂H] + ⁻OH, H₂O → PhCH₂CO₂⁻ + [aminothiazolidine carboxylate]

12. 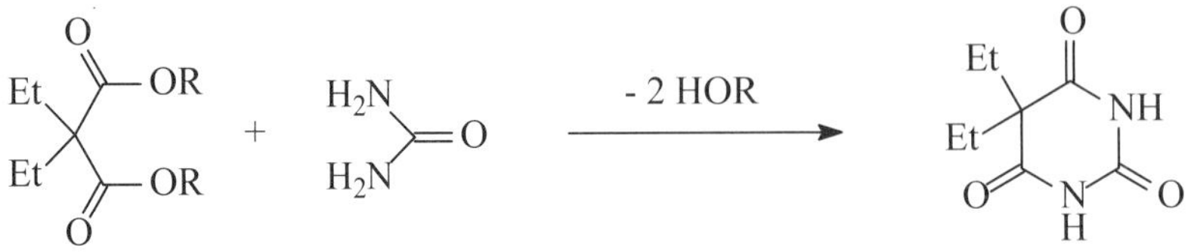 Et₂C(CO-OR)₂ + H₂N-C(=O)-NH₂ → (− 2 HOR) → [5,5-diethylbarbituric acid: Et₂C ring with (C=O)-NH-C(=O)-NH-C(=O)]

13. 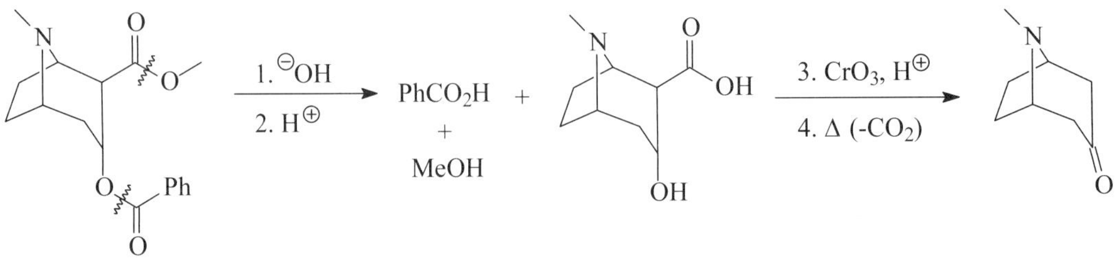 [N-methyl bicyclic with CO₂Me and O-C(=O)Ph ester] —1. ⁻OH; 2. H⁺→ PhCO₂H + MeOH + [N-methyl bicyclic with CO₂H and OH] —3. CrO₃, H⁺; 4. Δ (−CO₂)→ [N-methyl bicyclic ketone (tropinone-like)]

14. 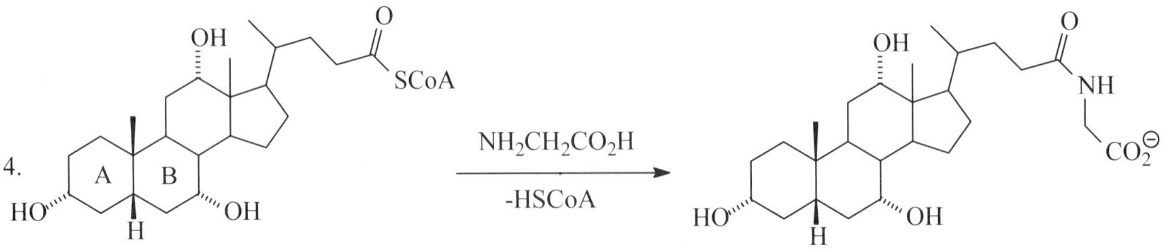 [cholic acid CoA thioester with OH groups, rings A, B] + NH₂CH₂CO₂H (−HSCoA) → [glycine conjugate, glycocholate]

15. 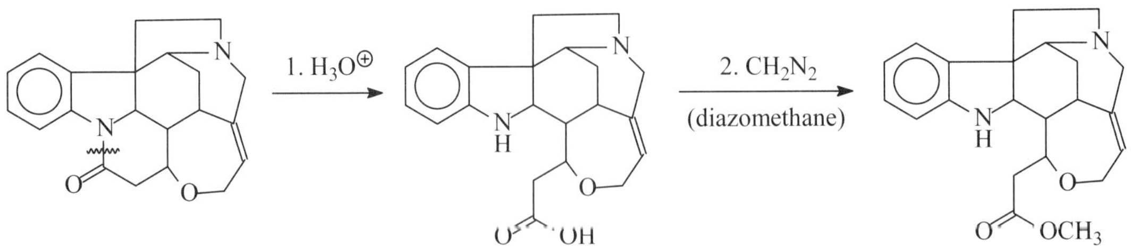 [strychnine-type lactam] —1. H₃O⁺→ [ring-opened carboxylic acid] —2. CH₂N₂ (diazomethane)→ [methyl ester, -OCH₃]

16. 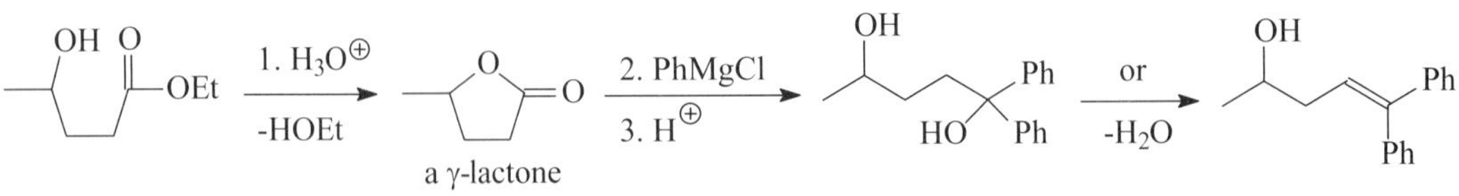 HO-CH(...)-CH₂CH₂-C(=O)-OEt —1. H₃O⁺, −HOEt→ [γ-butyrolactone] (a γ-lactone) —2. PhMgCl; 3. H⁺→ [HO-CH-CH₂CH₂-C(Ph)₂-OH] —or, −H₂O→ [HO-CH-CH₂-CH=CPh₂]

Solutions • 389

17. [structure: bicyclic anhydride] → 1. LiAlH₄ / 2. H⊕ → [diol structure] = [tetrahydrofuran diol: HO–C(Me)–CH₂–O–C(Me)–CH₂–OH in ring form]

18. MeNH–CH₂CH₂–NHMe → 1. ClC(O)Cl, −HCl → Cl–C(O)–N(Me)–CH₂CH₂–NHMe → −HCl → [cyclic urea: 1,3-dimethylimidazolidin-2-one] → 2. LiAlH₄ / 3. H⊕ → 1,3-dimethylimidazolidine

19. iPrO–P(=O)(Me)–F + protein–CH(OH)– → −HF → protein–CH(O–P(=O)(Me)–OiPr)–

20. [lysergic acid with CO₂H] → 1. SOCl₂ → [acid chloride] → 2. HNEt₂, −HCl → [amide with C(=O)NEt₂]

    *lys*ergic acid *d*iethylamide

21. tosyl chloride (p-MeC₆H₄–SO₂Cl) + H₂N–C(=O)–NH–n-Bu (more nucleophilic nitrogen indicated) → −HCl → p-MeC₆H₄–SO₂–NH–C(=O)–NH–n-Bu

22. [saccharin structure: benzisothiazol-3(2H)-one 1,1-dioxide] →⁻OH, H₂O→ 2-carboxylate benzenesulfonate (o-⁻O₂C–C₆H₄–SO₃⁻) + NH₃

23. HO–C₆H₄–NH₂ + Ac₂O (1 equiv) → HO–C₆H₄–NH–C(=O)Me + HOAc

24. H₂C=C=O + :NH₂Ph → H₂C=C(O⁻)(NH₂Ph⁺) → ∼H⊕ → H₂C=C(OH)(NHPh) → taut → CH₃–C(=O)–NHPh

17.1 Reactions

390 • Chapter 17 Carboxylic Acid Derivatives

25. 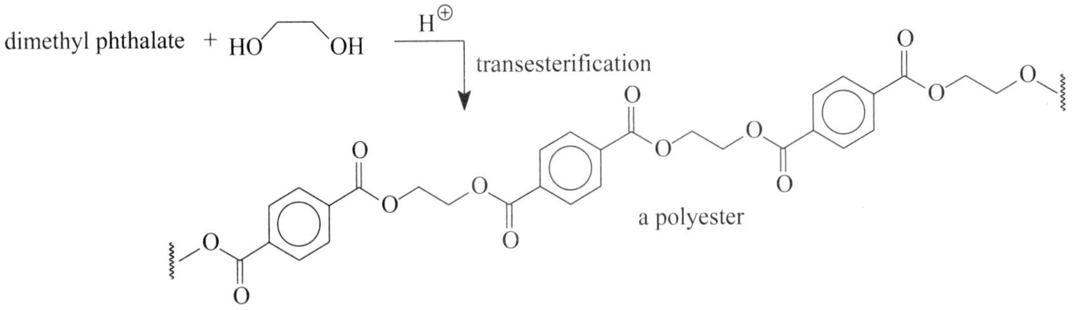 dimethyl phthalate + HO-CH₂CH₂-OH $\xrightarrow{H^\oplus}$ transesterification → a polyester

26. 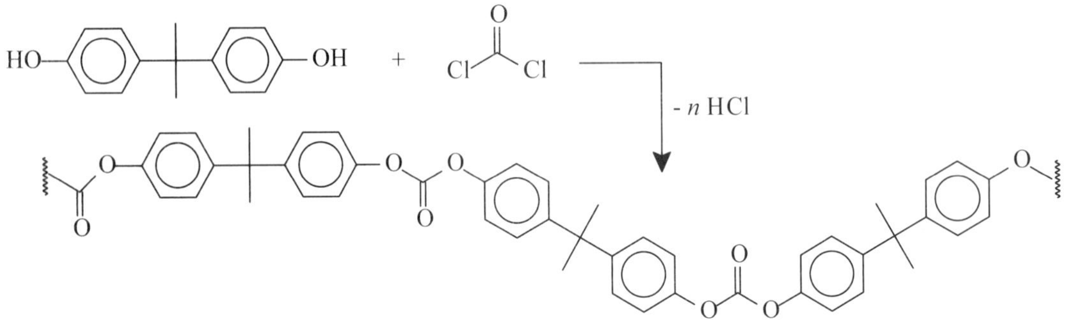 Bisphenol A + phosgene (Cl-CO-Cl) $\xrightarrow{-n\,HCl}$ polycarbonate

27. 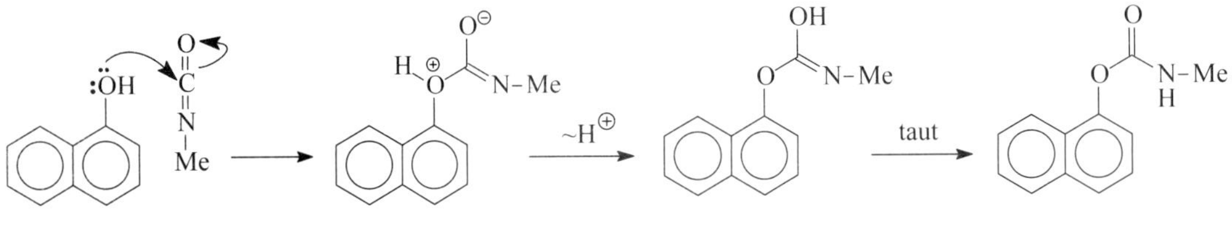 1-naphthol + MeN=C=O → addition → proton transfer (~H⁺) → tautomerization → naphthyl N-methylcarbamate

28. 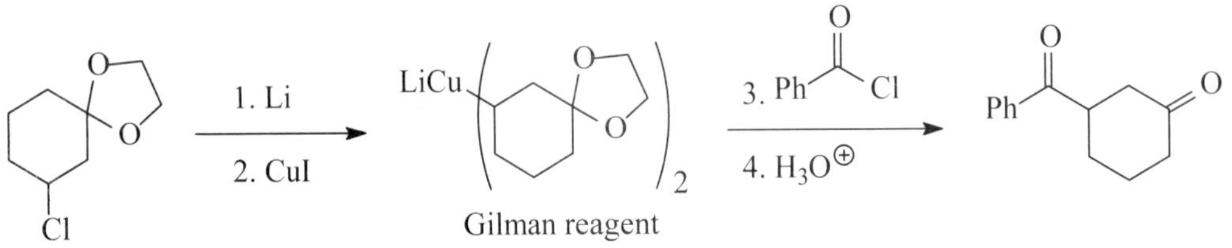 4-chloro-1,1-(ethylenedioxy)cyclohexane
    1. Li
    2. CuI
    → Gilman reagent (LiCu(R)₂)
    3. PhCOCl
    4. H₃O⁺
    → 3-benzoylcyclohexanone

29. 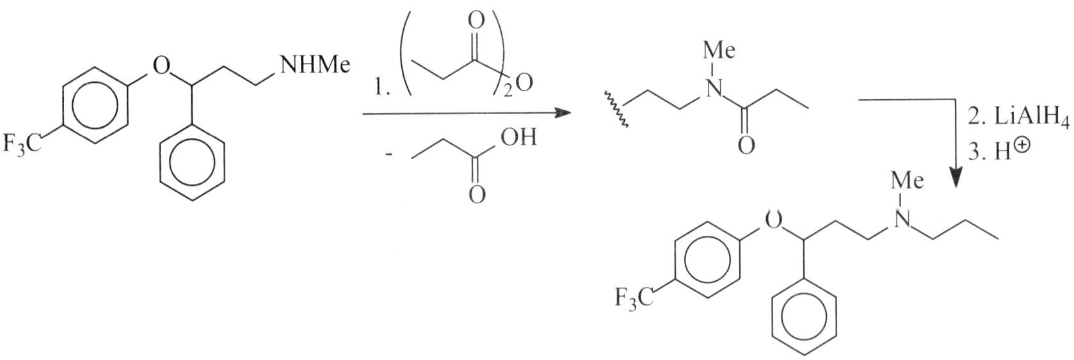 ArO-CH(Ph)-CH₂CH₂-NHMe (Ar = 4-CF₃-C₆H₄)
    1. (propanoic anhydride, (EtCO)₂O), −EtCOOH
    → amide
    2. LiAlH₄
    3. H⁺
    → ArO-CH(Ph)-CH₂CH₂-N(Me)(CH₂CH₂CH₃)

17.1 Reactions

17.1 Reactions

17.1 Reactions

17.1 Reactions

394 • Chapter 17 Carboxylic Acid Derivatives

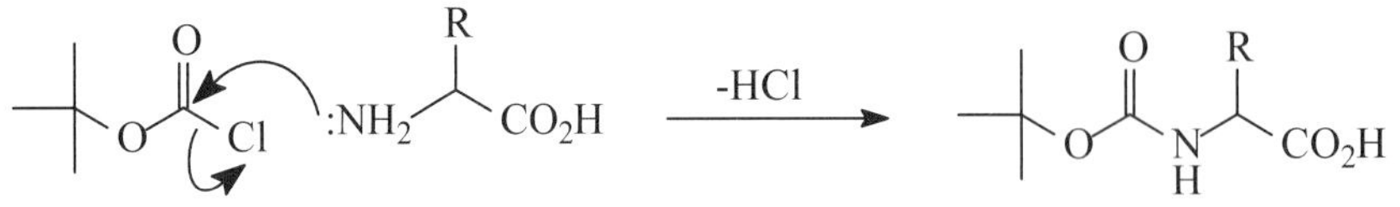

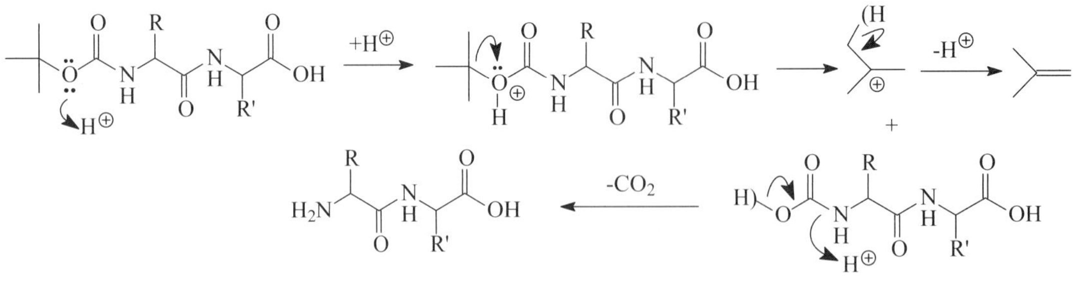

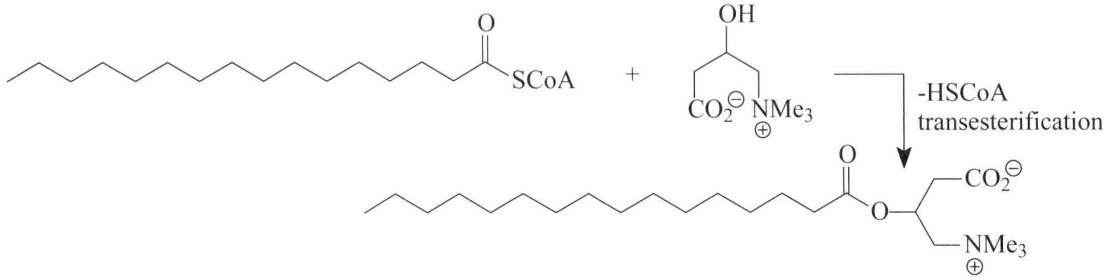

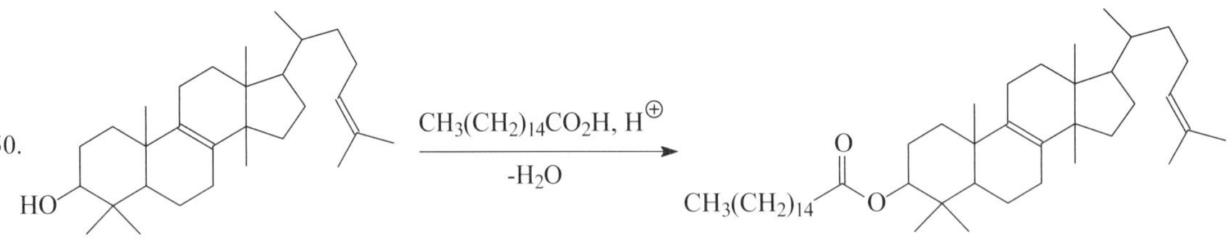

17.1 Reactions

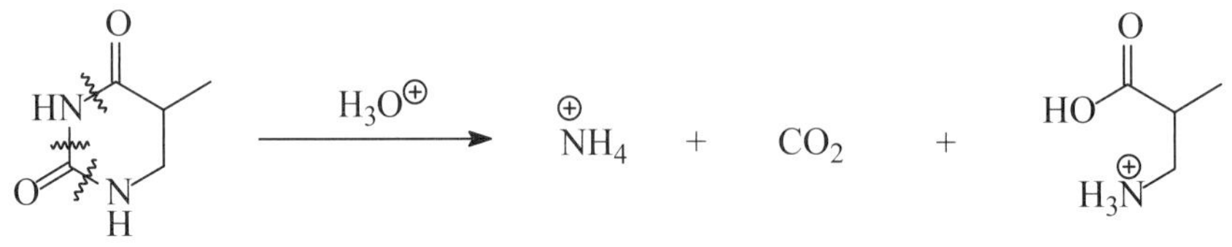

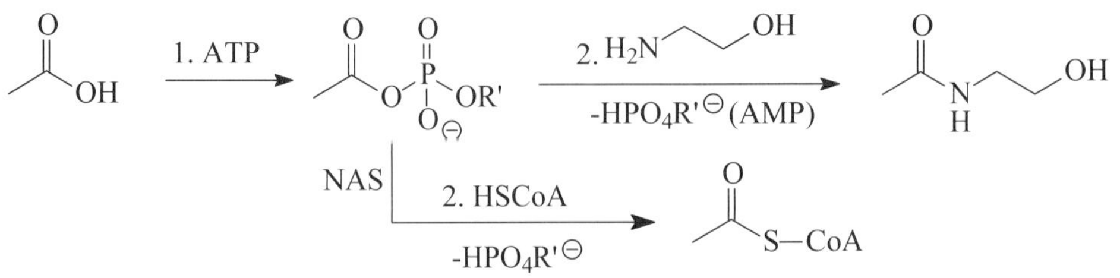

17.2 Syntheses

396 • Chapter 17 Carboxylic Acid Derivatives

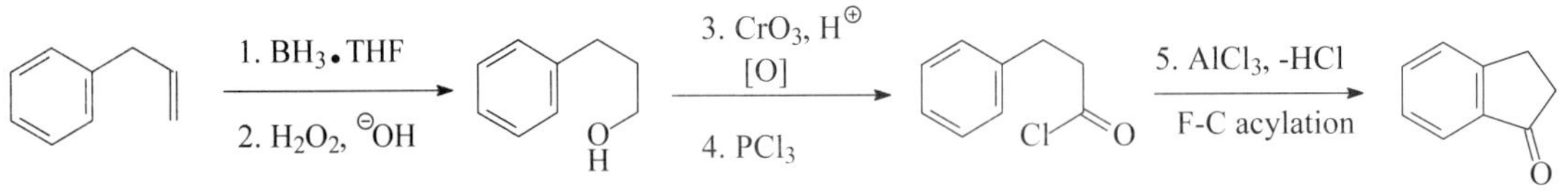

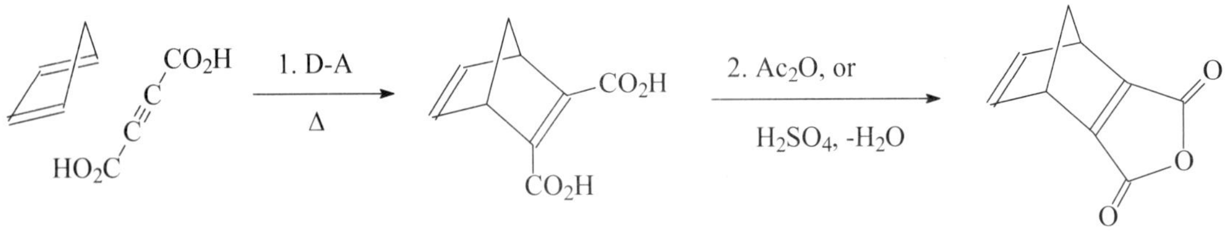

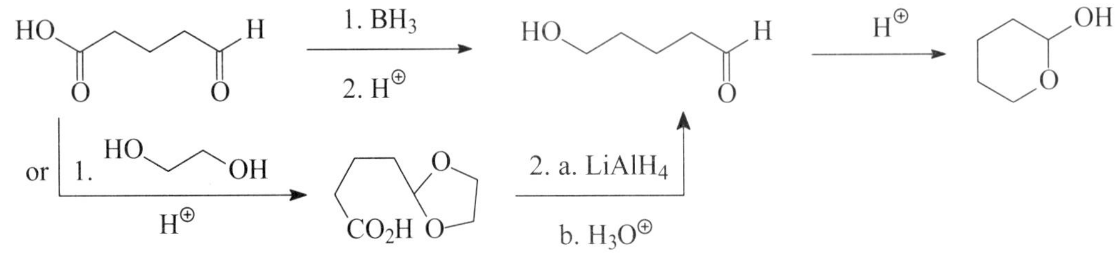

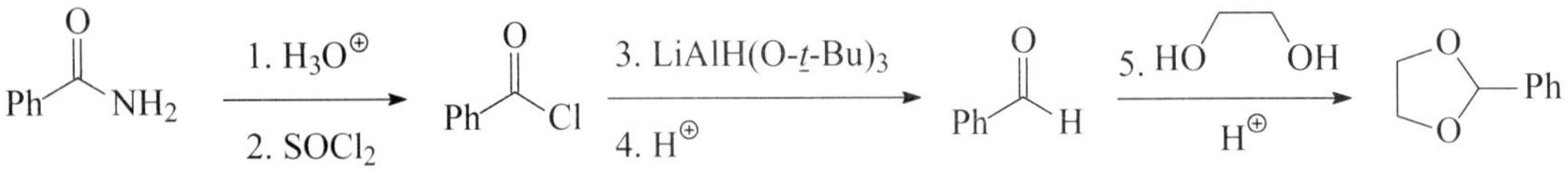

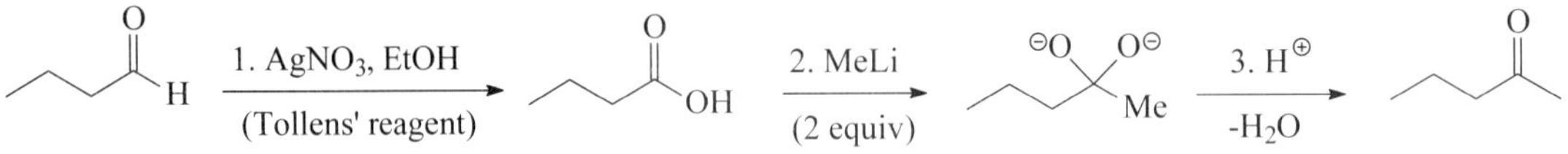

17.2 Syntheses

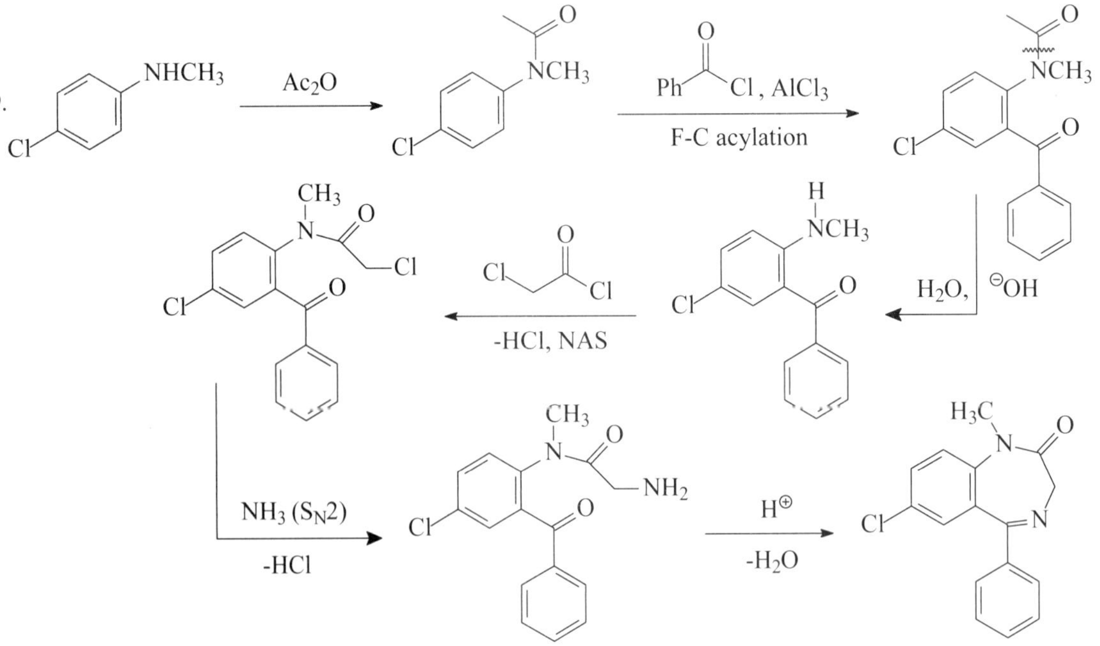

17.2 Syntheses

14. (*cont.*) Mechanism for step 5:

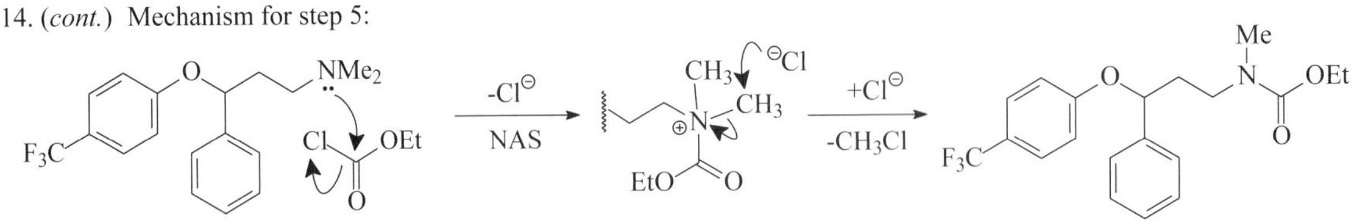

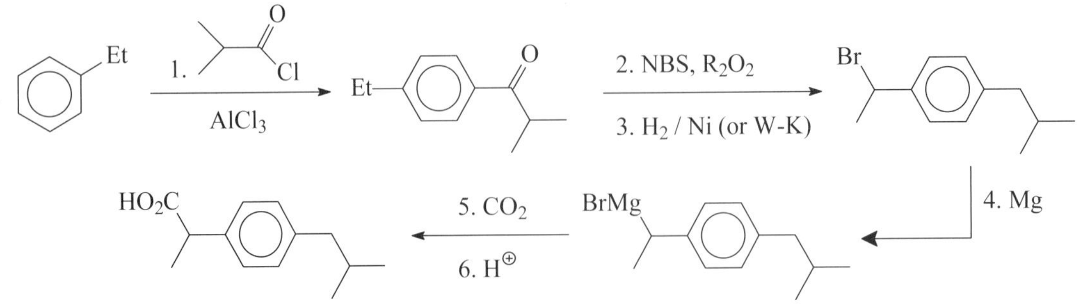

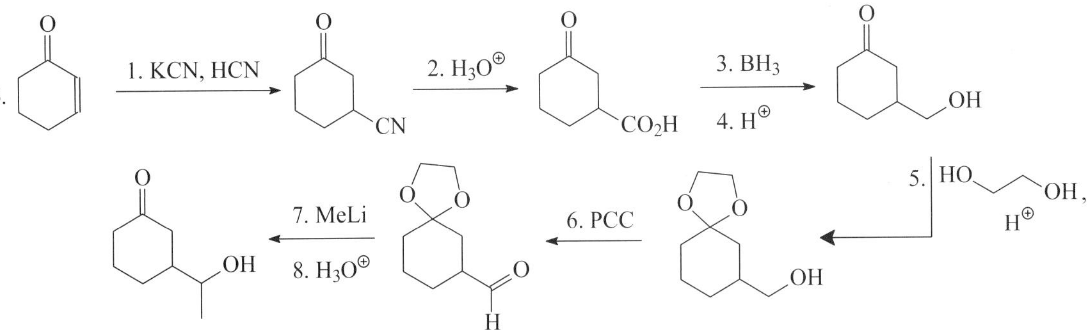

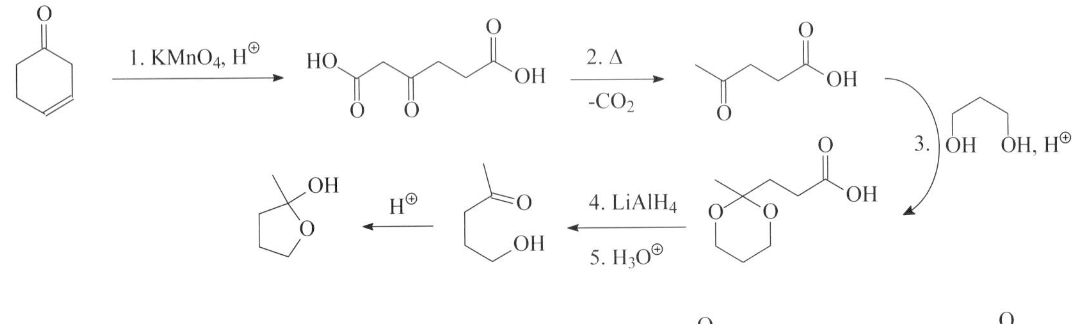

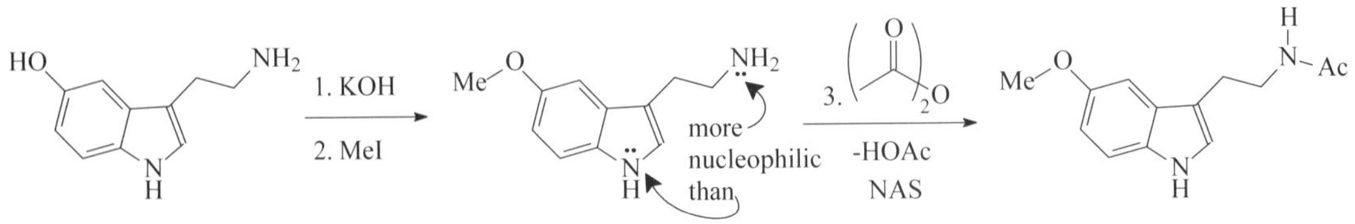

17.2 Syntheses

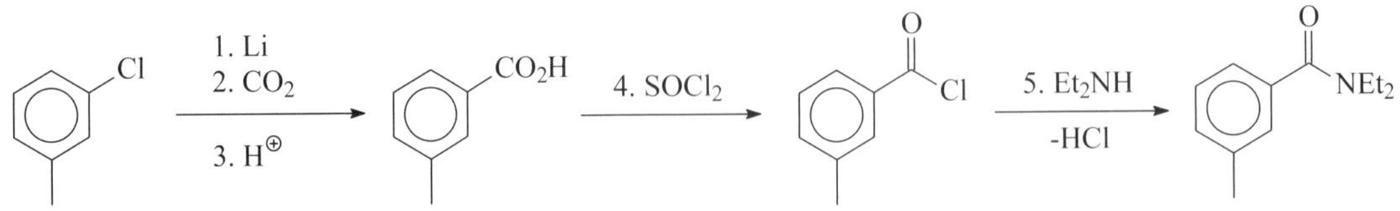

20.

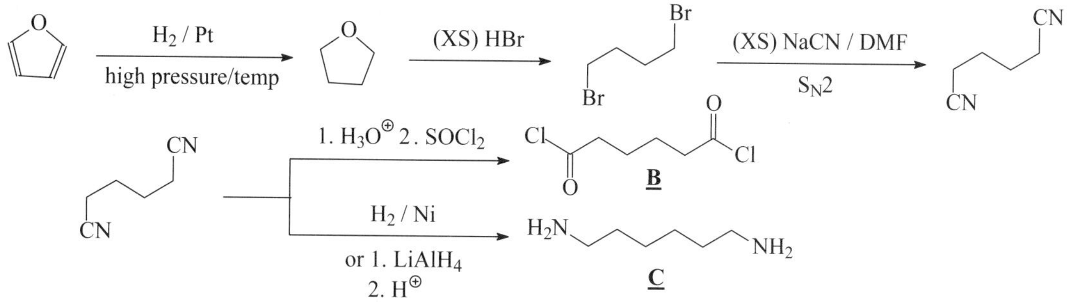

21. Poly(vinyl alcohol). Vinyl alcohol is unstable and rapidly tautomerizes to acetaldehyde:

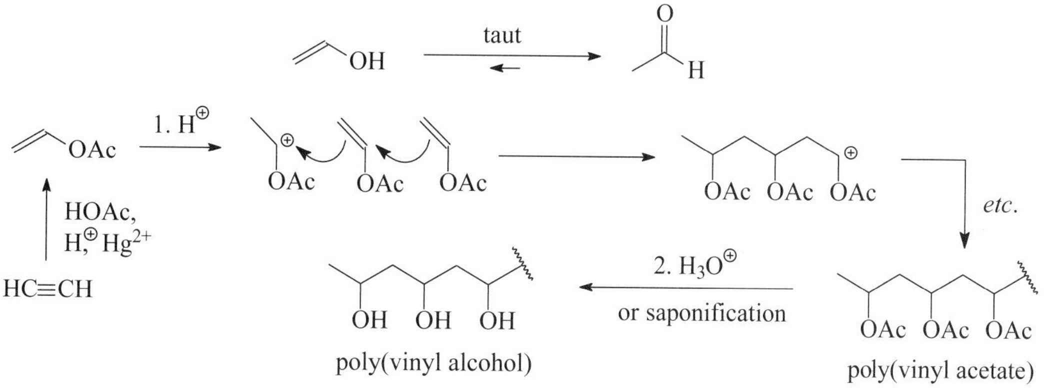

22. a.

b.

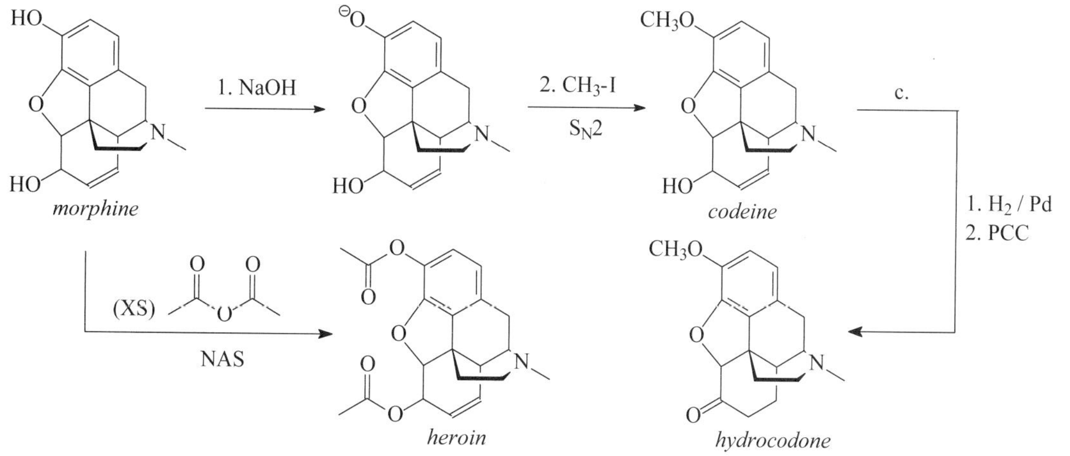

d.

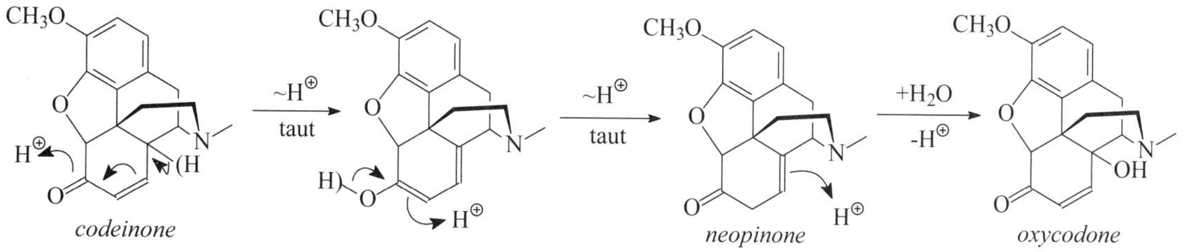

17.2 Syntheses

17.3 Mechanisms

1. 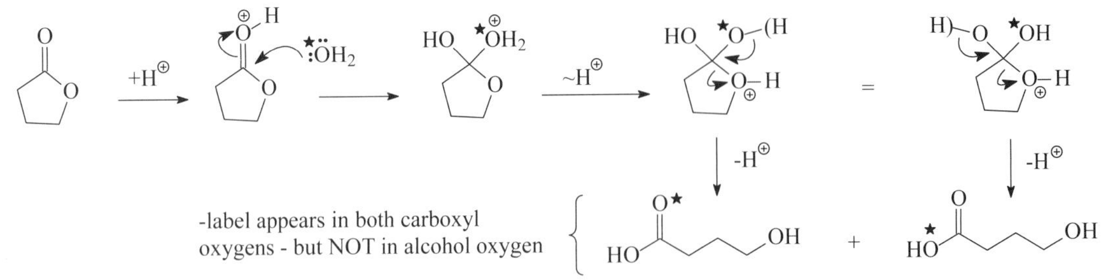 [Mechanism showing acid-catalyzed hydrolysis of γ-butyrolactone with ¹⁸O labeling]

   —label appears in both carboxyl oxygens - but NOT in alcohol oxygen

2. 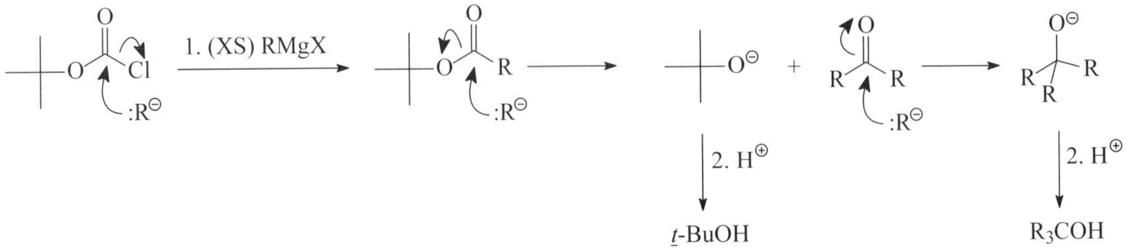 [Mechanism showing reaction of t-butyl chloroformate with excess RMgX giving t-BuOH and R$_3$COH]

3. [Mechanism showing lactic acid dehydration]

   α-lactone does NOT form because of ring strain

   intermolecular condensation → intramolecular condensation → much less ring strain

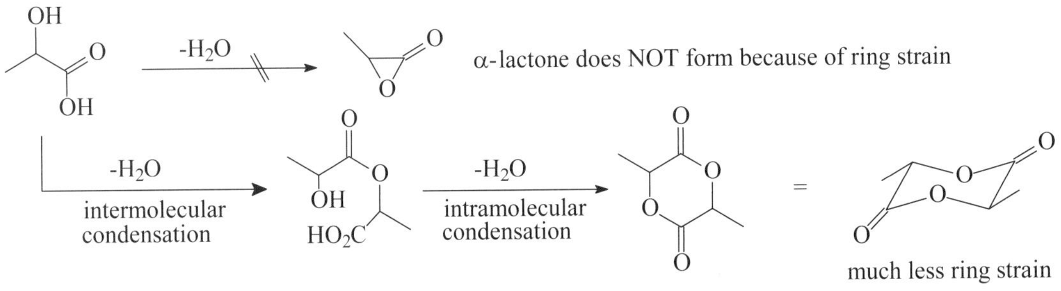

4. 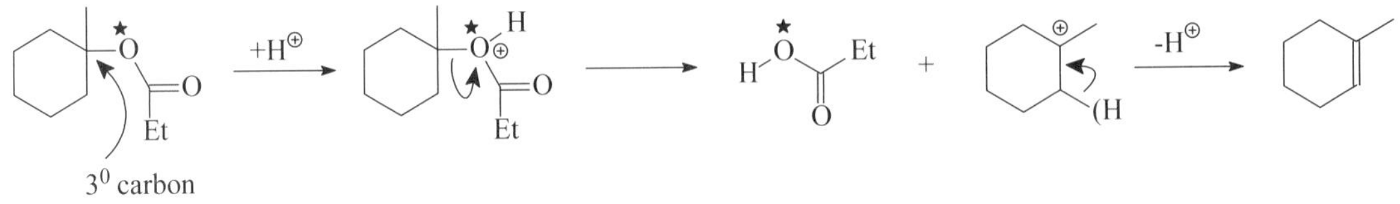 [Mechanism showing acid-catalyzed elimination of cyclohexyl ethanoate ester at 3° carbon to give cyclohexene]

   3° carbon

5. 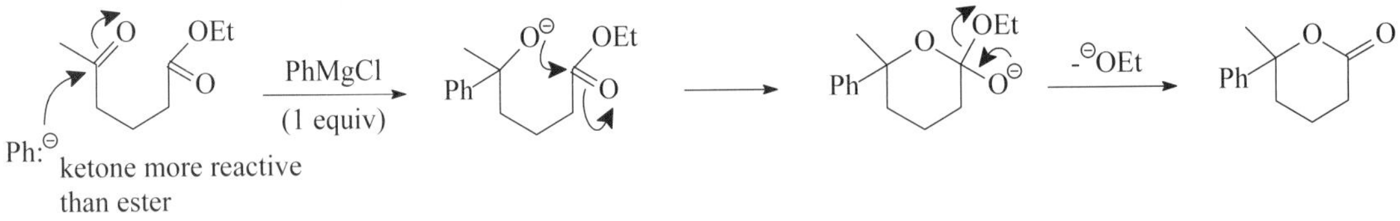 [Mechanism showing PhMgCl (1 equiv) addition to keto-ester forming lactone]

   Ph:⁻ ketone more reactive than ester

6. 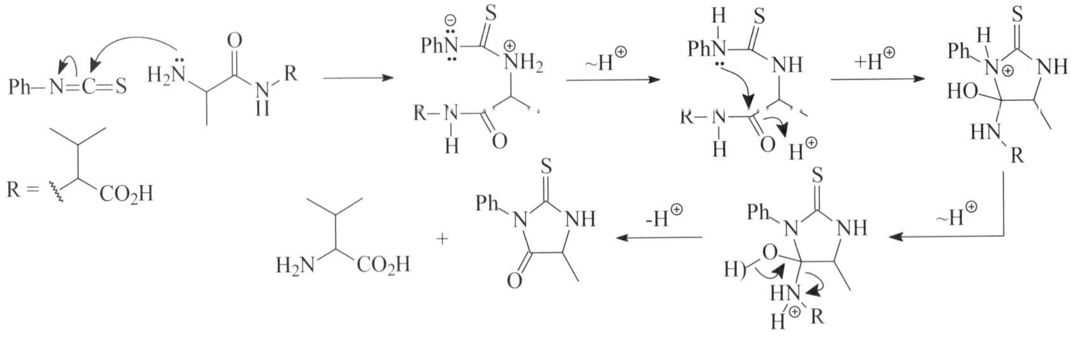 [Mechanism showing Edman degradation: Ph–N=C=S reacting with peptide H$_2$N-CHR-C(O)-NHR to give phenylthiohydantoin and H$_2$N-CHR-CO$_2$H]

   R = CH(CH$_3$)CH(CH$_3$)... CO$_2$H

17.3 Mechanisms

402 • Chapter 17 Carboxylic Acid Derivatives

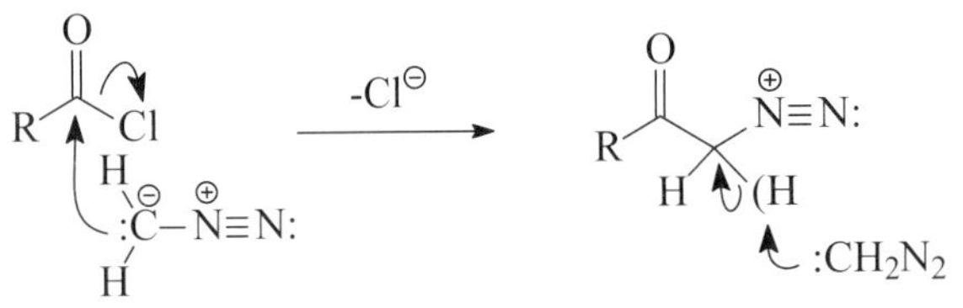

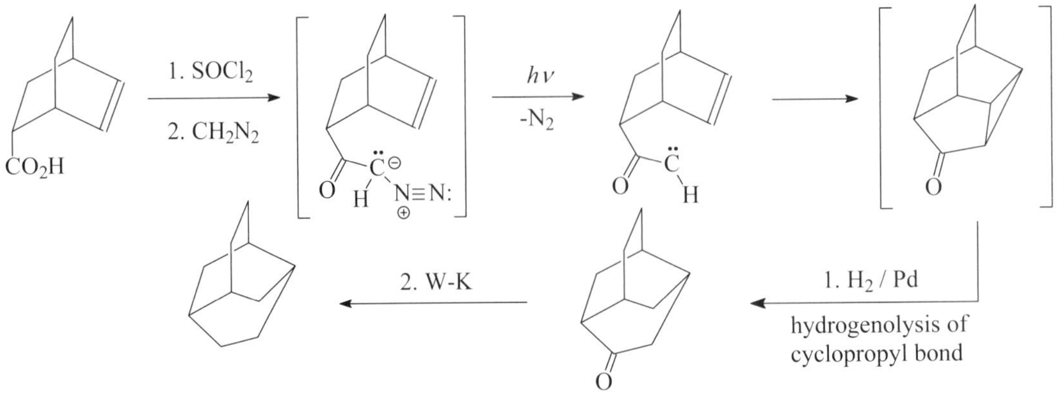

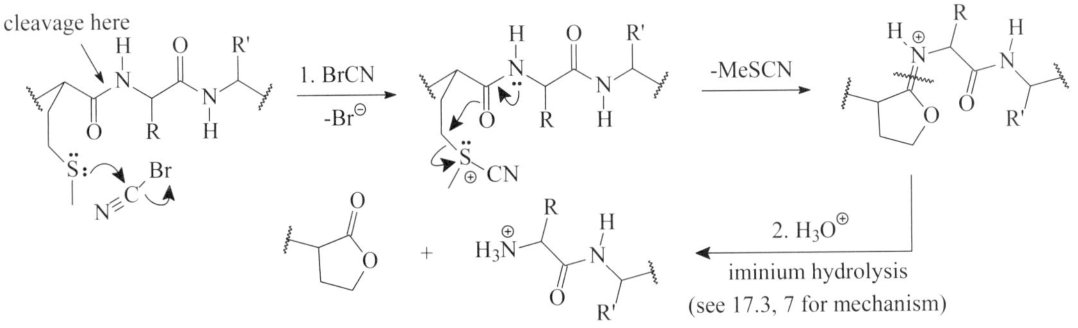

12. a.

b.

13.

14.

17.3 Mechanisms

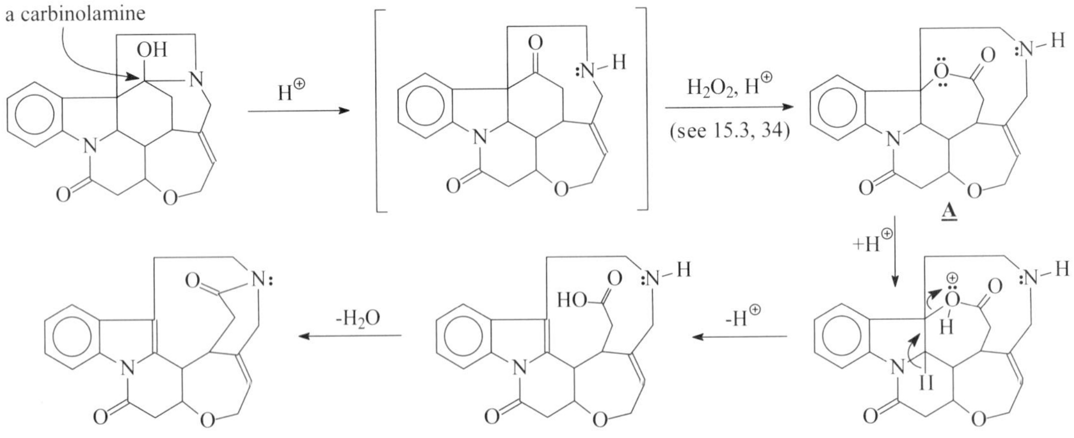

17.3 Mechanisms

15. [Mechanism scheme showing acetic anhydride reaction with an indole alkaloid carboxylic acid, proceeding via pyridine-mediated decarboxylation to form an enol acetate product.]

16. [Mechanism scheme showing DCC-mediated coupling, with R = cyclohexyl, forming dicyclohexylurea and a penicillin-type β-lactam with PhOCH₂CONH- side chain.]

17.

$$H-\overset{O}{\underset{OH}{P}}-OH \underset{\text{taut}}{\overset{\sim H^+}{\rightleftharpoons}} :\overset{OH}{\underset{OH}{P}}-OH$$

nucleophilic form

R = -CH₂CH₂NH₂

[Mechanism scheme: RCO₂H + PCl₃ → acyl chloride intermediate, then addition of P(OH)₃, loss of HCl, second PCl₃ addition, hydrolysis (+3 H₂O, −3 HCl), giving the bisphosphonate H₂N-CH₂CH₂-C(OH)(PO(OH)₂)₂.]

18. [Mechanism scheme: isopropenyl acetate + H⁺ → acetone + acylium ion Me—C≡O⁺; acylium ion then acylates a tertiary OH on a steroid-like methoxy-containing polycyclic structure, giving after −H⁺ the acetate ester.]

404 • Chapter 17 Carboxylic Acid Derivatives

19. a. Carboxylate as a *nucleophile*:

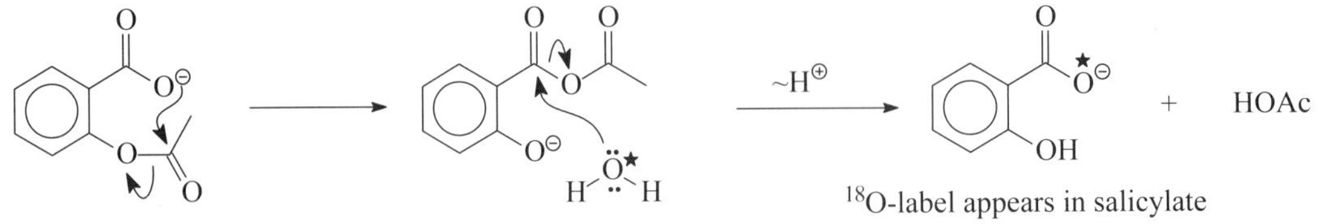

^{18}O-label appears in salicylate

b. Carboxylate as a *base*:

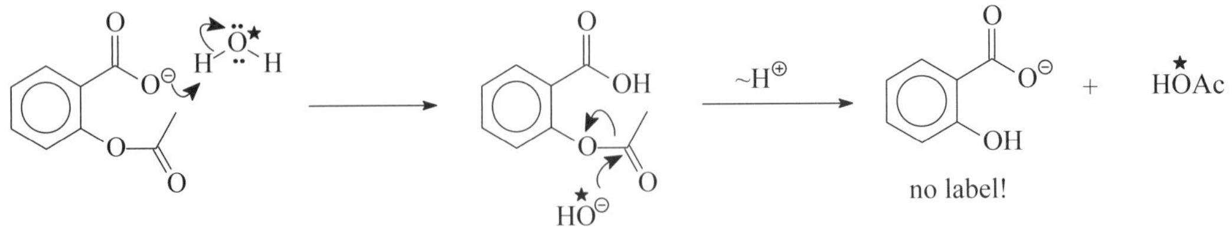

no label!

c. Therefore, *pathway b* is preferred.

20.

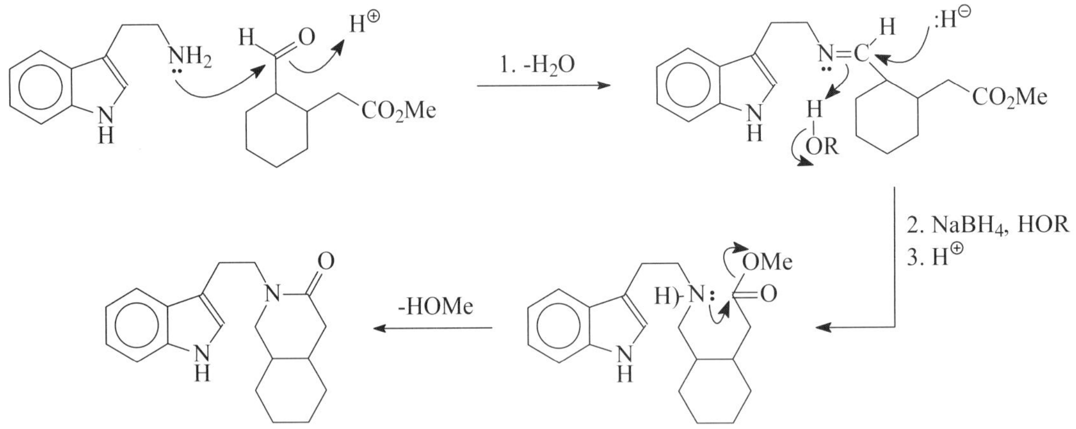

1. $-H_2O$

2. $NaBH_4$, HOR
3. $H^{\oplus}$

$-HOMe$

17.3 Mechanisms

CHAPTER 18
CARBONYL α-SUBSTITUTION REACTION AND ENOLATES

18.1 Reactions

1. 2,2-dimethylcyclopentanone —1. LDA→ enolate —2. n-PrBr→ 2,2-dimethyl-5-propylcyclopentanone

2. methyl 3-oxopentanoate —1. ⁻OMe / MeOH→ enolate (OMe, O⁻) —2. PhCOBr→ α-benzoyl-β-ketoester —3. H₃O⁺→ β-ketoacid —Δ, −CO₂→ 1-phenylpentane-1,3-dione

3. cyclohexanone —1. LDA→ enolate —EtO-CO-OEt (NAS)→ 2-(ethoxycarbonyl)cyclohexanone —2. ⁻OEt; 3. EtI→ 2-ethyl-2-(ethoxycarbonyl)cyclohexanone —4. H₃O⁺, Δ→ 2-ethylcyclohexanone

4. diethyl malonate (CO₂Et, CO₂Et) —1. base; 2. (PhCO)₂O→ Ph-CO-CH(CO₂Et)₂ —3. H₃O⁺, Δ, −CO₂→ PhCOCH₂CO₂H —−CO₂→ PhCOCH₃

5. NC-CH₂-CO₂Et —1. ⁻OEt→ ⁻:CH(CN)(CO₂Et) — + 1-Cl-2-CN-4-NO₂-benzene, −Cl⁻ (nucleophilic aromatic subst'n)→ aryl-CH(CN)(CO₂Et) with o-CN, p-NO₂ —2. H₃O⁺, Δ, −CO₂→ 2-(carboxymethyl)-4-nitrobenzoic acid (HO₂C-CH₂-Ar-CO₂H, NO₂)

6. cis-4-methylcyclohexyl tosylate —NaCH(CO₂R)₂, S_N2→ trans-1-methyl-4-[CH(CO₂R)₂]cyclohexane

7. (E)-pent-3-en-2-one —1. H₃O⁺; 2. CrO₃, H⁺→ pentane-2,4-dione —3. NaH, 2 equiv→ dianion (more reactive enolate) —4. PhCH₂Cl, 1 equiv→ benzylated diketone enolate —5. H⁺→ 1-phenylheptane-3,5-dione

18.1 Reactions

18.2 Syntheses

408 • Chapter 18 Carbonyl α-Substitution Reactions and Enolates

2. $\begin{array}{c}CO_2Me\\CO_2Me\end{array}$ $\xrightarrow[\text{2. Br}\diagup\diagup\diagup OH]{\text{1. }^\ominus OMe,\text{ MeOH}}$ $\diagup\diagup\diagup\underset{OH}{(CO_2Me)_2}$ $\xrightarrow[-CO_2]{\text{3. }H_3O^\oplus}$ $\diagup\diagup\diagup\underset{OH}{CO_2H}$ $\xrightarrow{-H_2O}$ δ-valerolactone

3. $\begin{array}{c}CO_2Me\\CO_2Me\end{array}$ $\xrightarrow[\text{2. BrCH}_2CO_2Me]{\text{1. }^\ominus OMe}$ MeO$_2$C–CH(CO$_2$Me)–CH$_2$CO$_2$Me $\xrightarrow[\text{4. PCl}_5]{\text{3. }H_3O^\oplus,\ -CO_2}$ ClC(O)CH$_2$CH$_2$C(O)Cl $\xrightarrow[\text{6. }H^\oplus]{\text{5. LiAlH(O-}t\text{-Bu)}_3}$ OHC–CH$_2$CH$_2$–CHO

4. $\begin{array}{c}CO_2Me\\CO_2Me\end{array}$ $\xrightarrow[\substack{\text{2. 2-chloropentane}\\\text{3. }^\ominus OMe\\\text{4. allyl chloride}}]{\text{1. }^\ominus OMe}$ (disubstituted malonate with sec-pentyl and allyl groups) $\xrightarrow[-2\text{ MeOH}]{\text{5. H}_2N\text{–CO–}NH_2\ (\text{urea})}$ 5-sec-pentyl-5-allylbarbituric acid

5. $\begin{array}{c}CO_2Me\\CO_2Me\end{array}$ $\xrightarrow[\text{2. EtI}]{\text{1. }^\ominus OMe}$ $\underset{H}{\overset{Et}{>}}C(CO_2Me)_2$ $\xrightarrow[\text{4. PhCH}_2Br]{\text{3. }^\ominus OMe}$ $\underset{Bn}{\overset{Et}{>}}C(CO_2Me)_2$ $\xrightarrow[-CO_2]{\text{5. }H_3O^\oplus,\ \Delta}$ 2-benzyl-butanoic acid (EtCH(CH$_2$Ph)CO$_2$H)

6. PhCH=CH$_2$ $\xrightarrow[\text{2. H}_2O_2,\ ^\ominus OH]{\text{1. BH}_3\cdot THF}$ PhCH$_2$CH$_2$OH $\xrightarrow[\text{4. SOCl}_2]{\text{3. CrO}_3,\ H^\oplus}$ PhCH$_2$COCl $\xrightarrow[\text{NAS}]{\text{5. (MeO}_2C)_2CHNa}$ PhCH$_2$COCH(CO$_2$Me)$_2$ $\xrightarrow[-CO_2]{\text{6. }H_3O^\oplus,\ \Delta}$ PhCH$_2$COCH$_2$CO$_2$H $\xrightarrow[\text{8. }H^\oplus(-H_2O)]{\text{7. NaBH}_4}$ PhCH=CHCH$_2$CO$_2$H (cinnamyl-type acid)

7. $\begin{array}{c}CO_2Me\\CO_2Me\end{array}$ $\xrightarrow[\text{2. CH}_2\text{=CHCO}_2Me]{\text{1. }^\ominus OMe}$ $\diagup\diagup\underset{(CO_2Me)_2}{CO_2Me}$ $\xrightarrow[\text{5. }H^\oplus]{\substack{\text{3. }H_3O^\oplus(-CO_2)\\\text{4. LiAlH}_4}}$ HOCH$_2$CH$_2$CH$_2$CH$_2$CH$_2$OH $\xrightarrow[\text{S}_N2]{\text{6. HBr}}$ BrCH$_2$CH$_2$CH$_2$CH$_2$CH$_2$Br

8. ethyl acetoacetate $\xrightarrow[\text{2. MeI}]{\text{1. NaOEt, HOEt}}$ ethyl 2-methylacetoacetate $\xrightarrow[\substack{^\ominus OH,\ ROH\\\text{W-K}}]{\text{3. H}_2NNH_2}$ ethyl 2-methylbutanoate $\xrightarrow{\text{4. NH}_3}$ 2-methylbutanamide

18.2 Syntheses

18.2 Syntheses

18.2 Syntheses

18.3 Mechanisms

1. [Mechanism diagram: dihydroxyacetone → glyceraldehyde via acid-catalyzed enolization, showing +H⁺, −H⁺, +H⁺, −H⁺ steps through an enediol intermediate]

2. [Mechanism diagram: ethyl acetoacetate with 1. ⁻OEt, −H⁺ then 2. propylene oxide, leading to lactone formation with loss of ⁻OEt]

3. [Mechanism diagram: dibenzoylmethane (PhCOCH₂COPh) with I₂/⁻OH, iodination twice, then cleavage giving PhCO₂⁻ + diiodoenolate; then I–I, I₃C–COPh, ⁻OH attack, then I₃C:⁻ departs with PhCO₂H to give I₃CH + PhCO₂⁻]

4. [Mechanism diagram: bicyclic ketone with α-H epimerization via enol through achiral intermediate: +H⁺/−H⁺ to protonated carbonyl, −H⁺/+H⁺ to enol, +H⁺/−H⁺ achiral, then −H⁺/+H⁺ back to epimerized ketone]

5. [Mechanism diagram: bicyclic β-ketolactone + MeO⁻ → MeOAc + cycloheptenone enolate; then H–OMe protonation, −MeO⁻ → 1,4-cycloheptanedione]

6. [Mechanism diagram: hexane-2,5-dione, taut/+H⁺ to enol, cyclization to oxocarbenium, ~H⁺ shift, −H₃O⁺/~H⁺ to give 2,5-dimethylfuran]

18.3 Mechanisms

18.3 Mechanisms

CHAPTER 19
CARBONYL CONDENSATION REACTIONS

19.1 Reactions

1. PhCHO + PhCOCH$_3$ $\xrightarrow[\text{-H}_2\text{O}]{\text{1. }^{\ominus}\text{OH}}$ PhCH=CHCOPh $\xrightarrow[\text{Michael}]{\text{2. }^{\ominus}\text{:CH(CO}_2\text{R)}_2}$ [RO$_2$C, CO$_2$R adduct] $\xrightarrow[\text{-CO}_2]{\text{3. H}_3\text{O}^{\oplus}}$ HO$_2$C–CH$_2$–CH(Ph)–CH$_2$–COPh

2. (isobutyraldehyde) $\xrightarrow[\text{taut}]{\text{H}^{\oplus}}$ (enol) $\xrightarrow{\text{aldol}}$ (β-hydroxy aldehyde product)

3. (1,3-cyclohexanedione) $\xrightarrow{^{\ominus}\text{OEt}}$ (enolate) $\xrightarrow[\text{Knoevenagel}]{t\text{-Bu-CHO}}$ (aldol with OH, t-Bu) or (enone with =CH-t-Bu)

4. (cyclopentane with CH$_2$CO$_2$Me and CH$_2$COCH$_3$) $\xrightarrow[\text{-H}^{\oplus}]{^{\ominus}\text{OMe}}$ (enolate attacking ester) $\xrightarrow{-^{\ominus}\text{OMe}}$ (spiro diketone)

5. (MeO$_2$C–(CH$_2$)$_4$–CO$_2$Me) $\xrightarrow{\text{1. }^{\ominus}\text{OMe}}$ (enolate) $\xrightarrow[\text{Dieckmann}]{-^{\ominus}\text{OMe}}$ (2-carbomethoxycyclohexanone) $\xrightarrow[\text{-CO}_2]{\text{2. H}_3\text{O}^{\oplus}}$ (cyclohexanone)

6. (2,3-dimethylcyclopentenone) $\xrightarrow[\text{conj. add'n}]{^{\ominus}\text{OH, ROH}}$ (β-hydroxy ketone) $\xrightarrow{\text{retro-aldol}}$ (2,5-heptanedione)

7. (diketone with Z-alkene, Et) $\xrightarrow[\text{-H}^{\oplus}]{^{\ominus}\text{OH}}$ (enolate) $\rightarrow$ (cyclopentanol with HO, Me, Et chain) $\xrightarrow{\text{-H}_2\text{O}}$ (cyclopentenone with Et side chain)

19.1 Reactions

19.1 Reactions

416 • Chapter 19 Carbonyl Condensation Reactions

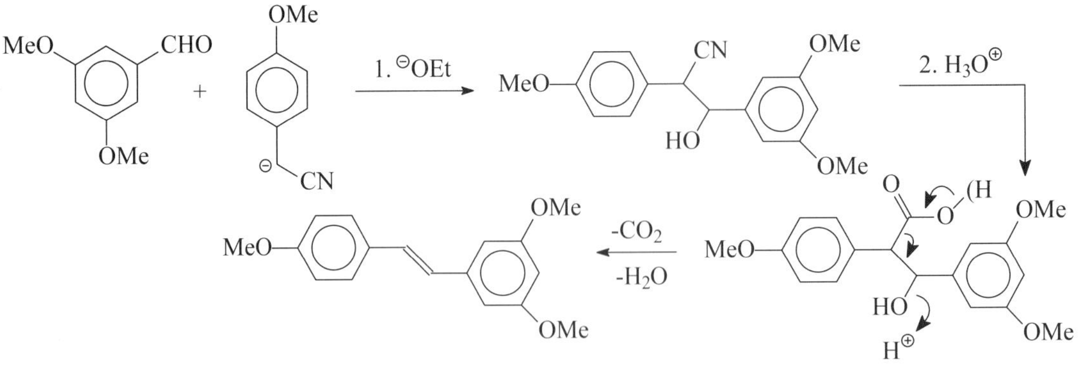

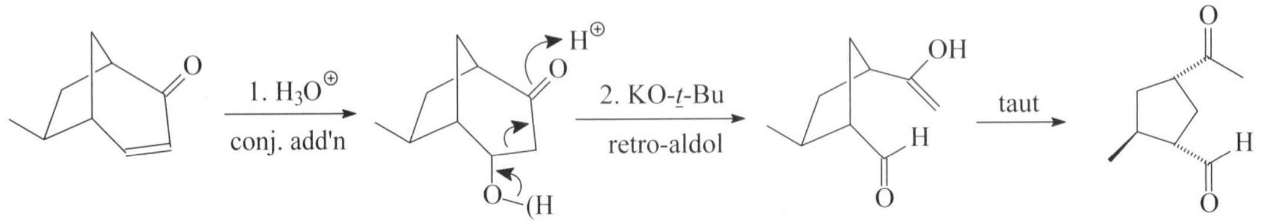

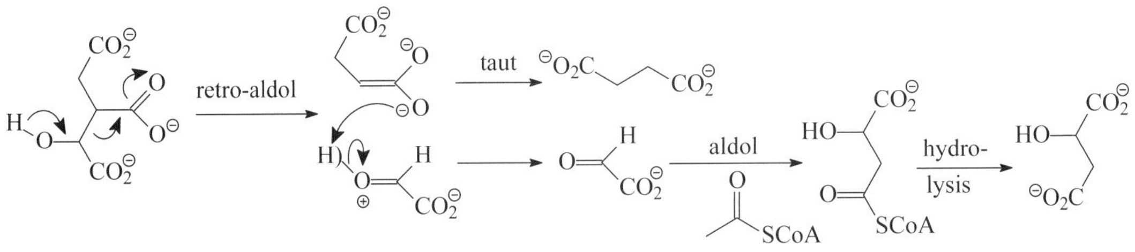

24.

25.

26.

27.

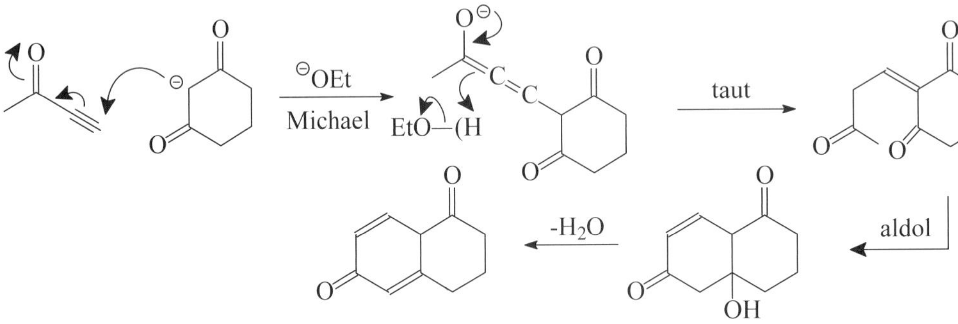

28.

19.1 Reactions

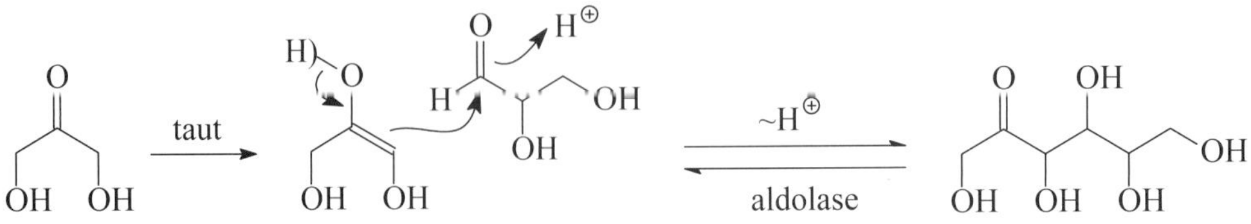

19.1 Reactions

418 • Chapter 19 Carbonyl Condensation Reactions

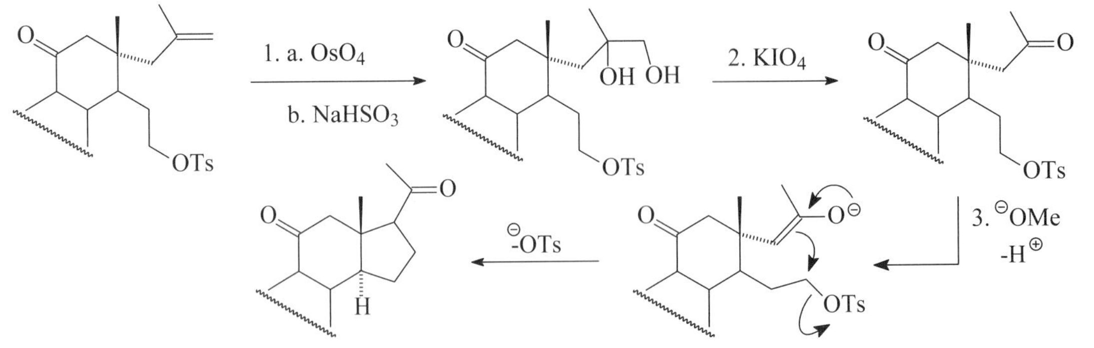

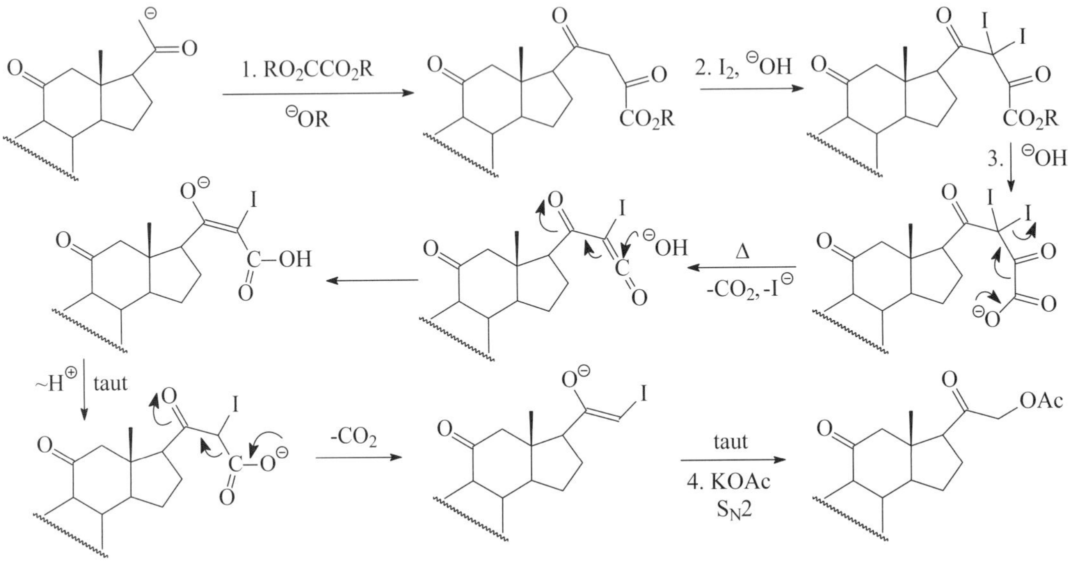

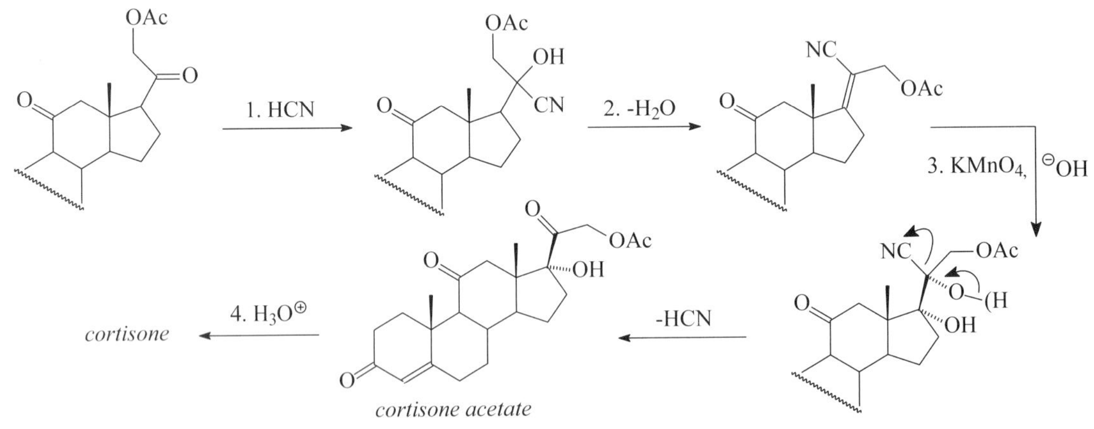

c.

d.

e.

33.

19.1 Reactions

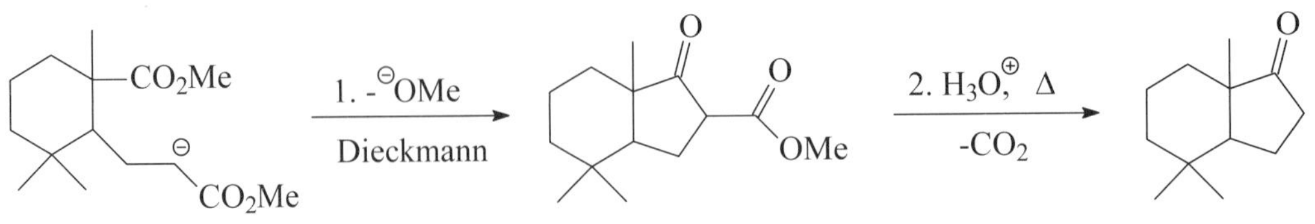

Solutions • 419

34. [Mechanism showing $O_2N-CH_2(H)$ with $^⊖OMe$ giving $O_2N-\ddot{C}H_2$, attacking 2-nitrobenzaldehyde to form the OH / nitro intermediate, then $-H^+$ giving the sodium nitronate product]

35. [Mechanism: bicyclic dione with Et$_3$N, $-H^+$ forms enolate, intramolecular Michael addition, then $+H^+$ tautomerization gives tricyclic diketone]

19.2 Syntheses

1. PhCHO + PhCOCH$_3$ $\xrightarrow[-H_2O]{1.\ H_3O^⊕}$ PhCH=CHCOPh $\xrightarrow{2.\ H_2/Pt}$ PhCH$_2$CH$_2$CH(OH)Ph

2. Cyclohexanone + piperidine $\xrightarrow[-H_2O]{1.\ H^⊕}$ 1-(1-piperidinyl)cyclohexene $\xrightarrow{2.\ \text{MVK}}$ iminium intermediate $\xrightarrow{3.\ H_3O^⊕}$ 2-(3-oxobutyl)cyclohexanone

3. Heptanedial $\xrightarrow{^⊖OH,\ ROH}$ enolate $\longrightarrow$ 2-hydroxycyclohexanecarbaldehyde

4. 2,6-heptanedione $\xrightarrow{H_3O^⊕}$ 1-methyl-2-acetyl-cyclopentanol $\xrightarrow{-H_2O}$ 1-acetyl-2-methylcyclopentene

5. 3-pentanone $\xrightarrow[-H_2O]{1.\ ^⊖OH\ (aldol)}$ enone $\xrightarrow{2.\ H_2/Pd}$ alcohol $\xrightarrow[4.\ H_2/Pd]{3.\ H_2SO_4\ (E1)}$ 3-ethyl-4-methylheptane

Chapter 19 Carbonyl Condensation Reactions

6. Acetone → (1. LDA, 2. MeI, 3. Cl₂, H⁺) → chloromethyl ethyl ketone → (4. HOCH₂CH₂OH, H⁺ protect) → dioxolane-CH₂Cl → (5. Ph₃P:, 6. MeLi) → ylide → (7. acetone, Wittig) → dioxolane alkene → (8. H₃O⁺) → 2-methyl-2-hexen-... ketone

- *via* a Wittig, not a mixed aldol!

7. Dimethyl phthalate → (1. NaCH₂CO₂R, Claisen) → intermediate → (2. ⁻OR, HOR, Dieckmann) → indan-1,3-dione-2-CO₂R

8. Cyclohexanone → (1. Cl₂, H⁺; 2. KO-t-Bu (E2)) → cyclohexenone → (3. H₃O⁺; 4. CrO₃, H⁺) → 1,3-cyclohexanedione → (5. NaH (2 equiv)) → dianion → (6. PhCH₂Br (1 equiv)) → benzyl dione anion → (7. H⁺) → 2-benzyl-1,3-cyclohexanedione

9. Pentene → (1. BH₃·THF, 2. H₂O₂, ⁻OH, 3. PCC) → pentanal → (4. pyrrolidine, H⁺, −H₂O) → enamine → (5. allyl chloride, 6. H₃O⁺) → 2-allylpentanal

10. PhCH₂CHO → (1. CH₂O, ⁻OH, mixed aldol) → PhCH(CH₂OH)CHO → (2. NaBH₄, 3. H⁺) → PhCH(CH₂OH)(CH₂OH)

11. 1,3-cyclohexanedione + methyl vinyl ketone (methylated) → (⁻OR, ROH, Michael) → Michael adduct → (⁻OR, ROH, −H₂O, aldol) → Wieland-Miescher-type enone

12. Decalin → (1. O₃, 2. Zn, H⁺) → cyclodecane-1,6-dione → (3. KOH, EtOH, aldol) → bicyclic β-hydroxy ketone → (−H₂O) → bicyclic enone

19.2 Syntheses

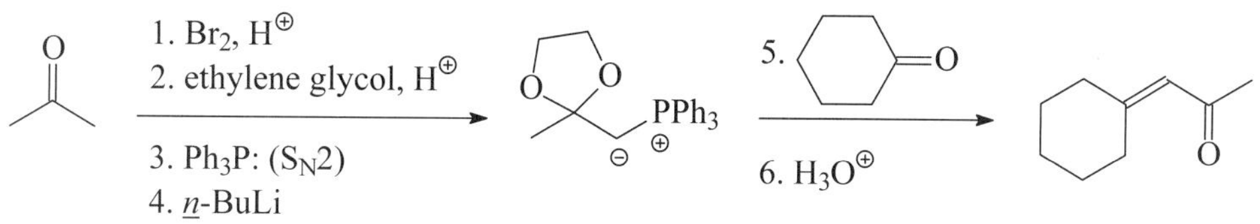

- the reaction of cyclohexanone with acetone *via* an aldol would yield four possible products!

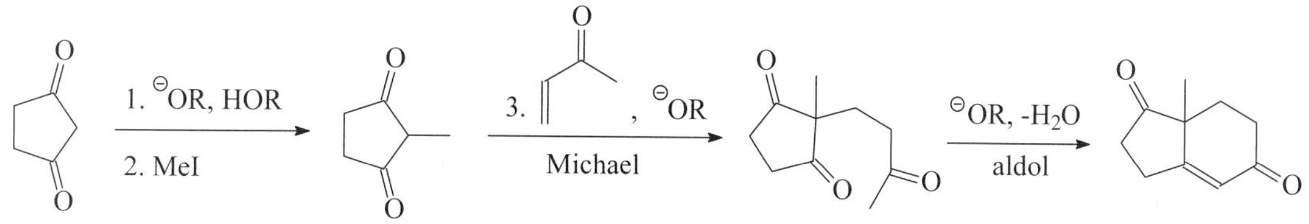

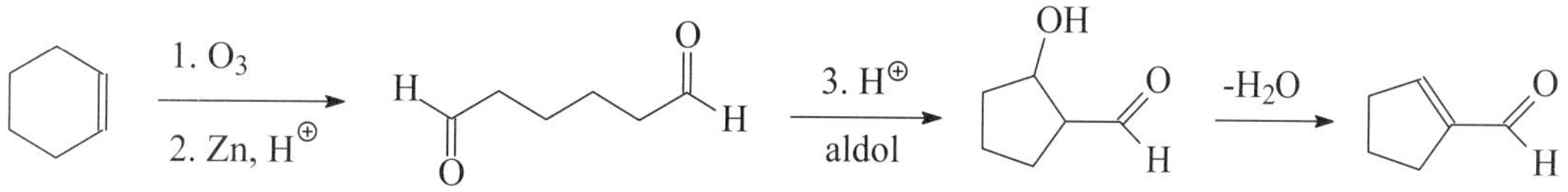

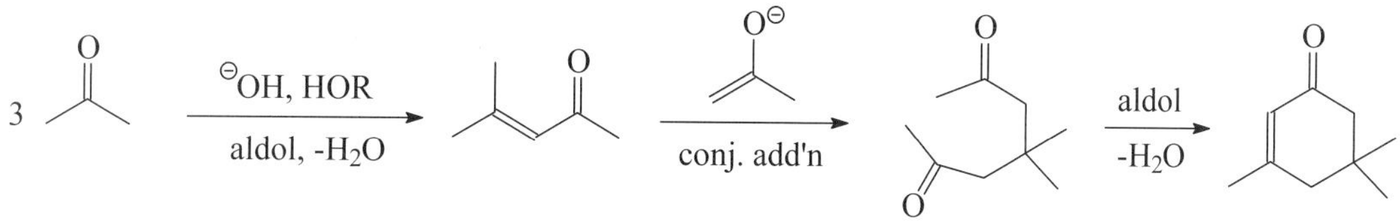

19.3 Mechanisms

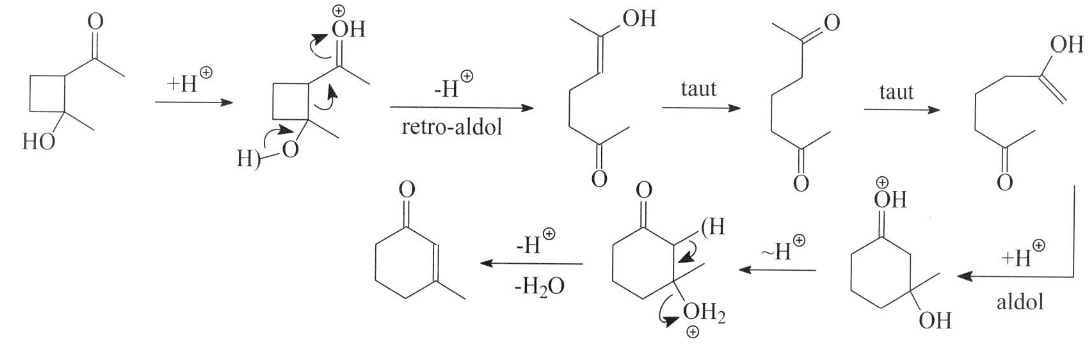

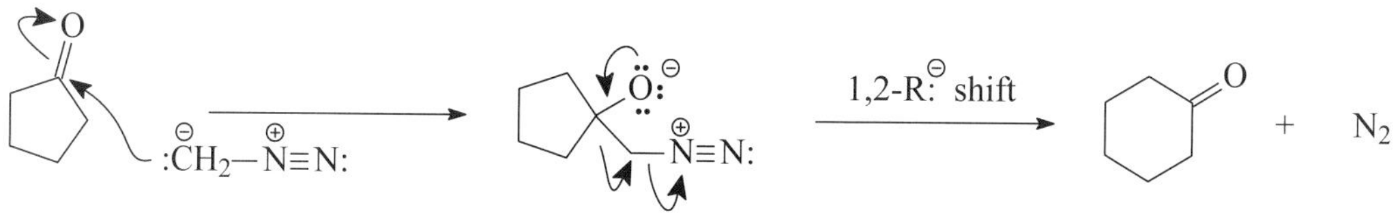

19.3 Mechanisms

19.3 Mechanisms

424 • Chapter 19 Carbonyl Condensation Reactions

13. *(cont.)*

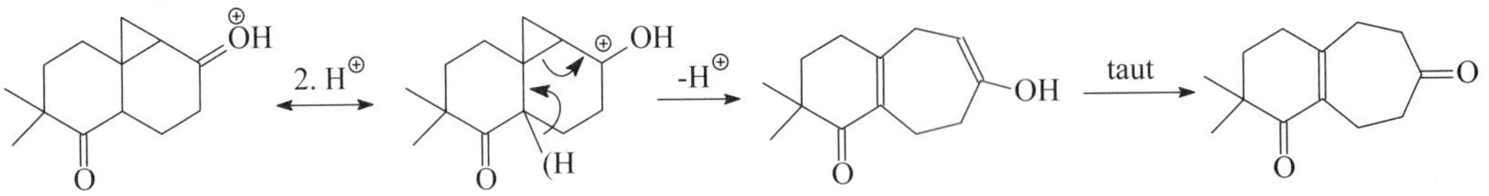

14.

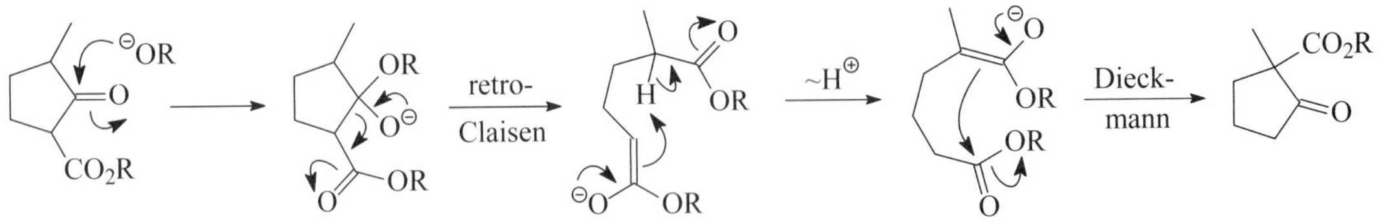

15.

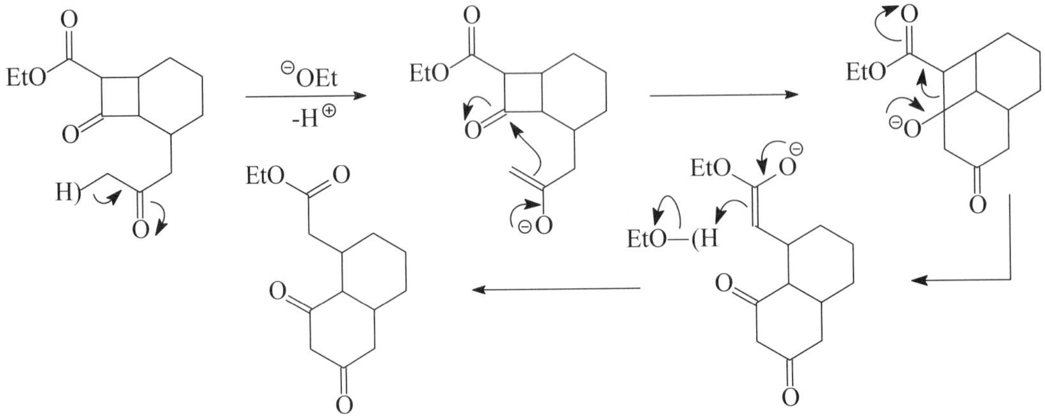

16.

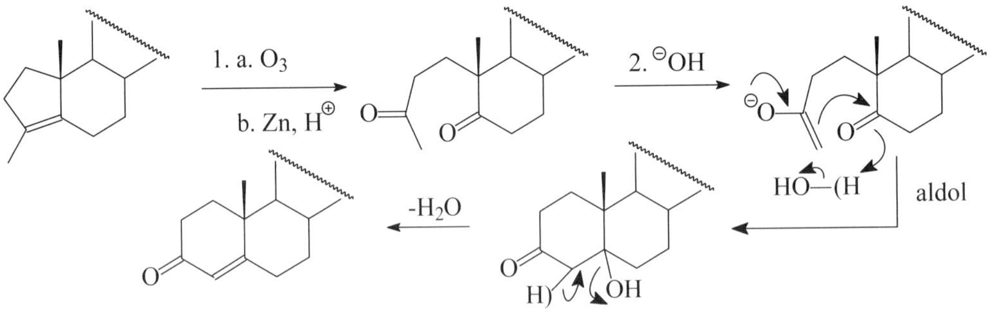

17.

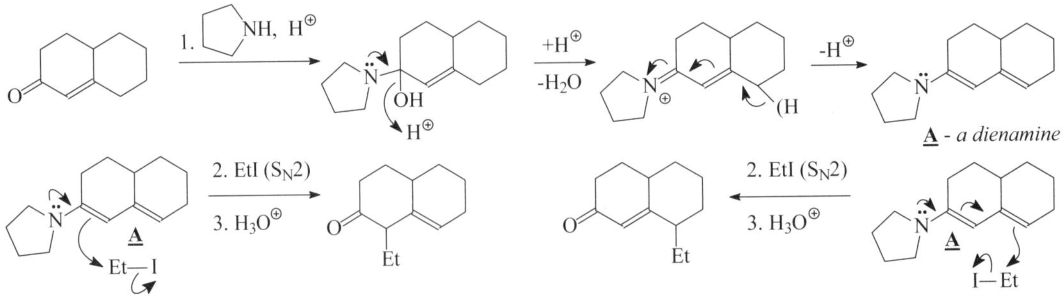

18.

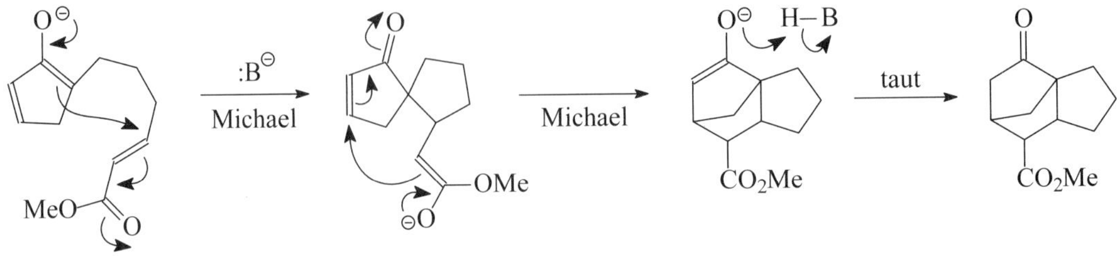

19.3 Mechanisms

19.3 Mechanisms

19.

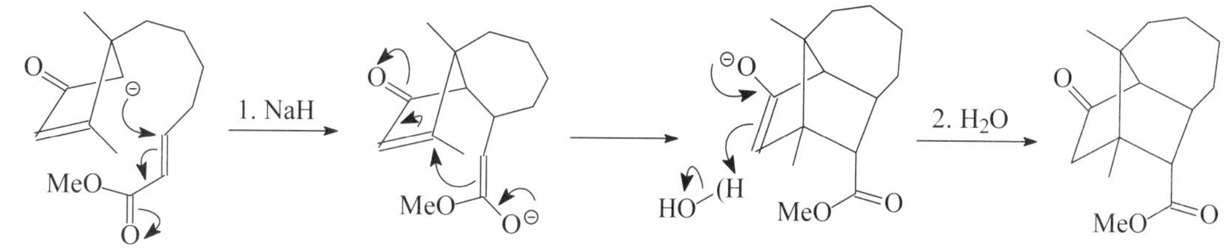

20.

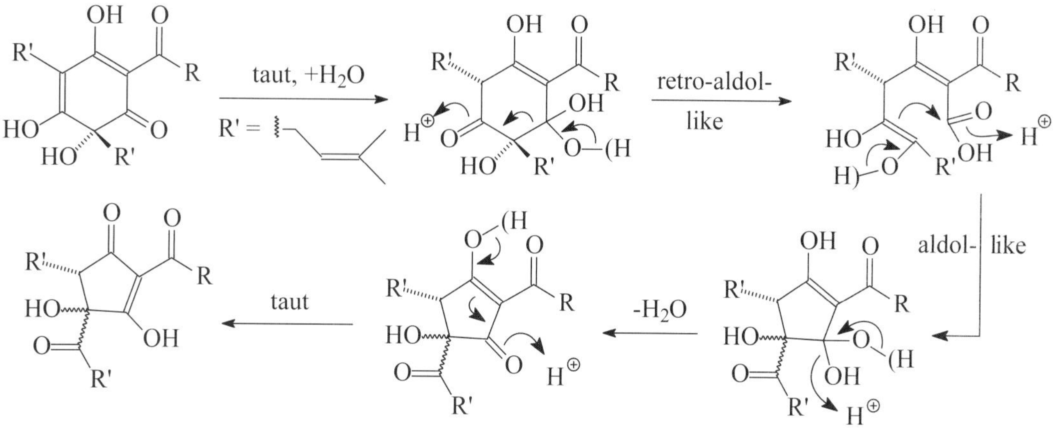

21. a. Mannich

b.

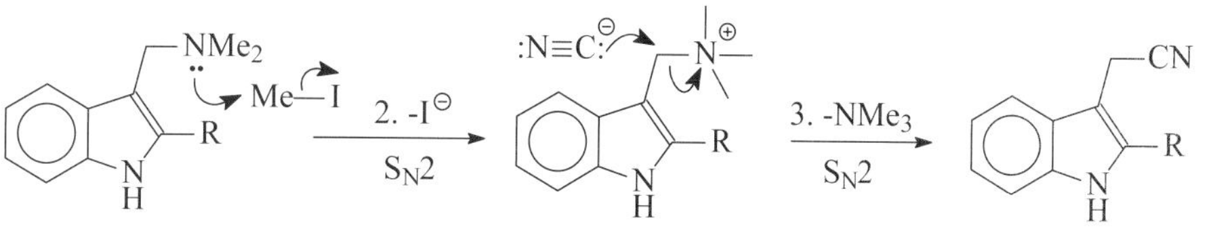

c.

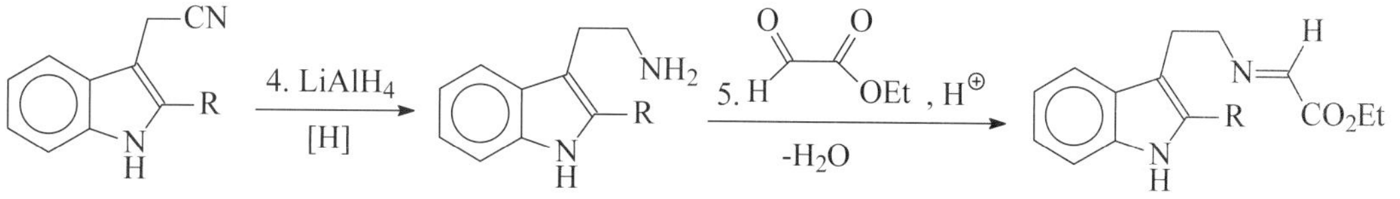

22. PEP

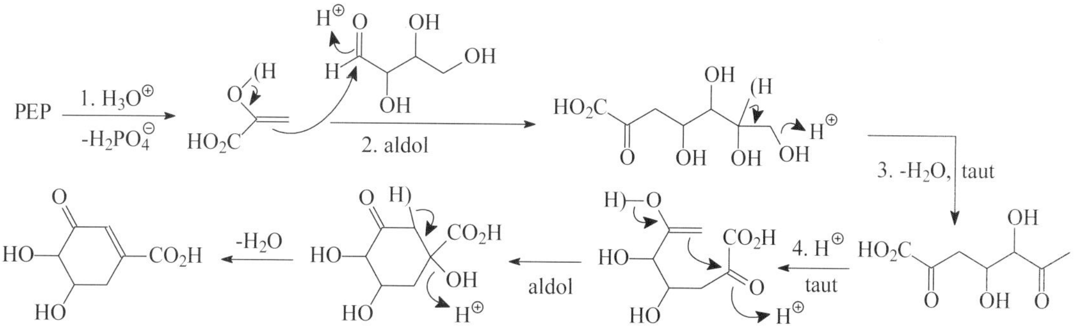

23.

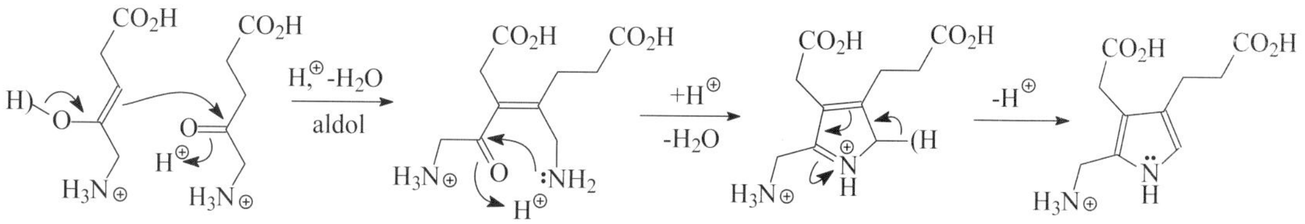

24. a.

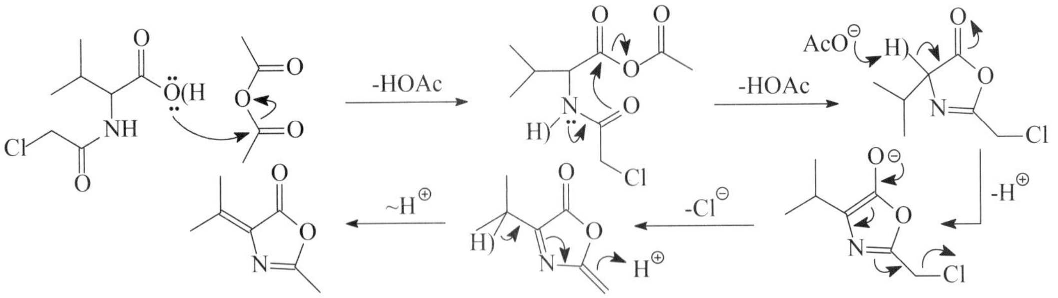

b.

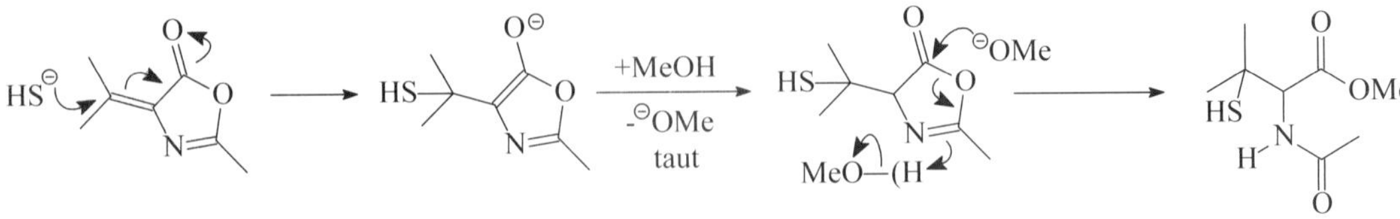

25.

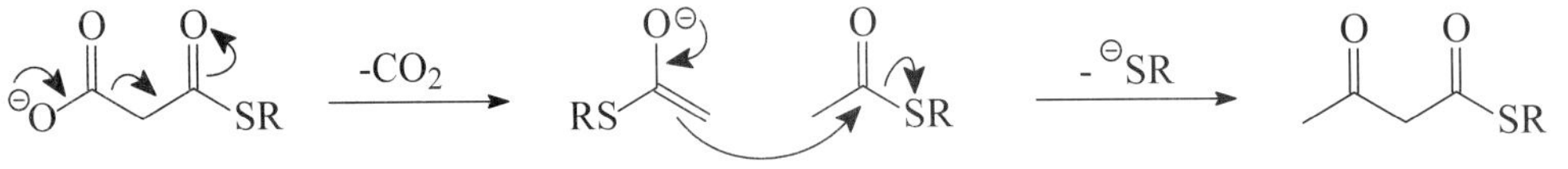

26.

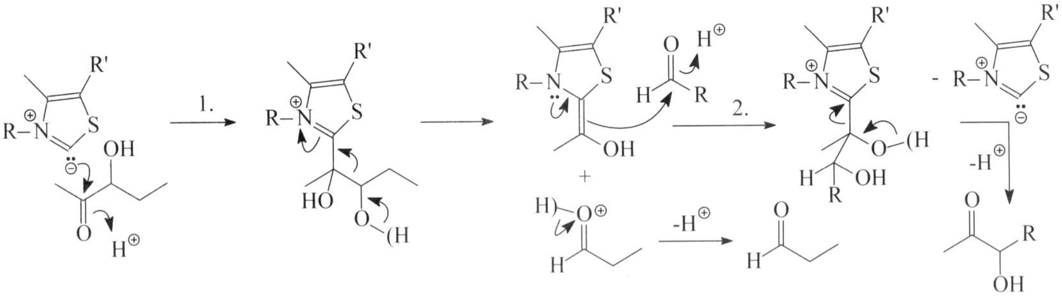

27. a.

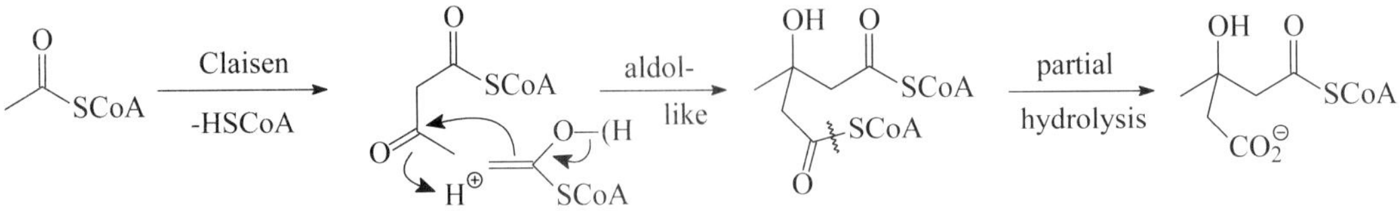

b.

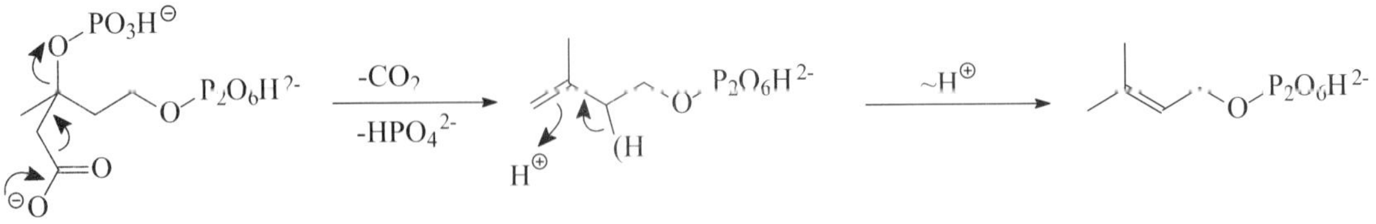

CHAPTER 20
AMINES

20.1 Reactions

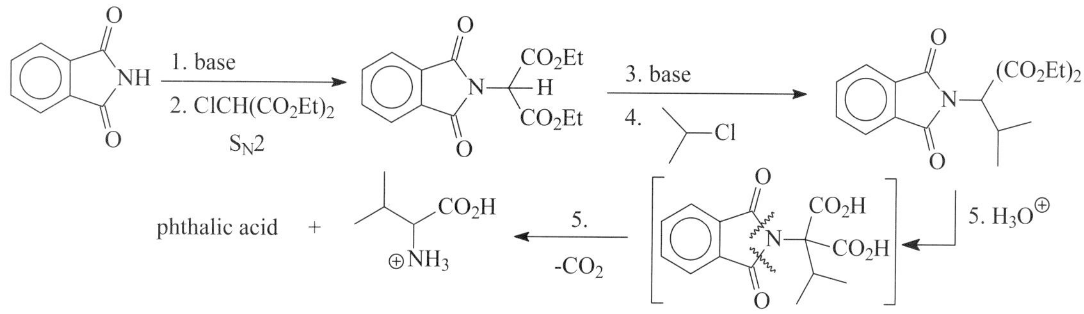

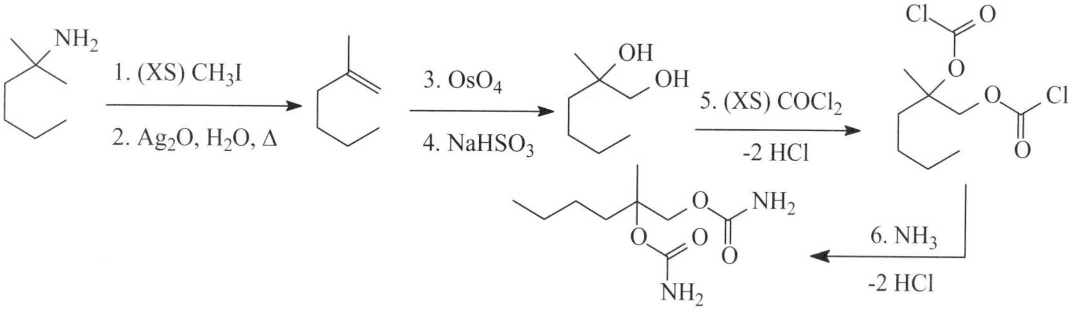

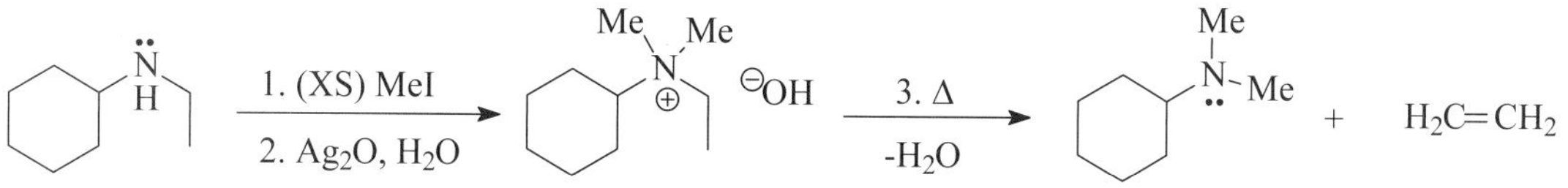

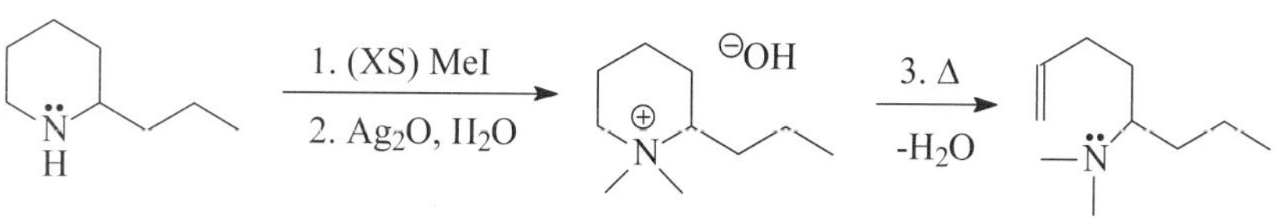

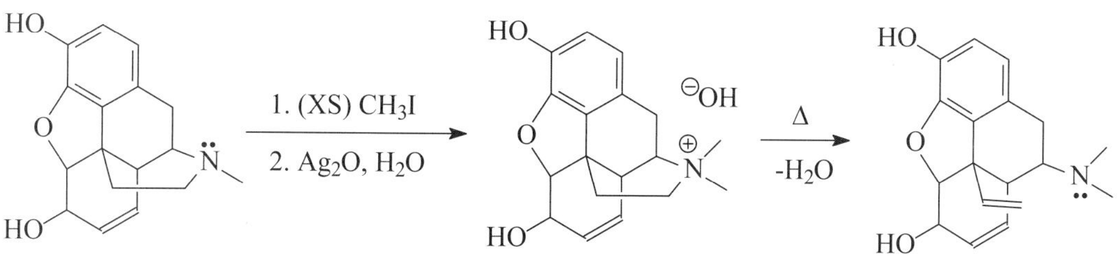

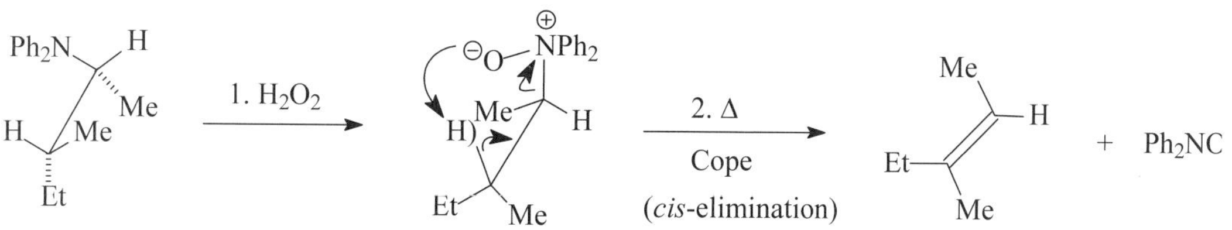

20.1 Reactions

20.1 Reactions

14. Ph$_2$CHOH $\xrightarrow{\text{1. NaNH}_2}{\text{2. }\triangle\text{O}}$ Ph$_2$CH-O-\~\~-O$^\ominus$ $\xrightarrow{\text{3. PBr}_3}$ Ph$_2$CH-O-\~\~-Br $\xrightarrow{\text{4. Me}_2\text{NH}}$ Ph$_2$CH-O-\~\~-NMe$_2$

15. PhCH=CHCH$_3$ $\xrightarrow[\text{ROOR}]{\text{1. HBr}}$ PhCH$_2$CH(Br)CH$_3$ $\xrightarrow[\text{3. H}_2\text{NNH}_2]{\text{2. phthalimide N}^\ominus}$ PhCH$_2$CH(NH$_2$)CH$_3$ + phthalhydrazide

16. (Me$_2$N-CH$_2$-C(OH)(cyclohexyl)(p-MeO-C$_6$H$_4$)) $\xrightarrow[\text{2. Ag}_2\text{O, H}_2\text{O, }\Delta]{\text{1. MeI}}$ (alkene-alcohol intermediate, H$^\oplus$ loss) $\xrightarrow[\text{pinacol-like rearrangement}]{\text{3. H}^\oplus}$ (methyl-aryl-cycloheptanone)

17. (3-aryl-3-phenyl-butanoic acid) $\xrightarrow[\text{2. NH}_3]{\text{1. SOCl}_2}$ (amide) $\xrightarrow[\text{H}_2\text{O}]{\text{3. Br}_2, {}^\ominus\text{OH}}$ (primary amine, Hofmann) $\xleftarrow{\text{4. 2 } i\text{-PrI}}$ (diisopropylamino product)

18. (4-methylpentanoic acid) $\xrightarrow[\text{2. KO-}t\text{-Bu (E2)}]{\text{1. a. Br}_2\text{, PBr}_3\text{ b. H}_2\text{O (H-V-Z)}}$ (α,β-unsaturated carboxylate) $\xrightarrow[\text{conj. add'n}]{\text{3. HCN, }^\ominus\text{CN}}$ (β-cyano acid) $\xrightarrow{\text{4. H}_2\text{/Pt}}$ (β-aminomethyl acid)

19. p-toluidine $\xrightarrow{\text{1. Br}_2}$ 2,6-dibromo-4-methylaniline $\xrightarrow[\text{HCl}]{\text{2. NaNO}_2}$ diazonium $\xrightarrow[-\text{N}_2]{\text{3. KI}}$ 2,6-dibromo-4-methyl-iodobenzene

20. PhCN $\xrightarrow[\text{FeCl}_3]{\text{1. Cl}_2}$ 3-chlorobenzonitrile $\xrightarrow[\text{(NAS via benzyne)}]{\text{2. NaNH}_2}$ 3-aminobenzonitrile $\xrightarrow[\text{4. CuCN}]{\text{3. KNO}_2\text{, H}^\oplus}$ 1,3-dicyanobenzene $\xrightarrow{\text{5. H}_3\text{O}^\oplus}$ isophthalic acid

21. toluene $\xrightarrow[\text{2. fuming nitric acid}]{\text{1. KMnO}_4\text{, H}^\oplus}$ 3-nitrobenzoic acid $\xrightarrow[\text{4. NaNO}_2\text{, HCl}]{\text{3. Fe, HCl}}$ 3-diazoniobenzoic acid $\xrightarrow{\text{5. HBF}_4}$ 3-fluorobenzoic acid

430 • Chapter 20 Amines

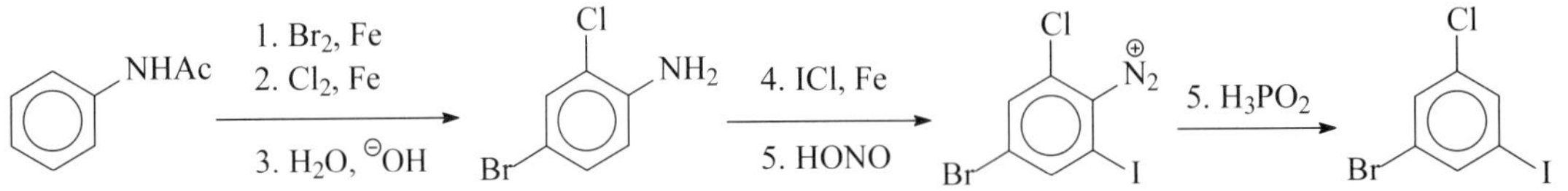

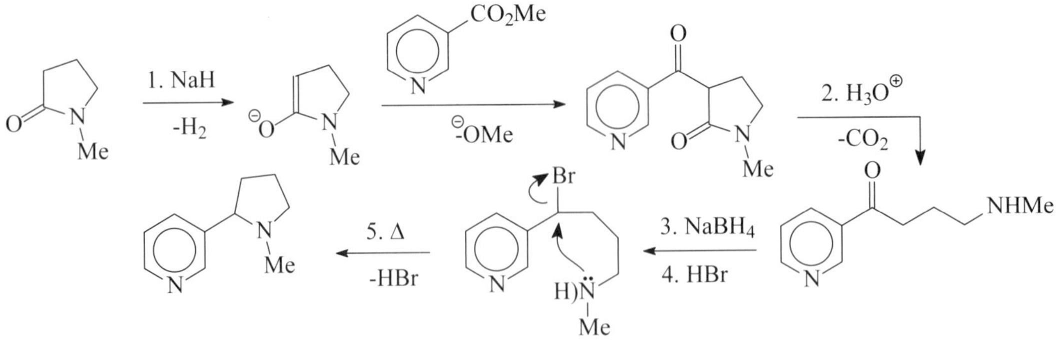

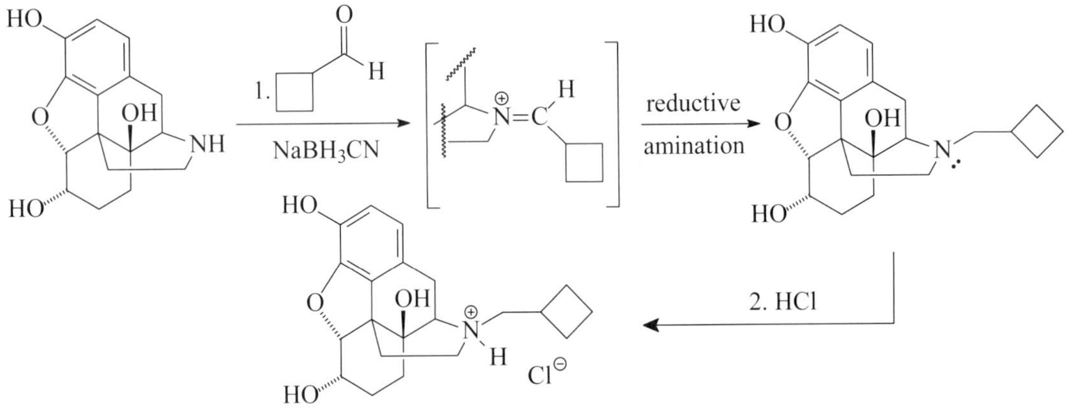

20.2 Syntheses

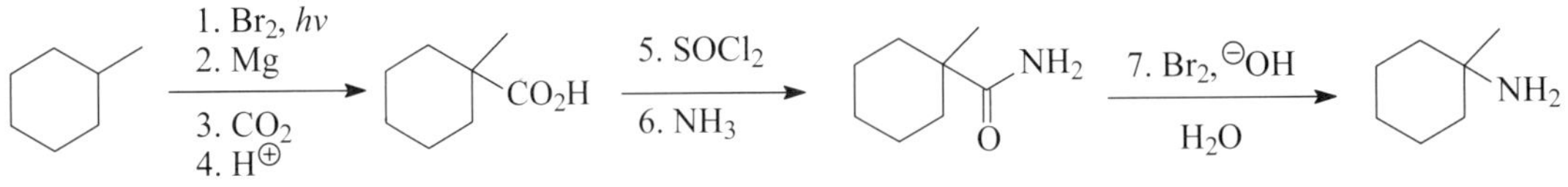

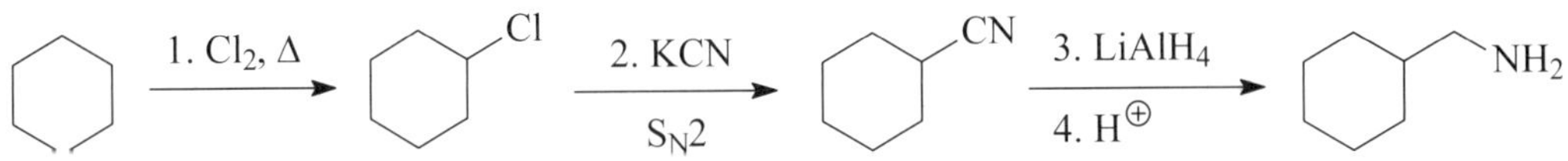

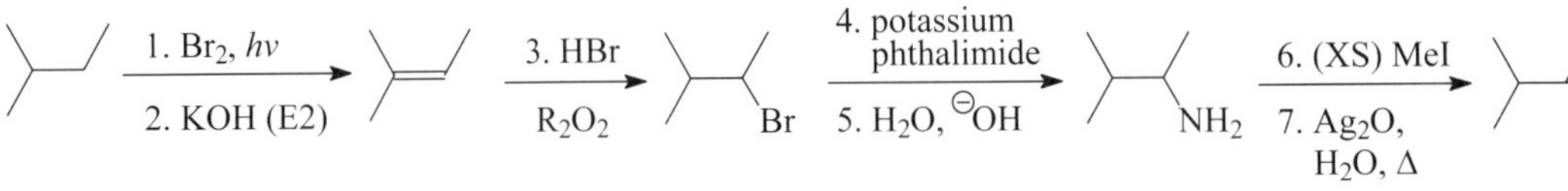

20.2 Syntheses

432 • Chapter 20 Amines

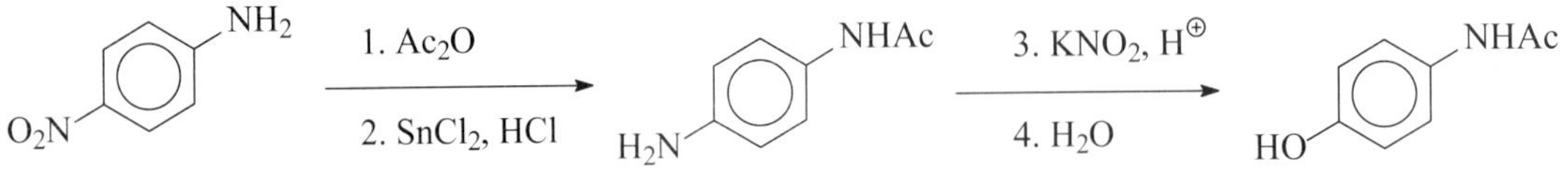

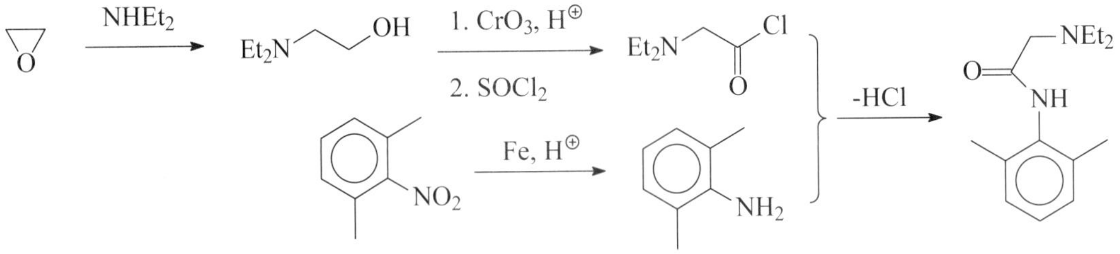

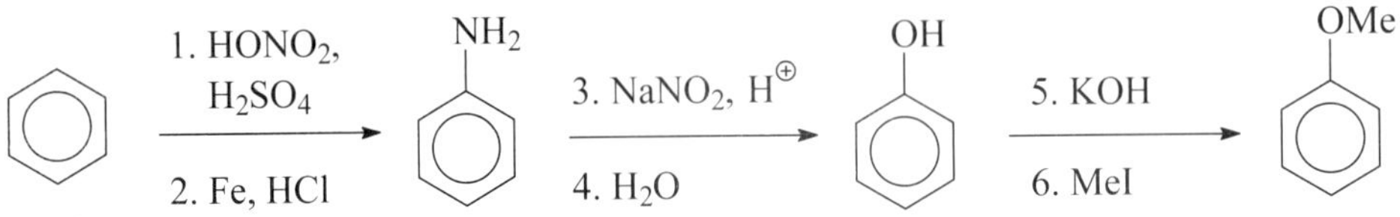

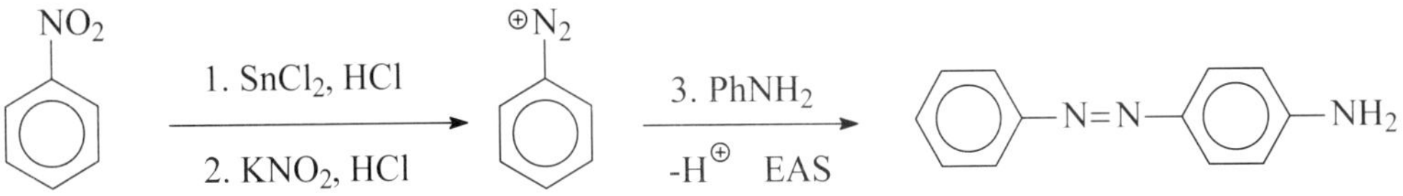

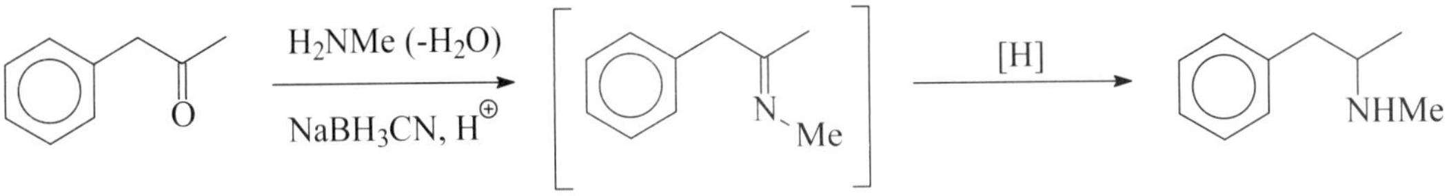

20.2 Syntheses

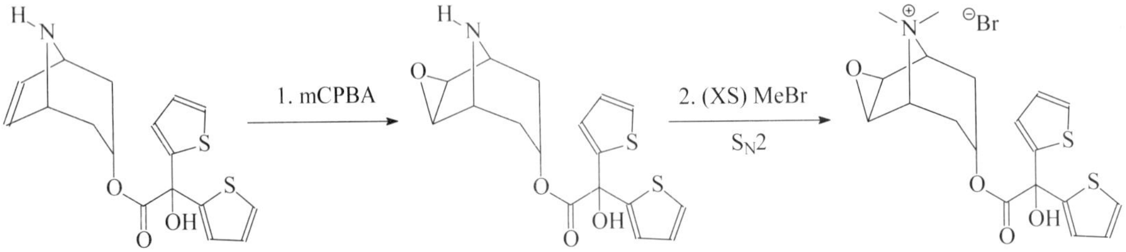

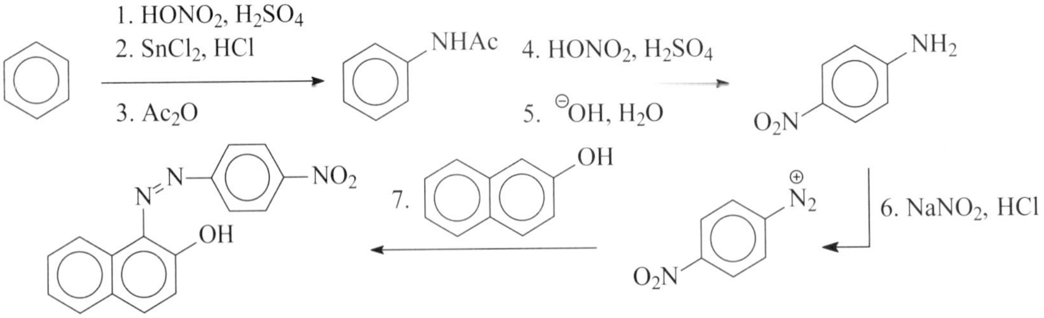

19. Ph-H →(1. HONO₂, H₂SO₄)→ PhNO₂ →(2. DCl, EAS)→ 3,5-dideutero-nitrobenzene →(3. Fe, HCl; 4. KNO₂, HCl; 5. H₃PO₂)→ 1,3-dideuterobenzene

20. 4-NHAc-phenol →(1. propylene, H⁺, F-C alkylation)→ 2,6-diisopropyl-4-acetamidophenol →(2. H₃O⁺, -HOAc)→ 2,6-diisopropyl-4-aminophenol →(3. KNO₂, HCl; 4. H₃PO₂)→ 2,6-diisopropylphenol

20.3 Mechanisms

1. Butyramide + ⁻OH → deprotonated amide + Br-Br → N-bromoamide (-HBr) → N-Br nitrogen anion (-Br⁻) → propyl isocyanate (PrN=C=O) + HO-Ph → tetrahedral intermediate → (~H⁺, taut) → PhO-C(=O)-NH-Pr (phenyl carbamate)

2. Phthalimide →(1. Cl₂, ⁻OH, H₂O)→ N-chlorophthalimide + ⁻OH →(-H⁺)→ ring-opened carboxylate with N-Cl →(-Cl⁻)→ 2-isocyanatobenzoate →(⁻OH, taut)→ 2-(carbamate)benzoate →(2. H⁺, -CO₂)→ anthranilic acid (2-aminobenzoic acid)

3. 2-aminobenzoic acid →(1. HONO)→ 2-diazoniobenzoic acid →(2. pH 8, -H⁺)→ carboxylate diazonium →(-CO₂, -N₂)→ benzyne + butadiene →(3. D-A)→ 1,4-dihydronaphthalene

20.3 Mechanisms

20.3 Mechanisms

20.3 Mechanisms